THE
OXFORD GUIDE TO
THE HISTORY OF PHYSICS
AND ASTRONOMY

THE
OXFORD GUIDE TO
THE HISTORY OF PHYSICS
AND ASTRONOMY

Edited by
J. L. Heilbron

OXFORD
UNIVERSITY PRESS
2005

OXFORD

UNIVERSITY PRESS

Oxford University Press, Inc., publishes works that further
Oxford University's objective of excellence
in research, scholarship, and education

Oxford   New York
Auckland   Bangkok   Buenos Aires   Cape Town   Chennai
Dar es Salaam   Delhi   Hong Kong   Istanbul   Karachi   Kolkata
Kuala Lumpur   Madrid   Melbourne   Mexico City   Mumbai
Nairobi   São Paulo   Shanghai   Taipei   Tokyo   Toronto

Copyright © 2005 by Oxford University Press, Inc.

Published by Oxford University Press, Inc.
198 Madison Avenue, New York, New York, 10016
http://www.oup.com/us

Oxford is a registered trademark of Oxford University Press

Library of Congress Cataloging-in-Publication Data

The Oxford guide to the history of physics and astronomy /
general editor, John L. Heilbron.
p. cm.
Includes bibliographical references.
ISBN-13: 978-0-19-517198-3
ISBN-10: 0-19-517198-5
1. Physics—History—Dictionaries. 2. Astronomy—
History—Dictionaries. I. Heilbron, J. L.
QC7.O94 2005
530'.03—dc22   2004020388

9 8 7 6 5 4 3 2 1

Printed in the United States of America
on acid-free paper

CONTENTS

PREFACE

The present *Guide* offers a coherent selection of articles from the *Oxford Companion to the History of Modern Science* (2002). Like the *Companion*, the *Guide* is modern in coverage, cosmopolitan in reach, agreeable in style, and particular in language. The first and last of these virtues need some elucidation.

Modernity. It may not be easy to say when modernity started or ended, if it is over, but few will dispute that Columbus sailed before it began, that it flourished during the nineteenth and twentieth centuries, and that it saw the creation and exploitation of natural science as we know it. Hence our coverage of the modern period runs from the sixteenth through the twentieth century. Apart from appropriate backward glances and a biography of Nicholas Copernicus, the *Guide* picks up around 1550 and dwells on the seventeenth and eighteenth centuries before turning to the development of the modern scientific disciplines in the nineteenth and twentieth centuries. The earlier period receives emphasis not to allow contributors to expatiate on the so-called Scientific Revolution, but to stress that modern science is a discovery as well as an invention. It was a discovery that nature generally acts regularly enough to be described by laws and even by mathematics; and it required invention to devise the techniques, abstractions, apparatus, and organization for exhibiting the regularities and securing their law-like descriptions.

The discovery of the regularities of planetary motions goes back to antiquity; but the discovery of principles that allowed for ever finer quantitative agreement between prediction and observation of astronomical phenomena dates from the seventeenth century. The means by which the agreement was secured—the instruments, computational techniques, and physical models—suggested that the phenomena of physics and chemistry might also be (or with suitable definitions and abstractions become) as law-like as astronomy. The systematic production of evidence favorable to the suggestion dates from the last few decades of the eighteenth century.

Answers to the questions when modern science began and how it grew express attitudes toward modernity. Some historians, put off by science's methods of aggrandizement, represent the secularly progressive path of science as a locally bumpy road along which scientists advance by fighting with one another over facts, interpretations, and authority. Others have located the fighting between forward-thinking philosophers of nature and reactionary powers of religion, state, and society. Still others couple the development of science tightly with practical applications, with pharmacy, medicine, mining, and manufacture.

The *Guide* speaks to all these possibilities and to more besides. Readers can choose among them or construct their own stories, aware that the choice matters—that placing the threshold of the modern science in the work of Galileo or the Industrial Revolution (to pick standard candidates) implies and conveys a worldview.

Propriety. The word "scientist" was not seriously propounded before 1840 and not used widely until well into the twentieth century ("man of science" or "scientific man" being preferred). Nonetheless, historians and others who should know better readily refer to doubtful "scientists" of the renaissance and "geologists" of the Middle Ages. This usage hides discontinuities in the subject matters of the sciences and the circumstances of their cultivation. Anachronistic vocabulary distorts understanding of the nature of modern sciences and their place in the societies that support them.

The *Guide* avoids using the names of modern sciences in reference to the sixteenth, seventeenth, and eighteenth centuries except for astronomy. Otherwise it recommends terms of the period, like "natural knowledge" or "natural philosophy." It shrinks from labeling as a "scientist" any student of nature active before the middle of the nineteenth century, preferring "scholar," "man of letters," "academician," "professor," and, where relevant, "astronomer" or "mathematician;" and it prohibits altogether honorifica and horrifica like "the father (or mother) of modern physics." Earlier students of nature should be allowed the roles they played. Few were, or would have wished to be, professional researchers of today's type.

The *Guide* abates this prissiness when enforcing it would result in pedanticism, which it abhors even more than bad language. Nevertheless, to paraphrase the editor of the *Saggi* or *Essays* issued in 1667 by the Accademia del Cimento of Florence, perhaps the earliest of scientific societies, "if sometimes there shall be inserted any hints of anachronism, we request that they be taken always for the thoughts, and particular sense of some one of the Contributors, but not imputed to the whole of the Company."

The guides. The *Companion* may be regarded as a history of scientific ideas and their immediate contexts embedded in a web of connections to epistemology, ethics, government, law, the mass media, the military, music, philanthropy, polite conversation, religion, scientific institutions, theology, and universities. This conception recommended a division of the *Companion* into free-standing volumes on neighboring sciences and a volume on the wider web. The first of these derived volumes, the present *Guide*, covers sciences that have ranked at the top of classifications of knowledge through most of the modern period. These are astronomy, physics, and their offspring, now called planetary science. A guide to the modern history of chemistry and the life sciences, and one on the historical setting of modern science, are under consideration.

Modern astronomy, physics, and planetary science have transformed mankind's understanding of the world; in industrialized societies, they mediate everyone's everyday experience of it. The *Guide*'s considered and focused accounts of this important material offer instructive surveys and entertaining browsing for readers with a good high-school knowledge of physics. The articles on the unavoidable subject of quantum theory, however, require a little more background. But the *Guide* as a whole is aimed at the casual and general reader who wants to know where science is and how it got there; students preparing papers and (there are some) merely expanding their minds; all who teach, study, apply, or analyze any science whatsoever; every expert in need of breadth and refreshment; and anyone who likes to see a difficult job conscientiously tackled.

In general, the *Guide* reprints articles in their entirety although it occasionally omits a sentence or paragraph to eliminate inappropriate overlap or unnecessary detail. Institutional and social history embedded in the articles remains there. The editor has corrected a few errors and updated a few bibliographical citations. New articles, all but one a biography, tighten connections within and between astronomy and planetary science: Agassiz, Bessel, Cassini Family, Deluc, Halley, Herschels, Hoyle and Ryle, Meteoritics, Werner. Since their authors also contributed to the *Companion*, the new articles enter seamlessly among the old.

A guide to the *Guide*. The articles are presented alphabetically but arranged hierarchically. Each comprehensive discipline has a primary article (Astronomy; Earth Sciences; Physics) that indicates general historical development, appropriate terminology, and main subdivisions. These subdivisions make the subject matter of a second tier of articles (Astronomy, Non-Optical; Paleontology; Mechanics), which, in turn, spin off tertiary notices of people, discoveries, concepts, and instruments (Cavendish and Coulomb; Pulsars and Quasars; Isostasy; Barometer). A system of cross-references, either stated explicitly or indicated by asterisk, leads down through the hierarchy. The Index, which should be considered an integral part of the *Guide*, offers further guidance.

Biographies presented a special problem and opportunity. The *Guide* has room for relatively few in comparison with standard compilations like the *Dictionary of Scientific Biography* (Scribners) in eighteen volumes or the *Oxford Dictionary of Scientists* with its 1600 entries. On the other hand, the *Guide* can scarcely send its readers elsewhere to learn where Isaac Newton or Marie Curie fit in time and space. The solution has been to include separate notices of the most famous and deserving, and a sprinkling of the less famous whose lives were representative and also remarkable. Where opportunities for useful comparisons exist, the *Guide* has parallel biographies in the style of Plutarch (Kurchatov and Oppenheimer; Lee, Yang, and Wu). Because the work of those given biographies is also covered in the *Guide*'s scientific articles, the lengths of their notices are not proportional to their standing in science. The democratic *Guide* allots 750 words to individuals, 1000 to pairs, and 1250 to triplets.

Each article has a short list of books to help readers follow up whatever captures their fancy. Although we have not taken over the *Companion*'s lengthy bibliographical essay, which points to general reference works and websites, we have reproduced its "Thematic Listing of Entries," which provides a systematic way to look at relations among sciences and technologies, and between them and their wider connections.

Acknowledgments. The Editor once again expresses his thanks to the contributors to the *Companion* whose work reappears here, and to the staff in the Trade Reference Department at Oxford University Press who have prepared the *Guide* for the press: David Bowers, Benjamin Keene, Ruth Mannes, and Anne Savarese.

DIRECTORY OF CONTRIBUTORS

Finn Aaserud, *Director, Niels Bohr Archive, Copenhagen, Denmark*
Bohr, Niels; Complementarity and Uncertainty

Theodore Arabatzis, *Assistant Professor of History and Philosophy of Science, University of Athens*
Thermodynamics and Statistical Mechanics

Jim Bennett, *Director, Museum of the History of Science, University of Oxford*
Instruments and Instrument Making

Joanne Bourgeois, *Associate Professor of Geological Sciences, University of Washington, Seattle, Washington*
Gaia Hypothesis; Glaciology; Mohole Project and Mohorovicic Discontinuity; Planetary Science; Seismology

Brian Bowers, *Retired Senior Curator (Electrical Engineering), Science Museum, London*
Lighting

Robert Brain, *Associate Professor of History of Science, Harvard University*
Exhibitions

William H. Brock, *Professor Emeritus of History of Science, University of Leicester*
Transuranic Elements; Urey, Harold

Laurie M. Brown, *Professor of Physics and Astronomy, Emeritus, Northwestern University*
Yukawa, Hideki

Jed Buchwald, *Dreyfuss Professor of History, California Institute of Technology*
Carnot, Sadi, and Augustin Fresnel

Joe D. Burchfield, *Associate Professor of History, Northern Illinois University*
Earth, Age of the; Universe, Age and Size of the

Regis Cabral, *International R & D Manager, Uminova Center, Umeå University, Umeå, Sweden*
Lattes, Cesar, and Jose Leite Lopes

David Cahan, *Professor of History, University of Nebraska-Lincoln*
Helmholtz, Hermann von, and Heinrich Hertz

Geoffrey Cantor, *Professor of the History of Science, University of Leeds*
Young, Thomas

Tian Yu Cao, *Associate Professor of Physics, Boston University*
Space and Time

David C. Cassidy, *Professor of Natural Science, Hofstra University*
Heisenberg, Werner, and Wolfgang Pauli

Allan Chapman, *Member of the Faculty of Modern History, University of Oxford*
Moon; Planet

I. Bernard Cohen, *Victor S. Thomas Professor of the History of Science, Emeritus, Harvard University*
Newton, Isaac

Patrick Curry, *Associate Lecturer, Centre for the Study of Cultural Astronomy and Astrology, Bath Spa University College, London*
Astrology

Olivier Darrigol, *CNRS, Paris*
Hydrodynamics and Hydraulics

Suzanne Débarbat, *Astronome Titulaire Honoraire, Observatoire de Paris*
Carte du ciel

Robert J. Deltete, *Professor of Philosophy, Seattle University*
Energetics

Michael Dettelbach, *Associate Director, Corporate and Foundation Relations, Boston University*
Geography

Isobel Falconer, *Research Associate, Open University, United Kingdom*
Electron; Mass Spectrograph; Thomson, Joseph John

Theodore S. Feldman, *Boston, Massachusetts*
Barometer; Climate; Climate Change and Global Warming; Hooke, Robert; Mathematization and Quantification; Meteorology; Thermometer

Maurice A. Finocchiaro, *Distinguished Professor of Philosophy, University of Nevada, Las Vegas*
Galileo

Paul Forman, *Curator, Modern Physics Collection, National Museum of American History, Smithsonian Institution*
Schrödinger, Erwin

Tore Frängsmyr, *Professor in History of Science, Uppsala University*
Agassiz, Louis; Werner, Abraham Gottlob

Alan Gabbey, *Professor of Philosophy, Barnard College, Columbia University*
Mechanics

Elizabeth Garber, *Associate Professor of History, State University of New York at Stony Brook*
Conservation Laws; Gibbs, J. Willard

Kostas Gavroglu, *Professor of History of Science, University of Athens, Greece*
Cold and Cryonics; Entropy; Solid State (Condensed Matter) Physics

Owen Gingerich, *Research Professor of Astronomy and History of Science, Harvard-Smithsonian Center for Astrophysics*
Brahe, Tycho; Copernicus, Nicholas

Gregory A. Good, *Associate Professor of History of Science, West Virginia University*
Atmospheric Electricity; Ionosphere; Lightning

Gennady Gorelik, *Research Fellow, Center for Philosophy and History of Science, Boston University*
Sakharov, Andrei, and Edward Teller

Loren R. Graham, *Professor of History of Science, Massachusetts Institute of Technology/Harvard University*
Kapitsa, Pyotr; Lomonosov, Mikhail Vasilievich

John L. Greenberg, *Paris*
Euler, Leonhard

Mott T. Greene, *John Magee Professor of Science and Values, University of Puget Sound*
Polar Science

Roger Hahn, *Professor of History, University of California, Berkeley*
Laplace, Pierre-Simon

P. M. Harman, *Professor of the History of Science, Lancaster University*
Maxwell, James Clerk

J. L. Heilbron, *Professor of History and the Vice Chancellor, Emeritus, University of California, Berkeley; Senior Research Fellow, Worcester College, Oxford*
Atomic Structure; Cassini Family; Cavendish, Henry, and Charles-Augustin Coulomb; Cosmic Rays; Curie, Marie, and Pierre Curie; Deluc, Jean André; Ether; Experimental Philosophy; Fire and Heat; Geodesy; Imponderables; Lightning Conductor; Magneto-optics; Noble Gases; Physics; Planck, Max; Pneumatics; Quantum Physics; Rainbow; Relativity; Röntgen, Wilhelm Conrad; Rutherford, Ernest; Sympathy and Occult Quality; Terminology; X Rays

Norriss S. Hetherington, *Director, Institute for the History of Astronomy, and Visiting Scholar, University of California, Berkeley*
Anthropic Principle; Astronomy; Astronomy, Non-Optical; Astrophysics; Black Hole; Chandrasekhar, Subrahmanyan; Cosmology; Extraterrestrial Life; Galaxy; Halley, Edmund; Hawking, Stephen, and Carl Sagan; Herschel, William, and John Herschel; Hoyle, Fred, and Martin Rye; Hubble, Edwin; Light, Speed of; Meteorics; Milky Way; Nebula; Parallax; Pulsars and Quasars; Steady-State Universe; Telescope

Frederic Lawrence Holmes, *Avalon Professor of the History of Medicine, Section of the History of Medicine, Yale University*
Homeostasis; Lavoisier, Antoine

Gerald Holton, *Mallinckrodt Professor of Physics and Professor of History of Science, Emeritus, Harvard University*
Einstein, Albert

R. W. Home, *Professor of History and Philosophy of Science, University of Melbourne*
Aepinus, F. U. T; Cohesion; Electricity; Franklin, Benjamin; Magnetism

David W. Hughes, *Professor of Astronomy, University of Sheffield, Department of Physics and Astronomy*
Aberration, Stellar; Eclipse; Orbit; Solar Physics; Star; Supernova

Bruce J. Hunt, *Associate Professor of History, University of Texas*
Michelson, A. A.

Myles W. Jackson, *Professor of the History of Science, Willamette University*
Optics and Vision

Frank A. J. L. James, *Reader in History of Science, The Royal Institution of Great Britain*
Faraday, Michael; Field; Spectroscopy

D. J. Kevles, *Stanley Woodward Professor of History, Yale University*
Cold Fusion; Sputnik

Alexei Kojevnikov, *Associate Professor of History, Department of History, University of Georgia, Athens*
Vavilov, Nikolai, and Sergey Ivanovich Vavilov

L. R. Lagerstrom, *Senior Lecturer, Electrical and Computer Engineering, University of California, Davis*
Constants, Fundamental; Standardization

Rachel Laudan, *Guanajuato, Mexico*
Cartography; Crystallography; Earth Science; Geology; Geophysics; Hutton, James; Ice Age; International Geophysical Year; Isostasy; Lyell, Charles; Mineralogy and Petrology; Mining Academy; Neptunism and Plutonism; Paleontology; Plate Tectonics; Stratigraphy and Geochronology; Terrestrial Magnetism; Uniformitarianism and Catastrophism

Marjorie C. Malley, *Cary, North Carolina*
Radium

William R. Newman, *Professor, Department of History and Philosophy of Science, Indiana University*
Boyle, Robert

Mary Jo Nye, *Thomas Hart and Mary Jones Horning Professor of the Humanities and Professor of History, Oregon State University*
Cathode Rays and Gas Discharge; Electrolysis; Ideal Gas; Radioactivity

Kathryn Olesko, *Associate Professor, Georgetown University*
Bessel, Friedrich Wilhelm; Error and the Personal Equation; Humboldt, Alexander von; Humboldtian Science

Giuliano Pancaldi, *Professor of History of Science, University of Bologna*
Galvani, Luigi, and Alessandro Volta

Manolis Patiniotis, *Lecturer, Department of History and Philosophy of Science, Athens University*
Boltzmann, Ludwig

Philip Rehbock, *Professor, Department of History, University of Hawaii*
Naturphilosophie; Oceanography

Michael Riordan, *Adjunct Professor of Physics, University of California, Santa Cruz*
Collider

Jessica Riskin, *Assistant Professor of History, Stanford University, California*
Newtonianism

Alan J. Rocke, *Henry Eldridge Bourne Professor of History, Case Western Reserve University, Cleveland, Ohio*
Atom and Molecule; Dalton, John; Periodic Table

Nicolaas A. Rupke, *Professor of the History of Science and Director of the Institute for the History of Science, Göttingen University*
Gauss, Johann Carl Friedrich

Arturo Russo, *Associate Professor of History of Physics, University of Palermo*
Fermi, Enrico; Satellite

Rose-Mary Sargent, *Professor of Philosophy, Merrimack College, North Andover, Massachusetts*
Matter; Mechanical Philosophy

Sara Schechner, *David P. Wheatland Curator, Collection of Historical Scientific Instruments, Harvard University*
Comets and Meteors

Silvan S. Schweber, *Professor of Physics and Richard Koret Professor in the History of Ideas, Brandeis University, Waltham, Massachusetts*
Bethe, Hans; Dirac, Paul Adrien Maurice; Elementary Particles; Feynman, Richard; Quantum Electrodynamics; Quantum Field Theory; Quark; Salam, Abdus; Tomonaga, Shin'ichirō

Robert Seidel, *Professor of History of Science and Technology, University of Minnesota*
Accelerator

H. Otto Sibum, *Associate Professor, Research Director, Max Planck Institute for the History of Science, Berlin*
Joule, James, and Robert Mayer

Ruth Lewin Sime, *Professor Emeritus of Chemistry, Sacramento City College*
Meitner, Lise

Crosbie Smith, *Professor of History of Science, University of Kent at Canterbury*
Thomson, William, Lord Kelvin

Abha Sur, *Lecturer, Department of Urban Studies and Planning, Massachusetts Institute of Technology*
Raman, C. V.

Anne van Helden, *Museum Boerhaave, Leiden, The Netherlands*
Air Pump and Vacuum Pump

Theo Verbeek, *Professor of Early Modern Philosophy, Utrecht University*
Descartes, René

Christiane Vilain, *Assistant Professor of Physics and Epistemology of Physics, University Denis-Diderot, Paris*
Huygens, Christiaan

James Voelkel, *Professor, Department of the History of Science, Johns Hopkins University*
Kepler, Johannes

Mike Ware, *Honorary Fellow in Chemistry, University of Manchester*
Photography

Peter Westwick, *Senior Research Fellow in Humanities, California Institute of Technology*
Acoustics and Hearing; Blackett, Patrick M. S., and Ernest O. Lawrence; Brain Drains and Paperclip Operations; Chaos and Complexity; Cherenkov Radiation; Cloud and Bubble Chambers; Electromagnetism; High-Energy Physics; Kurchatov, Igor Vasilyevich, and J. Robert Oppenheimer; Lee, T. D., C.S. Wu, and C.N. Yang; Low-Temperature Physics; Nuclear Magnetic Resonance; Nuclear Physics and Nuclear Chemistry; Plasma Physics and Fusion; Strangeness; Theory of Everything; Vacuum

Curtis Wilson, *Tutor Emeritus, St. John's College, Annapolis, Maryland*
Celestial Mechanics

THEMATIC LISTING OF ENTRIES IN THE *OXFORD COMPANION TO THE HISTORY OF MODERN SCIENCE*

Entries in bold face are included in the present *Guide*.
A dagger indicates entries commissioned for the Guide

HISTORIOGRAPHY OF SCIENCE

GENERAL CONCEPTS AND APPROACHES
Classification in Science
Discipline(s)
Gender and Science
Historiography of Science
History of Science
Modernity and Postmodernity
National Culture and Styles
Non-Western Traditions
Priority
Scientific Development, Theories of
Scientific Revolutions
Terminology

MAJOR PERIODS IN TIME
Enlightenment and Industrial Revolution
Long Fin-de-Siècle, the
Positivism and Scientism
Renaissance
Revolution, Restoration, and the Royal Society
Scientific Revolution
Shift of Hegemony
World War II and Cold War

MAJOR DIVISIONS
Aristotelianism
Baconianism
Darwinism
Hermeticism
Humboldtian Science
Mechanical Philosophy
Naturphilosophie
Neoplatonism
Newtonianism

ORGANIZATION AND DIFFUSION OF SCIENCE

THE SCIENTIFIC PROFESSION
Engineer
Internationalism and Nationalism
Scientist

GENERALIZED INSTITUTIONS
Academies and Learned Societies
Advancement of Science, National Associations for the
Botanical Garden
Bureaus of Standards
Cabinets and Collections
Hospital
Institute
International Organizations
Laboratory, Chemical
Laboratory, Industrial
Library
Meteorological Station
Military Institutions
Mining Academy
Multi-National Laboratories
National Parks and Nature Reserves
Observatory
Oceanographic Institutions
Professional Society
Schools, Research
Seminar
University
Zoological Garden

INDIVIDUAL INSTITUTIONS
Bell Labs
Bohr Institute
CERN
I. G. Farben
Institute for Scientific Information, Philadelphia
Kaiser-Wilhelm/Max-Planck-Gesellschaft
Lunar Society of Birmingham
Manhattan Project
RAND
Solvay Congresses and Institute
Third World Academy of Science

THE
OXFORD GUIDE TO
THE HISTORY OF PHYSICS
AND ASTRONOMY

A

ABERRATION, STELLAR. Because the observer on the earth is often moving across the path of the incoming light from a star, the observed direction of the star deviates from its true direction. This deviation, known as aberration, depends on the velocity of the observer on the earth and on the velocity of light. The maximum deviation owing to the earth's moving around its orbit is 20.5 seconds of arc. The earth's spin produces an additional much smaller diurnal aberration.

James Bradley, England's third Astronomer Royal, discovered stellar aberration serendipitously. He was looking for evidence of stellar parallax, a concept at the heart of the heliocentric solar system. Since the diameter of the earth's orbit is 300 million km, nearby stars should appear to move with respect to background stars as the earth orbits the sun. Since typical visual stars are twenty million times further away than the sun, the parallax angle usually is much less than a second of arc, and therefore extremely difficult to measure.

Robert *Hooke in London, Jacques Cassini in Paris, and Francesco Bianchini in Rome attempted to measure parallax angles, but with little success. Bradley and his friend Samuel Molyneux decided to check Hooke's observations of 1669 of the star Gamma Draconis, which passed overhead at the latitude of London.

The *telescopes of the day were long, cumbersome, and suffered considerably from tube flexure. Molyneux commissioned George Graham to construct a special vertical, 24-four-foot long, refracting telescope that could image Gamma Draconis once a day. Observations of stars at the zenith had the additional advantage that no correction for atmospheric refraction was required. Bradley and Molyneux observed deviations in the position of the star, but not of the sort expected for parallax. Other overhead stars shifted about in the manner of Gamma Draconis. After discounting the possibility that the axis of the earth might be changing direction, Bradley traced the phenomenon to the vector addition of the velocities of starlight and of the earth. He announced this result to the Royal Society in January 1729.

This observation confirmed the heliocentric *cosmology and helped to prompt Pope Benedetto XIV to remove the blanket proscription of Copernican treatises from the *Index of Prohibited Books*. Another removal—that of the effect of stellar aberration from recorded stellar positions—heralded a more accurate approach to positional astronomy. The fact that the aberration for all the stars in a specific direction had the same value, independent of the brightnesses (and thus the distances) of the stars, indicated the constancy of the velocity of the light. Bradley calculated that sunlight took an average of 8.2 minutes to reach the earth, about 0.1 minutes from the time accepted today. The precision of Bradley's instrument indicated that stellar parallax must be less than 1 second of arc.

Robert Grant, *History of Physical Astronomy* (1852). Colin A. Ronan, *Their Majesties' Astronomers: A survey of astronomy in Britain between the two Elizabeths* (1967).

DAVID W. HUGHES

ACCELERATOR. During the twentieth century physicists developed increasingly powerful artificial means to produce very high-energy particles to transform or disintegrate atoms. At the end of World War I, Ernest *Rutherford used alpha particles from naturally occurring sources of radiation to transform nitrogen into oxygen. He called for the development of more energetic sources of charged particles for nuclear experiments. Two of his students, John D. Cockroft and Ernest T. S. Walton, completed the first successful particle accelerator at the Cavendish Laboratory in Cambridge in 1932. By accumulating a potential of hundreds of thousands of volts, Cockroft and Walton accelerated protons to energies sufficient to disintegrate the nuclei of light elements. Owing to the repulsion between nuclei and protons, and the difficulty of creating and maintaining high potentials, their machine could not transform heavier elements.

Physicists soon turned to other means to accelerate particles. In the United States, Ernest Lawrence's magnetic resonance accelerator (the cyclotron), which applied energy to protons or deuterons in successive small steps rather than, as in the Cockroft–Walton machine, in one large jump, provided bombardments able to transform almost all nuclear species. His linear accelerator used the same principle of resonance acceleration to propel heavier nuclei. Another American, Robert Van de Graaff, returned to the one-jump method by a technique that allowed the accumulation of up to about ten million volts on a spherical conductor. The Van de Graaff, cyclotron, and linear accelerators were used at many universities and research institutes in the 1930s to explore the new field of nuclear physics. Financial support for the development of particle accelerators came largely from medical philanthropies. They hoped that their high-voltage X rays, neutrons, and other particles as well as the artificially radioactive products of nuclear interactions would be more effective against cancer and other diseases than the natural radiations from radium.

Lawrence was especially successful in generating support for his cyclotrons. The parameter most often employed to indicate their power—the diameter of the pole pieces of the magnet that retained the particles in their spiraling orbits as they accumulated energy—grew from a few inches to sixty. The increase in size gave a monumental increase in the energy with which the particles escaped from the magnet—from a few hundred thousand to thirty-two million electron volts. In 1939 Lawrence received the Nobel Prize in physics for his cyclotron and work done with it. The consequent enlargement of his prestige helped him to convince the Rockefeller Foundation to give the money to build a giant cyclotron with pole pieces 184 inches in diameter. Intended to be the last and largest of all particle accelerators and, in that way, the counterpart to the 200-inch telescope the foundation supported at Palomar, the 184-inch proved instead to be the first of a generation of much larger particle accelerators.

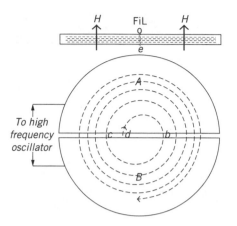

The cyclotron principle. Ions admitted at the center of the apparatus (at fil) are accelerated as they cross the gap (c, d, b) between the two D-shaped hollow cans (A, B) and describe circular orbits under a magnetic field when within the "dees"; the combination of circular arcs and horizontal accelerations produces the spiral paths shown.

The cyclotron did not accelerate electrons because their loss in energy and increase in mass owing to acceleration made them unattractive for work at high energy. Nonetheless, in 1939 Donald William Kerst invented a "betatron," which produced electrons or energies useful in nuclear investigations. After the war linear electron accelerators became competitive with proton accelerators for some purposes, and physicists turned the intense radiation of electrons maintained in circular orbits to advantage in "synchrotrons" whose "light" could be used in materials science and other applications.

The discovery of nuclear fission in uranium in 1939 provided a new role for particle accelerators and nuclear physicists. Cyclotrons at Berkeley and Los Alamos produced the first samples of plutonium, and cyclotroneering principles underlay the electromagnetic separation of isotopes in large banks of "calutrons" (after California University) that separated the fissile isotope of uranium for use in the first atomic bombs. Although other techniques eventually proved more efficient, these wartime successes opened the door to federal funding of nuclear physics by the Manhattan Engineer District and its successor, the Atomic Energy Commission.

The completion of the 184-inch cyclotron as a synchrocyclotron in 1946 was the first successful application of the synchrotron principle discovered by Edwin M. McMillan and Vladimir I. Veksler during the war. This principle enabled designers to accelerate particles in tight bunches by changing the frequency of the accelerating fields in step with relativistic changes in the mass of the particles. A series of accelerators built throughout the second half of the century, ranging from Brookhaven's 'Cosmotron' to the Tevatron, which produces trillions of electron-volts, incorporate the synchrotron principle and advances in magnetic technique that permit the confinement of the beam of accelerating particles to a narrow, evacuated pipe of fixed radius. The only limit to the size of these machines is financial. The first version of CERN had a diameter of 200 meters; the one now under construction extends to 27 kilometers. The largest current machine in the United States, at Fermilab, has a circumference of four miles, within which a herd of bison graze.

The linear accelerator also developed rapidly after World War II. At Berkeley, Luis Alvarez invented a type for protons using war surplus radar equipment. His student Wolfgang Panofsky applied the scheme to electrons at the Stanford Linear Accelerator (1962). Discoveries of new elements and particles by particle accelerators led to Nobel Prizes as well as to increasingly larger accelerators in the United States, western Europe, and the Soviet Union. The prestige and power of these machines made them political as well as physical icons. Only after the end of the Cold War did the enormous cost of accelerators prompt the United States to withdraw from the competition by canceling the Superconducting Super Collider.

See also ELEMENTARY PARTICLES; NUCLEAR PHYSICS.

Milton Stanley Livingston, *Particle Accelerators; a Brief History* (1969). Armin Hermann et al., *History of CERN*, 3 vols. (1987–1996). J. L. Heilbron and Robert W. Seidel, *Lawrence and His Laboratory: A History of the Lawrence Berkeley Laboratory* (1990).

ROBERT W. SEIDEL

ACOUSTICS AND HEARING. Acoustics, the science of sound, falls at the intersection of several fields, including mechanics, hydrodynamics, thermodynamics, and electromagnetism. Scientific interest in sound also derives from human hearing, and hence involves physiology and psychology; and acoustics engages fields outside science such as music and architecture. A distinct field of acoustics, including but not limited to hearing, gradually emerged from this disparate background in the eighteenth and nineteenth centuries, aided by the use of quantitative apparatus to produce and detect sound and the mathematical analysis of the results.

Ancient and medieval philosophers of nature studied acoustics mainly as a means to understand music. Mathematical theories of music dated at least to the Pythagoreans, who identified musical intervals as ratios of whole numbers and related musical pitches to lengths of vibrating strings. Although a few ancient writers speculated on the wave nature of sound and the propagation of compressions, Arabic and European natural philosophers through the Middle Ages and Renaissance continued to study acoustics only as part of music theory, if at all. In the early modern period, natural philosophers began to undertake systematic experiments and to extend their investigations to sound in general. Experiments with vibrating strings led *Galileo Galilei to posit the relation between pitch and frequency, elucidated around the same time by Giovanni Benedetti and Isaac Beeckman, and also to suggest that pitch depended on the tension and diameter of the string as well as its length. Marin Mersenne in the early seventeenth century used very long vibrating strings, some over one hundred feet long, to arrive at a quantitative relation between pitch and frequency. Mersenne also measured the speed of sound, as did his contemporary Pierre Gassendi, who asserted that soft and loud sounds traveled at the same speed.

The scientific academies that sprang up later in the seventeenth century made the speed of sound a prime program. Around the same time, experiments with improved air pumps, beginning with Otto von Guericke and extended by Robert *Boyle and Francis Hauksbee, convinced natural philosophers that sound did not travel in a vacuum. Experiments on the speed of sound tested in particular the theory of Isaac *Newton, whose work on fluid mechanics in the *Principia* included acoustics. But Newton arrived at a figure for the speed of sound, based on the pressure and density of the medium, at odds with contemporary estimates. Mathematicians in the eighteenth century, notably Leonhard *Euler, Jean d'Alembert, and Joseph Louis Lagrange, extended Newton's analytical mechanics and pneumatics to explain the discrepancy between theory and measurement, and Pierre-Simon *Laplace, Jean-Baptiste Biot, and Siméon-Denis Poisson at the start of the nineteenth century succeeded by suggesting that the passing sound wave heated the medium.

The quantification of acoustics accelerated in the nineteenth century, driven by the use of laboratory apparatus such as tuning forks, vibrating plates, and sirens to produce standardized tones, and sounding boards and the stethoscope to detect them. The new instruments helped bring acoustics into the realm of precision physics and also indicated the increasing quantification of physiology. Hermann von *Helmholtz combined knowledge of the physiology and physics of sound in his synthetic treatise *On Sensations of Tone* of 1862, which confirmed that the ear analyzes periodic sound waves into Fourier sums of simple harmonics and predicted the existence of nonlinear summation tones. John Tyndall helped bring Helmholtz's work to English-speaking audiences, elaborating it with his own experiments and generalizing beyond Helmholtz's particular interest in music. Lord Rayleigh then provided a systematic mathematical analysis of sound in his *Theory of Sound* of 1877–78; although Helmholtz provided mathematical details in appendices to his works, he had kept the main text nonmathematical. Rayleigh's two-volume work analyzed diverse phenomena of sound based on the vibration of air, liquids, and gases and of solid strings, plates, and rods under various perturbations; his books completed the edifice of classical acoustics.

New problems and programs were meanwhile emerging from the development of electromagnetism and its application to acoustics. Alexander Graham Bell, familiar with Helmholtz's research and inspired by his own work with deaf students, invented a means for transmitting speech over wires. Bell faced several coclaimants for the invention of the telephone and strong competition for its development, including an improved system designed by Thomas Edison. Edison in the meantime invented the phonograph, the foundation for the recording industry. The development of radio further spurred the invention of microphones, loudspeakers, amplifiers, vacuum tubes, and oscillators by the burgeoning electrical industry. Industrial engineers also agreed on an international standard for the intensity of sound, the decibel. Magnetostriction, piezoelectricity, and acousto-optics, three more developments of the late nineteenth century, provided new ways to produce and detect sound. Electroacoustics opened up a rich new field for the study of sound in the twentieth century.

Architectural acoustics, an ancient study, was extended in the seventeenth century by the Jesuit polymath Athanasius Kircher, among others. Around 1900 Wallace Sabine revived the subject with his work relating reverberation times to the volume and building materials of rooms. The introduction of electroacoustic technology, along with new sound-absorbing building materials, provided a means to active control of sound inside buildings, and in 1930 Carl Eyring revised Sabine's results to accommodate the new acoustic environments.

Sound does not just travel through air. The use of submarines in World War I spurred efforts to detect them, which soon focused on sound. The subsequent development of sonar for submarine warfare provided much support of acoustics research and also new tools for marine biology and oceanography. Sound also travels through matter. Ultra-high frequency sound, or ultrasound, became an important probe for solid-state research and found application in industrial materials and in medicine in the second half of the twentieth century.

Frederick Vinton Hunt, *Origins in Acoustics: The Science of Sound from Antiquity to the Age of Newton* (1978). Emily Thompson, *Soundscape of Modernity* (2002); Robert T. Beyer, *Sounds of Our Times: Two Hundred Years of Acoustics* (1999).

PETER J. WESTWICK

AEPINUS, Franz Ulrich Theodosius (1724–1802), German natural philosopher.

Aepinus's work *Tentamen theoriae electricitatis et magnetismi* (1759) revolutionized the study of *electricity and *magnetism, adding new conceptual rigor and bringing these subjects within the reach of mathematical analysis. Aepinus greatly strengthened and improved Benjamin *Franklin's controversial new theory of electricity and also developed, in close analogy to Franklin's theory, a highly successful new theory of magnetism. In the process he provided a model for the mathematization of other fields of experimental natural philosophy, which until then had been entirely qualitative studies.

Aepinus was born in Rostock, Germany, where his father was a professor of theology. He studied at Rostock and Jena universities, then taught mathematics and experimental physics at Rostock for several years. He also began a program of astronomical observation that led to his appointment as astronomer at the Berlin Academy of Sciences in 1755.

In Berlin, Aepinus joined the circle surrounding the mathematician Leonhard *Euler, who became a powerful patron. Electricity was then a leading subject of inquiry, following the discovery of the electrical nature of *lightning. Aepinus and his Swedish student Johan Carl Wilcke became highly skilled experimenters.

Aepinus and Wilcke adopted Franklin's idea that electrification consisted in the redistribution of a subtle "electric fluid" that supposedly pervaded all bodies. When a substance was rubbed some fluid transferred from one body to the other, leaving one with a surplus and the other with a deficiency of electric fluid. Aepinus and Wilcke systematically pursued this notion of two contrary electricities. Aepinus used this notion to make sense of the extraordinary behavior of tourmaline crystals, which, he showed, acquired opposite electrical charges on opposing faces when warmed. Struck by the analogy between an electrified tourmaline and a magnet, he conceived the idea that magnetization resulted from the redistribution within a piece of iron of a subtle magnetic fluid, different from but analogous to Franklin's electric fluid, the poles of a magnet being regions of "plus" and "minus" magnetic charge. Aepinus pursued this idea in a brilliant series of studies, which he brought together in 1759 in his *Tentamen*.

Using Isaac *Newton's discussion of gravity as a model, Aepinus reduced all electrical effects (and, by analogy, all magnetic effects) to unexplained forces acting at a distance between particles of ordinary matter and particles of electric (or magnetic) fluid. He transformed Franklin's inspired but rough ideas into a coherent and mathematically consistent theory of electricity and magnetism. He explained in detail a wide range of puzzling induction effects—including the most pressing problem confronting theorists of electricity at the time, the shock delivered by the Leyden jar—in which changing the electrical (or magnetic) condition of bodies induced changes in the electrical (or magnetic) condition of other, nearby bodies. The forces he invoked varied with distance. With the (inverse square) law of variation not yet established, however, Aepinus, being unwilling to base his work on speculation, could not progress beyond semiquantitative computations. Nevertheless, his work constituted a dramatic advance toward a fully mathematized physics. A generation later, it provided the starting point for the work of both Charles-Augustin Coulomb and Alessandro Volta (*see* CAVENDISH, HENRY AND CHARLES-AUGUSTIN COULOMB and GALVANI, LUIGI, AND ALESSANDRO VOLTA).

In 1757 Aepinus became professor of physics at the Academy of Sciences in Saint Petersburg. There, he caught the eye of the Grand Duchess, soon to become Russia's Empress Catherine the Great, and became her personal tutor in natural philosophy. For several years he published extensively on electricity and magnetism, mathematical questions, astronomy, and geography. However, further appointments—as director of studies at the Imperial Corps of Noble Cadets, tutor to Catherine's son (later the Emperor Paul), and head of the cipher department in the Russian Foreign Ministry—took him away from natural philosophy. In the early 1780s, at Catherine's request, he prepared detailed proposals for a national school system, which were only partly implemented. In 1798 he retired to Dorpat (Tartu), Estonia, where he died several years later.

J. L. Heilbron, *Electricity in the 17th and 18th Centuries* (1979). R. W. Home, *Aepinus's Essay on the Theory of Electricity and Magnetism* (1979).

R. W. HOME

AGASSIZ, Louis (1807–1873), geologist and zoologist, born in Switzerland.

Agassiz was educated at the universities of Zürich, Heidelberg and Munich and in 1829 obtained a doctorate in philosophy for a thesis on objects rare in Europe—Brazilian fish. He then took a medical degree before returning to natural history, under Georges Cuvier in Paris. He accepted a professorship at Neuchâtel in 1832 and published works on various species of fish, both living and fossil, notably *Recherches sur les poissons fossiles,* in five volumes (1833–1844), and *Monographies d'échinodermes vivans et fossiles*, in four volumes (1838–1842).

In 1836 Agassiz became interested in glaciers and learned of the observations of Ignace Venetz, civil engineer, and Jean De Charpentier, director of the salt mines at Bex. They had conjectured that the glaciers must once have been larger and that the huge erratic boulders that could be found in Alpine valleys had been transported by ice and not, as previously thought, by Charles *Lyell and others, by polar seas and currents. Agassiz had shared Lyell's interpretation, but his field studies under the direction of Venetz and Charpentier changed his mind. He quickly developed his own theory, which he presented in a paper to the Société Helvétique des Sciences Naturales in Neuchâtel in 1837. Within three years he had produced a more detailed study in book form, *Études sur les glaciers* (1840). In it he developed the idea that the whole of the northern hemisphere had been covered by an enormous sheet of ice, which had left traces in the form of erratic boulders, smoothed and striated rock surfaces, rounded stones and moraine. In addition Agassiz believed, unlike Charpentier, that the ice sheet in the Alps had been there before the mountains formed. By comparison with other regions, Agassiz worked out the effects of the ice and its geographical extent.

Agassiz's theory implied that the earth had developed from a colder to a warmer state, which contradicted prevailing belief. He explained that periods of colder climate were short episodes in the longer-term change from a warmer to a cooler state. Each such period had brought about the extinction of all plants and animals; no genetic connection existed between extinct and living species, even if they appeared to be similar. Each new warm period had seen a new Creation.

Agassiz's colleagues, particularly Leopold von Buch and Elie de Beaumont, and also geologists in Britain, severely criticized his ideas as far too speculative and based on too little evidence. It was not accepted until the 1870s, after it had been supplemented by several empirical studies. Despite certain weaknesses, like the idea that the ice sheet had been formed before the Alps, Agassiz's theory provided the basis for the development of modern quaternary geology.

In 1846 Agassiz went to the United States to lecture in Boston. He accepted the offer of a chair at Harvard and stayed there for the rest of his life. Although he returned to geology and studied glaciation from time to time, he devoted most of his attention to natural history in general. He became a great collector and in 1855 began to publish a gigantic work, *Contributions to the Natural History of the United States,* planned to comprise ten volumes although only four were published. In 1859, on Agassiz's initiative, the Museum of Comparative Zoology was founded at Harvard to connect field research with popular education. He was also instrumental in the founding of the National Academy of Sciences in 1863 and became an adviser to governments and various institutions. Agassiz never accepted Darwin's theory of evolution and remained a keen critic of evolutionary thinking all his life. Despite this he retained his position as the leading natural historian in the United States until his death.

Edward Lurie, *Louis Agassiz: A Life in Science* (1960). Jules Marcou, *Life, Letters, and Works of Louis Agassiz*, 2 vols. (1896).

TORE FRÄNGSMYR

AIR PUMP AND VACUUM PUMP. Air pumps exist in two variants: vacuum pumps to take gas from a vessel to create a (near) *vacuum, and compressors to supply gas to create a high pressure. Compressors have been important since the mid-nineteenth century for work on the thermodynamics of gases and for the cooling apparatus for low-temperature research. From about the same time, the vacuum pump became significant for research on *cathode rays. Production of sufficiently low pressures sometimes required state-of-the-art scientific knowledge, but in the modern period the vacuum pump has served merely as a technical tool. In the late seventeenth and early eighteenth centuries, however, it played a crucial role at a more fundamental level.

The idea of pumping air was not new when the Magdeburg Burgomaster Otto von Guericke adapted a fire syringe around 1647 to remove air from a vessel. Compressors had been known since antiquity. What was new was the application of a pump to a philosophical question (the possibility of a macroscopic vacuum) as well as the idea of evacuating rather than compressing. Guericke's work received attention through the dramatic demonstrations he gave of the weight and force sustainable by the pressure of the air. The vacuum pump caught on as a philosophical instrument only after Robert *Boyle published his pneumatic experiments in the 1660s. Boyle recognized the vacuum pump as an ideal tool for the experimental method he wished to apply. Rather than answering one specific philosophical question, Boyle wanted to map the properties of the vacuum and of air. His work met with serious criticism from people who did not believe in the experimental method, but within the circle of Baconian scholars Boyle's example reached an emblematic status. Pumps were depicted in frontispieces to works in natural philosophy, and savants had themselves portrayed with the instrument in order to express their commitment to the new science.

Early air pumps leaked. This fault proved significant both in using pumps and securing their acceptance as a reliable tool of philosophy. Correcting it proved far from easy. Hence the small number of air pumps initially made, some 15 in all before the instrument became available commercially in the 1670s. These early pumps were extremely expensive. Even in the eighteenth century, the air pump remained an expensive piece of philosophical equipment.

After 1750, the technical improvement of the air pump became a problem in engineering without much philosophical importance. Craftsmen introduced mechanically operated valves and diminished the dead space at the bottom of the cylinder. Makers made extravagant claims for the degree of exhaustion their pumps could reach, but in fact the vapor pressure of the lubricant barred access to pressure below 10 millibars (0.01 atmospheres). Only when the lubricants were improved or eliminated could further progress be made. In 1858, the German glassblower J. H. W. Geissler introduced a vacuum pump that employed a piston of liquid mercury. Variants of the pump reached pressures sufficiently low to allow the discovery of *X rays in 1895. In the early twentieth century, a series of entirely new pump techniques, developed by Wolfgang Gaede, relied on the molecular behavior of the gas. Thanks to these techniques, pressures of 10^{-9} bar became available for many fields of experimental research. By then, the air pump had an important place in the background technology of science.

See also BAROMETER; ETHER; VACUUM.

Gerard L'E. Turner, *Nineteenth-Century Scientific Instruments* (1983). Steven Shapin and Simon Schaffer, *Leviathan and the Air-Pump: Hobbes, Boyle and the Experimental Life* (1985). Anne C. van Helden, "The Age of the Air-Pump," *Tractrix, Yearbook for the History of Science, Medicine, Technology and Mathematics* 3 (1991): 149–172. Willem D. Hackmann, *Museo di*

Storia della Scienza, Catalogue of Pneumatical, Magnetical and Electrical Instruments (1995).

ANNE C. VAN HELDEN

ANTHROPIC PRINCIPLE. The anthropic cosmological principle, in its weak form, states that the universe must be such as to admit and sustain life. From Descartes's "I think, therefore I am," we proceed to "I am, therefore the nature of the universe permits me to be." Although a tautology, this has interesting implications. Hydrogen and helium formed nearly instantaneously in the primordial inferno of the cosmological Big Bang, but the building blocks of life, including carbon, oxygen, and nitrogen, form in the interiors of stars over long times. Hence the universe we observe must be billions of years old, older than the first generation of stars to spew forth the building blocks of life, and still young enough that our own sun has not expired, taking us with it. The physicist Robert Dicke at Princeton University in the 1950s noted this necessary connection between observers and the observed age of the universe. The universe is expanding. Not too slowly, or it would come to a halt and collapse. Nor too fast, lest it become too dilute for stars to coalesce. The rate of expansion depends on the density of the universe. Had the initial density differed by as little as one part in 10 to the 60th power, the result would have been either a big crunch or a big chill, and no life. Slight changes in the values of physical constants, including the strengths of fundamental forces such as the gravitational and electromagnetic, and the masses and charges of subatomic particles, would also render life impossible. The remarkable set of coincidences apparently necessary for human life has prompted many grandiose inferences. The strong version of the anthropic principle, articulated in the 1970s, asserts that the universe was created and fine-tuned so that intelligent life could evolve in it. Here we have passed from science to religion.

John D. Barrow and Frank J. Tipler, *The Anthropic Cosmological Principle* (1986).

NORRISS S. HETHERINGTON

ASTEROIDS. See METEORITES.

ASTROLOGY. Astrology is best defined as the set of theories and practices interpreting the positions of the heavenly bodies in terms of human and terrestrial implications. Although inextricably entangled with what are now demarcated as science, magic, religion, politics, psychology, and so on, astrology cannot be reduced to any of these. Its historical longevity and cultural spread guarantee a certain diversity; yet it has always managed to appear as much the same thing to practitioners, public, and opponents alike. Until recent decades historians of science conceived astrology anachronistically as a "pseudo-science," the human meanings of which could be derived from its lack of epistemological credentials.

Western astrology originated as Mesopotamian astral divination. The planets and prominent stars, identified with gods in ways that have since changed remarkably little, were considered celestial omens in which the divine messages, largely answering royal concerns, could be discerned. Many key elements of the astrological tradition—not only the planetary deities, zodiacal signs, risings, and settings, but also the effort to systematize divination through what we would now consider astronomical and empirical observations—developed before the fifth century B.C., when natal astrology first appeared. Following Alexander's conquest of Persia, natal astrology was transformed by Greek geometric and kinetic models, which added the aspects, or angles of separation between planets and points, and emphasized the importance of the *horoscopos* or Ascendent, the degree of the zodiacal sign rising on the Eastern horizon. Astrologers tended to develop increasingly flexible interpretive schemes, of which the most famous and influential was formulated by Ptolemy (c. A.D. 100–170) in his *Tetrabiblos*.

In the wake of Alexander's conquests Greek astrology spread to Persia and throughout Eastern Asia as far as India. In this way Greek astrology eventually became incorporated into, and benefited from, the learning of the Arabic world. It was introduced into medieval Europe in Latin translations, notably, from the mid-twelfth century onwards, of works by Abu

Ma'shar (787–886). These supplied a philosophical basis (largely Aristotelian) for astrology and popularized the idea that conjunctions of Jupiter and Saturn ("grand conjunctions") in particular regions of the heavens signify changes of political rulership. A complete revolution of the conjunctions around the zodiac indicated changes in the fortunes of entire religions. Pierre d'Ailly and Roger Bacon took up this astral historiography.

In the late fifteenth century, a series of influential translations by Marsilio Ficino made available rediscovered Greek texts, including much of Plato, Plotinus, and Iamblichus and the *Corpus Hermeticum*. These placed a renewed magical and/or mystical astrology at the heart of the *Renaissance revival of neo-Platonism and hermeticism. Typically, it managed to evade Giovanni Pico della Mirandola's powerful critique of astrology in his *Disputationes* (1494) by finding shelter elsewhere in the set of ideas that had inspired him (for example, occult *sympathy and antipathy).

Astrology survived the condemnations of St. Augustine and the early church fathers, who saw it as pagan and a transgression of both human free will and divine omnipotence. In the late thirteenth century, Thomas Aquinas arranged a compromise that secured for it a long-lived, if limited, niche. His synthesis of Christian theology and Aristotelian natural philosophy permitted "natural astrology" to influence physical and collective phenomena but not human souls directly; the individual judgments (and in particular predictions) of "judicial astrology" were therefore illusory in theory. Since Aquinas admitted that most people followed the promptings of their bodies, which felt the influence of the stars, he gave a tacit legitimation of astrology in practice. But the Reformation presented a serious new challenge. Luther and Calvin objected violently to astrology's idolatry, as they saw it, which they stigmatized as "superstition."

The seventeenth century was pivotal in the history of astrology. Contrary to the argument of Keith Thomas's influential *Religion and the Decline of Magic* (1971), the historical puzzle is not why so many intelligent people then believed in astrology (at a time when most people did), but why did they cease to believe in it?

Strong social and political forces abetted its fall from favor. In the English Revolution the pamphlets and almanacs of astrologers on both sides—but especially those of William Lilly for Parliament—played a major, and highly visible, role. In the late seventeenth and eighteenth centuries the new patrician and commercial alliance sought to put sectarian strife and upheaval behind it, and astrology became firmly identified as vulgar plebeian (rather than religious) superstition, to be contrasted with the spirit of rationalism and realism. A new set of opponents, the metropolitan literati, promoted the perception. Jonathan Swift's issue of a mock almanac in 1707 predicting the death of the prominent astrologer John Partridge, followed by another putatively confirming its fulfilment, epitomized the attack. Partridge became a laughing-stock in coffeehouse circles, although his almanacs continued to sell. Benjamin Franklin later employed the same tactic to promote *Poor Richard's Almanac*.

Increasing political centralization in France made astrologers' unlicensed prophecies unwelcome there too. After a short period of ambivalence, most prominent European natural philosophers also started to close ranks against astrology, ignoring or criticizing it as part of the old Aristotelian order, and/or as (plebeian or Platonic) magic. Isaac *Newton's success set the seal on this development. He borrowed the old idea of attraction at a distance, but substituted a single and quantifiable force for an astrological sine qua non: the planets as a qualitative plurality. Natural philosophy quietly absorbed natural astrology (including the moon's effects on tides), but judicial astrology, as a symbolic rather than mathematical system addressing merely "secondary" qualities and "subjective" concerns, had no place in a newly disenchanted world. In this context construal of astrology as "superstition" began to acquire its present meaning as a cognate of stupidity or ignorance.

Although belief in astrology declined in the 18th century, it did not die. It survived in largely rural strongholds dominated by farmers' almanacs, and into a relatively simple and magical set of beliefs. But early in the nineteenth century, as the middle classes grew in power and began to break away from patrician hegemony, a new urban astrology appeared that still remains. More individualistic than before, it succeeded in adapting to consumer capitalist society. And in the early twentieth century, through the work of Alan Leo and his commercially canny Theosophy, astrology secured a firm footing in both the popular press and the thriving middle-class market for psychology-cum-spirituality. At present it seems to meet a demand for (re-)enchantment that no amount of technical, technological, or purely theoretical progress can serve.

Astrology has managed to adapt to, and even exploit, every challenge history has thrown it. There is no reason to expect it will ever fail to do so, despite the outraged denunciations it continues to attract from contemporary guardians of scientific probity.

Eugenio Garin, *Astrology in the Renaissance* (1983). Patrick Curry, ed., *Astrology, Science and Society* (1987). Jim Tester, *A History of Western Astrology* (1987). Patrick Curry, *Prophecy and Power: Astrology in Early Modern England* (1989). Tamsyn Barton, *Ancient Astrology* (1994). Nicholas Campion, *The Great Year* (1994).

PATRICK CURRY

ASTRONOMICAL MEASURING INSTRUMENTS. See INSTRUMENTS, ASTRONOMICAL MEASURING.

ASTRONOMY. Astronomy, unlike modern sciences formed during the Scientific Revolution of the seventeenth century, has an ancient pedigree. The goals of astronomers, however—their theories, their instruments and techniques, their training, their places of work, and their sources of patronage—have undergone changes as revolutionary as those experienced by other sciences over the past four centuries.

For two millennia before the seventeenth century, a primary problem for astronomy in the Western world was to discover the true system of uniform circular motions believed to underlie the observed and seemingly irregular motions of the planets, Sun, and Moon. Astronomers observed and recorded a few planetary, solar, and lunar positions; attempted to fit geometrical models to the observations; and constructed tables of positions. In addition to the effort for its own sake, there were practical offshoots, including personal horoscopes, more general warnings of man-made and natural catastrophes, and calendars foretelling times of religious celebrations and the agricultural seasons. Solar and stellar navigation became important only later; fourteenth- and fifteenth-century explorers still found their way in sight of land. The rest of the universe scarcely existed for ancient and medieval astronomers other than as the limiting outer sphere of the stars. Nor did they concern themselves with the physical composition of the universe. Beyond the region of the Earth lay one unchanging element.

The Copernican Revolution, begun in the sixteenth century, switched the places of the Earth and the Sun in a geometrical model still composed of uniform circular motions. In other important ways, however, *Copernicus radically redefined the astronomical agenda. After new observations refuted the ancient assumption of circular motion, the physical nature and cause of orbital motion, previously outside the province of astronomers, became crucial. Interest in the physical composition of the heavens increased as well once the Earth left the center of the universe, and especially after *Galileo's revolutionary new telescopic observations.

In 1609 rumor from Holland of a device using pieces of curved glass to make distant objects on the Earth appear near reached Galileo in Italy. He constructed his own *telescope and turned it on the heavens. Among his discoveries were four moons circling Jupiter. These Medicean stars, as he named them, secured him a position at the court of the Grand Duke of Tuscany. Galileo also observed mountains on the Moon and sunspots. Henceforth the telescopic discovery of hitherto unknown planets, moons, asteroids, comets, and nebulae, as well as examination of their more prominent activities, was a standard occupation of astronomers. Amateurs wealthy enough to procure relatively large telescopes could excel.

The *telescope became an instrument of precise measurement through a happy accident. Noticing that a spider's web spun in the focal plane of his telescope was superimposed on the telescopic image, the Englishman William Gascoigne realized that crosshairs or wires could help center telescopes on objects and also help measure angles between them. By 1700, after some resistance, astronomers had accepted the telescope as the primary instrument of astronomical measurement. The cost of larger telescopes, their operation, and analysis of data speeded the transition from individual observers to organized observatories under government patronage. The Paris Observatory was founded as part of the new Paris Academy of Sciences in 1666, and the Greenwich Observatory began operations a decade later under the oversight of the Royal Society of London. In a new age of exploration, the task of the Greenwich Observatory, as stated in a royal warrant, was to rectify "the tables of the motions of the heavens, and the places of the fixed stars, so as to find out the so much-desired longitude of places for perfecting the art of navigation."

The last great achievement of pre-telescopic observations was Tycho *Brahe's body of positional measurements at the end of the sixteenth century. He enjoyed inherited wealth and also royal patronage, given in exchange for the glory his discoveries and fame cast over his patrons. Early in the seventeenth century, Johannes *Kepler used Brahe's positions to destroy faith in uniform circular motion, showing instead that ellipses more accurately describe planetary motions around the Sun. Aristotelian physics, referring all motion to a central Earth, did not work in a Sun-centered universe. An explanation of why the planets retrace their elliptical paths around the Sun became a central problem of astronomy.

Near the end of the seventeenth century, Isaac *Newton treated celestial motions as problems in mechanics governed by the same laws that determined terrestrial motions. Bodies remain at rest or move uniformly in straight lines unless external forces alter their state. A force of attraction toward the Sun continually draws the planets away from rectilinear paths and holds them in their orbits. Newton showed mathematically that, on his mechanical principles, Kepler's elliptical orbits result from a universal inverse-square law of gravity. The working out of details left undone by Newton focused the energies of many mathematical astronomers during the eighteenth century. Newtonians added quantitative success upon success, though not in the form Newton himself had used. Newton had employed geometry, the accepted medium of mathematical proof, in his demonstrations. His successors used new, more powerful algebraic methods. This change may help explain why Newton's loyal followers in England made little progress compared to mathematicians on the continent. Royal academies with generous support for astronomers, particularly at Paris, Berlin, and St. Petersburg, also made a difference. London's Royal Society, in contrast, neither received nor paid out royal emoluments, and the astronomers royal at Greenwich had to provide some of their own instruments.

Exact orbital calculations incorporate the influence of small perturbation effects. The Sun alters the Moon's motion around the Earth, and Jupiter and Saturn modify the motions of each other about the Sun. The Swiss-born mathematician Leonhard *Euler, at St. Petersburg and later at the Berlin Academy of Sciences, helped develop mathematical techniques to compute perturbation effects. He applied them first to the Moon, and in 1748 to Jupiter and Saturn, whose motions were the subject of that year's prize topic of the Paris Academy of Sciences. Euler, and Joseph Louis Lagrange and Pierre-Simon *Laplace, both members of the Paris Academy, applied the newly all-powerful calculus to the perturbations of planets and satellites, the motions of comets, the shape of the Earth, precession (a slow conical motion of the Earth's axis of rotation caused primarily by the gravitational pull of the Sun and the Moon on the Earth's equatorial bulge), and nutation (a smaller wobble superimposed on the precessional motion of the Earth's axis).

The wealthy brewer and tireless astronomer Johannes Hevelius (right) and his wife observing on the roof of their home in Danzig in the middle of the seventeenth century. The instrument, a brass sextant six feet in radius, was modeled on Tycho Brahe's instrument of a century earlier.

Orbital calculations now lie outside mainstream astronomy. NASA scientists use computers to calculate trajectories for their spacecraft. Astronomers would be willing, for additional government funding, to calculate whether various passing asteroids will safely miss the Earth.

The business of astronomy has at times been a family profession. Gian Domenico *Cassini, recruited from Italy as effective head of the Paris Observatory in 1669, was succeeded by a son, a grandson, and a great-grandson, before the French Revolution drove the family from the observatory in 1793. Friedrich Struve, who helped Czar Nicholas I chart his vast empire and also recorded the positions of double stars at the Pulkovo Observatory, opened in 1839 outside St. Petersburg, founded an astronomical dynasty spanning four generations. A son succeeded him at Pulkovo, a grandson directed the Königsberg Observatory, and a great-grandson, having fled Russia in 1921 after the revolution, directed the Yerkes Observatory of the University of Chicago and then, in the 1950s, the astronomy department of the University of California (see DYNASTY).

In the middle of the nineteenth century, the Pulkovo Observatory shared the honor of possessing the largest refracting telescope in the world. Refractors bend to a focus light passing through curved glass lenses. As late as the end of the eighteenth century, glass of the quality necessary for optical instruments could be cast only in small pieces, up to two or three inches in diameter. The English duty on manufacturing limited production and made further experimentation too costly. Progress occurred on the continent, but even there the largest lens achieved by 1824 was a 9.5-inch disc for the Dorpat Observatory (in what is now Tartu, Estonia). In 1847 the 15-inch refractors at Pulkovo and the Harvard College Observatory were the largest in the world.

British industry made possible the first large reflecting telescope (1780 to 1860), which used metal mirrors to reflect light to a focus. William *Herschel built a 48-inch metal mirror in 1789; in 1845 William Parsons, the Earl of Rosse, completed in Ireland his "leviathan," which had a metal mirror 72 inches in diameter. Large reflectors with their tremendous light-gathering power yielded remarkable observations of distant stellar conglomerations. But difficulty in aiming tons of metal and rapid tarnishing of mirrors rendered early reflectors unsuitable for observatories and professional astronomers, who could not harness consistently the instrument's raw power. Something new arrived with the twentieth century: a 60-inch reflecting telescope built and installed at the new Mount Wilson Observatory in 1908. The telescope had a reflective silver coating on a glass disc

ground to bring incoming light to a focus, and a mounting system and drive capable of keeping the multiton instrument fixed on a celestial object while the Earth turned beneath it. The mountain observatory, funded by Andrew Carnegie's philanthropic Carnegie Institution, was one of the first located above most of the Earth's obscuring atmosphere. Its reflecting telescope, specifically designed for photographic work, completed the revolution in astronomical practice—which had required the tedious drawing by hand of features seen through telescopes—begun with the invention of *photography.

With professional astronomers by definition already employed in research projects, amateur astronomers pioneered in applying photography. The American John Draper took the first known photograph of a celestial object, the Moon, in the 1840s. In 1851 a new process using plates exposed in a wet condition made possible a few photographs of the brightest stars, but not until the introduction of more sensitive dry plates after 1878 did *photography become common in astronomical studies. Draper's son Henry took the first photograph of a nebula, the Orion Nebula, in 1880. In England long exposures taken by Andrew Common and Isaac Roberts brought out details too faint for the eye to see. Photography facilitated comparisons over time, produced permanent recordings of positions suitable for more precise measurement, and was essential for exploitation of the other new astronomical tool of the nineteenth century, *spectroscopy.

The development of spectroscopy and the subsequent rise of the new science of *astrophysics created new activities for astronomers. When attached to telescopes, prisms splitting light into spectra opened to investigation the physical and chemical nature of stars. In 1859 Gustav Kirchhoff, professor of physics at Heidelberg, showed that each element produces its own pattern of spectral lines. In the first qualitative chemical analysis of a celestial body, he compared the sun's spectrum to laboratory spectra. The English amateur astronomer William Huggins seized on news of Kirchhoff's work. By 1870 he had identified several elements in spectra of stars and nebulae. He also measured motions of stars revealed by slight shifts of spectral lines. Early in the twentieth century Vesto M. Slipher at the Lowell Observatory in Arizona was the first to measure Doppler shifts in spectra of faint spiral nebulae, whose receding motions revealed the expansion of the universe. It required an extended photographic exposure over three nights to capture enough light for the measurement.

Astronomical entrepreneurship late in the nineteenth century saw the construction of new and larger instruments and a shift of the center of spectroscopic research from England to the United States. Charles Yerkes and James Lick put up the funds for their eponymous observatories, which came under the direction, respectively, of the University of Chicago and the University of California. Percival Lowell directed his own observatory in Arizona. All three observatories were far removed from cities; the latter two sit on mountain peaks.

A scientific education became necessary for professional astronomers in the later nineteenth century, as astrophysics came to predominate and the concerns of professionals and amateurs diverged. As late as the 1870s and 1880s the self-educated American astronomer Edward Emerson Barnard, an observaholic with indefatigable energy and sharp eyes, could earn a place for himself at the Lick and Yerkes observatories with his visual observations of planetary details and his discovery of comets and the fifth satellite of Jupiter, but he was an exception and an anachronism. A project begun in 1886 at the Harvard College Observatory and continued well into the twentieth century to obtain photographs and catalog stellar spectra furthered another social shift in astronomy. It employed women, for lower wages than men would have received, but at least made space for them in a male profession. Annie Jump Cannon was largely responsible for the *Henry Draper Catalogue,* published between 1918 and 1924, which gave spectral type and magnitude for some 225,000 stars. Also, she rearranged the previous order of spectra into one with progressive changes in the appearance of the spectral lines. Although she developed her spectral

sequence without any theory in mind, astronomers quickly realized that changes in the strength of hydrogen lines indicated decreasing surface temperature.

Spectral class became even more useful when, early in the twentieth century, it was related to luminosity. The relationship could be used to estimate distances. Students of stellar evolution asked what in the constitution of stars gave rise to dwarfs and giants and why brightness increased systematically with spectral type. The source of stellar energy, and with it the constitution of stars, became better known after the discovery of *radioactivity at the beginning of the twentieth century. As late as 1920, however, the English astrophysicist Arthur Eddington complained that the inertia of tradition was delaying acceptance of the most likely source of stellar energy: the fusion of hydrogen into helium. Eddington calculated how fast pressure increases downward into a star and how fast temperature increases to withstand the pressure. He used qualitative physical laws regarding the ionization of elements developed in the 1920s by Meghnad Saha, an Indian nuclear physicist. Saha's work also provided a theoretical basis for relating the spectral classes of stars to surface temperatures. World War II produced a deeper understanding of *nuclear physics and more powerful computational techniques. The practice of astrophysics has moved from observatories to scientific laboratories to giant computers running simulations.

*Cosmology, the study of the structure and evolution of the universe, only belatedly insinuated itself into modern mainstream astronomy. An inability to measure great distances limited cosmology for centuries to philosophical speculations, often focused on the nature of nebulae and the possible existence of island universes similar to our galaxy. Cosmology achieved an observational foundation early in the twentieth century, when Harlow Shapley and Edwin *Hubble at the Mount Wilson Observatory made the observations that revealed the size of our *galaxy, the existence of other galaxies, and the expansion of the universe.

At first these observations made little connection with *Einstein's *relativity theory. Astronomers, especially in the United States, possessed only limited mathematical knowledge, and were largely content to produce observations while leaving theory to theoreticians. In England, in contrast, interest in relativity theory relatively flourished. But the work was strictly mathematical, without observational input. In the 1930s Hubble attempted to bridge the gulf between observation and theory. Cooperation, he wrote, featured prominently in nebular research at the Mount Wilson Observatory, and he struck up a close collaboration with colleagues at the nearby California Institute of Technology. Not until the 1960s, however, after the discovery of quasars (see PULSARS AND QUASARS) and the intense theoretical effort to find a new energy source to explain them, did relativity theory secure a place in mainstream astronomy.

During the 1970s and 1980s scientists realized that important cosmological features could be explained as natural and inevitable consequences of new theories of *elementary particle physics, and particle physics now increasingly drives cosmology. Also, particle physicists, having exhausted the limits of particle *accelerators and public funding for yet larger instruments, now turn to cosmology for information regarding the behavior of matter under extreme conditions, such as prevailed in the early universe.

The greatest change in astronomy during the twentieth century in understanding the universe and also in the backgrounds of astronomers and their activities followed from observations beyond visual light. In the 1930s the American radio engineer Karl Jansky and the radio amateur Grote Reber pioneered detection of radio emissions from celestial phenomena, and radar research during World War II helped develop *radio astronomy, especially in England. X-ray astronomy also took off after the war, first aboard captured German V-2 rockets carrying detectors above the Earth's absorbing atmosphere. NASA subsequently funded a rocket survey program and then satellites to detect X rays. Most of the new X-ray astronomers came over from experimental physics with expertise in designing and building instru-

ments to detect high-energy particles. Their discoveries followed from technological innovations. Gamma rays and infrared and ultraviolet light provided further means of non-optical discoveries in the space age. Unlike the relatively quiescent universe open to Earth-bound astronomers' visual observations, the universe newly revealed to engineers and physicists observing at other wavelengths from satellites is violently energetic.

See also ASTRONOMY, NON-OPTICAL; ASTROPHYSICS; COSMOLOGY.

Otto Struve and Velta Zebergs, *Astronomy of the 20th Century* (1962). Martin Harwit, *Cosmic Discovery: The Search, Scope, and Heritage of Astronomy* (1981). Michael Hoskin, *Stellar Astronomy: Historical Studies* (1982). Dieter B. Hermann, *The History of Astronomy from Herschel to Hertzsprung* (1984). John North, *The Norton History of Astronomy and Cosmology* (1994). John Gribbin, *Companion to the Cosmos* (1996). *American Astronomy: Community, Careers, and Power, 1859–1940* (1997). Michael Hoskin, *The Cambridge Illustrated History of Astronomy* (1997). John Lankford, ed., *History of Astronomy: An Encyclopedia* (1997).

NORRISS S. HETHERINGTON

ASTRONOMY, NON-OPTICAL. Many recent astronomical discoveries have been made without using visible light. The newly revealed universe is fascinating in its activity. Stars are born, galaxies collide, neutron stars collapse into black holes, quasars vary in hours through luminosities greater than that of our entire galaxy, and pulsars rotate gigantic radio beams in fractions of seconds.

Radio astronomy is the oldest branch of the new astronomy. In 1932 Karl Jansky, an American radio engineer with the Bell Telephone Company, detected electrical emissions from the center of our galaxy while studying sources of radio noise. Neither optical astronomers nor Jansky's practical-minded supervisors cared. Grote Reber, an ardent radio amateur and distance-communication addict, took an interest, and built for a few thousand dollars a pointable radio antenna 31 feet in diameter in his backyard in Wheaton, Illinois. In

1940 he reported the intensity of radio sources at different positions in the sky.

Fundamental knowledge underlying radio astronomy techniques increased during World War II, especially with research on radar, and especially in England. After the war, research programs at Cambridge, at Manchester, and at Sydney, Australia, dominated radio astronomy for the next decade.

Manchester's large steerable radio telescope at Jodrell Bank was rescued from financial disaster in 1958 by its ability to track *Sputnik* (*see* SATELLITE). Other uses floated had included mobilizing the telescope as part of a ballistic missile tracking system and for long-range bomber navigation. As an astronomical instrument it detected radar signals bounced off ionized meteor trails in the earth's atmosphere. Other scientists bounced radar signals off the Moon and developed planetary radar astronomy to map the surfaces of planets.

The program at Cambridge was led by Martin Ryle (*see* HOYLE AND RYLE), who received a Nobel Prize in 1974 for his overall contributions to radio astronomy. He completed a survey of almost two thousand radio sources, most of them extragalactic, in 1955. These sources had a bearing on Fred Hoyle's steady-state cosmological theory. Within a few years radio data and their interpretation argued convincingly against Hoyle's *cosmology, which allowed for fewer faint radio sources than were detected.

The major blow to Hoyle's theory came in 1965 with discovery of the cosmic microwave (short radio wave) background radiation. It had been predicted in 1948 by nuclear physicists exploring the consequences of the cosmological big bang, but no astronomer looked for it. In 1963 Arno Penzias and Robert Wilson with the Bell Telephone Laboratories detected what they first regarded as noise in an antenna, but soon realized it must be excess radiation of cosmic origin. In 1965 Robert Dicke at Princeton interpreted the work for which Penzias and Wilson received a Nobel Prize as a remote consequence of the origin of the universe.

Radio sources were identified in the early 1960s with star-like objects, now called quasi-stellar objects, or quasars. They have luminosities a thousand times that of our entire galaxy. The only known source for so much energy from such a small volume is a *black hole. Quasars also have extremely large red shifts, probably a manifestation of an expanding universe. Ryle shared the nobel prize with his cambridge colleague anthony hewish, who was credited with discovering pulsating radio sources, although his student jocelyn bell made the actual observation, in 1967 (*see* PULSARS AND QUASARS).

Radio astronomy is increasingly threatened by modern society's growing use of garage-door openers, microwave ovens, wireless telephones, and other sources of interference with radio signals from astronomical objects. The commercial use of radio frequencies is light pollution's invisible cousin.

Like radio astronomy, X-ray astronomy took off after World War II, first aboard captured German V-2 rockets. Herbert Friedman at the U.S. Naval Research Laboratory found that the Sun weakly emits X rays, as expected. Astronomers did not expect to find strong X-ray sources, and doubted that brief and expensive rocket-borne experiments could contribute much to observational astronomy. In the immediate post-*Sputnik* period, however, more government money existed for astronomical research than imaginative scientists could spend.

Help came from Italian-born Riccardo Giacconi, who worked at the private company American Science and Engineering and then at the Harvard–Smithsonian Observatory before becoming director of the Space Telescope Science Institute. In 1960, with funding from the Air Force Cambridge Research Laboratories, Giacconi discovered the first cosmic X-ray source, Sco X-1 (the strongest X-ray source in the constellation Scorpius). In 1963 Friedman's group detected a second strong X-ray source, associated with the Crab nebula, the remnant of a supernova explosion.

NASA now funded a rocket survey program and a small satellite devoted exclusively to X-ray astronomy. Launched into an equatorial orbit from Kenya on its independence day in 1970, the satellite *Uhuru*, Swahili for freedom, discovered binary X-ray pulsars, neutron stars whose energies arise from infalling matter from companion stars. The Einstein X-ray telescope, launched in 1978, revealed that individual sources account for much of the X-ray background radiation. Giacconi had wanted to name this satellite *Pequod*, from Melville's *Moby Dick*, as a reminder of Massachusetts and the American Indian. The inevitable comparisons of Giacconi and the egomaniacal Captain Ahab indicate the qualities needed to drive a large and complex scientific project to completion. NASA declined to associate its satellite with a white whale.

A single scientific paper was published on X-ray astronomy in 1962, and 311 a decade later. Only 4 out of 507 American astronomers participated in X-ray astronomy in 1962, compared to 170 out of 1,518 in 1972. Over 80 percent of the participants came from experimental physics with expertise in designing and building instruments to detect high-energy particles.

More energetic than X rays, gamma rays arise from nuclear reactions, including those that create elements in stars and bombs. A satellite watching for nuclear tests in the atmosphere made the first observation of cosmic gamma ray sources in 1973. The Compton Gamma Ray Observatory, deployed from a U.S. space shuttle in 1991, has recorded bursts of gamma radiation, about one a day, perhaps from mergers of extremely distant neutron stars into black holes. Gravitational waves from these events may be detectable by the Laser Interferometric Gravitationwave Observatory, constructed in 1999. The final seconds of such an event would be more luminous than a million galaxies. If it occurred in our galaxy, the burst of gamma radiation would destroy the earth's ozone layer, kill all life on Earth, and leave its surface radioactive for thousands of years.

The ultraviolet is relatively quiet. The Extreme Ultraviolet Explorer Satellite, launched in 1992, has observed hot plasma flares in the outer regions of a few nearby stars. Much of the interstellar medium, however, stops ultraviolet light. The Hub-

ble Space Telescope spans the spectral region from far-ultraviolet through visible to near-infrared.

Cool stars, planets, grains of dust in interstellar space, and also animals radiate most of their energy at infrared frequencies. Frank Low at Texas Instruments developed an infrared detector in 1961 so sensitive that under the impossible ideal condition of no other interfering infrared source, and hooked up to the world's largest telescope, it could have detected the body heat of a mouse on the Moon. Carl *Sagan wanted to use it to beat the Russians in the race to detect life on Mars. In 1967 Low, then at the Kitt Peak National Observatory in Arizona, discovered a giant cloud of gas and dust in the constellation of Orion invisible at optical wavelengths but emitting enormous energy in the infrared. The cloud might be an interstellar nursery in which stars are born and grow in their violent way. The Infrared Astronomical Satellite during ten months in 1983 before all its cooling helium evaporated found more extensive dust tails for comets than are visible optically, rings of dust in our solar system, and an extensive ring or shell of gas and dust around the star Vega, perhaps left over from the original cloud of gas and dust from which Vega condensed.

The recent revolution in non-optical astronomy has important policy lessons. Should bureaucrats follow Thomas Kuhn's theory of the structure of scientific revolutions and act as keepers of the paradigm, directing research funds into observations intended to extend and consolidate mature theories? Or should they favor the early stages of development of new fields, when practitioners fumble along without any fixed conceptual framework? Non-optical astronomical discoveries have followed largely from technological innovations, with little advance prediction or justification for their search. Is it, then, nobler of mind and more cost-efficient to serve old paradigms or new technologies?

See also ASTRONOMY; RELATIVITY.

Sir Bernard Lovell, *The Story of Jodrell Bank* (1968). David O. Edge and Michael J. Mulkay, *Astronomy Transformed—The Emergence of Radio Astronomy in Britain*

(1976). Martin Harwit, *Cosmic Discovery: The Search, Scope, and Heritage of Astronomy* (1981). Richard F. Hirsch, *Glimpsing an Invisible Universe: The Emergence of X-Ray Astronomy* (1983). Woodruff T. Sullivan, *The Early Years of Radio Astronomy: Reflections Fifty Years after Jansky's Discovery* (1984). Wallace Tucker and Riccardo Giacconi, *The X-Ray Universe* (1985). Wallace Tucker and Karen Tucker, *The Cosmic Inquirers: Modern Telescopes and Their Makers* (1986). David H. Devorkin, *Race to the Stratosphere: Manned Scientific Ballooning in America* (1989). Andrew J. Butrica, *To See the Unseen: A History of Planetary Radar Astronomy* (1996).

NORRISS S. HETHERINGTON

ASTROPHYSICS. Telescopes equipped with prisms that split starlight into rainbow-like spectra made possible investigations of the physical and chemical nature of astronomical objects. In 1802 the English chemist William Wollaston found several dark lines in the solar spectrum, and in 1814 the German optician Joseph Fraunhofer observed and catalogued hundreds of solar lines. In 1859 Gustav Kirchhoff, professor of physics at the University of Heidelberg, showed that each element produces its own pattern of spectral lines. In the first qualitative chemical analysis of a celestial body, Kirchhoff compared laboratory spectra from thirty elements to the Sun's spectrum and found matches for iron, calcium, magnesium, sodium, nickel, and chromium.

The English astronomer William Huggins, tired of making drawings of planets and timing the meridian passage of stars, likened news of Kirchoff's work to "coming upon a spring of water in a dry and thirsty land." Huggins suggested to his friend and neighbor, William Allen Miller, a professor of chemistry at Kings College, London, that they commence observations of stellar spectra. Initially Miller doubted the wisdom of applying Kirchhoff's methods to the stars, because of their faint light; but by 1870 Huggins and Miller had identified several elements in spectra of stars and nebulae.

Spectroscopic techniques were also employed to measure motion in the line of sight. In 1842, Johann Christian Doppler, an Austrian physicist, argued that motion of a source of light should shift the lines in its spectrum. A more correct explanation of the principle involved was presented by the French physicist Armand-Hippolyte-Louis Fizeau in a paper read in 1841 but not published until 1848. Not all scientists accepted the theory. In 1868, however, Huggins found what appeared to be a slight shift for a hydrogen line in the spectrum of the bright star Sirius, and by 1872 he had more conclusive evidence of the motion of Sirius and several other stars. Early in the twentieth century Vesto M. Slipher at the Lowell Observatory in Arizona measured Doppler shifts in spectra of faint spiral nebulae, whose receding motions revealed the expansion of the universe (*see* Cosmology).

Instrumental limitations prevented Huggins from extending his spectroscopic investigations to other galaxies. Astronomical entrepreneurship in America's gilded age saw the construction of new and larger instruments and a shift of the center of astronomical spectroscopic research from England to the United States. Also, a scientific education became necessary for astronomers, as astrophysics predominated and the concerns of professional researchers and amateurs like Huggins diverged.

George Ellery Hale, a leader in founding the *Astrophysical Journal* in 1895, the American Astronomical and Astrophysical Society in 1899, the Mount Wilson Observatory in 1904, and the International Astronomical Union in 1919, was a prototype of the high-pressure, heavy-hardware, big-spending, team-organized scientific entrepreneur. While an undergraduate at the Massachusetts Institute of Technology in 1889, he invented a device to photograph outbursts of gas at the Sun's limb. He continued studying the Sun at his home observatory and then at the Yerkes Observatory of the University of Chicago. In 1902 Andrew Carnegie established the Carnegie Institution of Washington with a $10,000,000 endowment for research, exceeding the total of endowed funds for research of all American colleges combined. With grants from the Carnegie Institution, Hale built the Mount Wilson Observatory with a 60-inch reflecting telescope in 1908 and a 100-

inch completed in 1917. In his own research, Hale in 1908 detected the magnetic splitting of spectral lines from sunspots (see MAGNETO-OPTICS).

Stellar spectra were obtained at the Harvard College Observatory and catalogued in a project begun in 1886 and continued well into the twentieth century. Women did much of the tedious computing work. Edward Charles Pickering, the newly appointed director of the Harvard College Observatory in 1881 and an advocate of advanced study for women, declared that even his maid could do a better job of copying and computing than his incompetent male assistant. And so she did. And so did some twenty more women over the next several decades, recruited for their steadiness, adaptability, acuteness of vision, and willingness to work for low wages. Initially the stars were catalogued alphabetically, beginning with A, on the strength of their hydrogen lines, but Annie Jump Cannon found a continuous sequence of gradual changes among them: O, B, A, F, G, K, M, R, N, S (mnemonically: Oh, be a fine girl; kiss me right now, Sweet.). Antonia Maury added a second dimension to the classification system by noting that spectral lines were narrower in more luminous, giant stars of the same spectral class. Cecilia Payne-Gaposkin, a graduate student at Harvard, determined spectra relative abundances of elements in stellar atmospheres. Her Ph.D. thesis of 1925, published as *Stellar Atmospheres*, has been lauded as the most brilliant thesis written in astronomy. She received her degree from Radcliffe College, however—Harvard did not then grant degrees to women—and later when employed at Harvard she was initially budgeted as "equipment."

Ejnar Hertzsprung in Copenhagen and Henry Norris Russell at Princeton University independently, in 1911 and 1913 respectively, related spectral class to luminosity. The Hertzsprung–Russell diagram was used to estimate distances by determining spectral class, reading absolute brightness off the diagram, and comparing that to the observed brightness diminished by distance. Once the source of stellar energy became better known, evolutionary tracks of stars would be drawn on the H–R diagram.

By the middle of the nineteenth century, discussions of the source of solar energy had rejected chemical combustion, which would have burnt away a mass as large as the Sun in only 8,000 years. From 1860 on, William *Thomson (Lord Kelvin) switched from an influx of meteoritic matter as the Sun's energy source to gravitational contraction. Thomson estimated the age of the Sun at twenty million years, less than a tenth required by Darwin's theory of evolution. In 1903 Pierre and Marie *Curie measured the heat given off by a gram of radium. This hitherto unknown source of energy opened up vast spans of time for geological and biological evolution, though astronomers were slow to abandon gravitational contraction (see EARTH, AGE OF).

In 1938 the German-born Hans Bethe, then at Cornell University, proposed a plausible mechanism for energy production in stars. He knew much about atoms but little about stellar interiors when he attended a conference in April 1938 reviewing the problem of thermonuclear sources in stars. Shortly thereafter, Bethe worked out the carbon cycle. It begins with a carbon-12 nucleus; adds four protons in stages, converting the carbon to nitrogen and then to oxygen; and ends in nuclei of carbon-12 and helium-4 plus energy. A bit later Bethe also envisioned a proton-proton reaction: two protons (hydrogen nuclei) form a nucleus of deuterium (one proton becoming one neutron); the addition of a third proton creates a helium-3 nucleus (two protons, one neutron), two of which collide to form helium-4 (two protons, two neutrons) while ejecting two protons. The net result in both cases is the fusion of four hydrogen nuclei into one helium nucleus plus a release of energy. The proton-proton reaction provides the main source of energy in stars about the mass of our Sun, 70 percent of which is hydrogen and 28 percent helium. In more massive and hotter stars with more heavy elements, the carbon cycle is more important. Bethe received the Nobel Prize in 1967 for his work on the mechanisms of energy production in stars.

See also ASTRONOMY; ASTRONOMY, NON-OPTICAL; COSMOLOGY.

Cecilia Payne, *Stellar Atmospheres* (1925). A. S. Eddington, *Stars and Atoms* (1929). Otto Struve and Velta Zebergs, *Astronomy of the 20th* Century (1962). Helen Wright, Joan N. Warnow, and Charles Weiner, *The Legacy of George Ellery Hale: Evolution of Astronomy and Scientific Institutions, in Pictures and Documents* (1972). J. B. Hearnshaw, *The Analysis of Starlight: One Hundred and Fifty Years of Astronomical Spectroscopy* (1986). John Gribbin, *Companion to the Cosmos* (1996).

NORRISS S. HETHERINGTON

ATMOSPHERIC ELECTRICITY. In the early eighteenth century, noticing the similarity between lightning and static electric discharges produced in the laboratory, Francis Hauksbee and William Wall suggested that there was electricity in the atmosphere. Lightning, aurora borealis, the odor of ozone on a mountain ridge, and St. Elmo's fire had not yet been associated with electricity.

In 1752, following a suggestion of Benjamin Franklin, the Comte de Buffon arranged for a test of the electrical character of *lightning. That same year, astronomer Pierre Charles Lemonnier discovered that the clear atmosphere exhibits an electrical charge. This was a transforming moment: air as well as clouds could be electrified. In 1757 Giambattista Beccaria, Franklin's first defender in Italy, began systematic research into atmospheric electricity using rockets and other probes. He published the first extended treatise on the subject in 1775. Charles Augustin *Coulomb investigated the electrical conductivity of the atmosphere. He showed that inadequate insulation could not account for most of the charge lost by an electrified metal plate and that losses owing to atmospheric conductivity increased with the charge on the test body.

The early nineteenth century saw natural historical studies of atmospheric electricity and speculations about its processes. Around 1812 Gustav Scöbler studied the diurnal period of atmospheric electricity, a phenomenon later investigated by François Arago and Alexander von *Humboldt. In the early 1840s Jean Charles Athanase Peltier discovered that the earth has a negative charge. Theorists pondered how processes on the earth's surface such as evaporation and chemical changes might produce atmospheric electricity.

Since Coulomb, scientists had believed that dust or moisture in the atmosphere, and not the air itself, dissipated charges. Early-nineteenth-century chemists Jöns Jacob Berzelius and Humphry Davy, among others, suggested that chemical compounds are held together by electrical force. After F. Linss discovered in 1887 that dry air can conduct electricity, scientists began to apply the theory of ions, introduced by Svante Arrhenius the same year to explain electrolytic and other phenomena in liquids, to gases. Joseph John *Thomson's discovery of the *electron in the late 1890s strengthened this understanding. Around 1900, Julius Elster and Hans Geitel, gymnasium teachers in Germany, explained charge dissipation as a result of the movement of ions and electrons. The investigation of ionized gas flow formed a central part of atmospheric research throughout the twentieth century.

Because the ground is usually charged negatively with respect to the air, a potential gradient (or voltage) amounting to at least hundreds and often thousands of volts per meter exists between the two. During the late nineteenth and early twentieth centuries, scientists measured the daily, seasonal, and other regularities of the gradient at land observatories, at high altitude using balloons, and at sea, especially on the research vessel *Carnegie* between 1909 and 1929. They developed ingenious instruments: William *Thomson's water-drop electroscope and flame electrometer, Franz Exner's gold-leaf electroscope with attached oil lamp, and Hans Benndorf's device that used *radium to bring itself to the ambient potential of the air. Victor Hess summarized the research in *Electrical Conductivity of the Atmosphere* (1926 in German, 1928 in English). *Radioactivity, discovered in 1896, the physics of various "rays" (*cathode and *X rays, for example), and the discovery by Henri Becquerel and Ernest *Rutherford that rays from radioactive substances ionize gases suggested exploration of the effects of naturally occurring radioactivity on the atmosphere. Elster and Geitel, C. T. R. Wilson, John Joly, H. Gerdien, Victor Hess, and John Satterly all investigated radioactivity in the atmosphere long before nuclear fallout brought it to public attention. Their work resulted in the discovery of *cosmic rays. From the 1920s on, Hess, Werner Kolhörster, Robert Millikan, Arthur Holly Compton, Scott Forbush, and Oliver Gish linked investigations of gamma radiation with atmospheric electricity through the effects of altitude on the ionization of gases in closed vessels.

In the later twentieth century, as some scientists concentrated on modeling the global electrical circuit, others examined the physics of lightning discharge closely, including the electrical fields produced, and discovered a wide assortment of new electrical phenomena, including red sprites (extensive, brief light flashes above large thunderstorms), trolls (Transient Red Optical Luminous Lineament—red discharges occurring just above cloud tops after an especially strong sprite), blue jets (conical discharges traveling upward from cloud tops to 45 km at 100 km/sec), and elves (Emissions of Light and Very low frequency perturbations from Electromagnetic pulse Sources—brief, expanding disks of red light at about 100 km altitude). Many of the newly discovered events occur in the stratosphere and above it.

Park Benjamin, *A History of Electricity (The Intellectual Rise in Electricity) from Antiquity to the Days of Benjamin Franklin* (1898). Paul Fleury Mottelay, *Bibliographical History of Electricity and Magnetism* (1922). W. J. Humphreys, *Physics of the Air*, 3d ed. (1964). Peter E. Viemeister, *The Lightning Book* (1972). J. L. Heilbron, *Elements of Early Modern Physics* (1982). Donald R. MacGorman and W. David Rust, *The Electrical Nature of Storms* (1998).

GREGORY A. GOOD

ATOM AND MOLECULE. Greek philosophers in the pre-Socratic era first expressed the belief that matter is not infinitely divisible, and that there exist invisibly small ultimate particles called "atoms" (Greek for "unsplittable") that constitute all perceptible matter. The Stoic philosophers and to some degree Aristotle and his followers opposed this opinion, and the debate continued sporadically into the early modern era. The idea of atoms became increasingly popular in the seventeenth and eighteenth centuries, but natural philosophers could do little more than speculate about them.

Chemists began vicariously to explore the world of ultimate particles soon after the death of Antoine-Laurent *Lavoisier in 1794. Lavoisier's definition of a chemical "element"—a chemically irreducible species of matter—provided for the first time a basis for an empirical approach to atoms, for each element might be thought to consist of a characteristic kind of irreducible particle. If the atoms of the various elements had truly distinguishable characteristics, such as weight, for example, then that fact ought to be discernible in the combinations of the various elements with each other. And so it was, for the laws of "stoichiometry" (regularities in the weight proportions in which elements combine with each other) were soon discovered, by Joseph Louis Proust, Jeremias Richter, and others.

These laws could only have been discerned by starting with the assumption of atoms. For example, carbon and oxygen combine in two different ways to make carbonic oxide and carbonic acid gas. The first has 43 percent carbon and 57 percent oxygen by weight; the second, 27 percent carbon and 73 percent oxygen. These numbers seemed to suggest nothing about integral ratios associated with presumed integral atoms. The Englishman John *Dalton in 1803 set out under the assumption that the substance that is less rich in oxygen consists of collections of identical molecules, each molecule consisting of one atom of carbon united to one atom of oxygen. In such a case, the carbon atom would have to weigh 12 to the oxygen atom's 16 units (or any other equal ratio), for 12 is to 16 as 43 is to 57. Then the second substance could easily be understood as consisting of molecules containing one carbon atom united to *two* of oxygen, for 12 is to 32 (16 + 16) as 27 is to 73. In this fashion, Dalton used the regularities of elemental weight relations to justify the theory of chemical atoms.

Assumptions about the composition of the molecules were required to determine the calculation of the relative weights of their component atoms. Eleven years after Dalton's calculation, Joseph Louis Gay-Lussac chose to believe that the more highly oxygenated gas consisted of a "volume" of carbon united with a "volume" of oxygen, with relative weights of 6 and 16 respectively (reflecting the 27/73 percentage composition). Then the other compound would have two carbons and one oxygen, for 12 (6+6) is to 16 as 43 is to 57. Thus, even if stoichiometry implied chemical atoms, it did not dictate *unambiguous* relative weight relations among them. The terms "atom" and "molecule" are used here as they are defined today, definitions not uniformly adopted until the 1860s.

Gay-Lussac preferred to reason in terms of "volumes" because he had discovered in 1808 that chemical reactions between gases take place in integral volumes. In 1811 and 1814, respectively, Amedeo Avogadro and Andre-Marie Ampère independently concluded from this law of combining volumes that the ultimate particles of gases must all have the same volume, so that equal volumes of any two gases under the same temperature and pressure must contain equal numbers of particles. "Avogadro's hypothesis," as it came to be known, suggested a simple physical method of determining relative atomic weights, namely by measuring relative vapor densities. But Avogadro's hypothesis did not seem to be reconcilable with certain facts, such as that water vapor was less dense than one of its constituents, oxygen. To solve this problem, Avogadro suggested that oxygen particles split in two in combining with hydrogen. This seemed both ad hoc and highly improbable to most of Avogadro's contemporaries.

Another disputed point was whether all atomic weights were integral multiples of that of the lightest element, hydrogen—a thesis defended as early as 1815 by the English physician William Prout. So many atomic weights were close to integers (when, by convention, hydrogen's was set equal to one) that Prout's hypothesis seemed probable. The hypothesis led to the notion that there might be a fundamental particle whose aggregation accounted for the various elemental atoms. But contrary cases such as chlorine (35.5 times that of hydrogen) gradually drew most chemists away from Prout's hypothesis.

Chemists working their way through the puzzles of atomic theory distinguished between atoms and molecules by thinking of the former as the smallest integral particle of an element, and the latter as the smallest integral particle of a compound; the first was homogeneous and unsplittable, the second a heterogeneous compound particle that could be formed, transformed, or dismembered. Every chemist applied his own distinct terminology to these concepts. Avogadro complicated the scheme, for his theory required a distinction between atoms and molecules *of elements*. Since most chemists supposed that the force uniting atoms into molecules was electrical, the idea that identical atoms of an element could attract each other seemed absurd.

Gradually there developed other physical approaches to understanding the atomic-molecular level: in addition to Gay-Lussac's law of combining volumes, there was Alexis Petit and Pierre Dulong's law of atomic heats, and Eilhard Mitscherlich's study of isomorphic crystalline compounds. The great Swedish chemist Jöns Jacob Berzelius gave the fullest early development of atomic theory between 1813 and 1826. Berzelius used all of the physical methods known to him, as well as sensitive and broad-ranging chemical data and analogies, to construct a system of atomic weights and formulas that most European chemists accepted from the late 1820s through the early 1840s.

That did not stop new proposals for deriving atomic weights and formulas, each system implying its own conception of the atoms and molecules it represented. By 1850 four versions including Berzelius's competed for European dominance. One of the two leading variants relied on chemical "equivalents," whose claimed advantage was being purely empirical, hence permanently defensible and unalterable; the other used a revised system of atomic weights advertised as being ontologically true, even though theoretically derived.

The latter had been developed by the Frenchmen Auguste Laurent and Charles Gerhardt. The Laurent-Gerhardt system resembled Berzelius's, but used Avogadro's hypothesis more consistently. Laurent proposed definitions of atom and molecule in 1846 that were subsequently adopted into all of the major European languages and are current today: the atom is the smallest *chemically* active unit of an element, while the molecule is the smallest freely existing and *physically* active unit of an element or compound.

During the 1850s a consensus quietly developed among leading German and British theorists in favor of the Laurent–Gerhardt system. The evidence driving this shift came primarily from ingeniously designed (and interpreted) chemical reactions, not from evidence from physics. Chemists had long puzzled over the phenomenon called isomerism (first named by Berzelius in 1830): the word applied to cases of two or more unequivocally distinct substances that contained the same elements combined in the same proportions. Many had conjectured that the differences in properties might be traced to differing arrangements of the atoms in the molecules of isomeric compounds. By the 1850s chemists had begun to get purchase on the question of intramolecular atomic arrangements, and by the end of the decade August Kekulé's theory of "chemical structure" (as his Russian rival Aleksandr Butlerov called it) was beginning to gain adherents. These novelties were made possible only by manipulating the newer formulas and atomic weights.

To gain fuller support for the new system, a small coterie of reformers organized an international chemical symposium, the first of its kind, which took place in Karlsruhe, Germany, in September 1860. Although the meeting itself had little drama, it widened the consensus after participants had digested the persuasive exposition of the Italian theorist Stanislao Cannizzaro. By the 1860s, most chemists had adopted the Laurent–Gerhardt system, which is nearly identical to that used today. Only in France were equivalents preferred by the majority of chemists until they, too, adopted atomic weights in the 1890s. The capstone of the chemical atomic theory, the periodic system of the elements (*see* PERIODIC TABLE), began with the work of Lothar Meyer and especially Dmitrii Mendeleev in the late 1860s.

In the late 1850s physicists were also penetrating the invisible world of atoms and molecules. The early development of the kinetic theory of gases by Rudolf Clausius and James Clerk *Maxwell led to a reaffirmation of Avogadro's hypothesis. As the kinetic theory gained in stature and success, Avogadro's hypothesis took on the status of an accepted fact of nature. This sense was reinforced when a number of estimates of the sizes of molecules, inferred from experimental physical data, emerged in the late 1860s and early 1870s, all fairly consistent among themselves.

Still, a current of skepticism toward the physical reality of atoms and molecules ran throughout the nineteenth century (*see* ENERGETICS). This skepticism was finally and permanently removed at the beginning of the twentieth century, as the result of the discovery of *radioactivity, subatomic particles, light quanta, and a new interpretation of Brownian motion.

A. G. van Melsen, *From Atomos to Atom* (1952). Frank Greenaway, *John Dalton and the Atom* (1966). W. H. Brock, *The Atomic Debates* (1967). David M. Knight, *Atoms and Elements* (1967). Stephen G. Brush, *The Kind of Motion We Call Heat* (1976). Mary Jo Nye, *The Question of the Atom* (1984). Alan J. Rocke, *Chemical Atomism in the Nineteenth Century* (1984).

A. J. ROCKE

ATOMIC STRUCTURE. By 1890, much evidence had accumulated that the atom of chemistry and the molecule of physics must have parts. The chemical evidence included analogies between the behavior of dilute solutions and of gases, as developed in the ionic theory of Svante Arrhenius (Nobel Prize in chemistry, 1903), and the implication from the *periodic table that the elements must have some ingredient in common. The physical evidence included the emission of characteristic spectra, which was likened to the ringing of a bell, and the formation of ions in gas discharges.

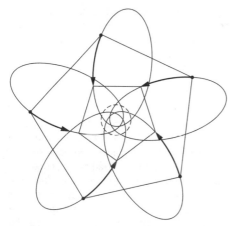

Sommerfeld's "Ellipsenverein" showing the linked motions of five electrons describing similar precessing elliptical orbits; at any instant the electrons occupy the vertices of a rotating, pulsating pentagon.

Study of the rays emanating from the cathode in these discharges prompted Joseph John *Thomson (Nobel Prize in physics, 1906) to assert in 1897 that the rays consisted of tiny charged corpuscles, which made up chemical atoms and constituted their common bond, and also emitted spectral lines and proffered the key to ionization. This hazardous extrapolation quickly received confirmation. Thomson showed that the *ions liberated from metals by ultraviolet light had the same low mass-to-charge ratio (m/e) as corpuscles, around one one-thousandth the corresponding ratio for hydrogen atoms in *electrolysis. Thomson's student Ernest *Rutherford (Nobel Prize in chemistry, 1908) determined that the rays from radioactive substances consisted of two sorts, one, alpha, unbendable, the other, beta, bendable, by a magnetic field; and he and the discoverer of *radioactivity, Henri Becquerel (Nobel Prize in physics, 1903), showed by the degree of bending that beta rays also had a very small m/e.

Indication of the presence of corpuscles within atoms came from the magnetic splitting of spectral lines accomplished by Pieter Zeeman (Nobel Prize in physics, 1902) as elucidated by Hendrik Antoon Lorentz (Nobel Prize in physics, 1902).

Lorentz traced spectra to the vibrations of "ions" whose m/e, as determined from Zeeman's magneto-optical splitting, was very close to those of the cathode-ray, photoelectric, and beta-ray particles. The omnipresent corpuscle became the *electron around 1900 when experiments at Thomson's laboratory at Cambridge indicated that the charge on corpuscles was about the same as that on a hydrogen ion.

Thomson was a leader of the British school, largely Cambridge trained, which, to paraphrase another of its lights, William *Thomson, Lord Kelvin, understood a problem best after constructing a physical model of it. J. J. Thomson supposed that the corpuscle-electrons constituting an atom circulated within a spherical space that acted as if filled with a homogeneous, unresisting, diffuse, weightless positive charge. On this last assumption, an atom of mass A would contain around $1,000A$ electrons. In 1904 Thomson exhibited analogies between a swarm of electrons circulating in concentric rings and the chemical properties of the elements. He thus introduced the important idea that one element differed from another only in its electronic structure. In 1906, having calculated the capacity of his model atoms to scatter *X rays and beta rays, he deduced that the number of scattering centers was more nearly $3A$ than $1000A$, thus giving weight and reality to the positive charge.

In 1910, Rutherford and Hans Geiger showed that alpha particles could be scattered through large angles. Rutherford inferred that Thomson's latest estimate was still too high. The scattering experiments indicated that the charges making up an individual scatterer acted as if assembled together at one point. Such an assemblage needed a smaller total charge to deflect an alpha particle through a given angle than a diffuse scatterer. Thus the atomic nucleus, which in Rutherford's original formulation could be either positive or negative, entered physics. He soon chose a positive center carrying the entire mass A and a charge of around $Ae/2$. Rutherford's student Henry G. J. Moseley then (in 1913–1914) deduced from his examination of characteristic x-ray spectra that the nucleus could be characterized by a whole number Z, beginning at 1 with hydrogen and increasing by a unit thereafter through the periodic table. Rutherford's school associated Z (the "atomic number") with the positive charge on the nucleus and recognized it to be a more reliable indicator of chemical nature than atomic weight. Classification by Z allowed an explanation of the few places in the periodic table where chemical properties did not follow weight and provided space for the accommodation of isotopes according to the ideas put forward by Rutherford's former colleagues Frederick Soddy (Nobel Prize in chemistry, 1921), Georg von György Hevesy (Nobel Prize in chemistry, 1943), Kasimir Fajans, and others around 1913.

During a study trip to Rutherford's laboratory in 1911, Niels *Bohr (Nobel Prize in physics, 1922) convinced himself that the nuclear atom was the route to a theory of atomic structure much better than Thomson's. One cause of his conviction was that ordinary electrodynamics required that a nuclear hydrogen atom with one electron could destroy itself instantly by radiation, whereas atoms stuffed with electrons could last a very long time. By presenting the problem of the existence of atoms in its strongest form, the nuclear hydrogen atom appealed to Bohr's dialectical mind. He solved the problem by fiat, by declaring that electrons could circulate in atoms only on orbits restricting their angular momenta to integral multiples of the *quantum of action. With this condition Bohr derived the wavelengths of the spectra of hydrogen and ionized helium and gave indications of the nature of molecular bonding in (1913–1914). Arnold Sommer-

feld extended the scheme to spectra containing doublets and triplets, and also to the x-ray spectra investigated by Moseley and Manne Siegbahn (Nobel Prize in physics, 1924).

Physicists returning from World War I found that Bohr (who as a Dane had been neutral), Sommerfeld (who was too old to fight), and their coworkers had made Bohr's combination of ordinary mechanics and quantum conditions into a hybrid, inconsistent, vigorous theory where none had existed before. Reasoning about electrons orbiting nuclei in quantized orbits, they could explain many features of atomic spectra, ionization and electron impact, and molecular bonding. The high point of the hybrid approach, known subsequently as "the old quantum theory," came in 1922, when Bohr claimed to be able to derive the lengths of the periods of the table of elements. But by then quantitative mismatches between theory and data on atoms more complicated than hydrogen had begun to undermine confidence in the internally structured mechanical atom. During the years between 1925 and 1927, it was superseded by ascribing non-mechanical spin, unsociability (Pauli Exclusion), and exchange forces to the electron, and by replacing the Bohr-Sommerfeld equations with matrix and wave mechanics (see *QUANTUM PHYSICS). Physicists declared that they knew everything about the atom "in principle" and moved their model-making into the recesses of the nucleus.

J. L. Heilbron, *Historical Studies in the Theory of Atomic Structure* (1981). Abraham Pais, *Inward Bound* (1986).

J. L. HEILBRON

B

BAROMETER. The barometer grew out of practical hydrostatics in the early seventeenth century. Italian mining and hydro-engineers had noticed that pumps would not raise water more than about thirty feet. *Galileo proposed that "the force of the vacuum" could hold up a column of water only so tall in a pump before it broke, as if the *vacuum were a rope holding up a weight. Isaac Beeckman, Giovanni Baliani, and others argued that the weight of the air outside the pump balanced the water column. Around 1641 Gasparo Berti attached a forty-foot lead pipe to the side of his house, filled it with water, sealed the top, and opened a cock at the bottom, which stood in a large vessel of water. Ten feet of water flowed out, leaving a column suspended some thirty feet high and a space above it that posed a difficult puzzle, since the reigning Aristotelian physics held the vacuum to be an impossibility.

Galileo's disciples repeated Berti's experiment with different liquids until, in 1644, Evangelista Torricelli filled a glass tube with mercury, inverted it in a bowl of the same liquid, and watched the silver liquid fall. Torricelli reasoned from the equality of the ratios of the heights of the mercury and water columns and their specific weights that the atmosphere indeed balanced the standing column. He suggested that his instrument "might show the changes of the air," but the meteorological possibilities of the instrument were largely ignored during two decades' debate over the nature of the space above the mercury and the balancing act of the air. The "Torricellian experiment" remained an experiment for the demonstration and investigation of the vacuum. Blaise Pascal pushed the experiment further towards a measuring instrument by fitting the tube with a paper scale and watching the mercury travel "up or down according as the weather is more or less overcast." This arrangement he designated a "'continuous experiment,' because one may observe, if he wishes, continually." The word "barometer" appeared in 1663 in the work of Robert *Boyle, who with Robert

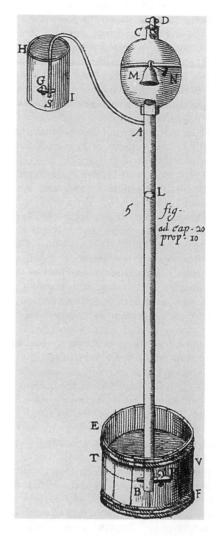

The water barometer standing over thirty feet high set up by Galileo's disciple Gasparo Berti in Rome around 1640; a hammer moved by a magnet outside the experimental space above the water struck a bell to determine whether sound could propagate in the space.

*Hooke had set a tube in a window for weather observation.

Soon many new types of barometers appeared: siphon instruments, in which a recurved lower end replaced the barometer's cistern; double and triple barometers, in which successive liquids magnified the motions of the column; diagonal barometers, whose inclination had the identical function; and others. Many designs aimed at a growing market for philosophical and mathematical instruments among the well-to-do. None increased the barometer's accuracy, which did not concern natural philosophers of the seventeenth and early eighteenth centuries.

Precision became important after the Seven Years' War. European states inaugurated national cartographic projects to provide accurate topographic information for military campaigns, taxation, agricultural reform, and other programs of the late Enlightenment. Barometers provided a convenient, if not the most authoritative, method for measuring heights on these surveys, and a demand arose for their precision. The instruments also proved useful to scientific travelers and for investigations into the properties of the air at a time when philosophers were discovering the different gases constituting it and puzzling over water vapor and its pressure in the atmosphere. The Genevan natural philosopher Jean André *Deluc, an avid Alpine explorer, designed the first barometer capable of precise measurement. Deluc's exhaustive *Recherches sur les modifications de l'atmosphère* (1772) included such reforms as boiling the mercury to remove dissolved air, methods for leveling the barometer and for reading the mercury meniscus, and corrections for the expansion with temperature of mercury, glass, and the ambient air. Equally important, Deluc showed how to deploy the instrument in repeated series of exhaustive measurements—a revolutionary technique. Deluc's rule for converting barometric measurements to heights was quickly adopted in England, where a national geodetic survey was underway.

In England instrument makers, moving away from craft-based organization, implemented industrial techniques such as division of labor, machine manufacture, and research and development financed out of profits from government trade. By 1770 Jesse Ramsden, the most advanced among them, was manufacturing barometers accurate to one-thousandth of an inch. In France the guilds hobbled efforts by the enlightened monarchy to modernize the instrument trade, though a few excellent instrument makers worked there towards the end of the ancien régime. The dissolution of the guilds during the Revolution and the need for instruments for military purposes and for metrification strengthened the instrument trade in France. Jean Nicolas Fortin's barometer, accurate to two-thousandths of an inch, became the standard model in France (see INSTRUMENTS AND INSTRUMENT MAKING).

In the nineteenth century the popularity of scientific travel and the increasing reach of European imperialism favored the development of the aneroid barometer. At the end of the seventeenth century Leibniz had suggested an instrument that would balance the pressure of the atmosphere against a spring-loaded, flexible bellows or box. In 1844 Lucien Vidie developed a practical barometer based on this principle. The aneroid became popular among mountaineers and mariners; in the twentieth century aviation pushed the development of highly accurate aneroid instruments.

The requirements of networks of weather observers organized by national governments in the late nineteenth century helped create a new class of barometers: the high-precision "primary" barometer, against which observers' instruments were calibrated. Manufacturers employed sophisticated methods, including the Sprengel mercury vacuum pump, to clean, fill, and evacuate these instruments.

See also AIR PUMP AND VACUUM PUMP.

W. K. Middleton, *History of the Barometer* (1964). Anthony J. Turner, *Early Scientific Instruments, Europe 1400–1800* (1987). Theodore S. Feldman, "Late Enlightenment Meteorology," in *The Quantifying Spirit in the 18th Century*, ed. Tore Frängsmyr, J. L. Heilbron, and Robin E. Rider (1990): 143–178. Jan Golinski, "Barometers of Change: Meteorological Instruments as Machines of Enlightenment," in *The Sciences in Enlightened Europe*, ed. William Clark, Jan Golinski, and Simon Schaffer (1999): 69–93.

THEODORE S. FELDMAN

BESSEL, Friedrich Wilhelm (1784–1846), German astronomer and mathematician.

Bessel was born in Minden, West Prussia, one of nine children of a justice department civil servant, Carl Friedrich Bessel, and his wife, Friederike Ernestine (nee Schrader), who came from a family of theologians and teachers. An inconsistent student, he left secondary school at 14, three years before graduating, to take up an unpaid apprenticeship with the trade firm Kuhlenkamp & Sons in Bremen. To pursue his interest in international trade, Bessel studied French, Spanish, English, tax regulations, economics, geography, mathematics, navigation, and astronomy. Heinrich Wilhelm Matthäus Olbers, a Bremen physician and astronomer with a special interest in comets, promoted Bessel's astronomical interests and talents.

In 1805 Bessel became director of the Lilienthal observatory where he primarily studied the paths of comets and planets. He accepted a royal offer to direct a new observatory at the University of Königsberg in 1809, but the faculty objected to his lack of a doctorate. Carl Friedrich *Gauss thereupon arranged for Bessel to receive an honorary doctorate from the University of Göttingen. Bessel arrived at Königsberg in 1810 to oversee the building of his cruciform observatory, completed in 1813 with the most precise instruments in the apse, the moveable instruments in the transept, and a lecture room, his office, and living space in the nave. He bought his first instruments, which included many of British manufacture, second-hand, from the estate of Count Friedrich von Hahn. Bessel later acquired exemplars of the first generation of German-made precision instruments, including meridian circles by G. F. von Reichenbach and J. G. Repsold and a heliometer by Joseph Fraunhofer.

Bessel and his colleagues the mathematician Carl Gustav Jacobi and the physicist Franz Ernst Neumann made Königsberg a center for the study of the exact sciences. Bessel was beloved by his students, many of whom became professional astronomers, notably Carl August Steinheil. In 1812 Bessel married Johanna Hagen, the daughter of a colleague, which made him a relative by marriage to several scientists including Neumann. Devoted to his family, Bessel was devastated by the death of his only son Wilhelm in 1840. Bessel died shortly before the announcement of the discovery of the planet Neptune, whose existence he had predicted based on perturbations in Uranus's orbit.

Bessel's astronomical work began with a computation of the orbit of Halley's comet. Orbital analysis remained one of his strong specialties. He also measured or calculated the positions of stars—almost 100,000 of them; between 1810 and 1813, when the Königsberg observatory was being built, he reduced James Bradley's measurements of 3,222 stars, taken between 1750 and 1762, to a common year, 1755, and eliminated errors in the measurements, including the effect of the density and temperature of the air. His results, compiled in *Fundamentae Astronomiae,* appeared in 1818. In 1821 he began a map of stars down to the ninth magnitude. His measurements of 61 Cygni in 1837–38 resulted in the first determination of stellar parallax (0.314″) and distance (10.4 light years).

Laced throughout Bessel's nearly 400 publications are theories of measurement and data reduction. Caught up in contemporary epistemology, he was intensely interested in the limits of reliability of measurements. He believed that instruments were constructed twice: first in the workshop and then on paper through error analysis. He provided theories of, and corrections for, precession, nutation, *aberration, refraction, temperature, and air pressure and other material factors affecting measurement. In addition to creating novel ways of treating constant errors and providing a fresh analysis of the method of least squares, Bessel discovered the human effect on measurement, the so-called personal equation.

State officials sought Bessel's expertise for Prussian state projects on navigation, the reform of weights and measures, and triangulation. Between 1823 and 1839 he served on the Prussian Ministry of Trade's examination board at the navigation school in Pillau. For the redetermination of the Prussian unit of length Bessel calculated the length of the simple seconds pendulum for both Königsberg and Berlin in the 1820s and 1830s in a study regarded as exemplary until the end of the century. Long courted for state cartography projects, especially after Prussia's reacquisition of the Rhine area in 1815, Bessel finally agreed to collaborate with General Johann Jakob Baeyer on the triangulation of East Prussia in 1830. He used the opportunity to link Prussian measurements to Danish and Russian ones. His measurement of the baseline for the triangulation by means of 4 iron bars of 3.9 meters each remained a metrological standard until the end of the century. In 1838 he calculated the earth's shape as a spheroid with an ellipticity of 1/299.15.

Although heavily involved in state projects, Bessel sometimes expressed political beliefs and observations at odds with prevailing views. He criticized legal cases where the rights of the accused were not protected, believed that a certain degree of dissent was necessary for strengthening norms and regulations, opposed the state-sanctioned neohumanist curriculum based on the classics, and held that strong religiosity and state service were incompatible. Bessel used his contact with state ministers (and two kings, Friedrich Wilhelm III and IV) to promote personal causes and support younger academic colleagues.

Sir J. F. W. Herschel, *A Brief Notice of the Life, Researches, and Discoveries of Friedrich Wilhelm Bessel* (1847). Kasimir Lawrynowicz, *Friedrich Wilhelm Bessel, 1784–1846* (1995). Besselgymnasium der Stadt Minden, ed., *Friedrich Wilhelm Bessel: Beiträge über Leben und Werk des bekannten Astronomen* (1996). Agnes M. Clerke, *Popular Astronomy during the Nineteenth Century* (2003).

KATHRYN OLESKO

BETHE, Hans Albrecht, theoretical physicist, born 2 July 1906 in Strasbourg, then a part of Germany. Bethe's father, a respected physiologist, transferred to the newly founded Frankfurt University in 1915. His mother had been raised in Strasbourg, where her father had been a professor of medicine. Bethe's father was Protestant. His mother, born Jewish, had become a Lutheran before she met Hans's father. Bethe grew up in a Protestant household in which religion did not play an important role.

Bethe attended a traditional Humanistisches in Gymnasium Frankfurt that emphasized Greek and Latin. Although instruction in mathematics and the sciences at gymnasia was in general poorer than in the humanities, several of the mathematics instructors at the Goethe Gymnasium that Bethe attended stimulated his interest. He left school with a fair amount of mathematics, a good deal of science, much Latin and Greek, and a knowledge of the German classics—Kant, Goethe, and Schiller—and the French and English languages.

By the time Bethe had finished Gymnasium, he knew he wanted to be a scientist, and his poor manual dexterity steered him first into mathematics and then into theoretical physics. After completing two years at the University in Frankfurt, he went to Arnold Sommerfeld's seminar in Munich. He obtained his doctorate in 1928 summa cum laude. After a brief stay in Stuttgart as Paul Ewald's assistant, he returned to Munich to habilitate under Sommerfeld. During the academic year 1930–1931, Bethe was a Rockefeller fellow at the Cavendish Laboratory in Cambridge, then directed by Ernest *Rutherford, and in Rome in Enrico *Fermi's Institute. In 1932, he again spent six months in Rome working with Fermi. By 1933, Bethe was recognized as one of the outstanding theorists of his generation. His book-length articles in the *Handbuch der Physik* on the quantum theory of one- and two-electron systems and on the quantum theory of solids immediately became classics. In April 1933, after Hitler's accession to power, he was removed from his position in Tübigen because he had two Jewish grandparents. He went to England and, in February 1935, to Cornell University, where he has been ever since.

Bethe's explanation of the mechanism for energy generation in stars grew out of his participation in the third Washington conference on theoretical physics in April 1938. These conferences, organized by George Gamow, Edward Teller, at the time one of Bethe's closest friends (*see* SAKHAROV, ANDREI AND EDWARD TELLER), and Merle Tuve, had become annual events. Most of the participants at that conference conflated the problem of nucleosynthesis with the problem of energy generation. Bethe returned to Cornell and separated the two problems. He advanced two sets of reactions—the proton-proton and the carbon cycle—to account for energy production in stars such as the sun. The carbon cycle depended on the presence of carbon in the star. At that time, physicists knew no mechanism to account for the abundance of carbon in stars. However, the presence of carbon in stars had been corroborated by their spectral lines in stellar atmospheres. Bethe accepted this fact and proceeded to compute the characteristics of stars nourished by the two cycles. He found that the carbon-nitrogen cycle gave about the correct energy production in the sun.

During World War II, Bethe worked on armor penetration, radar, and atomic weaponry. After a stint at the Radiation Laboratory at M.I.T, he joined Robert Oppenheimer (*see* KURCHATOV, IGOR VASILIEVICH, AND J. ROBERT OPPENHEIMER) at Los Alamos in 1943 as the head of the theoretical division. His ability to translate his understanding of the microscopic world to the design of macroscopic devices made his services invaluable at Los Alamos. Bethe, Fermi, and others converted their knowledge of the interaction of neutrons with nuclei into diffusion equations, and the solutions of the equations into reactors and bombs.

After the war, Bethe became deeply involved in the peaceful applications of nuclear power, as well as studying the feasibility, and participating in the design, of fusion bombs and ballistic missiles. He served on the President's Science Advisory Committee (PSAC) and other government committees. In 1967, he won the Nobel Prize for his theoretical investigations in 1938 explaining the mechanism of energy production in stars.

Bethe's life can be divided into well-delineated stages. Between 1906 and 1933, German culture and German institutions molded him. From 1934 until 1940, Cornell was his haven. He then published "Bethe's Bible" (in the *Reviews of Modern Physics*,

1935–37), a synthesis and concise presentation of all the extant knowledge of nuclear structure and reactions, and his solution of the problem of energy generation in stars. During the third period, World War II, he acquired new sorts of powers at M.I.T. and Los Alamos. Capitalizing on them, he played major national and international roles in the first postwar decade. He was at the center of developments in *quantum electrodynamics and meson theory, helped Cornell become one of the outstanding universities in the world, consulted with private industries trying to develop atomic energy for peaceful purposes, and exerted great influence on national security. His membership in PSAC and its subcommittees, consultancies with Avco, GE, and other industrial firms, and crises within the home help explain the routine character of his scientific production between 1955 and 1970. From the mid 1970s, in collaboration with Gerald E. Brown, Bethe returned to creative and productive work. The life cycle of supernovas and the properties of the neutrinos involved in solar fusion particularly engaged his attention.

See also ASTROPHYSICS.

J. Bernstein, *Hans Bethe: Prophet of Energy* (1980). *Hans Bethe: A Life in Science* [video-recording: 9 videocassettes (VHS NTSC)] (1997). S. S. Schweber, *In the Shadow of the Bomb. Bethe, Oppenheimer and the Moral Responsibility of the Scientist* (2000).
SILVAN S. SCHWEBER

BIG BANG. See SPACE AND TIME.

BLACKETT, Patrick M. S. (1897–1974), British experimental physicist, co-detector of pair production of positron, and opponent of nuclear weapons, and **Ernest O. LAWRENCE** (1901–1958), American experimental physicist, inventor of the cyclotron, and proponent of nuclear weapons.

Blackett and Lawrence were among the leading experimental physicists of their time and worked in the preeminent physics labs, Blackett at the Cavendish Laboratory in Cambridge, Lawrence at the Radiation Laboratory he founded at the University of California at Berkeley. Both won Nobel Prizes for their work in nuclear physics, as much for their development of instruments as for the discoveries they made with them. And both applied physics to military problems in World War II, although afterward Blackett spoke out against nuclear weapons while Lawrence strongly supported their development.

Blackett was born and raised in cosmopolitan London, Lawrence in the American Midwest. As youths both tinkered with radio sets, thus gaining valuable familiarity with electronics. Blackett trained in naval schools and saw action in the British Navy in World War I. After the war he went to Cambridge, where he studied mathematics and in 1921 joined the Cavendish under Ernest *Rutherford. Rutherford had just achieved the disintegration of a nitrogen nucleus by alpha particles, and Blackett undertook to observe the disintegration in a *cloud chamber. In 1924, after culling some 23,000 cloud chamber photographs of particle tracks, he identified eight that showed disintegration; the famous photos would adorn generations of nuclear physics textbooks.

Blackett then turned the cloud chamber to the study of *cosmic rays, incorporating a coincidence circuit using *Geiger counters to trigger the expansion of the chamber when cosmic rays passed through it. In 1932 Blackett and Giuseppe P. S. Occhialini obtained photos of positive electrons in cosmic ray showers and explained their appearance by pair production. Blackett won the Nobel Prize in physics in 1948, "for his development of the Wilson [cloud chamber] method and his discoveries, made by this method, in nuclear physics and on cosmic radiation."

Unlike Blackett, who was educated in elite naval schools, Lawrence studied in small midwestern colleges before finishing his Ph.D. in physics at Yale in 1924 with a thesis on the photoelectric effect. He stayed on at Yale until Berkeley hired him away in 1928. He took up the problem of accelerating subatomic particles to energies comparable to those of particles produced by natural *radioactivity, such as the alphas used by Rutherford and Blackett in their disintegration experiments.

Instead of trying to maintain a very high voltage, a difficult task with available

electrical technology, Lawrence in 1929 thought to turn a particle through a lower voltage many times. By bending particles into a circular path using a magnetic field and sending them through an alternating electric field provided by a radio-frequency oscillator, Lawrence could give them successive pushes and pulls to reach high voltages. The device, dubbed the cyclotron, promised ever-higher energies, which Lawrence pursued through the 1930s in a series of ever-larger machines and an ever-widening circle of funding sources. The cyclotrons provided copious sources of high-energy particles for nuclear physics and new radioactive isotopes for chemistry, biology, and medicine. Lawrence won the Nobel Prize in physics in 1939, "for the invention and development of the cyclotron and for results attained with it, especially with regard to artificial radioactive elements."

Lawrence pursued a pragmatic approach to the art of *accelerator building, often resorting to cut-and-try instead of theory to achieve higher energies. The Cavendish cultivated a similar tradition of string-and-sealing-wax experiment. But both labs increasingly adopted characteristics of big science, including large collaborative teams of engineers and scientists, complex organization, and connections to industrial equipment-makers.

World War II diverted both men, like many of their colleagues, to war work. By 1935 Blackett was involved in the British project to develop radar for air defense; after the outbreak of war he started work on bomb-sights and joined the so-called MAUD committee on the British atomic bomb project. He spent most of the war itself engaged in operations research, a new field that applied quantitative methods to military tactics, such as the optimum placement of antiaircraft batteries and the coordination of antisubmarine warfare. Lawrence lent several of his top scientists to the American radar project, but his main contribution was to the Manhattan Project for the production of atomic bombs, where he turned his cyclotrons into giant mass spectrographs to separate uranium isotopes. As one of the scientific leaders of the project, Lawrence helped advise the government on the use of atomic bombs.

Blackett became involved in left-leaning politics in the 1930s and began to explore relations between science and society. During the war he had opposed in vain Britain's strategic bombing of German cities, both on moral grounds and on the practical military consideration, backed up by operations research, that bombers would be more useful in antisubmarine warfare. His opposition to strategic bombing underlay his postwar criticism of atomic weapons, again on both moral and military grounds: he discounted the effectiveness of atomic weapons and urged Britain not to develop them and to stay neutral in the deepening cold war. Blackett publicized his views in a widely read book, *Military and Political Consequences of Atomic Energy* (1948; published in 1949 in the United States as *Fear, War, and the Bomb*), which also introduced the argument that the United States had used atomic bombs not so much as a military weapon against Japan but as a political weapon against the Soviet Union. His activism earned him an undeserved reputation as a communist fellow-traveler, but he continued to speak out against nuclear weapons and the arms race.

The pragmatic Lawrence professed to keep politics out of science and vice versa, but his ascent into influential circles made that impossible. He displayed no moral scruples after Hiroshima; he felt atomic bombs had ended the war and further bloodshed, and might prevent future wars. He became a staunch advocate of nuclear-weapons development by the United States, arguing in favor of a crash program to build a hydrogen bomb in 1949 and mobilizing his laboratory to aid the effort. A consequence of his advocacy was the establishment of an offshoot of the Berkeley lab that became the Livermore weapons laboratory. Although Lawrence did not testify in person at the security hearing of J. Robert Oppenheimer in 1954, he opposed the continued influence of Oppenheimer, whom he thought insufficiently hawkish on nuclear policy (*see* Kurchatov, Igor Vasilievich, and J. Robert Oppenheimer).

The scientific careers of Lawrence and Blackett also diverged. Lawrence was the exemplar of the entrepreneurial lab director, his attention after the war devoted more to administration than to his own research. Blackett turned down the directorship of the Cavendish in 1953. He remained immersed in research but left *nuclear physics for problems in geomagnetism, where he exercised his instrumental dexterity free of the trappings of big science.

Some parallels persisted. Both Blackett and Lawrence propagated their opinions in the postwar decade not from official advisory positions, but as physicists of high reputation with personal connections to political and scientific leaders. Their respective governments honored their achievements. The two national laboratories Lawrence established in Berkeley and Livermore bear his name. In the 1960s Blackett re-entered high-level advising under the Labor government, served as president of the Royal Society, and was elevated to the peerage as Baron Blackett of Chelsea.

See also Accelerator.

Herbert Childs, *An American Genius: The Life of Ernest Orlando Lawrence* (1968). Bernard Lovell, *P. M. S. Blackett: A Biographical Memoir* (1976). J. L. Heilbron and Robert W. Seidel, *Lawrence And His Laboratory: A History of the Lawrence Berkeley Laboratory* (1989). Mary Jo Nye, *Blackett: Physics, War, and Politics in the Twentieth Century* (2004). Peter J. Westwick, *The National Labs* (2003).

Peter J. Westwick

BLACK HOLE. An early intellectual precursor to the concept of black holes may be found in John Michell's dark stars. Michell noted in 1783 that the gravity of a sufficiently large sun would prevent its light particles from reaching us. The existence of such an object might be inferred from the motions of luminous bodies around it. Michell's idea was ignored even before wave theory, in which gravity does not act on light, overthrew the particle theory from which he reasoned.

Collapsed stars make another intellectual precursor to black holes. After Einstein's general theory of *relativity (1915), gravity again could act on light. In the 1930s, Subrahmanyan *Chandrasekhar, recently arrived from India to study in England, modeled stellar structures. He found that stars of less than 1.4 solar mass shrink until they become white dwarfs, but more massive stars continue contracting. The British astrophysicist Arthur Eddington noted that at high compression, gravity would be so great that radiation could not escape, a situation he regarded as absurd. Others accepted Chandrasekhar's mathematics but believed that continuous or catastrophic mass ejection would bring stars below the critical mass. Also, massive stars might evolve into stars composed of neutrons. In 1939, however, the American physicist J. Robert Oppenheimer established a mass limit for neutron stars (see KURCHATOV, IGOR VASILIEVICH, AND J. ROBERT OPPENHEIMER). When it has exhausted thermonuclear sources of energy, a sufficiently heavy star will collapse indefinitely, unless it can reduce its mass.

In the 1960s, computer programs that simulated bomb explosions were modified to simulate implosions of stars. A renewed theoretical assault followed on "black holes," as they were named by the American nuclear physicist John Wheeler in 1967. In contrast to collapsed or frozen stars, black holes are now known to be dynamic, evolving, energy-storing, and energy-releasing objects.

Because no light escapes from black holes, detection of them requires observing manifestations of their gravitational attraction. From a companion star, a black hole captures and heats gas to millions of degrees, hot enough to emit *X rays. Because the Earth's atmosphere absorbs X rays, devices to detect them must be lofted on rockets or *satellites. A few black holes probably have been found, although other explanations of the observational data are possible.

In another predicted manifestation, two black holes spiral together, gyrate wildly while coalescing, and then become steady. Outward ripples of curvature of spacetime, also called gravitational waves, would carry an unequivocal black-hole signature.

Gravitational waves should propagate through matter, diminishing in intensity with distance. On Earth, they should create tides the size of an atom's nucleus, in contrast to lunar tides of about a meter. Gravitational-wave detectors may be operational early in the twenty-first century.

Meanwhile, without benefit of prediction and intent, we may already have observed manifestations of black holes. Extraordinarily strong radio emissions from both the centers of *galaxies and from quasars (see PULSARS AND QUASARS) may be powered by the rotational energy of gigantic black holes, either coalesced from many stars or from the implosion of a single supermassive rotating star a hundred million times heavier than our sun. Other possible explanations for radio galaxies and quasars, however, do not require black holes.

S. W. Hawking and W. Israel, eds., *Three Hundred Years of Gravitation* (1987). Kip S. Thorne, *Black Holes and Time Warps: Einstein's Outrageous Legacy* (1994).

NORRISS S. HETHERINGTON

BODE'S LAW. See METEORITES.

BOHR, Niels (1885–1962), physicist, inventor of the quantum theory of the atom. Niels Bohr's mother Ellen, née Adler, belonged to a flourishing Danish–Jewish banking family. His father, Christian Bohr, was an internationally renowned physiologist at the University of Copenhagen. Around the turn of the century, Christian Bohr hosted an informal discussion group with three other prominent Copenhagen professors that young Niels was allowed to attend. What he heard helped shape his mind.

At the University of Copenhagen, Bohr's teacher was Christian Christiansen, the university's only physics professor and a member of Christian Bohr's discussion group. Niels Bohr's first major contribution to science—which was to guide the development of physics for years to come—derived from a postdoctoral stay at Manchester, where Ernest *Rutherford and his international group of collaborators had just established that the mass of the atom was concentrated in a small nu-

Bohr lecturing at Columbia, 1937

cleus with electrons swirling around it at relatively large distances. According to classical physics, such a system would be unstable. As a remedy, Bohr postulated, in a trilogy of papers published in 1913, that the quantum of action introduced by Max *Planck in 1900 set a condition on the nature of the orbits that an electron can occupy. This seemingly arbitrary postulate proved to have impressive predictive power, accounting for hitherto inexplicable spectroscopic data as well as providing a theoretical basis for the *periodic table of the elements. Bohr received the Nobel Prize for this work in 1922.

At this time Bohr had secured himself a solid institutional base at the University of Copenhagen, where a new Institute for Theoretical Physics was inaugurated in 1921. The very best of the younger theoretical physicists from all over the world flocked to Copenhagen. Bohr's celebrated need for a "helper" in developing his ideas contributed to the institute's unique atmosphere, subsequently described as the "Copenhagen spirit."

After the formulation of quantum mechanics in 1925, which completed the development of quantum theory, Bohr resumed the philosophical interest of his youth, taking a leading role in interpreting the new theory and pondering its philosophical implications outside the field of

physics. In 1927, he introduced the complementarity argument, which he continued to refine and promote for the rest of his life and which still constitutes the basis for the "Copenhagen interpretation" of quantum mechanics (see COMPLEMENTARITY AND UNCERTAINTY). Yet his emphasis on an experimental basis for theoretical work did not subside, and beginning in the early 1930s he changed the object of experimental and theoretical research at the institute from the atom as a whole to its nucleus. In so doing, Bohr kept his institute in the forefront of contemporary international physics research.

In 1931 the distinguished Danish philosopher Harald Høffding, one of Christian Bohr's discussion circle, died. Høffding had been the first occupant of the honorary residence at Carlsberg, which was conferred for life by the Royal Danish Academy of Sciences and Letters on the most prominent intellectual in Danish society. The academy chose Bohr as the second occupant. In 1932 he settled in with his wife, Margrethe, and their six sons.

Because several of the institute's guests came from the Soviet Union and Germany, Bohr learned early on about the lack of openness of Soviet society and Hitler's persecution of Jews. Virtually overnight Bohr's institute became a sanctuary for young German physicists unable to return to their homeland—until Bohr was able to find permanent placement for them, most often in the United States. At the end of 1938, physicists at the Copenhagen Institute provided a theory for the recent discovery of fission based on Bohr's liquid drop model of the nucleus, and in 1939, during a stay of several months in the United States, Bohr contributed important insights about the mechanism of the fission process. Yet, as he announced publicly in several lectures and publications, he did not believe in the feasibility of an atomic bomb within the foreseeable future.

Bohr held this view until he arrived in England in October 1943 after escaping from Nazi-occupied Denmark. Once acquainted with the Allied efforts, he came to consider the atomic bomb project feasible and took part in it for the rest of the war. At the same time he carried on his own personal crusade to convince Churchill and Roosevelt that Stalin needed to be informed about the project in order to retain mutual confidence among nations after the war as well as to avoid a nuclear arms race. Bohr's continued insistence on an "open world" among nations, the necessity of which he brought to the attention of statesmen at every opportunity, came to public expression in 1950 in an Open Letter to the United Nations.

Bohr served as a mentor and guide to several generations of theoretical physicists during a particularly important and exciting period for the field. He died peacefully at his home in Copenhagen.

Stefan Rozental, ed., *Niels Bohr: His Life and Work as Seen by His Friends and Colleagues* (1967). Abraham Pais, *Niels Bohr's Times, in Physics, Philosophy, and Polity* (1991).

FINN AASERUD

BOLTZMANN, Ludwig (1844–1906), theoretical physicist.

Born in Vienna to a comfortable middle-class family, Boltzmann studied at the university there from 1863 to 1866, when he received his doctorate on the kinetic theory of gases. In 1869 he became professor of mathematical physics at the University of Graz, where he remained until 1873. During this period he spent some months with Robert Bunsen and Leo Königsberg in Heidelberg and with Gustav Kirchhoff and Hermann von *Helmholtz in Berlin. In 1873 he returned to Vienna to the chair of mathematics at the university, which he held for the next three years. In 1876 he relocated again to Graz as professor of experimental physics and began his acquaintance with his future friend but persistent opponent in scientific matters, Wilhelm Ostwald.

The peripatetic Boltzmann went to the University of Munich as professor of theoretical physics in 1890 and then, four years later, to the University of Vienna in the same capacity, to take the chair vacated by the death of his teacher Joseph Stephan. At about the same time, Ernst Mach, who would become both a philosophical and personal adversary, became professor of history and theory of the inductive sciences at Vienna. The friction between the two prompted Boltzmann to accept the offer of Ostwald to move to Leipzig. The hoped-for heaven turned into a hell, owing to disagreements with Ostwald over atomism (*see* ENERGETICS). Upon Mach's retirement in 1901, Boltzmann returned to his previous post. Emperor Francis Joseph asked him to give his word of honor that he would never accept a position outside the Empire again.

During much of the nineteenth century the belief in the strictly deterministic character of the physical laws was the cornerstone of the physicists' worldview. Boltzmann's work seriously undermined it. In 1877, he published a famous paper, *On the relation between the second law of the mechanical theory of heat and the probability calculus with respect to the theorems on thermal equilibrium*, which ascribed only a probabilistic value to the second law of *thermodynamics. A system tends to the state of thermodynamic equilibrium as the most probable, but by no means the only, state the system can reach.

Having realized that the second law could not be interpreted via mechanical principles alone, Boltzmann studied James Clerk *Maxwell's approach to the kinetic theory of gases. In a paper on thermal equilibrium, Boltzmann extended Maxwell's theory of distribution of energy among colliding gas particles, treating the case when external forces are present. He deduced that the average energy of a molecule was roughly equal to kT (where k is "Boltzmann's constant" and T the absolute temperature); larger or smaller energies could occur, but with proportionately lower probability.

Boltzmann next turned his attention to nonequilibrium systems. How could kinetic theory account for the process through which a gas tends towards an equilibrium state? In 1872 he formulated "Boltzmann's H-theorem," which states that H (as the negative of *entropy) always decreases, except when the distribution of molecular velocities complies with Maxwell's law. Boltzmann was the first to show that the increase of entropy corresponds to an increasing randomness of molecular motion, as required by Maxwell's distribution law.

These results gave rise to a paradox. If Newtonian mechanics held on the molecular level, interactions between particles had to be reversible, whereas thermodynamic changes on the macroscopic level were irreversible. The answer to this "reversibility paradox" lay in the statistical character of the second law. Nevertheless, the community of physicists became apprehensive about the statistical approach. The reversibility paradox—which was initially pointed out by William *Thomson—formed the basis of a controversy between Boltzmann and his friend and colleague Josef Loschmidt.

The main difficulty lay in accepting the implications of the use of the theory of probability in the formulation of a fundamental law of physics. Boltzmann took on the task of persuading his colleagues that the statistical approach could account legitimately for the macroscopic phenomena of the real world. In developing his position he reached one of his major results, $S = k\log W$, which connected the entropy S of a system in a given state with the probability W of the state. The formula connects a thermodynamic or macroscopic quantity, the entropy, with a statistical or microscopic quantity, probability.

Boltzmann aggressively defended his belief in the atomic structure of matter and tried to reconcile this perspective with the statistical description of macroscopic phenomena. In 1903, he started offering a university course on *Methods and General Theory of the Natural Sciences*. It appeared that, at last, he could defend his views in a wider setting. But ill health and recurrence of the depression that sometimes plagued him caused him to take his own life in October 1906, while on vacation with his family at the Bay of Duino, a resort near Trieste.

Engelbert Broda, *Ludwig Boltzmann: Man, Physicist, Philosopher* (1983). John Blackmore, ed., *Ludwig Boltzmann, His Later Life and Philosophy, 1900–1906*, 2 vols. (1995). Carlo Cercignani, *Ludwig Boltzmann: The Man Who Trusted Atoms* (1998). David Lindley, *Boltzmann's Atom: The Great Debate That Launched a Revolution in Physics* (2001).

MANOLIS PATINIOTIS

BOYLE, Robert (1627–1691), chemist, natural philosopher.

Robert Boyle was born in Lismore, Ireland, the fourteenth child of Richard Boyle, the Earl of Cork. As the son of the immensely wealthy "Great Earl," Boyle had a privileged upbringing. He was initially educated at home by a tutor, and attended Eton from 1635 to 1638. Soon thereafter he came under the tutorship of Isaac Marcombes, a native of Auvergne, with whom Robert and his brother Francis toured parts of Europe from 1639 until 1642. When the Great Earl experienced a sudden reversal of fortune owing to the Irish Rebellion, Robert went to live with Marcombes in Geneva until 1644, when he returned to England.

Ensconced in the family estate at Stalbridge in Dorset, Boyle bent himself to devotional writing. He composed early versions of *Seraphic Love, The Martyrdom of Theodora*, and other pious reveries. During the second half of the 1640s, Boyle made contact with several members of the loosely organized group of technical and utopian writers inspired by Francis Bacon and clustering around the expatriate "intelligencer" Samuel Hartlib. Although much attention has been focused on the role that Benjamin Worsley, a member of the group, played in Boyle's scientific formation, his first exposure to systematic experimentation occurred at the hands of George Starkey, an emigré from New England who wrote immensely popular chrysopoetic treatises under the pseudonym Eirenaeus Philalethes. From 1650 until the middle of the decade, Boyle acquired from Starkey a full experimental knowledge of Helmontian "chymistry," a discipline that fused mundane chemical pursuits with the quest for such "grand arcana" as the universal dissolvent or alkahest and the philosophers' stone.

After an interlude in Ireland, Boyle moved in 1655–1656 to Oxford, where he came into contact with a group of physicians and natural philosophers who encouraged his pursuit of natural philosophy. Before or around this time, Boyle became a corpuscularian thinker, committed to the idea that matter is composed of discrete particles rather than making up an infinitely divisible continuum. Already in his treatise *Of the Atomicall Philosophy* (c. 1654–1655), Boyle attempted to ground his corpuscular doctrine on the phenomena of the laboratory, an approach that would develop into a lifelong quest. Although Boyle's matter theory pitted him against the mainstream of school philosophy, his most important early source for the experimental verification of corpuscles at the microlevel was Daniel Sennert, a scholastic medical professor and iatrochemist at the University of Wittenberg.

In later works such as *The Sceptical Chymist* (1661) and *The Origine of Formes and Qualities* (1666–1667), Boyle would fuse his experimentally based corpuscular theory with continental versions of mechanism to arrive at "the *mechanical philosophy"—a programmatic attempt to reduce sensible phenomena to the two "catholick principles" matter and motion. In these and other works, Boyle constructed highly effective experimental means of debunking such scholastic entities as substantial forms and "real qualities" in favor of the interaction of unseen particles. This Baconian program would exercise a major influence on subsequent philosophers and scientists as diverse as John Locke and Isaac *Newton.

During the early 1660s Boyle began publishing a succession of experiments with the *air-pump designed by Robert *Hooke at his request. These experiments, which led to the formulation of Boyle's law on the inverse proportionality of volume and pressure in gases, became an important venue for the justification of the Royal Society's experimentalist program against the aprioristic rationalist stance of Thomas Hobbes. Boyle's pneumatic experiments by no means represent the major thrust of his continuing research, however, which focused rather on the justification of the mechanical philosophy by means of experiment.

Boyle's pursuit of the philosophers' stone, which seems to have reached a climax in the late 1670s, was probably related to his increasing discomfort with the uneasy relationship between his religion and the mechanical philosophy. Other "chymical" pursuits also continued to engage him, such as the use of indicator tests to divide salts into the three "tribes" of acidic, "urinous" (ammonia and its compounds), and "alcalizate" (fixed alkalies such as potassium carbonate). This project, mediated by Starkey, led Boyle to some of his most fruitful discoveries in the realm of chemistry, and may have helped set the stage for the increasing emphasis on the chemistry of salts in the subsequent century.

R. E. W. Maddison, *The Life of the Honourable Robert Boyle, F.R.S.* (1969). Steven Shapin and Simon Schaffer, *Leviathan and the Air Pump: Hobbes, Boyle, and the Experimental Life* (1985). Michael Hunter, ed., *Robert Boyle Reconsidered* (1994). Rose-Mary Sargent, *The Diffident Naturalist: Robert Boyle and the Philosophy of Experiment* (1995). Lawrence M. Principe, *The Aspiring Adept: Robert Boyle and his Alchemical Quest* (1998). William R. Newman and Lawrence M. Principe, *Tried in the Fire: Starkey, Boyle, and the Fate of Helmontian Chymistry* (2002).

WILLIAM R. NEWMAN

BRAHE, TYCHO (1546–1601), astronomer.

Tycho once boasted that his observatory on the island of Hven had cost King Frederick II of Denmark more than a ton of gold. Born into a noble Danish family, Tycho was accustomed to walk in corridors of wealth and power, yet he chose a commoner wife and devoted his energies to the construction of an unrivaled scientific establishment. A volume published in 1667 dramatically illustrates the scope of his achievement: in an attempt to catalog all known astronomical observations of planetary and stellar positions, the book devotes 92 pages to pre-Tychonic observations and 65 pages to three decades of post-Tychonic data, compared to an overwhelming 912 pages to Tycho's own measurements.

As a teenager Tycho was impressed by the ability of astronomers to predict an eclipse accurately, but dismayed by their errors in predicting the time of the great conjunction of Jupiter and Saturn in 1563. The appearance of a brilliant new star in 1572 gave him an occasion to make his mark in astronomy. Besides giving the usual astrological interpretations, he demonstrated that the nova lay beyond the moon, that is, in the realm of the eternal and incorruptible stars and planets. His lectures and small booklet on the star, *De nova stella* (1573), procured for him several fiefdoms, including Hven, a small island in the strait north of Copenhagen. There, as he was beginning to build his observatory, he observed the conspicuous comet of 1577, and showed that it too lay in the supralunary sphere.

On Hven Tycho built Uraniborg, his castle of the stars, containing observing decks for his elaborately graduated naked-eye instruments as well as a cellar full of alchemical furnaces. His extensive observations of the sun, moon, planets, and stars led to revised and improved solar and lunar theories as well as to a star catalogue that would finally supersede Ptolemy's. Flushed with his success in establishing large minimum distances to the nova and comet, Tycho undertook an even more ambitious project: to determine the distance to Mars by using evening and morning observations, that is, with a triangulation baseline approximately equal to the diameter of the earth. The parallactic angle is too tiny for naked-eye detection, but since Tycho and his predecessors assumed a distance twenty times too small and therefore a parallactic angle twenty

times too large, he had good reason to believe that his project was feasible. After an initial failure at the close approach of Mars in 1582, he began a major overhaul of his instruments. He built a subterranean observatory, Stjerneborg, on Hven near Uraniborg to give his devices greater stability and protection from the wind.

While making his Martian observations, Tycho discovered the critical role of refraction for observations made with altitudes under 30 degrees. Correcting for refraction with an erroneous table, in 1587 Tycho convinced himself that he had finally found a parallax for Mars. The calculated distance ruled out the traditional earth-centered arrangement of the planets, but rather than adopting the Copernican system, which, according to Tycho, violated both physics and sacred scripture, he postulated an alternative geo-heliocentric system in which the moon and sun circled a fixed central earth while the sun in turn carried a retinue of planets in orbit around it. This he published in his *De mundi aetherei recentioribus phaenomenis* (1588). Subsequently he realized that his refraction correction was erroneous, but he never withdrew his unsubstantiated claim about the Mars parallax or his cosmological Tychonic system. Although he kept silent about his failed campaign to detect the distance to Mars, his treasure trove of detailed observations would provide the precious database for Johannes *Kepler's study of the Martian orbit.

In the mid-1590s Tycho grew increasingly concerned about the future of his observatory, since by Danish law his children by his commoner wife could not inherit his property. Appeals to the young new king Christian IV not only failed, but Tycho lost much of the lavish support to which he had grown accustomed. Deciding to leave Denmark for the Prague court of Emperor Rudolf II, where the inheritance laws were more liberal, in 1597 Tycho packed up his instruments, library, and printing press for the journey to Bohemia. He paused outside the Danish border, but when no call came for his return, he printed up a well-illustrated book, *Astronomiae instauratae mechanica* (*Instruments for the Reform of Astronomy*, 1598), as his calling card for the nobility of the Holy Roman Empire. Welcomed in Prague by Rudolf II, he began setting up his great instruments at Benatky Castle, some distance north of the capital, and there he was joined by Kepler, a young, precariously employed apprentice. A renewed observational program had barely begun when it was interrupted by Rudolf's call to relocate in Prague. Tycho did not last long in Prague, his life possibly shortened by mercurial elixirs brewed in his alchemical laboratory. His partially printed study of the 1572 supernova, *Astronomiae instauratae progymnasmata* (*Exercises for the Reform of Astronomy*), was completed by Kepler and published posthumously in 1602.

J. L. E. Dreyer, *Tycho Brahe* (1890). Victor Thoren, *Lord of Uraniborg* (1990). Owen Gingerich and James Voelkel, "Tycho Brahe's Copernican Campaign," *Journal for the History of Astronomy* 29 (1998): 1–34. John Robert Christianson, *On Tycho's Island: Tycho Brahe and His Assistants, 1570–1601* (2000).

OWEN GINGERICH

BRAIN DRAINS AND PAPERCLIP OPERATIONS. Like other forms of labor and capital crucial to economic development in the modern world, scientists and their intellectual capital have moved from nation to nation in the global economy of science. The phenomenon appeared in the early modern period, when proliferating scientific academies under state sponsorship recruited leading scientists from abroad: the Academy of Berlin, for example, in the early eighteenth century lured Pierre de Maupertuis from Paris and Leonhard *Euler from St. Petersburg; and the St. Petersburg Academy drew most of its initial members from Germany and Switzerland.

In the twentieth century flows of intellectual capital increased from trickles to torrents, measured not just in individuals but in dozens and hundreds of scientists. Some migrated to take advantage of professional opportunity, for instance abandoning the backlog of academic jobs in Germany for the growing academic and industrial research system in the United States early in the century. But many scientists were uprooted either as victims of political persecution or as spoils of war claimed by victorious nations. The rise of fascism in the 1930s drove hundreds of scientists from Germany, Austria, Hungary, and Italy, including many of the leading lights of European science. Over thirty nations took in émigrés, but most went to Britain or the United States. Their colleagues tried to find academic jobs for them, whether out of obligation or opportunity, and often succeeded despite the Great Depression and anti-Semitism.

The victors of World War II engaged in a form of intellectual reparations. The United States pursued Project Paperclip, which aimed to bring leading German scientists with their families to America and put them to work on problems of military or economic importance. Paperclip would also deny America's competitors, especially the Soviet Union, the services of the Germans it recruited. Between 1945 and 1952, the United States imported over six hundred technical specialists under Paperclip. Similarly, Great Britain undertook Operation Matchbox to spirit away scientists from the Soviet sphere in Germany; the Soviets meanwhile took Germans east. Some of these postwar expatriates went voluntarily, either out of political conviction or pragmatic recognition that working conditions would be better in the host country than in their ruined homelands.

The prewar brain drain and postwar Paperclip operations helped shift the scientific center of gravity away from central Europe. European émigrés contributed to many important scientific developments, from the electronic computer to molecular biology. Émigrés helped the Allies develop the atomic bomb in World War II, and expatriated scientists after the war brought German rocket technology to the United States and Soviet Union.

Another brain drain flowed from Asia to the United States. The Nationalist Chinese government sent promising students to the United States for graduate training in the 1930s and 1940s; some of them stayed there to take advantage of greater resources or to escape the world war and ensuing civil war in China. The rapprochement between the United States and China in the early 1970s led to another influx of Chinese scientists; by the 1990s some eight hundred scientists and engineers from China had settled in America. The flow of scientific capital from developing countries to the developed world continued through the end of the twentieth century in the labor market for high-tech industry, which relied heavily on foreign workers, especially from South and East Asia.

Donald Fleming and Bernard Bailyn, eds., *The Intellectual Migration: Europe and America, 1930–1960* (1969). Clarence G. Lasby, *Project Paperclip: German Scientists and the Cold War* (1971).

PETER J. WESTWICK

C

CARNOT, Sadi (1796–1832), and **Augustin FRESNEL** (1788–1827), physicists.

On 10 May 1788, the year before the French Revolution began, Fresnel was born in Broglie in Normandy. His father, an architect, and his mother, a daughter of the manager of the local chateau, had religious and Royalist sympathies. Revolutionary turmoil soon forced them to move. Carnot was born at the Palais de Luxembourg to Lazare Carnot, one of the post-Thermidorean directors, a surviving member of the Committee on Public Safety, and an organizer of the successful military campaign in 1793 against Austrian and Dutch forces.

Fresnel entered the new Central School at Caen in 1799, where he learned mathematics and the sciences. Sadi Carnot was educated at home by his father until he was sixteen. Both Fresnel (in 1804) and Carnot (in 1812) did well in the admissions examinations for the École Polytechnique; Fresnel placed seventeenth and Carnot twenty-fourth, each out of nearly two hundred applicants. Both entered the engineering corps after graduation; Fresnel joined Ponts et Chausées (Bridges and Roads) and Carnot (after two years at the Artillery and Engineering School at Metz), the Royal Corps of Engineers. True to family sympathies (and like many others at Ponts et Chausées), Fresnel resisted the return of Napoleon from Elba in 1815 and was sent into home exile. Both men spent much of their careers in Paris and both died young: Fresnel, always physically weak, of tuberculosis, and Carnot probably of cholera.

They produced novel though unequally influential physical theories. Beginning in 1816 and extending over nearly seven years, Fresnel developed the mathematical, theoretical, and experimental foundations of wave optics. By the early 1830s his principles and devices had spread widely. Carnot's book, *Reflections on the Motive Power of Fire* (1824), provided for the first time a general theoretical system for understanding the conditions under which the efficiency of heat engines could be max-

imized. His scheme lay fallow for over two decades until William *Thomson, during a sojourn at Victor Regnault's laboratory in Paris, learned of it in 1845. Carnot's theory, suitably adjusted, became central to the development of thermodynamics by Thomson and Rudolf Clausius.

When Fresnel began his innovative work in 1815, several *physiciens* had been exploring novel optical phenomena for about half a decade as a result of Étienne Louis Malus's discovery of polarization in 1809. François Arago and Jean-Baptiste Biot (like Malus, graduates of the École Polytechnique) had exchanged bitter words over priority issues concerning the colors produced by polarized light that passes through thin crystals. This topic (chromatic polarization) lay at the center of Parisian interest in optics. Fresnel knew nothing about the dispute or the details of polarization. His interest in optics had been stimulated by difficulties he perceived in the contemporary French understanding of heat and light as fluids consisting of minute particles subject to various forces (*see* IMPONDERABLES).

Fresnel probably met Arago at the École Polytechnique, but his first substantial encounter with him took place in Paris at a dinner party given by Fresnel's maternal uncle, Léonor Merimée. At home in the countryside, and apparently ignorant of previous work in the area by Christiaan *Huygens and Thomas Young, Fresnel developed a theory of light as a wave in a universally present medium (or *ether), according to which striped patterns would form when light passes through narrow slits or near edges. This phenomenon of diffraction, known since the seventeenth century and discussed by Isaac *Newton in the *Opticks*, inspired Fresnel to create a mathematical system tightly connected with persuasive experiments that eventually embraced the influential topic of polarization.

Arago acted as Fresnel's patron and collaborator, and, through his access to the centers of power, stimulated ongoing interest in Fresnel's work, which culminated in 1818 in Fresnel's winning an Academy of Sciences prize contest on diffraction. Supported by impressive new experiments, quantitative, bound to a subject (polarization) of intense contemporary interest to powerful people like Pierre-Simon *Laplace, and pushed forward by the aggressively political Arago, Fresnel's new wave optics spread rapidly in French circles and soon thereafter throughout Europe.

Carnot's theory of heat engines had a different fate. Despite or perhaps because of the fame and power his father had wielded before Napoleon, Carnot did not have the access to the centers of Parisian science that Arago (despite his republican sympathies) and Merimée gave Fresnel. Moreover, Carnot's subject did not connect with the main interests of his contemporaries. His system was not based on any novel physical concept of heat (which might have excited curiosity), but rather on an obscure analogy between the transfer of heat by means of an engine from a hot to a cold source and the fall of water from a high to a low place. Carnot's father had considered the requirements that ordinary machines must satisfy to produce the greatest possible effect under given circumstances and had clarified how that effect should be estimated. His son appropriated several of Lazare Carnot's notions and produced a theory based on the principal axioms that, for maximum effect, no heat must flow directly from hot to cold but must rather be transferred through an engine; and that the total quantity of heat is always conserved. Carnot doubted the truth of this second requirement, but could never restructure his theory to do without it.

The principles of Fresnel's wave optics endure essentially unchanged, despite the photon, because many optical instruments are based upon them. Carnot's system has no comparable legacy, except indirectly through the principles of thermodynamics, which stabilized during the late 1850s and 1860s. This difference may be explained as follows. Unlike Fresnel, Carnot had neither a patron nor the skill to turn his scientific knowledge to practical success. Fresnel invented the layered lens and associated lighting elements that transformed coastal illumination in France and elsewhere.

Carnot did not develop his system to a high mathematical level, perhaps because he wished to reach an audience of working engineers, who, however, gleaned little new from it. Nonetheless Carnot, like Fresnel, was a signal contributor to the creation of *physics during Napoleonic and Restoration France.

D. S. L. Cardwell, *From Watt to Clausius* (1971). Sadi Carnot, *Reflexions on the Motive Power of Fire: A Critical Edition with the Surviving Manuscripts*, ed. Robert Fox (1986). Jed Buchwald, *The Rise of the Wave Theory of Light* (1989).

JED Z. BUCHWALD

CARTE DU CIEL. Astronomical *photography, for which François Arago foresaw a promising future in January 1839, was soon afterwards inaugurated by Henry Draper, Armand-Hippolyte-Louis Fizeau, and Léon Foucault. At the Cape of Good Hope in 1882, David Gill succeeded in photographing a very bright comet with many stars clearly visible in the field of view. He sent his photograph to Ernest Barthélémy Mouchez, director of the Paris Observatory, who, when presenting it to the Académie des Sciences, proposed a scheme for the photographic production of celestial charts, an idea that had been advanced twenty-five years previously by Warren De la Rue.

At the Paris Observatory, the brothers Paul and Prosper Henry installed a successful astrograph in 1885. Mouchez, with the help of Gill, organized an international congress in 1887 to establish a project for a general photographic chart of the heavens. Sixteen observatories participated; the number rose to eighteen in 1889. In 1896 a further agreement was reached on a unified system of astronomical constants.

The work of the Carte du Ciel (the French name being used by all the participants) aimed at establishing standard observations for future generations and began with two sets of plates of the sky around 1900: "catalogue" plates including stars up to magnitude 11, and "carte" plates up to magnitude 14. In the 1950s new plates were taken to determine proper motions by comparisons with the original photographs, but the comparisons did not prove satisfactory, and the whole operation ceased in 1970.

The success of the astrometric satellite *Hipparchos* between 1989 and 1992 resulted in the Hipparchos and Tycho catalogues of stars and led astronomers at the United States Naval Observatory to reconsider the original project. By comparison with the satellite observations, they assigned new values to the proper motions of the stars of the general catalogue of the Carte du Ciel within 0.002 s. Two CD-ROMs were issued from data based on an interval of about one hundred years, an appropriate outcome for those who, at the end of the nineteenth century, proposed a photographic map of the heavens for the benefit of their successors.

See also PHOTOGRAPHY.

Suzanne De'barbat et al., *Mapping the Sky—Past Heritage and Future Directions* (1988).

SUZANNE DÉBARBAT

CARTESIANISM. See COPERNICUS; DESCARTES, RENÉ; MECHANICAL PHILOSOPHY.

CARTOGRAPHY, the art and science of making representations of areas of the earth and other spatially extended objects, has been connected with the history of modern science both as a field of inquiry and as a tool. As a field, it has supplemented and stimulated *geodesy and, later, *planetary science. As a tool, it has been indispensable to *geology, *geophysics, *oceanography, *meteorology, and biology, all of which rely on thematic maps to represent and analyze spatially distributed phenomena such as magnetic variation, rock outcrops, or warm and cold fronts.

Three major events around 1500 triggered an explosion of cartographic activity in Europe. First, cartographers assimilated the lessons of Claudius Ptolemy's *Geographia*, the greatest cartographic achievement of the ancient world. Brought to Italy from Constantinople, the complete text was translated into Latin early in the fifteenth century. By the century's end, at least seven different editions had been published. From Ptolemy, cartographers learned to arrange geographical knowledge according to a coordinate system of parallels (latitudes) and meridians (longitudes) and to project the system onto flat surfaces. Second, thanks to the invention of printing and advances in engraving, cartographers found ways to reproduce maps, first as wood engravings and, in the sixteenth century, by clearer copper engravings. Third, the voyages of discovery and increased international trade that followed produced a demand for more accurate maps of the world. Thus the chief focus of cartographers became the production of atlases.

The Low Countries were the center of atlas making in the sixteenth century. There, Gerardus Mercator invented the famous projection that showed lines of constant direction (rhumb lines) as straight lines, just as in the portolan charts of the late Middle Ages. Mercator used this projection in the marine chart he published in 1569. In 1599, Edward Wright used the newly invented logarithms to lay out the mathematical basis of this projection. In 1570, Abraham Ortelius published one of the most comprehensive early atlases, the *Teatrum Orbis Terrarum*, with 53 maps by various authors. Numerous other atlases followed. By the eighteenth century, France had become the center of cartography. Guillaume Delisle published an atlas of 98 maps, and Jean-Baptiste Bourguignon d'Anville included 211 maps in his *Atlas général* (1737–1789).

Part of France's lead in cartography came from its research in geodesy, which led not only to a determination of the earth's shape and the length of a degree, but also to the first detailed topographical map of any nation, the *Carte géométrique de la France* (or *Carte de Cassini*). Its 182 sheets were finally completed and published by 1793, the culmination of decades of work.

Across Europe, nations founded surveys, mostly run by the military: the Ordnance Survey of Great Britain in 1791, the Institut Géographique National in France, the Landestopographie in Switzerland. The Spanish set up their topographic survey in the eighteenth century, the Austro-Hungarians, Swiss, and Americans in the nineteenth century. The Germans combined the existing state surveys into a national survey following unification in 1871. In 1888, Japan instituted its Imperial Land Survey. European nations also surveyed their overseas territories. The most ambitious of these surveys, the Great Trigonometrical Survey of India, established early in the nineteenth century, not only mapped the subcontinent but raised gravimetric problems that quickly led to the theory of *isostasy.

Already in the seventeenth century, scientists had designed maps for specific scientific purposes. Between 1698 and 1700, Edmund Halley sailed the Atlantic, measured the variations in magnetic declination, and charted them on a pioneering map that appeared in different editions between 1701 and 1703. By drawing lines between points of equal variation, he pioneered the technique of isogonic lines that would be used often in later thematic maps. In the early nineteenth century, geologists invented the stratigraphic map that represented each kind of rock that outcropped at the earth's surface by a different color. Constructing such maps became the major goal of newly instituted national geological surveys. So important was stratigraphic mapping, so urgent its need for good underlying base maps, that in the United States the government gave the Geological Survey, established in 1878, the responsibility for producing the prerequisite topographic maps.

As national topographic surveys designed to promote national military and commercial interests began to publish their results, entrepreneurs founded private companies to exploit the information in maps for the use of the general public. Among the more important of the companies they founded were Bartholemew in Great Britain, Justus Perthes in Germany (an enterprise begun in the late eighteenth century), and Rand McNally in the United States.

From the 1880s to World War I, cartographers, like many other scientists, created international institutions. In 1875, the Convention of the Meter, attended by twenty nations, accepted the metric system, initially proposed in France in 1791, as the universal system of measurement. This aided the systematization of map scales (see STANDARDIZATION). In 1884, the International Meridian Conference in Washington, D.C., decided on Greenwich, London, as the site of the prime meridian; the choice was universally accepted by World War I. At the International Geographical Congress of 1891, the German geographer Albrecht Penck proposed an International Map of the World on a scale of 1:1,000,000 in 1,500 sheets. Work proceeded slowly. The most significant accomplishment of this still incomplete project was the 107-sheet *Map of Hispanic America* published by the American Geographical Society in 1945.

World War I brought a halt to international cooperation and ushered in the most productive century ever in cartography. In succession, the airplane, aerial photography, seismic techniques, echo sounders and sonar, radar, satellites, and computers made possible the mapping of features of the land, ocean bottoms, and extraterrestrial objects with an accuracy, ease, and variety hitherto undreamed of.

Without a growth industry in aeronautical charts, aviation would have foundered. Conversely, airplanes made aerial photography and reconnaissance mapping possible once cartographers had developed techniques to translate photographs into maps. Meteorologists responded to the needs of the aviation industry, and benefited from it as they developed theories of high- and low-pressure systems and hot and cold fronts, all of them represented on maps. Today, meteorology is a map-intensive enterprise. *Seismology, the echo sounder, and sonar displaced sounding as methods of mapping underwater. The unexpected topographic features detected with these new tools helped bring about the *plate-tectonic revolution.

In the 1960s, the already important remote-sensing systems came into their own. Electronic measuring helped extend continental surveys to oceanic areas. Soon thereafter, satellite triangulation helped to tie continental grids together into a single system. Innovations multiplied. For example, scientists developed SLAR (side-looking airborne radar) in the 1970s. After SLAR had been tested in Ecuador and Nicaragua, RADAM (Radar Amazon Commission) of

Brazil and Aero Service of Philadelphia mapped the previously inaccessible Amazon Basin. Wide-angle cameras and improved film allowed the mapping of the earth and the moon and planets.

Computers were quickly harnessed for the purposes of mapping by institutions such as the Laboratory for Computer Graphics and Spatial Analysis at Harvard. They made the previously tedious business of laying out projections routine. After experiments in the 1970s, computers by the 1980s could translate aerial photographs into topographic maps complete with contours, allowing rapid and accurate contour mapping for the first time. Statistical mapping also became much easier.

The greatest change brought about by computers was a shift from considering the map as a completed object to the map as a manipulable tool. In the 1960s, geographers developed computerized information systems. Instead of entering data on a two-dimensional surface, they digitalized it. They could easily superimpose different sets of information and, even more important, draw conclusions based on the superimposition. With this new tool in hand, cartographers were better equipped to prepare specialized maps for business, law-enforcement, natural-disaster prediction, or other specialized interests.

Today, cartography flourishes as never before. The military remains heavily involved, and the private companies multiply. Most nations have their surveys. Non-governmental groups such as the American Geographical Society and the National Geographic Society promote cartography. Professional societies such as the American Congress on Surveying and Mapping, the American Society of Photogrammetry, and the American Society of Civil Engineers (focusing here on the United States) have multiplied. They publish journals such as the *Manual of Photogrammetry*, *Photogrametria*, and *Surveying and Mapping*. Finally, a variety of international organizations attempt to coordinate mapping worldwide. Among them are the U.N. Office of Cartography, the Inter-American Geodetic Survey, the Pan-American Institute, and the International Hydrographic Organization in Monaco.

During most of the twentieth century, historians of cartography described how their discipline had achieved increasing scope and precision. At the end of the century, the critical cartography movement led by J. B. Harley and other British cartographers emphasized instead that, however exact maps may be, they are far from neutral. Cartographers' choice of projections, scales, symbols, and units reflect their own and their patrons' interests. According to these historians, maps depend not only on scientific knowledge but also on political interests.

Arthur Robinson, *Early Thematic Mapping in the History of Cartography* (1982). Leo Bagrow, rev. R. A. Skelton, *History of Cartography*, 2d ed. (1985). J. B. Harley and David Woodward, eds., *History of Cartography*, 4 vols. (1987–1999). Timothy Foresman, ed., *The History of Geographic Information Systems: Perspectives from the Pioneers* (1998). Mark Monmonier, *Air Apparent* (1999). J. B. Harley, Paul Laxton, and J. H. Andrews, eds., *The New Nature of Maps* (2001).

RACHEL LAUDAN

CASSINI FAMILY: Gian Domenico (1625–1712); **Jacques** (1677–1756); **César-François Cassini de Thury** (1714–84); **Jean-Dominique** (1748–1845); astronomers and geodecists.

The Cassini dynasty, which dominated the Paris Observatory from its foundation in 1666 until the French Revolution, was almost extinguished before birth. The progenitor, Gian Domenico Cassini, the greatest observational astronomer of his time, had felt a great temptation to join the Jesuits, who educated him at their college in Genoa, near his hometown Perinaldo. But though a man of deep religiosity, he lacked the calling of a priest. His teachers deflected his youthful preoccupation with theology to mathematics, which then included astronomy and astrology. Attracted by the exactness of the one and the complexity of the other, he mastered both; which, with the recommendation of his teachers, brought him to the Marchese Cornelio Malvasia, a former soldier then (in the late 1640s) setting up an observatory in his villa near Bologna. Comparison of their astrological forecasts with subsequent events persuaded both of them that the art that had drawn them together was fallacious.

Among the visitors to Malvasia's villa were the Jesuit astronomers Giambattista Riccioli and his collaborator Francesco Maria Grimaldi, who has enduring fame in the history of science as the discoverer of optical diffraction. Through them Cassini came to the attention of the faculty of the University of Bologna, which offered him a chair in mathematics that had become available in 1648 by the death of its incumbent, a distinguished disciple of Galileo. Cassini took up his professorship in 1651.

His most notable achievement in Bologna was the installation of a meridian line (*meridiana*) in the cathedral of San Petronio in 1654. Ostensibly built to obtain data to update the ecclesiastical calendar, the *meridiana* served Cassini to confirm that *Kepler's model of the orbit of the sun agreed better with observation than Ptolemy's. San Petronio remained in use as a solar observatory for 75 years. The first harvest of the ongoing observations was a vast improvement, made by Cassini, in tables of atmospheric refraction.

Cassini's combination of exact science, practical ability, and influential acquaintance recommended him to highly placed people in need of technical advice. In 1657 he advised Bologna (and hence the Pope) about the hydraulics of the river Reno. The Pope, Alexander VII, let the Bologna Senate know his high opinion of Cassini's expertise; the Senate immediately appointed him superintendent of the water works of Bologna. Cassini advised the Pope's brother about the fortifications of the Papal stronghold at Urbano and the Pope himself about water rights against the Grand Duke of Tuscany. The Grand Duke tried to win Cassini for himself; but Cassini preferred to retain his professorship in Bologna, his control of the Papal waters, and his easy access to the Vatican.

In June 1665, using the good lenses made by Giuseppe Campani, Cassini saw spots move across the surface of Jupiter, which he interpreted, rightly, as shadows cast by its moons. Cassini's sharp eyes also spied a spot that did not move relative to Jupiter's surface. From observations of this blemish,

he argued that Jupiter spins and gave the rate of rotation. He also detected the rotations of Mars and (erroneously) Venus. The Paris astronomers followed these feats closely. Christiaan *Huygens, then the leading member of the Paris Academy, confirmed the shadows of Jupiter's moons and the revolution of its body. Like everyone else, he wanted Cassini as a colleague. As a further attraction, Cassini had calculated an *Ephemeris* of Jupiter's moons, which gave their orbits and periods with sufficient accuracy to make possible reliable predictions of their eclipses, and thence the determination of the longitude of any place with a competent astronomer. The method succeeded well on land; using Cassini's tables, Parisian geodesists found that the channel coasts of France had been placed too far West by many leagues.

Louis XIV's all-seeing minister, Jean Baptiste Colbert, recognized Cassini's value as an ornament and also as the likely inventor of a practical means of finding the longitude at sea. The negotiation to win him for Paris involved two sovereign states and the Senate of Bologna. Cassini would accept only with the Pope's blessing—and only with a salary half again as large as Huygens'. A direct appeal from the Sun King to the Pope and the Senate procured the release of the solar astronomer, and then only on the understanding that the relocation would not be permanent.

Cassini fitted out the Paris Observatory with his relatives, beginning with his nephew Giacomo Filippo Maraldi, an astronomer and father of astronomers from fertile Perinaldo, and continuing with his own family. Cassinis lived in the Observatory for over 120 years; whence, no doubt, has arisen the common assumption that the Cassinis directed it from its early days. In fact, it had no official head, budget, or organization before Cassini's grandson was named director in 1771, and no paid assistants before his great-grandson took over in 1784.

The heavens opened to Cassini in France as they had to *Galileo half a century earlier in Italy. Cassini discovered two moons of Saturn in the 1670s and another two in the 1680s, a division in Saturn's ring, a comet or two, and many lunar features. In 1672, coordinating his measurements in Paris with those made by Jean Richer in Cayenne, he deduced a good value for the parallax of Mars and thence an excellent measure of the size of the solar system. Cassini thus provided the means of correctly correcting observations for refraction and parallax—data essential to the further progress of astronomy.

Cassini did not return to Italy except for a tour in 1695, accompanied by his son Jacques, with whose help he refurbished the *meridiana* in San Petronio. Jacques built a similar line in the Paris Observatory in 1729; but his most important act of filial piety was to complete the measurement of the arc of longitude from Dunkirk through the Observatory and down to Perpignan begun by his father. Jacques computed that the length of a degree along the arc near Perpignan exceeded the length of one near Dunkirk, which implied that the earth did not have the shape of a pumpkin, as *Newton had deduced from his theory of gravitation, but of an American football. The Cartesians who dominated the Paris Academy eagerly accepted this empirical result; but after many more measurements (*see* GEODESY) the Academy found for Newton. As an astronomer Jacques Cassini made many observations of the planets and their satellites, but contributed little qualitatively new apart from the revelation of the proper motion of the stars.

The primary purpose of the Cassinis' ongoing triangulations was to provide the framework for an exact map of France. Cassini de Thury brought the business—which under him literally became a business—almost to completion by the time of his death. It lacked only Brittany, to which his son Jean-Domique attended. The whole of France was thus mapped at a scale of 1:86400 against a grid determined by the stars and 40000 terrestrial triangles. With Jean-Dominique's appointment as director of the Observatory in 1784, the Cassinis seemed to be more entrenched than ever in their family fief. The king provided munificently for new instruments. The engraving of the invaluable Carte de France neared completion. The Paris and Greenwich observatories cooperated to join the geodetic grids of France and Britain. Only a revolu-

tion could remove the Cassinis. And so it happened. The revolutionary government turned Jean-Dominique out in 1793, confiscated his map (and completed the engraving of the last 15 of the 180 sheets), and put him in jail as a reward for his family's service to the ancien regime. On his release he returned to the family chateau in Thury. He spent the rest of his life combating liberal ideas and defending the scientific work of his ancestors.

J. L. Heilbron, *The Sun in the Church* (1999). John Lankford, ed., *History of Astronomy. An Encyclopedia* (1997).

J. L. HEILBRON

CATHODE RAYS AND GAS DISCHARGE. The study of cathode rays and gas discharge laid the groundwork for the elucidation of the properties of the *electron and its role in determining physical and chemical properties of matter.

As early as 1833 Michael *Faraday investigated glows produced by electrical discharge through gases. Subsequently the invention of a mercury pump that could reduce gas pressures in glass tubes to as low as 10^{-6} atmosphere multiplied investigations of the phenomena that Eugen Goldstein in 1876 called "Kathodenstrahlen" since the radiations appeared to flow from the negative, or cathode, pole of the *vacuum tube. William Crookes, editor of *Chemical News*, found evidence in the 1870s and 1880s that the "rays" are negatively charged particles of matter. He proposed that these particles constituted a "fourth state" of matter that was neither solid, liquid, nor gas. Heinrich Hertz in 1883 and Philipp Lenard in the early 1890s reported results inconsistent with the particle hypothesis and supportive of an elec-

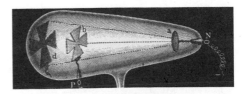

Demonstration by William Crookes that the invisible cathode rays follow straight lines (because they throw sharp shadows).

tromagnetic-wave interpretation. There appeared to be an English interpretation and a German interpretation of the agent of the Kathodenstrahlen.

After assuming the directorship of the Cavendish Laboratory in Cambridge in 1884, Joseph John *Thomson directed much of the laboratory's research toward studies of ionization and electric gas discharge and, later, to *X rays and *radioactivity. Working on the hypothesis that the cathode radiation consists of fast-moving charged particles that produce pulses of X rays when they hit glass or metal in the vacuum tube, Thomson studied the velocity of the suppositious particles and their response to electrical and magnetic fields. Here he drew on recent work by Jean Baptiste Perrin in Paris to help counter the evidence that the rays could not be charged. In 1897 Thomson calculated the mass-to-charge ratio for the cathode-ray particles and made a case that the cathode radiation comprises negatively charged corpuscles a thousand times smaller in mass than the hydrogen *ion.

A beam of negative particles in the electric discharge tube must be associated with an oppositely directed beam of positive particles. Physicists obtained evidence that these counterparts existed in the late 1880s. Wilhelm Wien reported investigations in 1902 on a positive radiation that he could draw through holes or canals bored in the cathode ("Kanalstrahlen"). Canal rays consist of gaseous ions with a mass-to-charge ratio identical to that of the ions made of the residual gas in the electric discharge tube. J. J. Thomson and his assistant Francis Aston devised ways of manipulating the various constituents of canal rays that led to the physical separation of isotopes and, in the 1920s, to the rapid deployment of *mass spectrometry.

See also AIR PUMP AND VACUUM PUMP; ATOMIC STRUCTURE.

E. A. Davis and I. J. Falconer, *J. J. Thomson and the Discovery of the Electron* (1997). Per Dahl, *Flash of the Cathode Rays* (1997). Mary Jo Nye, *Before Big Science: The Pursuit of Modern Chemistry and Physics, 1800–1940* (1999).

MARY JO NYE

CAVENDISH, Henry (1731–1810), and **Charles Augustin COULOMB** (1736–1806), the two most important contributors to the quantification of physical science during the eighteenth century. Both derived their major scientific questions, and the keys to answering these questions, from Isaac *Newton's conception of distance *forces. Otherwise, the two men were altogether different.

Cavendish, a grandson of dukes on both sides of his family, became one of the wealthiest men in England. Educated at an upper-class school and at Cambridge, and lacking no personal comfort, he nonetheless developed a painful shyness from which his solitary scientific pursuits provided a refuge. He spent his entire life in England and did not marry. Coulomb also came from a wealthy family and received his early education in Parisian colleges. But his father's business speculations eventually ended in bankruptcy and Coulomb had to make his way in the world. He became a military engineer. Neither shy nor home-bound, he spent eight years directing the building of a huge fortress in Martinique. Following his return to France in 1764 he wrote papers on mechanics and magnetic compasses, which won prizes from the Académie des Sciences of Paris. On the strength of these works, the academy admitted Coulomb as a member in 1781, at the age of forty-five. In contrast, Cavendish was elected a Fellow of the Royal Society before he was thirty, not because of prize-winning work but because his father, a leading light of the society, wished it. Academy membership allowed Coulomb to withdraw from active service and to raise a family. Membership in the Royal Society gave Cavendish a social life.

Cavendish's first important investigation, which he did not publish, aimed at producing a Newtonian theory of heat based on corpuscular motions and distance forces. While pursuing these ideas he came naturally to the study of airs, the expansion of which provided a simple and convenient model for the interaction of heat and matter (see PNEUMATICS). In 1762 Cavendish discovered that on treatment with acids, metals released an inflammable air (H_2). He was one of several experimenters who reached the implausible conclusion (in his case in 1784) that inflammable air combined with "eminently respirable air" (so named by Joseph Priestley) to make water. In Cavendish's explanation, which preserved the old chemistry against the innovations of Antoine-Laurent *Lavoisier, inflammable and eminently respirable air were water plus phlogiston and water minus phlogiston, respectively.

Cavendish's deepest work concerned electricity. Conceiving electrical matter to be compressible in the manner of a pneumatic fluid (a gas), and supposing further that the particles of electrical matter acted on one another and on particles of common matter by a distance force diminishing as some unknown power of the distance between them, he described the electrical states of systems of charged and uncharged conductors exposed to one another's influence. His long and difficult memoir of 1771

introduced the concepts of electrical capacity and, as represented by the compression of the gas-like electrical matter, electrical potential. Cavendish's interpretation of the results of electrical induction in terms of potential was well beyond the grasp of almost all his contemporaries. Unfortunately, his thoroughness, combined with the profligacy of nature, defeated his intention of producing a *Principia electricitatis*. As he learned more about the range of inductive capacities and conductivities of bodies, he came to doubt the possibility of describing all these phenomena with the aid of only mathematics and a single undifferentiated electric fluid.

Cavendish's unpublished papers on electricity contain many results that would have been important if released when he obtained them. One was a demonstration of the inverse-square repulsion between particles of electric fluid, based on a mathematical consequence of the law of squares—that

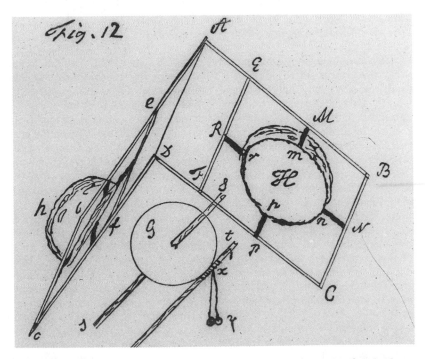

Cavendish's Rube-Goldberg null experiment. The large exterior insulated sphere (which opens like a book) is shut to enclose the previously charged insulated internal sphere; a wire briefly connects the two internally, whereupon the outer sphere opens and the experimenter shows that no charge remains on the inner.

a perfect conductor can maintain an electric charge only on its surface. He managed this demonstration with the rickety device pictured here and enriched it by estimating how far the exponent in the force law could differ from –2 and remain consistent with his measurements (answer: by < 0.02). This quantitative judgment of possible error in determining an empirical "law" was probably unique in its time. Another important investigation concerned the division of current in parallel circuits. Cavendish made his results public and palpable in 1776 by constructing a model of a torpedo (an electric fish) that made clear to his colleagues how an animal living in salt water (a passable conductor) could store electricity and appear to direct it to numb its prey.

In his last significant published work Cavendish returned to old Newtonian problems. In 1798 he used a torsion balance to detect the gravitational force between masses in the laboratory. He did not derive a value of the universal constant of gravitation from the tiny deflection of the balance, but rather estimated from the measurements a likely value for the density of the earth—that too being a Newtonian problem.

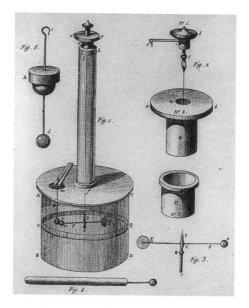

Coulomb's well-engineered torsion balance, in which the repulsive force between two charged pith balls (one fixed to the housing and the other to an end of the mobile horizontal insulated arm) is balanced by the twist in the vertical wire suspending the arm.

Coulomb's most important contribution to physics also concerned the forces between particles of electric fluids, of which, in contrast to Cavendish, he admitted two. Like Cavendish, in designing his method of measurement he made use of a Newtonian theorem: that a gravitating spherical shell acts on bodies outside it as if its entire matter resided at its center. If the particles of each electric fluid repelled one another by an inverse-square force, a charged conducting sphere should act on bodies outside it as if its charge were concentrated at its center. Coulomb adopted a torsion balance he had invented to test the strength of wires and to investigate magnetism to measure "the force of electricity." The apparatus with which he claimed to demonstrate the law of squares (Coulomb's law, 1785) has become an icon in physics. It is a carefully engineered device for obtaining numbers (the degree of twist of a wire), in striking contrast to Cavendish's homemade rig for registering zeros (the lack of a charge on a previously charged sphere).

Coulomb's contemporaries had trouble reproducing his measurements (keeping the balance rod horizontal is difficult), and modern attempts have succeeded only by using precautions that Coulomb did not specify. Nonetheless, Coulomb's academic colleagues accepted his numbers and inferences, and his similar results for magnetism, no doubt because they had already enlisted electricity among Newtonian *imponderable fluids.

Coulomb followed up his measurements of electrical forces with determinations of the distribution of electricity on the surfaces of charged, touching, insulated spheres. Powerful mathematicians, notably Simeon-Denis Poisson and Carl Friedrich *Gauss, confirmed Coulomb's determinations, and in the early nineteenth century they built up a fully quantitative electrostatics and magnetostatics.

The difficulty historians have found in repeating Coulomb's experiments has given rise to doubts about the veracity of his reports and to meditations on the meaning of "replication" in science. One answer to the doubters is that Coulomb perfected his technique over many years and that the coordination of eye and hand acquired from his long engineering practice is not a common attainment of historians.

A. J. Berry, *Henry Cavendish* (1960). C. Stewart Gillmor, *Coulomb and the Evolution of Physics and Engineering in 18th Century France* (1971). J. L. Heilbron, *Electricity in the 17th and 18th Centuries* (1979; 2d ed. 1999). Christine Blondel and Matthias Dörries, eds., *Restaging Coulomb* (1994). Christa Jungnickel and Russell McCormmach, *Cavendish* (1996).

J. L. HEILBRON

CELESTIAL MECHANICS. The study that Pierre-Simon Laplace named celestial mechanics originated in Isaac *Newton's *Principia* (1687). Newton assembled impressive empirical evidence for the inverse-square law of universal gravity and gave closed solutions for the one- and two-body problems (to determine the motion of a body under gravitational attraction towards a fixed center, and to determine the motions of two bodies attracting each other gravitationally). The three-body problem (to determine the motions of three mutually attracting bodies) admits of no such solution, and Newton's geometrical approach to it did not provide for systematic refinement. Celestial mechanics emerged as an algebraic attack on these problems, first treating the bodies as points, and then taking into account their extension in space. Thus the Moon's attraction for the Earth's equatorial bulge causes a precessional motion of the Earth's axis ("the precession of the equinoxes"); Newton's derivation of this motion was badly flawed, posing for the new science one of its major challenges.

Advance came through formulating these problems as differential equations and solving them with trigonometric functions. Having systematized the calculus of these functions in 1739, Leonhard *Euler in 1748 introduced trigonometric series and statistical procedures for the differential correction of orbital elements. In 1749 Jean Le Rond d'Alembert carried out the first valid derivation of the precession of the equinoxes. Thus prodded, Euler completed a formal theory of the dynamics of rigid bodies. Alexis-Claude Clairaut in 1752 gave the first satisfactory derivation of the moon's apsidal motion. In 1758–1759, by carrying out the first extended numerical integration, he achieved a prediction of the return of Halley's Comet accurate to a month.

Lunar theories—both Newton's and the newer analytic ones—remained insufficiently accurate to permit determination of longitude at sea to within one degree. To achieve the necessary accuracy, Tobias Mayer combined analytic theory with statistical correction of coefficients, thus providing the basis for the British *Nautical Almanac* (published from 1767 onwards). Semiempirical lunar theories would remain at the heart of national almanacs until 1862.

Joseph Louis Lagrange enriched orbital calculations of the "perturbing function." Assume a planet or satellite moving in an ellipse about its primary in accordance with the solution of the two-body problem while the attraction of a third body disturbs the motion. Lagrange's new function, when differentiated with respect to any coordinate, yielded the force disturbing the motion in the direction of the coordinate. Lagrange also completed and systematized the method of variation of orbital parameters. Laplace in 1773 showed that, to a good approximation, the planetary mean motions are immune from secular variation. In 1787 he provided an apparently cogent gravitational explanation, later fundamentally revised, for the moon's observed secular acceleration.

Although Laplace's *Mécanique céleste* remained the prime textbook of celestial mechanics until the 1890s, its treatment of second- and higher-order approximations relative to the perturbing forces, however, was unsystematic. Resolutions of this difficulty came from the method of variation of orbital parameters as developed by Simeon-Denis Poisson and Lagrange in 1808–1809. Peter Andreas Hansen made all perturbations additive to the mean motion; his lunar theory was used for the national ephemerides from 1862 to 1922. Charles-Eugène Delaunay successively removed perturbations from the perturbing function, incorporating them in progressively refined canonical variables. The slow convergence of his method led George William Hill to found lunar theory anew, starting from periodic solutions of a restricted form of the three-body problem. Hill's work stimulated the fundamental topological researches of Henri Poincaré into the stability of dynamical systems.

René Taton and Curtis Wilson, eds., *Planetary Astronomy from the Renaissance to the Rise of Astrophysics, Part B: The Eighteenth and Nineteenth Centuries* (1995): chs. 20–22, 28. Martin C. Gutzwiller, "Moon-Earth-Sun: The Oldest Three-Body Problem," *Reviews of Modern Physics* 70 (1998): 589–639.

CURTIS WILSON

CHANDRASEKHAR, Subrahmanyan (1910–1995), mathematical astrophysicist.

Chandrasekhar showed that stars up to a particular mass limit evolve into white dwarfs, implying that more massive stars collapse to form black holes. He wrote thorough mathematical monographs inspired by a variety of astrophysical problems, including *relativity and *black holes, and elevated the *Astrophysical Journal* to world-class status during his nineteen-year editorship.

In 1930 on a boat from India to England, where he would pursue graduate studies in stellar structure, Chandrasekhar modified stellar models for relativistic and quantum effects and found a limit—the Chandrasekhar limit—to the mass of a star (approximately 1.4 solar masses) that can evolve into a white dwarf (an extremely dense star). This startling result won little attention. In 1934 Chandrasekhar was advised that to persuade astronomers he would have to compute the masses of a representative sample of white dwarfs and demonstrate that all fell below the limit. Chandrasekhar presented his results before the Royal Astronomical Society in 1935. The famous English astrophysicist Arthur Eddington pointed out that in Chandrasekhar's mathematical model heavier stars would keep contracting and gravity would become strong enough to hold in radiation, which, in Eddington's opinion, was an absurd way for stars to behave. Devastated, Chandrasekhar abandoned this line of research, resuming it again only in the 1960s with a mathematical study of *relativity and *black holes. In 1983 he received a Nobel Prize for the work.

In 1937 Chandrasekhar took up a position at the University of Chicago's Yerkes Observatory, where his theoretical work could be combined with observations. Yerkes quickly built an outstanding graduate program in astronomy and astrophysics with Chandrasekhar as its most active faculty member. In 1952, however, the astronomy department made an abrupt change in emphasis, from theoretical to observational courses, and Chandrasekhar shifted much of his teaching to the physics department. Also in 1952 he became managing editor of the *Astrophysical Journal*, a position he exercised autocratically and energetically.

After the debacle with Eddington in 1935, Chandrasekhar, a classical applied mathematician in the Cambridge University tradition, chose mathematically beautiful problems usually neglected because of their difficulty, solved fundamental equations about a problem, and eventually published a formal, logical, rigorous treatise about it. Then he would move on to a new problem. Stellar dynamics was followed in the 1940s by radiative transfer, particularly with regard to stellar and planetary atmospheres. Next Chandrasekhar explored hydrodynamic and hydromagnetic stability, and then ellipsoidal figures in equilibrium, particularly the stability of rotating and vibrating stars within the framework of general relativity. This work, culminating in 1969, turned out to be relevant to pulsars, rapidly rotating neutron stars discovered in 1967.

In the 1960s Chandrasekhar finally felt free to indulge his early interest in relativity theory, a field he had long perceived as a graveyard for theoretical astronomers. In 1963, however, a relativistic astrophysics revolution began with the discovery of the first quasar (quasi-stellar object). Its tremendous energy output exceeded that obtainable from thermonuclear combustion of all its mass. In the intense theoretical effort to find a new energy source, gravitational collapse was the most likely mechanism, and relativity theorists suddenly became respectable. No doubt a factor freeing Chandrasekhar to pursue research in a field he earlier had shied away from as too risky was his now firmly established scientific reputation and professional position. He had been elected to the Royal Society in 1944 and received its Gold Medal in 1952, and would receive the U.S. National Medal of Science in 1967. The Chandra X-ray Observatory, launched into space in 1999, was named in his honor. Relativity and black holes remained the consuming scientific passions of the remaining three decades of Chandrasekhar's life. Black holes were the perfect object for him since their construction depended solely on mathematical concepts of space and time. The beauty of the equations drew him to the mathematics, irrespective of any astrophysical relevance. Chandrasekhar's intellectual legacy includes lectures in 1983 in memory of Arthur Eddington; a collection of essays published in 1987 on aesthetics and motivations in science, especially the quest for beauty in mathematical equations, which he likened in 1992 to a sequence of paintings by Claude Monet; and an analysis of Isaac *Newton's *Principia* in 1995.

See also PULSARS AND QUASARS.

Kameshwar C. Wali, *Chandra: A Biography of S. Chandrasekhar* (1991). G. Srinivasan, ed., *White Dwarfs to Black Holes: The Legacy of S. Chandrasekhar* (1999). www.phys-astro.sonoma.edu/BruceMedalists.

NORRISS S. HETHERINGTON

CHAOS AND COMPLEXITY. Nature is complicated. Scientists trying to understand it have to simplify and approximate in order to discern regularity in phenomena and describe it mathematically. In the late twentieth century a new field called chaos theory emerged that instead embraced complexity and the nonlinear mathematical equations that expressed it.

Scientists and mathematicians had previously addressed the topic of complexity and nonlinear equations, notably Henri Poincaré, who worked on the theory of differential equations and dynamical systems. In 1908 Poincaré pointed out that small differences in the initial conditions of a system could result in large changes in their long-term evolution, and noted as an example the unpredictability of the weather. But Poincaré's work did not immediately spark a new line of research; physicists at the time were fruitfully exploiting linear differential equations in the development of *relativity and quantum theory.

Chaos theory first emerged from the increasing use of computers in *meteorology after World War II. In 1961 Edward Lorenz, a meteorologist at the Massachusetts Institute of Technology (MIT), was running simplified atmospheric models through his computer. He decided to retrace a run, but instead of starting at the beginning he started halfway through, typing in the numbers for the initial conditions from the printout for the first run. The printout had rounded off the six decimal places used by the computer to just three, but Lorenz assumed a difference of one part in a thousand would be inconsequential. Instead he found that the second run, from almost the same initial conditions, diverged wildly from the first. He first thought he had blown a vacuum tube, but then recognized the importance of the small difference in initial conditions.

Using a still simpler system of three nonlinear equations modeling convection, Lorenz demonstrated sensitive dependence on initial conditions and cast doubt on the prospects of long-range weather forecasts. But he also revealed a sort of abstract order within the disorderly behavior that resulted: a plot of the results in three dimensions traced a complex double spiral, nonintersecting and nonrepeating yet with distinctive boundaries and structure. The image, later called a Lorenz attractor, appeared with Lorenz's results in 1963, in a paper entitled "Deterministic Nonperiodic Flow" in the *Journal of the Atmospheric Sciences*. The title asserted the persistence of determinism; the avenue of publication indicated the source of Lorenz's interest in the problem and ensured that most physicists and mathematicians would miss its initial appearance.

In the meantime Stephen Smale and several other mathematicians at the University of California at Berkeley were developing ways to model dynamical systems through topology, folding and stretching plots in phase space to reproduce the unpredictable histories of nonlinear systems; two points on the plot could be close together or far apart depending on the sequence of folds, thus exhibiting sensitivity to initial conditions. In the early 1970s mathematician James Yorke came across Lorenz's paper and publicized it in a mathematics journal, in which he applied the term "chaos" to the subject. Yorke drew on the work of Robert May, a mathematical physicist who had turned to population biology. May had found that nonlinear equations describing cyclic changes in populations could begin doubling rapidly in period before giving way to apparently random fluctuations; but within the random behavior stable cycles with different periods would reappear, then start doubling again toward randomness. Yorke explained May's results with chaos theory. A review article in *Nature* in May 1976 brought chaos to a still wider audience.

Yorke learned that Soviet mathematicians and physicists had been pursuing similar lines of research, starting with the work of A. N. Kolmogorov in the 1950s. A. N. Sarkovskii arrived at the same conclusions as Yorke, and Yakov Grigorevich Sinai developed the theory in the framework of thermodynamics. Physicists in the Soviet Union, the United States, and Europe saw in chaos a way to tackle longstanding problems in fluid dynamics, especially turbulence and phase transitions. The appearance of periodic order within longer-term disorder found visual expression in the geometry developed by Benoit Mandelbrot and other mathematicians in the 1970s. Mandelbrot coined the term "fractal" to describe the new class of irregular shapes that seemed to duplicate their irregularity when viewed at different scales and dimensions. In 1976 Mitchell Feigenbaum found that a single constant described the scaling or convergence rate—that is, the rate at which cycles doubled in period on the way to chaos—no matter the type of physical system or mathematical function (quadratic or trigonometric). Shortly after Fei-

genbaum announced the single universal scaling law, the first conference on chaos convened in Como, Italy. That was in 1977, fifty years after another conference of physicists there had considered the competing interpretations of quantum mechanics.

Chaos theory emerged from diverse disciplinary and institutional origins: Lorenz was an academic meteorologist; Mandelbrot worked at International Business Machines (IBM) on mathematical economics; Feigenbaum was a theoretical physicist at Los Alamos National Laboratory. Los Alamos eventually created a Center for Nonlinear Studies, and other centers for chaos theory emerged at the University of California at Santa Cruz, at Gorky in the Soviet Union, and elsewhere. Chaos theory served to bridge disparate disciplines dealing with apparent disorder: biology, ecology, economics, meteorology, and physics. Digital electronic computers were central to all of the work. Chaos also connected abstract mathematics with real-world problems. The theory provided tools for astronomers studying the red spot on Jupiter and galactic structure, population biologists modeling the fluctuations of species, epidemiologists charting the cycles of disease, physiologists investigating cardiac fibrillations, and urban engineers tracking traffic flows. A few enterprising chaos theorists sought to predict the stock market and make investors, and themselves, rich. Some practitioners viewed chaos theory as a subset of a wider field called complexity, which studied neural nets, cellular automata, spin glasses, and other exotic systems exhibiting complex interconnections among individual components.

Chaos theory emerged during a time of general cultural ferment often manifested in antiscientism. The theory itself seemed to reject reductionism and determinism in favor of a holistic embrace of complexity and flux, even if it still found rules and regularity buried deeper in disorder, and it thus resonated with critics of deterministic science. Several chaos pioneers themselves drew inspiration from the romanticism of Goethe. The emergence of centers for the field in Santa Cruz and Santa Fe, towns with New Age reputations, suggest the countercultural component in the chaos community.

Benoit Mandelbrot, *The Fractal Geometry of Nature* (1983). James Gleick, *Chaos: The Making of a New Science* (1987). David Ruelle, *Chance and Chaos* (1991). George A. Cowan, David Pines, and David Meltzer, eds., *Complexity: Metaphors, Models, and Reality* (1994). Ilya Prigogine, *The End of Certainty: Time, Chaos, and the New Laws of Nature* (1997).

PETER J. WESTWICK

CHERENKOV RADIATION. Scientists working on radioactivity around 1900 noticed a faint bluish light emanating from transparent substances, such as liquids, near their sources. The strange phenomenon seemed to pose less of a mystery than radioactivity itself, and inspired comment but not further investigation at the time. The effect got its name from Pavel Cherenkov, a Soviet physicist who in 1934, while investigating luminescence, found that clear liquids excited by gamma rays from a radium source emitted faint blue light. An intense examination of the spectrum, lifetime, and polarization of the radiation and the characteristics of the source convinced Cherenkov that the phenomenon was not luminescence but a new type of radiation. His colleague Sergey Ivanovich *Vavilov suggested that electrons knocked from atoms by gamma rays caused the radiation, which led Cherenkov to more experiments showing that most of the radiation was emitted in the direction of the forward path of the particle that stimulated it. Igor Tamm and Ilya Frank explained the phenomenon in 1937 as a consequence of classical electrodynamics: a charged particle, such as an *electron, traveling through a transparent medium faster than the phase velocity of light in that medium, will emit directed radiation analogous to the sonic boom produced by a plane flying faster than the speed of sound. Cherenkov, Frank, and Tamm shared the Nobel Prize in physics for 1958 for the work; Vavilov had died by that time and thus was ineligible.

Cherenkov radiation later reached popular audiences in images of nuclear reactors, where radiation from fuel rods excited a blue glow in pools of cooling water. It found scientific application in detectors for high-energy experiments at particle *accelerators. The Cherenkov effect converted particles to light and thus offered a way to detect and count them via photomultipliers in combination with electronic logic circuits. Furthermore, since the radiation emerged sharply collimated, and since the emission angle depended only on the incident particle's velocity and the refractive index of the medium, Cherenkov radiation also provided the speed and direction of high-energy particles. Cherenkov counters would figure in several prominent discoveries in postwar particle physics, including the antiproton.

J. V. Jelley, *Cherenkov Radiation and Its Applications* (1958). V. P. Zrelov, *Cherenkov Radiation in High-Energy Physics*, trans. Y. Oren, 2 vols. (1970).

PETER J. WESTWICK

CLIMATE. The mid-eighteenth-century *Encyclopédie* of Denis Diderot and Jean Le Rond d'Alembert offered three definitions of climate. A climate, first, is a latitude-band around the earth, of such a width that the longest day along its polar circle exceeds that along its equatorial circle by some set amount, say one half-hour. Second, climate denotes a region characterized by its seasons, the quality of the soil, "or even the manners of the inhabitants." Third, climates are synonymous with the temperatures or "degrees of heat" proper to them. Climate, according to these definitions, is a region rather than a pattern of typical weather; the term belonged to *geography rather than *meteorology.

Geographers had discussed climate and its relationship to culture since ancient times. The source of this tradition, the Hippocratic treatise *Airs, Waters, and Places*, attributed a population's character to the winds ("airs"), water sources ("waters"), and soil and orientation ("places") of its locale, as well as to diet, sanitation, occupational patterns, and so on. So much did the Hippocratic tradition flourish in the Enlightenment that the *Encyclopédie* took Montesquieu to task for his famous discussion of climate in *The Spirit of the Laws*, which, they claimed, added nothing to "such familiar topics."

Before the late eighteenth century the geographic tradition had little to do with meteorology. Geographers' pronouncements on climate remained general and qualitative, benefiting little from the increasing availability of good meteorological observations. Meteorologists for their part showed little interest in climate. Their chief goal was to discover recurring patterns they could use to predict the weather and its influence on agriculture and health. In the late eighteenth century, using precise instruments then being developed and organized by learned societies, observers began to collect a significant body of reliable weather data. But they did not construct from it an understanding of climates: they did not integrate observations of the weather at many locales into perceptions of the unity of the weather over periods of time and extents of space. Kant's criticism of contemporary natural history applies well to this approach, which placed objects "merely beside each other and ordered in sequence one after another," rather than integrating them into a "whole out of which the manifold character of things is derived."

Medical topography came closest to uniting meteorology, precise measurement, and geography in the late eighteenth century. The largest undertaking occurred in France, where the Royal Society of Medicine dispatched physicians to report on environmental conditions throughout the realm in Hippocratic fashion. The members gathered weather observations, descriptions of "airs, waters, and places," and information on local populations in order to publish a "medical and topographical map" of France. Before their efforts could be integrated in this way into what would have been a climatology of the nation, the Revolution closed the Society.

A true climatology required a new vision of nature. This vision emerged in the first decades of the nineteenth century, integrating geography with meteorology, the Romantic vision of the unity of nature with late Enlightenment methods of precise instrumentation and measurement. Alexander von *Humboldt is the best known representative of this synthesis. Schooled in late-eighteenth-century experimental natural philosophy, Humboldt also enjoyed close relations with the leaders of the German Romantic movement. Drawing on available observations, he constructed a unity among all aspects of nature, seeking in a region's geology, climate, flora, fauna, and human culture a coherent, interacting whole, which he called "physique générale." Climate, in Humboldt's formulation, grew out of the manifold relations of physical geography: the size and orientation of landforms, their height and geological constitution, the relations of land and water, plant and snow cover, and so on. Suddenly Humboldt and his contemporaries discovered regional and continental breadth in meteorological patterns that their predecessors had perceived as merely local. Humboldt's famous isotherms, lines of equal mean temperature drawn on a map, were a visual representation of one aspect of this breadth, uniting global observations of temperature into a coherent whole.

Climatologists of the age of Humboldt studied the distribution of climates over the globe, applied climatic considerations to biogeography, and speculated on the climates of historical and geological time periods, drawing on the evidence of written and fossil records. After the 1850s imperialists and racists increasingly applied the old Hippocratic arguments to buttress claims of European racial superiority and to justify African and Asian conquests. Following Hippocrates' dictum that "races are the daughters of climate," geographers and anthropologists ascribed Victorian virtues of intelligence, industry, sobriety, and more to the influence of temperate European and American climates. The enervating climates of colonial regions, on the other hand, had given birth to feeble races fit only to be ruled. Tropical medicine addressed closely related questions of the climatological basis of diseases that threatened Europeans' ability to govern their possessions and the physical and moral degeneration of those resident in the colonies. Among the more notorious climatological racists was Ellsworth Huntington, who also investigated the extent to which workers' performance depended on their climatological environment, particularly temperature. Experiments inspired by Huntington and carried out in the 1920s by the National Research Council of the United States included taking rectal measurements of body temperature of workers at ten-minute intervals while they labored under heavy loads at environmental temperatures of up to 115° F. Eugenicists and Nazi ideologues continued to propagate theories of racial climatology as late as the mid-twentieth century.

The more empirical and practical branches of climatology in the nineteenth century focused particularly on biogeographical and agricultural problems. Typical questions included the influence of climatological elements such as temperature, humidity, and precipitation on the life-cycles of food crops and on human settlement. Climatologists developed classifications that integrated both meteorological and biogeographical elements in their schemes. The best-known of these was presented in 1884 by Wladimir Köppen and refined over a half century. In the last third of the nineteenth century, so-called "classical climatology" devoted much energy to standardizing measurements and insuring their reliability.

Particularly in the universities and schools, the dominant role of climatology led many to understand meteorology as part of geography; the association reflected Europeans' pursuit of knowledge of exotic regions in an age of imperial expansion. After the turn of the century, as meteorologists like Vilhelm Bjerknes pushed to ally their field with physics, climatology receded from center stage while developing in new directions. Especially for the amphibious landings of World War II, military planners drew on climatology for knowledge of times and seasons offering best conditions for operations. Earlier in the century agricultural climatology had developed sophisticated criteria, such as "critical periods," during which plants particularly need moisture, but it was not until after World War II that meteorologists applied detailed climatological and microclimatological studies to agriculture, precisely quantifying the climatic environment and integrating climatology and plant physiology. Since the middle of the twentieth century, issues of climate change—present in climatological discourse since the Enlightenment—have become prominent.

See also CLIMATE CHANGE AND GLOBAL WARMING; STANDARDIZATION.

Aleksandr Khristoforovivh Khrgian, *Meteorology. A Historical Survey*, 2d ed., vol. 1, ed. Kh. P. Pogosyan (1959). Theodore S. Feldman, "Late Enlightenment Meteorology," in *The Quantifying Spirit in the Eighteenth Century*, eds. Tore Frängsmyr, John L. Heilbron, and Robin E. Rider (1990): 143–178. David N. Livingstone, "The Moral Discourse of Climate: Historical Considerations On Race, Place, And Virtue," *Journal of Historical Geography* 17 (1991): 413–434. Richard Grove, "The East India Company, the Raj and the El Niño: The Critical Role Played by Colonial Scientists in Establishing the Mechanisms of Global Climate Teleconnections, 1770–1930," in *Nature and the Orient: The Environmental History of South and Southeast Asia* (1998): 301–23. David Livingstone, "Tropical Climate and Moral Hygiene: the Anatomy of a Victorian Debate," *British Journal for the History of Science* 32 (1999): 93–110.

THEODORE S. FELDMAN

CLIMATE CHANGE AND GLOBAL WARMING. Theories of climate change date from very early times; Theophrastus (371–287 B.C.) wrote on desiccation wrought by deforestation. His work, revived in the Renaissance, helped fuel concern over deforestation in European colonies, and from the late eighteenth century onward, colonial governments established forest reserves that were among the earliest measures of environmental conservation. Meanwhile, Enlightenment students of the classical world uncovered literary evidence for climate change since ancient times, while Thomas Jefferson argued that deforestation from European settlement had moderated the climate, rendering it fit for civilization. Nineteenth-century climatologists, applying more exacting historical and scientific analysis to these questions, found no convincing evidence of climate change in historical times.

Climate change on astronomical and geological time-scales on the other hand was more nearly certain. From the Comte de Buffon, Immanuel Kant, and Pierre Simon de *Laplace to Lord Kelvin (William *Thomson) and beyond, cosmologists speculated about the long-term cooling of the earth from its origin as a molten ball and about the longer-term cooling of the sun. Nineteenth-century geologists found indisputable evidence in fossils of a warmer climate in ancient times, and around the middle of the century discovered the *Ice Ages. In order to reconcile these discoveries with his belief in a uniform state of the earth, Charles Lyell developed a theory of cyclical climate change, according to which "all...changes are to happen in future again, and iguanodons...must as surely live in the latitude of Cuckfield as they have done so."

Several eighteenth- and early nineteenth-century natural philosophers and mathematicians, including Horace Bénédict de Saussure, Joseph Fourier, and Claude Pouillet, had noted the atmosphere's selective transmission of heat; Fourier compared the atmosphere to the glass of a greenhouse. In 1859, John Tyndall began experiments on the radiative properties of atmospheric gases, and speculated that variations in their amounts might have altered the earth's climate on the geological time scale. In 1895, Svante Arrhenius, trying to explain the Ice Ages, calculated temperature changes from variations in carbon dioxide. Two decades later, he predicted that industrial generation of CO_2 would protect the globe from recurring ice ages and allow increased food production for a larger world population. Arrhenius's ideas were welcomed by the American geologist T. C. Chamberlin, who, in the first three decades of the twentieth century, developed a theory of the atmosphere as a large-scale geological agent based on a carbon cycle. Crustal uplifts expose large surface areas to weathering, a process that absorbs CO_2; global cooling and glaciation follow. The cycle turns about when mountain ranges reduce to nearly base levels.

By the 1950s, alteration in insolation (solar radiation received by the earth), orbital changes, mountain building, and volcanism had all been identified as agents of climate change. Water vapor eclipsed CO_2 as an agent of global warming until G. S. Callendar published a series of articles from 1938 to 1961 emphasizing anthropogenic influences on the amount of carbon dioxide in the atmosphere. Political tension fed fears of climate degradation, and the Cold War intensified anxiety over both global cooling and global warming. Proposals were floated for large-scale interventions such as damming the Bering Strait, orbiting fleets of mirrors in space, and spraying sulfur dioxide into the upper atmosphere. Today, the scientific community has reached a consensus that CO_2 levels have increased owing to industrial activity and that warming is taking place on a global scale. The relation between these two phenomena remains in doubt, as well as the role of other factors and the future course of climate change.

See also CLIMATE; EARTH, AGE OF; OZONE HOLE.

Spencer Weart, *The Discovery of Global Warming* (2003). James Rodger Fleming, *Historical Perspectives on Climate Change* (1998).

THEODORE S. FELDMAN

CLOUD AND BUBBLE CHAMBERS. C. T. R. Wilson built the first cloud chamber in 1895 to satisfy his interest in the weather. Previous work by John Aitken had examined the role of dust as a nucleating agent for water vapor in air, and hence as a source of fog and clouds. Like Aitken, Wilson built a chamber to reproduce in the laboratory the condensation of clouds: sudden expansion of the volume of a closed vessel containing saturated air produced a temperature drop and supersaturated the air. Unlike Aitken, Wilson brought a background in physics and the program of the Cavendish Laboratory at Cambridge University, especially in ion physics and discharge tubes, to bear on the subject. Instead of dust, Wilson employed *ions as nuclei for water droplets. He also began photographing the formation of drops. Then in 1910 Wilson thought to use the device as a detector of charged particles, whose passage through the chamber would leave a trail of ions and hence water droplets. The next year Wilson obtained his first photographs of the tracks left by alpha and beta rays, as well as evidence of X and gamma rays through the beta rays they produced, and thus provided compelling visual evidence of individual atoms and electrons and their interactions.

The cloud chamber then became a popular tool for the study of particles given off by the nuclei struck by *cosmic rays. Commercial firms helped propagate the device by providing affordable, experiment-ready Wilson chambers. P. M. S. *Blackett, also at the Cavendish, used a Wilson chamber in the early 1920s to confirm Ernest *Rutherford's transmutation of nitrogen into oxygen. In the early 1930s Blackett and Giuseppe P. S. Occhialini built a countercontrolled chamber, with a Geiger counter wired to trigger the expansion of the chamber whenever a cosmic ray passed through it. They used the chamber in 1932 to confirm the existence of the positron, which Carl Anderson of the California Institute of Technology (Caltech) had just detected in his own cloud chamber.

The Wilson cloud chamber suffered from a slow cycle time, and its diffuse gas offered few chances for interaction with incoming particles. The solid film of nuclear emulsions yielded more interactions and hence grew in popularity in the 1930s; but emulsions also constantly recorded tracks and hence complicated the resolution of occurrence times. In 1952 Donald Glaser, a physicist at the University of Michigan, tried to solve these problems by turning to a liquid analogue of the cloud chamber: instead of supersaturated gas, Glaser used a superheated liquid. He filled a small glass bulb, a couple of centimeters wide, with liquid diethyl ether held under pressure above its boiling point, then suddenly released the pressure, superheating the liquid. A charged particle passing through the chamber further heated the liquid and left a line of vapor bubbles in its wake. A high-speed camera filmed the first tracks in late 1952, and Glaser announced his results in 1953.

Glaser's bubble chamber provided sufficient density for numerous interactions and a fast cycle time. He intended to use his bubble chamber to detect cosmic rays and hence tried various means of triggering the expansion as Blackett had done with the cloud chamber (see ELEMENTARY PARTICLES). The bubble chamber instead became an important particle detector for high-energy *accelerators, whose predictable output dispensed with the need for countercontrolled expansion. A particular implication of the bubble chamber intrigued accelerator physicists: the possibility of using liquid hydrogen, which, because of its simple nucleus, already served as a target for interaction experiments at accelerators. A hydrogen bubble chamber would combine target and detector in one device.

Accelerator laboratories soon began building bubble chambers, including a group at the University of California at Berkeley under Luis Alvarez. The Berkeley group produced the first tracks in a hydrogen chamber in late 1953 and soon scaled up to larger chambers, culminating in a 180-cm. (72-in.) version to accommodate the greater interaction lengths of hyperons and other strange particles. The massive chamber, completed in 1959, presented new problems in cryogenics (see COLD AND CRYONICS), optical systems, and computerized data analysis. The coordination of the physicists, engineers, and technicians building the chamber produced a complex, corporate organizational structure.

The Berkeley bubble chamber signaled the growing importance accorded the detector in accelerator experiments. It also marked a decisive shift from the table-top device of Glaser, built for cosmic-ray physics, to the big science of high-energy physics. Bubble chambers at several accelerator laboratories paid dividends in evidence of new particles and resonances that supported the SU(3) theory for classifying strange particles in the early 1960s and of neutrino interactions that helped confirm the electroweak theory in the early 1970s. Both Glaser and Alvarez won Nobel Prizes in physics for their work with bubble chambers, and both eventually drifted away from the field, disillusioned with the routinization, specialization, and automation that the bubble chamber had brought to particle detection.

See also CATHODE RAYS AND GAS DISHARGE; ELECTRON; RADIOACTIVITY.

R. P. Shutt, ed., *Bubble and Spark Chambers: Principles and Use* (1967). Peter Galison, *Image and Logic: A Material Culture of Microphysics* (1997). Peter Westwick, "Chamber, Bubble," and Jeff Hughes, "Chamber, Cloud," in *Instruments of Science: An Historical Encyclopedia*, eds. Robert Bud and Deborah Jean Warner (1998): 98–100, 100–102.

PETER J. WESTWICK

COHESION, the property of sticking together, became a pressing theoretical issue in the seventeenth century, for with the rise of "corpuscular" or "mechanical" philosophy, it became the key to understanding what distinguished solid from fluid matter.

Corpuscular philosophers, who held that nature consisted solely of material particles of different shapes and sizes that variously moved, faced a challenge in explaining how particles hold together. There could be no glue, since an adhesive, too, could consist only of particles in motion; nor could the philosopher invoke an inherent tendency among particles to cohere, since this would amount to the same empty verbalism for which corpuscular philosophers derided their Aristotelian opponents, and which they prided themselves on having transcended (*see* SYMPATHY AND OCCULT QUALITY).

One line of thinking linked cohesion with the traditional notion, itself unacceptable to doctrinaire mechanists, that nature abhors a vacuum. Galileo *Galilei adopted this view in his *Discourse on Two New Sciences* (1638). He argued, however, that because a siphon could not raise water more than 18 *braccia* (about 10 m [33 ft]) before the water column broke under its own weight, the "attraction of the vacuum" could not exceed the weight of the breaking column. The cohesion of solid bodies resulted, he supposed, from their being composed of a multitude of minute particles separated by minute vacua, each of which exerted its attraction.

Pierre Gassendi, the great reviver of atomism, supposed that atoms that cohered did so because they possessed various cavities and protruberances that could catch together like the hooks and eyes used by dressmakers. René *Descartes, on the other hand, attributed cohesion to a state of relative rest between adjacent particles. Isaac *Newton in his *Opticks* dismissed both suggestions out of hand. Invoking hooked atoms was begging the question, he said, while saying that bodies adhere by rest made them stick together by nothing at all. As part of his wider challenge to the mechanical philosophy, Newton suggested that the particles of hard bodies attract one another with a force that is exceedingly strong at contact, but weakens very rapidly with separation. He claimed that experiments performed by Francis Hauksbee involving drops of oil creeping between sheets of glass slightly inclined to each other demonstrated the existence in nature of powerful, short-range forces such as Newton invoked.

With the widespread acceptance of the Newtonian program during the eighteenth century, natural philosophers concurred in explaining cohesion on the basis of a short-range interatomic force. They came to draw a distinction between the force of cohesion, now restricted to an attraction between corpuscles of the same kind, and the forces of affinity binding corpuscles of different chemical species together in compounds. In the late nineteenth century, molecular theorists invoked such an attraction, acting between molecules of a gas, to explain the deviation of real gases from ideal gas behavior. The twentieth century identified the force in question as ultimately electrical in nature, deriving from the electronic charge distributions within neighboring atoms.

Richard S. Westfall, *The Construction of Modern Science: Mechanisms and Mechanics* (1971).

R. W. HOME

COLD AND CRYONICS. In his *New Experiments Touching Cold* (1665), Robert *Boyle reported the first systematic researches on the experimental subject he found "the most difficult" of all. The ingenious experiments of Joseph Black to determine the latent heat and specific heat of water involved ordinary ice (*see* FIRE AND HEAT). Cryogenics received a boost from Michael *Faraday's liquefaction of nearly all the gases known in the 1820s. Among the gases he could not liquefy were oxygen, nitrogen, and hydrogen. By the end of the nineteenth century all gases except helium had been liquefied. Raoul-Pierre Pictet in Geneva and Louis Paul Cailletet in Paris first obtained small droplets of oxygen and nitrogen in 1877. Zygmunt Florenty von Wróblewski and Karol Stanisław Olszewski liquefied oxygen in appreciable quantities in 1883. Carl von Linde and William Hampson made significant improvements to the apparatus for reaching low temperatures. James Dewar liquefied hydrogen in 1898 at the Royal Institution in London. Heike Kamerlingh Onnes managed to liquefy helium on 10 July 1908 at the Physical Laboratory of the University of Leiden. Using the regeneration method and starting from liquid hydrogen temperatures, he made liquid helium and found its boiling point to be 4.25°K and its critical temperature 5 °K.

The development of *thermodynamics, especially James Prescott Joule's and William *Thomson's proofs that the temperature of a gas dropped when it expanded very quickly, provided the necessary background for the investigation and the understanding of the properties of matter in the very cold. Thomas Andrews's experiments determining the critical point—the temperature at which a gas whose pressure is increased at constant volume liquefies—and Johannes Diderik van der Waals's discussion of the continuity of the gaseous and liquid states brought further insight into the characteristics of very cold fluids.

The nineteenth century saw remarkable developments in the large-scale production of cold, especially through the development of the vapor compression process that led to different types of refrigerating machines and refrigeration processes. The plentiful availability of artificial cold transformed the preservation, circulation, and consumption of food. By the end of the nineteenth century the Linde Company had sold about 2600 gas liquefiers: 1406 were used in breweries, 403 for cooling land stores for meat and provisions, 204 for cooling ships' holds for transportation of meat and food, 220 for ice making, 73 in dairies for butter making, 64 in chemical factories, 15 in sugar refining, 15 in candle making, the rest for other purposes.

In 1911 the Institut International du Froid was founded in Paris to regulate the industry and to formulate directions of further research on cold. Delegates reached consensus about the preservation and transport of agrarian, fish, and dairy products, the standardization of the specifications for home refrigerators, the construction of trains and ships with large refrigerators, the installation of special refrigerators in mortuaries and slaughterhouses, the building of new hotels with air cooling systems, the design of breweries, the manufacture of transparent ice, and the possibilities of medical benefits from cold.

Also in 1911, Kamerlingh Onnes observed that certain metals become superconductors, losing all resistance to electrical current, below 4°K. In recent decades materials have been made that reach the superconducting state at much higher temperatures. Another bizarre bit of cold behavior, which came to light in the 1930s, is the superfluidity that liquid helium acquires below 2.19°K in virtue of which it does not display any of the features of classical fluid. These two phenomena turned out to be explicable only on the principles of *quantum mechanics. The explanation forced quantum mechanics to negate one of its basic methodological and historical tenets—that its effects show only in the microscopic world. Superconductivity and superfluidity showed that macroscopic quantum phenomena exist.

The availability of liquefied gases made the variation of the electrical resistance of metals with temperature a popular subject. Dewar and John Ambrose Fleming made the first systematic measurements in 1896. Their results derived at liquid oxygen temperatures suggested that electrical resistance would become zero at absolute zero. But the same measurements at liquid hydrogen temperatures showed that the resistances after reaching a minimum started increasing again. In 1911 Kamerlingh Onnes measured the resistance of platinum and that of pure mercury at helium temperatures. At 4.19°K the value of the resistance dropped abruptly and became 0.0001 times that of solid mercury at 0°C. Impurities did not affect the superconductivity of mercury, but a high magnetic field could destroy it.

The first successful quantum mechanical theory of electrical conduction, proposed by Felix Bloch (1928), predicted that superconductivity was impossible. He found that the most stable state of a conductor, in the absence of an external magnetic field, had no currents. But since superconductivity was a stable state displaying persistent currents without external fields, his theory did not explain how superconductivity could come about in the first place. At the beginning of November 1933 there appeared a short letter in *Naturwissenschaften* by Walther Meissner and Robert Ochsenfeld. It presented strong evidence that, contrary to every expectation and belief of the previous twenty years, a superconductor expelled the magnetic field after the transition to the superconducting state and the magnetic flux became zero (the Meissner effect). Superconductors were found to be diamagnetic and, hence, superconductivity a reversible phenomenon, thus allowing the application of thermodynamics.

On the assumption that diamagnetism must be an intrinsic property of an ideal superconductor, and not merely a consequence of perfect conductivity, Fritz London and his brother Heinz proposed in 1934 that superconductivity involved a connection not with the electric, but with the magnetic field. Their assumption led to the electrodynamics of a superconductor consistent both with zero resistance and the Meissner effect. Fritz London, in his discussion of superconductivity in 1936, formulated for the first time the notion of a macroscopic quantum phenomenon.

Because ionic masses are so much larger than the electron mass, physicists doubted that ions played an important role in the establishment of the superconducting state. Herbert Frohlich in 1950 asserted the opposite and found that the interaction of the electrons in a metal with the lattice vibrations would lead to an attraction between the electrons. Experiments confirmed his assertion. The mass became an important parameter when the motion of the ions was involved, and this, in turn, suggested that superconductivity could be derived from an interaction between the electrons and zero-point vibrations of the lattice. Soon after learning about these results, John Bardeen showed that superconductivity might arise from a new attraction between the electrons and the phonons resulting from lattice vibrations, thus laying the foundations for the electron pair theory. Based on these ideas, in 1957 Bardeen, Leon Cooper, and John R. Schrieffer worked out the details of a microscopic theory of superconductivity, and shared the Nobel Prize in physics of 1972 for their successful explanation of this elusive phenomenon.

All liquids solidify under their own pressure at low enough temperatures. Helium can only be solidified under a pressure of 26 atmospheres. The densities and specific heats of all liquids follow a continuous change and increase as the temperature goes down. In the case of helium, however, these parameters display a maximum at 2.19°K and then decrease. The two methods for measuring the viscosity of any liquid—rotating a disk in it or forcing it through very small capillaries—ordinarily give identical results. Not so for liquid helium below 2°K. The first of these methods gives a value a million times larger than the second. Finally, all normal liquids can be deposited in open containers, kept in containers with extremely small holes through which they cannot flow, and remain at rest when exposed to light. Liquid helium does not tolerate any such constraints. It goes over open containers, leaks through the smallest capillaries, and springs up in a fountain when light falls on it. Below 2.19°K liquid helium becomes a superfluid.

In 1938, Fritz London proposed that the transition to the superfluid state can be understood in terms of the Bose–Einstein condensation mechanism, first discussed by Albert *Einstein in 1924. For an ideal Bose–Einstein gas, the condensation phenomenon represented a discontinuity in the derivative of the specific heat. London argued that the sudden changes in the properties of helium at 2.19°K could result from such discontinuities. Below a certain temperature and depending on the mass and density of the particles, a finite fraction of them begins to collect in the energy state of zero momentum. The remaining particles fly about as individuals, like the molecules in a normal gas. Laszlo Tisza proposed to regard superfluid helium as a mixture of a normal and a superfluid. These two components had different hydrodynamical behaviors as well as different heat contents. At absolute zero, the entire liquid became a superfluid consisting of condensed atoms, while at the transition temperature the superfluid component vanished.

Excluding some applied fields, low-temperature physics became the high point of Soviet physics, especially during World War II. After receiving his doctorate under the supervision of Ernest Rutherford, Pyotr Kapitsa served as an Assistant Director of magnetic research at the Cavendish Laboratory, before becoming the director of the Mond Laboratory in Cambridge. There he liquefied helium in 1934. Though he was not allowed to return to England by the Soviet authorities after a trip in 1934, he was, by 1935, appointed as director of a new Institute of Physical Problems within the Academy of Sciences in Moscow. That is where he conducted his experiments with liquid helium in 1941 and coined the term "superfluid" when he discovered the remarkable characteristics of its viscosity. In 1941 Lev Landau developed a quantized hydrodynamics that explained the transition to the superfluid state in terms of rotons and phonons. The ground states and the excitations played the roles of the superfluid and normal state, respectively. The excitations were the normal state because they could be scattered and reflected and, hence, exhibit viscosity. The ground state described the superfluid because it could not absorb a phonon from the walls of the tube or a roton unless it met some conditions on the velocity. Landau's formalism led to two different equations for the velocity of sound. One was related to the usual velocity of compression, while the other depended strongly on temperature. Landau named it "second sound." Victor Peshkov demonstrated the existence of

standing thermal waves in 1944 for the first time, and in 1949 the experiments of Maurer and Herlin settled the issue of the temperature dependence of the second sound velocity below 1°K. By 1956 Richard Feynman could show that some of Landau's assumptions could be justified quantum mechanically and that the rotons were a quantum mechanical analog of a microscopic vortex ring.

Oscar Edward Anderson, Jr., *Refrigeration in America. A History of a New Technology and Its Impact* (1953). Kurt Mendelssohn, *The Quest of Absolute Zero* (1977). Roger Thevenot, *A History of Refrigeration Throughout the World* (1980). Kostas Gavroglu and Yorgos Goudaroulis, *Methodological Aspects of the Development of Low Temperature Physics 1881–1956. Concepts out of Context(s)* (1989). Kostas Gavroglu and Yorgos Goudaroulis, eds., *Through Measurement to Knowledge. The Selected Papers of Heike Kamerlingh Onnes 1853–1926* (1991). Per Fridtjof Dahl, *Superconductivity. Its Historical Roots and Development from Mercury to the Ceramic Oxides* (1992). Ralph Scurlock, ed., *History and Origins of Cryogenics* (1992). Kostas Gavroglu, *Fritz London, A Scientific Biography* (1995). Tom Shachtman, *Absolute Zero and the Conquest of Cold* (1999).

KOSTAS GAVROGLU

COLD FUSION. On 23 March 1989 in a press conference in Salt Lake City, B. Stanley Pons, a professor of chemistry at the University of Utah, and Martin Fleischmann, his collaborator from the University of Southampton, proclaimed that they had achieved the fusion of deuterium nuclei—the type of reaction that fuels the sun—in a laboratory experiment. They reported that when they passed an electrical current through palladium metal immersed in a beaker of heavy water with a bit of lithium, the cell produced an excess of heat—enough at one point to melt a cube of the metal and far more than could be accounted for by current running through the electrode or by ordinary chemistry. The heat, they said, had to be a product of cold fusion, so-called because it had been achieved at room temperature.

Their scientific miracle won wide coverage in the world press. It promised cheap, clean energy, which, in the wake of the fuel shortages of the 1970s, governments had been spending hundreds of millions of dollars to find. It captured attention in the White House, the U.S. Congress, and the government of Utah, which promptly appropriated five million dollars for a new National Cold Fusion Institute. Chase Peterson, the president of the university, enthused that the breakthrough "ranks right up there with fire, with cultivation of plants and with electricity."

Many scientists, however, greeted cold fusion with skepticism. It defied the known laws of physics. Moreover, Pons and Fleischman revealed few details of the experimental apparatus and methods that produced their astonishing result and persistently ducked requests from scientists for more information. Some scientists wryly observed that fusion cells running as hot as Pons and Fleischmann's should have been producing enough neutron radiation to kill them.

Laboratories in the United States, Europe, and Japan geared up to witness the miracle firsthand or debunk it. Theoretical calculations demonstrated that the fusion of deuterium was impossible with Pons and Fleischman's apparatus, and experimental tests in various laboratories showed that the heat generated with it did not arise from fusion. On 1 May 1989, at the meetings of the American Physical Society in Baltimore, Maryland, the evidence for cold fusion was authoritatively reviewed and found wanting at a special session that drew two thousand people.

Some observers wondered whether Pons and Fleischman had attempted to perpetrate a scientific fraud. At the least, one physicist remarked, cold fusion was a result of their "incompetence and perhaps delusion." Cold fusion indelibly tarred Fleischmann's reputation and forced Pons's departure from the University of Utah in the fall of 1989. Inspired by odd bits of seemingly encouraging evidence and more than a dash of hope, research into cold fusion continued at several laboratories in the United States, Japan, and Italy at least until the mid-1990s, some of it funded by the Japanese government, but none of the efforts yielded compelling results.

David L. Goodstein, "Pariah Science—Whatever Happened to Cold Fusion?" *The American Scholar* 63 (Fall 1994): 527–541. Gary Taubes, *Bad Science: The Short Life and Weird Times of Cold Fusion* (1993).

DANIEL J. KEVLES

COLLIDER. During the final quarter of the twentieth century, particle colliders emerged as the preferred instruments in high-energy physics. Their defining characteristic, besides their great size—the largest are measured in kilometers—is their manner of generating collisions. They accelerate two beams of subatomic particles and bring them together at interaction points, where a particle in one beam can collide with a particle in the other. Surrounding each interaction point, a particle detector records the tracks, energies, and other characteristics of particles emanating from these collisions.

The great advantage of colliders over conventional "fixed-target" machines (such as cyclotrons), in which particle beams strike stationary objects, is that essentially all the energy of the individual colliding entities can be used to create new subatomic particles. The available "center-of-mass" energy, or total collision energy, grows in proportion to the beam energy rather than to its square root, as in fixed-target experiments. All discoveries of massive new subatomic particles since 1975 have been made using colliders, while fixed-target experiments have excelled at examining the structures of known particles such as protons. To permit meaningful experiments, colliders must attain sufficient luminosity, a key measure of the rates of interaction between particles in the opposing beams.

The idea of particle colliders occurred to Rolf Widerøe and Donald Kerst in the mid-1950s, and the first significant work on developing such an instrument began at Stanford University in 1958. Led by Gerard O'Neill of Princeton University, a small group of physicists built two evacuated "storage rings" in a figure-eight configuration. Beams of electrons circulated in opposite directions within these rings at energies of up to 500 million electron volts (MeV); collisions occurred in the shared segment where the rings touched.

In parallel with this effort, physicists at the Frascati National Laboratory in Italy, led by Bruno Touschek, built a single-ring collider in which electrons circulated one way and positrons (their antimatter opposites) the other. Following the success of this prototype, the Italian physicists developed a full-scale electron-positron collider called ADONE, with beam energies of up to 1,500 MeV, or 1.5 billion electron volts (GeV). Experiments using this instrument began in 1968, recording electron-positron annihilations that usually created other subatomic particles.

Physicists at the European Center for Nuclear Physics (CERN), led by Kjell Johnsen, pioneered proton-proton colliders. In 1971 they successfully operated the Intersecting Storage Rings, in which beams of protons circulated at energies of up to 28 GeV. Collisions occurred at six interaction points where the interlaced rings crossed.

The most productive electron-positron collider was the SPEAR facility built at the Stanford Linear Accelerator Center (SLAC) under the direction of Burton Richter. Completed in 1972, SPEAR generated collisions at combined energies of up to 8 GeV. It yielded the discoveries of the massive psi particles and tau lepton, and Nobel Prizes for Richter (in 1976) and SLAC physicist Martin Perl (in 1995).

Following these advances, physicists built colliders at all leading high-energy physics laboratories. Especially noteworthy was a proton-antiproton collider built at CERN as an upgrade of its existing Super Proton Synchroton, stimulated by ideas and inventions of Peter McIntyre, Carlo Rubbia, and Simon Van der Meer. By observing proton-antiproton collisions at total energies of up to 540 GeV in 1982–1983, two teams of physicists discovered the massive W and Z bosons, the mediators of weak interactions and key elements of the *standard model.

These significant discoveries and the development of superconducting magnets for the Tevatron proton-antiproton collider at the Fermi National Accelerator Laboratory (Fermilab) encouraged U.S. physicists to design the Superconducting Super Collider (SSC), a 40,000 GeV proton-proton collider

that was to have a circumference of 86 km (54 miles), several interaction points, and a cost of $5.9 billion. Those were its parameters in 1989 when construction began south of Dallas, Texas. Congress terminated the project in 1993 owing to cost overruns, lack of major participation from other countries, and a concern to reduce budget deficits after the Cold War.

Since the SSC's demise, the development of particle colliders has continued largely through upgrades of existing instruments at CERN, Cornell University, Frascati, Fermilab, and SLAC, and at national laboratories in China and Japan. A prime example was the conversion of the Stanford Linear Accelerator into a linear electron-positron collider. The new machine accelerated individual "bunches" of the two types of particles and brought them together after a single pass through the linear accelerator. This approach contrasts with that of storage-ring colliders, in which the bunches of particles circulate continuously in fixed, intersecting orbits.

See also ACCELERATOR.

Burton Richter, "The Rise of Colliding Beams," in *The Rise of the Standard Model*, ed. L. Hoddeson et al. (1997): 261–284.
Michael Riordan, "The Demise of the Superconducting Super Collider," *Physics in Perspective* 2, no. 4 (2000): 411–425.

MICHAEL RIORDAN

COMETS AND METEORS. Until the seventeenth century, comets and meteors were classified as related natural phenomena and heavenly wonders that heralded calamity. The roots of these views reached back to antiquity. Aristotle saw comets as a type of fiery meteor that formed when terrestrial exhalations ascended into the upper atmosphere, below the moon's sphere, and began to burn. Other fiery meteors included shooting stars, fireballs, and the aurora borealis. Comets and meteors augured windy weather, drought, tidal waves, earthquakes, and stones falling from the sky because both the meteors and the portended disasters derived from hot, dry exhalations that had escaped from the earth. Romans came to view comets and showy meteors as monsters, contrary to nature.

Medieval chronicles recorded meteorological apparitions that heralded the death of holy men and kings, and augured wars of religion and civil strife. According to some early church fathers and later theologians, these heavenly signs demarcated critical periods in the history of the world and religion. Thus Origen and John of Damascus thought that the Star of Bethlehem had been a comet, whereas Saint Jerome, Thomas Aquinas, Martin Luther, and Thomas Burnet expected comets and fiery meteors to precede the Day of Judgment and consummation of all things.

Medieval and Renaissance natural philosophers agreed that comets and meteors prefigured calamity. John of Legnano and Johannes *Kepler looked for causal connections. Many more wrote guides, both in Latin and in vernacular languages, to interpret the meaning of these celestial hieroglyphs. Their tracts served astrologers and propagandists who used fiery meteors to legitimate political authority and to fortify conspirators up through the eighteenth century.

In the Renaissance, observations of the parallax, tails, and motion of comets by Tycho *Brahe and others convinced astronomers that comets were not sublunar meteorological phenomena, but celestial bodies traveling through interplanetary regions. The separation of comets from meteors was completed by the end of the seventeenth century when astronomers agreed with Isaac *Newton and Edmond *Halley that comets traveled in elliptical orbits around the sun. In the mid-eighteenth century, philosophers began to consider extraterrestrial origins for meteors such as the aurora and shooting stars.

How far these theoretical developments caused the decline of divination from comets and meteors has been much debated. Whatever the answer, in the late seventeenth century the learned of England and France (followed later by those in central and eastern Europe) rejected as vulgar the notion that comets and meteors were miraculous signs sent by God, or as causes of murder, rebellion, drought, flood, or plague. Nevertheless, neither the celestial locus nor the periodic orbits of comets required believers to give up their faith in the eschatological or prophetic functions of comets.

Newton suggested that comets transported life-sustaining materials to the earth and fuel to the sun. He, Halley, and William Whiston argued that comets had key roles to play in the earth's creation, Noachian deluge, and ultimate destruction. The final conflagration would be ignited by a comet, many theologians believed, and natural philosophers concurred that a blazing star could serve this function by immersing the earth in its fiery tail, by dropping into the sun and causing a solar flare, or by kicking the earth out of its orbit and transforming it into a comet. Forced to travel in a much more elongated circuit around the sun, the old earth would be scorched and frozen in turns. The new periodic theory of comets did not destroy the belief in comets as agents of upheaval or renewal, nor as tools God might use to punish the wicked or save the elect.

In the eighteenth and early nineteenth centuries, Georges-Louis LeClerc, Comte de Buffon, William Herschel, and Pierre-Simon *Laplace, continued to connect comets to the creation and dissolution of planets, but separated astrotheology from celestial mechanics. Unlike their predecessors, they neither hoped nor expected to find the moral order reflected in the natural world. When catastrophism (see UNIFORMITARIANISM) went out of style in the mid-nineteenth century, comets appeared to pose little risk or benefit to the earth. In recent years, however, the tide has turned, and the stage may be set for a new theological interpretation of comets and meteors. Most scientists now believe that comets (and their meteoric debris) may have been both the agents of death (most notably of the dinosaurs) and the conveyors of life's building blocks.

See also ASTROLOGY.

Roberta J. M. Olson, *Fire and Ice: A History of Comets in Art* (1985). John G. Burke, *Cosmic Debris: Meteorites in History* (1986). Donald K. Yeomans, *Comets: A Chronological History of Observation, Science, Myth, and Folklore* (1991). Sara J. Schechner, *Comets, Popular Culture, and the Birth of Modern Cosmology* (1997).

SARA J. SCHECHNER

COMPLEMENTARITY AND UNCER-TAINTY constitute the foundation of the "Copenhagen interpretation" of quantum physics, an acausal understanding of physics that remains predominant today. Both concepts were introduced in 1927, by Niels *Bohr and Werner *Heisenberg, respectively. They arose as part of the development of *quantum physics when the field was in tremendous flux.

Bohr had been a leader in the development of quantum physics since he published his revolutionary atomic model in 1913 (see ATOMIC STRUCTURE). In Bohr's model, atomic electrons could exist only in orbits determined by the quantum of action and emit electromagnetic radiation only when jumping from one orbit to another. During the "old quantum theory" (1913–1925), theorists invoked a mixture of Bohr's "correspondence principle" (which specified a numerical connection between quantum and classical physics) and arguments based on the quantum of action. As one of a long line of increasingly radical attempts to arrive at an overarching theory, in 1924 Bohr, his assistant Hendrik Kramers, and an American postdoctoral researcher, John Slater, published a paper based on a denial of the well-established principle of energy conservation (see CONSERVATION LAWS). In their view, the principle held only statistically and not for individual atomic processes. Experimental results showing energy conservation in collisions between individual photons and atomic electrons in the Compton effect (see X RAYS) quickly forced the abandonment of this view, and others took the lead in seeking to formulate a quantum theory. In the fall of 1925, Heisenberg, then working as Bohr's assistant, devised a means to calculate spectral data without explicit appeal to the correspondence principle. Whereas Heisenberg's severely operationalist, as well as particle-oriented, theory involved complicated matrix calculations, Erwin *Schrödinger's wave-oriented version of "quantum mechanics," published in the fall of 1926, involved mathematics with which the average physicist felt more comfortable. In spite of this, and although Schrödinger's "wave mechanics" seemed at first to allow the visualization of atomic processes by emphasizing continuity and retaining causality, Schrödinger and others soon showed that his approach was mathematically equivalent to Heisenberg's.

In September 1926 Schrödinger paid a now famous visit to Copenhagen. Bohr stubbornly sought to convince him of the reality of quantum jumps. In this tense environment Heisenberg wrote the article containing his "uncertainty principle," which stated that in the atomic domain the quantum of action set a limit to the precision with which two conjugate variables, such as a particle's position and momentum, or the time and energy of an interaction, could be measured. Since the present therefore cannot be fully specified, Heisenberg argued, neither can the future. By explaining the indeterminism of quantum physics in this way, Heisenberg tried to make his presentation more visualizable (anschaulich) and hence more acceptable to his fellow physicists.

Heisenberg overstepped common practice by submitting his article from Copenhagen without asking Bohr's permission. It turned out that Bohr disagreed so strongly with Heisenberg's presentation of the quantum that Heisenberg felt compelled to add a correction in the proofs that allowed a greater role for the wave picture. Bohr was then perfecting his own formulation of the foundations of quantum theory. At the end of a lecture surveying the general situation given in Como in September 1927, he proposed his notion of complementarity for the first time. The new notion provided an understanding of quantum mechanics in general and of Heisenberg's uncertainty principle in particular. Bohr maintained that, unlike in classical theory, a description of processes in space-time and a strictly causal account (by which Bohr meant an account recognizing conservation laws) of physical processes excluded one another. This meant in practice that the investigator could choose which aspect of microphysical reality he wished to see expressed by his choice of experimental setup. Although the setup required for realizing one aspect excluded the realization of the other—for example, an apparatus for exhibiting light with particulate properties cannot also show it as a wave—both sets of properties had to be invoked to obtain a complete description of the microphysical reality. While presented as a direct result of quantum mechanics, Bohr's interpretation and his subsequent elaborations of it resonated with philosophical views with which he had struggled in his youth. Only with Bohr's complementarity of 1927 did his work and that of Heisenberg, Max Born, Wolfgang Pauli, Pascual Jordan, and others begin to converge into what came to be seen as the unified "Copenhagen interpretation" of quantum mechanics.

The group surrounding Bohr soon came to perceive complementarity and uncertainty as so closely intertwined that in 1928 Heisenberg gave Bohr's concept precedence over his own. In 1935 Albert *Einstein and two collaborators challenged Bohr's interpretation for being inherently incomplete; they thought that they could obtain more information by experiment than complementarity allowed. Bohr repelled their attack by widening the divide between classical and quantum ideas. In the larger physics community, however, the uncertainty principle became inseparable from any presentation of quantum mechanics, while complementarity figured little in the teaching of the new physics. It tended to be regarded as overly philosophical, vague, and irrelevant.

In recompense, complementarity took on a life beyond physics. Bohr sought to generalize its application, first to psychology, then to biology, and ultimately beyond the scope of natural science. Although he did not complete the book on the topic that he had hoped to write, Bohr conceived complementarity as a general epistemological argument of great import for humanity. It constituted a guiding principle for his own activities, inside and outside physics. At the same time, Bohr's disciples sought to spread their understanding of Bohr's word, sometimes—as in the case of Jordan, who tried to use it to save the freedom of the will—to Bohr's embarrassment. Variations of complementarity became part of severe ideological struggles in Nazi Germany and the Soviet Union.

While extremely devoted, the audience for Bohr's philosophical statements was never large, and today consists largely of a specialized set of philosophers. Nevertheless, complementarity played an important role in providing a conceptual basis for the early work on quantum mechanics.

Max Jammer, *The Conceptual Development of Quantum Mechanics* (1966). Jørgen Kalckar, ed., *Niels Bohr Collected Works*, vols. 6 and 7 (1985, 1996). Mara Beller, *Quantum Dialogue: The Making of a Revolution* (1999). David Favrholdt, ed., *Niels Bohr Collected Works*, vol. 10 (1999).

FINN AASERUD

CONSERVATION LAWS. Isolated physical and chemical systems possess certain unchanging properties, for example, mass and energy, and, if in thermal equilibrium, temperature. Conservation laws refer to a subset of these properties conserved when these systems interact (conservation of energy, conservation of mass). Natural philosophers first explicitly set out such rules in the eighteenth century. Conservation laws have guided theory in the physical sciences ever since. Many instructive conflicts have erupted over the identity of the property conserved and the conditions of its conservation.

The first of these conflicts, fought out in the early eighteenth century, concerned the "force" of a particle or set of particles. "Force" could mean a particle's mass m multiplied by its velocity v (momentum), mv^2, or $mv^2/2$ (vis viva). Colin Maclaurin, Gottfried Leibniz, and Johann Bernoulli I put forward conflicting claims for the conservation of "force" based in metaphysical principles logically developed and eventually expressed mathematically. Experimental data was incorporated into various metaphysical schemes. These arguments intensified with the 1724 prize competition of the Paris Royal Academy of Sciences. Other laws and controversies followed including conservation of angular momentum (Jean d'Alembert, 1749). These quarrels died with their adherents after d'Alembert rooted rational mechanics in virtual velocity rather than conservation. The physical conditions governing momentum, vis viva, energy, force, and so on were disentangled only in the nineteenth century.

Another conserved quantity of the eighteenth century was weight, which became an important guide to chemical theory with the discovery and identification of the several sorts of air (see PNEUMATICS). Conservation of weight became a foundation of the reformed chemistry of Antoine-Laurent *Lavoisier. A third conservation law developed during the eighteenth century had to do with static electricity. Benjamin *Franklin's theory of positive and negative electricity (1747) explicitly conserved charge and, moreover, made good use of the law in explaining the operation of the Leyden jar. Later natural philosophers, who used two electrical fluids where Franklin had made do with one, also practiced, if they did not make explicit, the conservation of electricity. Most theories of caloric, the weightless matter supposed to cause the phenomena of heat (see IMPONDERABLES) also supposed its conservation. Conservation laws in physical sciences were thus well established by 1800.

During the nineteenth century, conservation became a tool for discovery. In 1824 the military engineer Sadi Carnot (1824) applied the principle of conservation to caloric considered as the fuel of steam engines. The work extracted from the engine came from the cooling of the caloric from the temperature of the boiler to that of the environment just as the fall of water works a mill. Carnot's analysis, which resulted in the important insight that no engine more efficient than a reversible one can exist, was put into mathematical form by Benoit-Pierre-Émile Clapeyron in 1837 and largely ignored. Meanwhile, Michael *Faraday, William Grove, and others explored the conservation of force, including *electricity and *magnetism.

In the 1840s this work changed direction and several men from various backgrounds became "discoverers" of the conservation of energy. William *Thomson developed Clapeyron's work, which led him to the definition of absolute temperature. In 1847 Thomson heard James Joule present an account of his measurements of the heat produced by an electrical current (Joule's law) and by mechanical motion. Joule had concluded that the forces of nature were not conserved but transformable one into another in accordance with an exact calculus. A certain amount of heat will always gener-

ate the same amount of mechanical work (mechanical equivalent of heat).

In *The Conservation of Force* (1847), Hermann von *Helmholtz announced a general principle of nature that he extracted from a representation of matter as a collection of atoms held together by central forces. He equated the change in vis viva of a particle moving under the influence of a center of force to the change in the "intensity of the force." He identified the latter with the potential function introduced earlier by Carl Friedrich *Gauss. Helmholtz showed how the results of experiments, like Joule's measurements of the production of heat in current-carrying wires, supported his principle of the interconvertibility of force.

In 1850 Rudolf Clausius put forward the clearest statement of the conservation of energy. He redid Carnot's analysis, replacing the conservation of caloric by the conservation of the "energy" of the perfect gas he assumed as the working substance of his heat engine and gave a mathematical expression for the conservation of energy, the first law of thermodynamics. Later he presented heat as the vis viva of gas molecules and the raising of a weight by the engine as the transformation of one type of mechanical energy (kinetic) into another (potential). In the second half of the nineteenth century the conservation of energy became a mainstay of the physical sciences.

The conservation of energy had no prominent place in James Clerk *Maxwell's kinetic theory or statistical mechanics but was central to Ludwig *Boltzmann's work on both mechanics and *thermodynamics. J. Willard *Gibbs extended thermodynamics from physics into chemistry and developed, along with Helmholtz, other conservation laws (enthalpy and free energy) useful in physical chemistry. Energy conservation underwrote a new philosophical approach to physics, developed by Wilhelm Ostwald and Georg Helm (see ENERGETICS). Also, Maxwell reworked his theory of electromagnetism and light within an energy framework using Thomson and Peter Guthrie Tait's *A Treatise on Natural Philosophy* (1867) as a guide.

Despite these substantial acquisitions, physicists had trouble making energy conservation fit certain phenomena of heat and radiation, and its applicability to *radioactivity at first appeared doubtful. The bleak situation, which Thomson described as clouds over the otherwise sunny landscape of physics, was saved by the quantum theory of Max *Planck and the demonstration of the conservation of weight and energy in radioactive decay through measurements by Marie Sklodowska *Curie, Ernest *Rutherford, and others and the mass-energy law ($E = mc^2$) of *relativity.

Conservation of energy was an integral part of Niels *Bohr's theory of *atomic structure (1913). As the problems of his quantized atoms mounted in the early 1920s, however, he limited conservation of energy to the average of all the interactions of atoms with the electromagnetic field, and freed individual interactions from the necessity of obeying the first law of thermodynamics (1924). This *lèse majesté* played a part in Werner *Heisenberg's quantum mechanics (1925), which changed the place of conservation laws in physics. Conservation laws now sprang from the mathematical symmetries inherent in the expressions for the matrices representing the operations that take a physical system from one state to another. *Symmetry here required that under geometrical change or time reversal the mathematical form remains the same: a rotation in space implied conservation of angular momentum; time reversal, conservation of energy; and linear translation, conservation of momentum. In addition, there was a nonclassical symmetry associated with the intrinsic angular momentum (the spin) of a particle at rest. In the 1930s an associated concept, isospin, was introduced and developed into a method of classifying the known nuclear particles. During the 1950s, isospin helped in classifying and predicting antiparticles and in generating a new conservation law, the conservation of nucleons.

To explore the nucleus physicists had to incorporate light into their theories of the atom and nucleus. The simplest problem, the interaction of the electron and the electromagnetic field (*see* QUANTUM ELECTRODYNAMICS), included a synthesis of quantum mechanics and the special theory of relativity. Techniques developed to bring convergence (renormalization) often forced changes in the conception of the nucleus and its constituents. P. A. M. *Dirac's derivation of the wave equation for the electron (1928) implied the existence of negative energy states, which he interpreted as the domain of an "antiparticle" with the same mass as the electron but with positive charge. In Dirac's theory the two sorts of electrons could be created and annihilated together and contradicted an implicit assumption in quantum mechanics—the conservation of particles. Dirac's interpretation gained credence through Carl David A. Anderson's observation in 1931 of the positron (positive electron) in tracks made in his cloud chamber by *cosmic rays.

The problem of beta decay further undermined assumptions as basic as the conservation of energy and momentum. Wolfgang Pauli suggested in 1931–1932 that an undetected particle of zero mass and electrical charge, named the neutrino by Enrico *Fermi, carried away the missing energy and momentum. The neutrino first revealed itself directly in experiments done by Frederick Reines and Clyde L. Cowan in 1956.

The development of particle *accelerators resulted in the discovery or manufacture of more and more "fundamental" particles and graver and graver problems for conservationists. Novelties included particles produced in associated pairs under circumstances so improbable that physicists gave them a quantum number named strangeness. Scientists postulated a new force within the nucleus, the weak force. The neutrino became a left-handed particle and the antineutrino right-handed. Conservation laws again needed revision, including those springing from the assumption of parity (P) conservation in weak interactions. (Parity requires that a device and its mirror image, if made of the same materials, function in the same way.) Nonconservation of parity led to investigations of other symmetry relations and their conservation laws, including charge conjugation (C), in which all the charges entering an equation become their opposites, and time reversal (T), in which the time variable t is changed to $-t$. The strongest result that physicists could produce was the conservation of CPT, in which the transformations C, P, and T simultaneously take place, with the corollary that if T conserves the relations, so does CP.

Experiments to test nonconservation under P demonstrated that some particles are intrinsically right-, and others intrinsically left-handed. By the 1970s a generally accepted model for particle behavior emerged, the so-called *Standard Model, whose fundamental building blocks, the *quarks, have fractional charges 1/3, 2/3, –1/3 of the electron's. In this model, however, all hadrons (protons, neutrons, etc.) should have the same mass, which they do not. The theory "breaks" this unwanted symmetry by introducing different sorts or "flavors" of quarks. Further symmetries and conservation laws emerged from the requirement that quarks be confined by the strong force. This led to more symmetries and new conservation laws. The dependence of physical interpretation on difficult mathematics seemed justified by the experimental identification of the different quarks in the 1970s and 1980s.

Thus conservation laws, at first intuitive expressions of physical regularities and lately of esoteric mathematical symmetries, have guided physics over the past 250 years.

J. L. Heilbron, *Electricity in the Seventeenth and Eighteenth Centuries* (1979). Abraham Pais, *Inward Bound: Of Matter and Forces in the Physical World* (1986). Necia Grant Cooper and Geoffrey B. West, *Particle Physics, A Los Alamos Primer* (1988). Silvan S. Schweber, *QED and the Men Who Made It: Dyson, Feynman, Schwinger, and Tomonaga* (1994). Crosbie Smith, *The Science of Energy: A Cultural History of Energy Physics in Victorian Britain* (1998).

ELIZABETH GARBER

CONSTANTS, FUNDAMENTAL. The belief that numbers constitute the essence of the universe runs deep in the human experience. The Greeks in particular anchored thought in number, both in the metaphorical, as in the speculations of the Pythagoreans and the Platonists, and in the practical, as in the exact geometrical astronomy of Hipparchus and Ptolemy.

These two approaches came together from time to time, notably in the work of Johannes *Kepler, who combined numerological beliefs in the literal harmonies of the celestial spheres with Tycho *Brahe's precision measurements to discover that the cube of a planet's average orbital diameter divided by the square of its orbital period was the same value, no matter the planet.

*Galileo Galilei confined his numerical endeavors to terrestrial bodies. He measured the rate of fall of various weights and found that the ratio of the distance traveled and the time of fall squared was the same for each body examined. Galileo's falling bodies and Kepler's fruitful numerology came together in Isaac *Newton's theory of universal gravitation. Newton's theory implied the existence of a fundamental constant (later labeled G) that specified the force of attraction not only between a planet and the Sun but also between a falling object and Earth.

Not much attention was paid to these constants of gravity. The mathematical methods of the time, which focused on the form of ratios between quantities and not their proportionality constants, veiled the importance of the constants themselves. Even in the late eighteenth century, Henry *Cavendish devised his famous torsion balance experiment not to measure the force between two weights and thus what modern physicists call the gravitational constant, but to measure the density of the earth. A significant exception to the lack of interest in the natural constants prevalent in early modern science was the measurement of the speed of light by Ole Rømer in 1675. A possible explanation of this exception is that, in contrast to gravitational acceleration, speed was conceptually familiar and, in the case of light, would be determined by astronomical phenomena—Rømer used eclipses of Jupiter's satellites—frequently subjected to measurement.

Not until the middle of the nineteenth century did the modern interest in fundamental constants in physics evolve. A broad-based quantifying spirit that had arisen during the eighteenth century supplied the general motivation, and the burgeoning telegraph industry, in dire need of well-defined electrical units and standards, supplied the immediate requirement. In 1851, Wilhelm Weber proposed a system of electrical units founded on the metric system. A decade later, the British Association for the Advancement of Science, under the leadership of William *Thomson, Lord Kelvin and James Clerk *Maxwell, took up the challenge of promulgating an international system of electrical units and standards that could meet the needs of both science and industry. In such a system, certain fundamental quantities of nature played a key role, such as the magnetic permeability and electrical permittivity of the *ether. The work also raised the possibility of defining a "natural," non-arbitrary system of units, perhaps based on the wavelength, mass, and period of vibration of the hypothesized atoms.

Meanwhile, certain key numbers were appearing in pathbreaking physical theories. James *Joule demonstrated the mechanical equivalent of heat and calculated its value. Ludwig *Boltzmann reframed thermodynamics in terms of statistics and an important constant, later called k, relating a molecule's average energy to temperature. Max *Planck introduced another key constant related to molecular energy, h, in his radiation law. Maxwell's electromagnetic theory emphasized that the speed of light c was actually the speed of electromagnetic radiation in general. Albert *Einstein's theory of special relativity and his mass-energy equivalence, $E = mc^2$, further established the fundamental status of c. And Joseph John *Thomson's discovery of the electron introduced its mass and charge (m_e and e) as candidates for the fundamental quantities of matter and electricity.

As the recognized physical constants multiplied into the twentieth century, they raised the question, how fundamental? Although they could be categorized by type, such as properties of objects or factors in physical laws, and some clearly possessed deeper and broader significance than others, it became apparent that many of them were interrelated and that the term "fundamental" always hid some arbitrariness.

A second question was, how precise? The 1920s and 1930s saw an informal international effort to identify not only the best extant value of each fundamental constant but also its precision. The technologies of World War II, and offspring like the laser and atomic clock, assisted by enabling revolutionary increases in decimal places. In the 1960s the project gained a formal footing with the establishment by the International Council of Science of a Committee on Data for Science and Technology (CODATA) and its Task Force on Fundamental Constants. It surveyed the literature and produced a set of "best" values for fundamental constants in 1973, 1986, and 1998.

The hard-eyed quest for the next decimal place did not eradicate interest in numerology, however. Certain combinations of e, h, and c seemed to contain deeper magic. The dimensionless constant e^2/hc, for example, was revealed by *quantum electrodynamics to be the constant that defined the strength of the electromagnetic force. Dimensionless constants held great allure because their value did not depend on the system of units chosen, but seemed to be pure numbers of the universe. Moreover, some simple combinations of fundamental constants yielded very large dimensionless numbers all on the order of 10^{40}. In his "large number hypothesis," P. A. M. *Dirac proposed that this coincidence hinted at an undiscovered law of the universe. The large number hypothesis also raised the question, how constant?, as it implied that some fundamental constants, such as G, might vary as the universe evolves. On some cosmological theories, small changes in the values of the constants can trigger large consequences for the development of the universe. Only for values close to those observed could complex life develop. Thus a consideration of the fundamental constants renewed interest in the anthropic principle and raised hopes that nature's numbers could be derived from the fact of human existence.

See also CONSERVATION.

B. W. Petley, *The Fundamental Physical Constants and the Frontier of Measurement* (1985). John D. Barrow and Frank J. Tipler, *The Anthropic Cosmological Principle* (1986). M. Norton Wise, ed., *The Values of*

Precision (1995). National Institute of Standards and Technology, *The NIST Reference on Constants, Units, and Uncertainty* (2000), online at physics.nist.gov/cuu.

LARRY R. LAGERSTROM

COPERNICUS, Nicholas, (1473–1543), astronomer.

Copernicus stands at the crossroads to modern science because his radically new plan for the cosmos, a heliocentric system to replace the time-honored, earth-centered cosmology, required a radically new physics. Only with a sun-centered arrangement does a causal scheme of gravitational physics follow, but in Copernicus's own work such a consequence was at best simply in an embryonic form. He had no empirical evidence to support the motion of the earth; his sixteenth-century successors for the most part suspended judgement on his cosmology.

Copernicus was born in Torun, a Hanseatic town that had shortly before transferred its allegiance to the Polish monarchy. His education at the University of Cracow was underwritten by his maternal uncle, Lucas Watzenrode, who, after becoming bishop of the northernmost Catholic diocese in Poland, provided a canonry for his nephew. Copernicus continued his education in Italy, in canon law at the University of Bologna (1496–1500) and then in medicine at Padua (1501–1503). He used these skills as a personal secretary and physician to his uncle and as a cathedral administrator.

Copernicus had begun to develop the heliocentric hypothesis by 1514, when a manuscript of his so-called *Commentariolus* existed in Cracow. In this small work Copernicus outlined the postulates behind his new theory and expressed his doubt about Ptolemy's use of the equant, a point within an orbit but outside its center from which the orbiting body appears to move with constant angular velocity. Copernicus went to considerable effort to eliminate the equant, which he felt violated the fundamental cosmological principle of representing celestial appearances by uniform circular motions. A major part of his final work, the replacement of the equant with combinations of circles, was greatly ad-

mired by his immediate successors. However, as some of his Islamic predecessors had already demonstrated, the equant can be eliminated without supposing a heliocentric framework.

In examining the arrangement of the major circles in Ptolemy's epicyclic theory (the deferent or carrying circle and the epicycle carried on it), Copernicus must have noticed that the directional line to the planet was preserved when the circles were interchanged. If he did not discover this for himself, he could have found it in Regiomontanus's *Epitome of the Almagest* (1496). This could have led Copernicus to a rescaling of the circles to allow one for each planet to be placed around the sun. Although very few working notes survive to show the progress of his thinking, the preserved fragments document this step.

The heliocentric arrangement led to a beautiful discovery: the swiftest planet, Mercury, took up the orbit closest to the Sun, whereas the slowest planet, Saturn, automatically fell farthest away. "In this arrangement, therefore, we discover a marvelous commensurability of the universe and a sure harmonious connection between the motion of the spheres and their size, such as can be found in no other way," Copernicus proclaimed in his great treatise, *De revolutionibus orbium coelestium*. It also explained naturally why the retrogression (or occasional apparent westward motion) of Mars, Jupiter, and Saturn always occurs when these three superior planets are opposite the sun in the sky, and why the retrogression of Mars exceeds Jupiter's, and Jupiter, Saturn's. Consideration of harmony and unity apparently drove Copernicus to his new vision of the cosmos.

Copernicus carried out occasional planetary observations to confirm the parameters used by Ptolemy. In order to check slow-moving Saturn, he made these observations over fourteen years, between 1514 and 1527. But because his final scheme amounted to a geometrical transformation of Ptolemy's models with essentially the same numerical parameters, his system had little quantitative advantage over the earlier ones. Both the Ptolemaic and Copernican tables give maximum errors exceeding four degrees in the worst cases for the planet Mars.

After 1530, with the basic observations in hand, Copernicus worked slowly to make the numerical work of his book consistent. It is unlikely that he would have succeeded under his own steam in getting the work ready for publication. The necessary catalyst, a young Lutheran astronomy professor from Wittenberg, George Joachim Rheticus, came to stay with Copernicus in Poland in 1539. The twenty-six-year-old Rheticus was so enthusiastic about the novel cosmology that Copernicus gave him permission to publish a short "first report" or *Narratio prima* (1540). When no explosion ensued, Copernicus allowed Rheticus to take a copy of the manuscript to Johannes Petreius in Nuremberg, who published *De revolutionibus* in 1543. Although Copernicus must have proofread finished pages as they arrived from Nuremberg, he received the completed book only on his deathbed, in May 1543.

While the great majority of readers understood the heliocentric cosmology simply as an astronomical hypothesis and not as physical reality, all perceived Copernicus's book as an important text. Additional printings, in Basel in 1566 and Amsterdam in 1617, supplied the continuing demand for this classic work.

Edward Rosen, *Three Copernican Treatises: The Commentariolus of Copernicus, The Letter against Werner, The Narratio prima of Rheticus*, 3d ed. (1971). N. M. Swerdlow and O. Neugebauer, *Mathematical Astronomy In Copernicus's De Revolutionibus* (1984). Nicholas Copernicus, *On the Revolutions*, trans. Edward Rosen (1992). Owen Gingerich, *The Eye of Heaven: Ptolemy, Copernicus, Kepler* (1993).

OWEN GINGERICH

COSMIC RAYS. The first explorers of *radioactivity found it in air and water as well as in the earth. Shielded electroscopes placed out of doors lost their charges as if they were exposed to penetrating radiation. Since leak diminished with height, physicists assigned its cause to rays emanating from the earth. As they mounted ever higher, however, from church steeples to the Eiffel Tower to manned balloons,

the leak leveled off or even increased. In 1912–1913, Victor Hess of the Radium Institute of Vienna ascertained that the ionization causing the leak declined during the first 1,000 m (3,280 ft) of ascent, but then began to rise, to reach double that at the earth's surface at 5,000 m (16,400 ft). Hess found further, by flying his balloon at night and during a solar eclipse, that the ionizing radiation did not come from the sun. He made the good guess—it brought him the Nobel Prize in physics in 1936—that the radiation came from the great beyond.

The need to know the meteorological state of the upper atmosphere for directing artillery during the Great War improved balloon technology. Robert Millikan, who would receive the Nobel Prize in 1923 for his measurement of the charge on the electron, was a powerful organizer of American science for war. His observations made with Army balloons seemed to show that the ionization in the atmosphere declines continually from the earth's surface. Experiments in lakes at different heights showed him his error and he became the champion of what he called "cosmic rays." He regarded them as the "birth cries of infant atoms" since, by his calculations, the relativistic conversion of mass into energy during the formation of light elements from hydrogen would produce high-frequency radiation (photons) of the penetrating power of Hess's rays.

The advance of electronics transformed the study of cosmic rays from a guessing game to an exact, expensive, and productive science. In 1929 Werner Kolhörster, who had confirmed Hess's measurements, and Walther Bothe placed two Geiger counters one above the other, separated them by a lead block, and arranged a circuit to register only when both counters fired simultaneously. They detected too many coincidences for photons to produce and they inferred that cosmic-ray primaries must be charged particles with at least a thousand times more energy than the hardest rays from radioactive substances. After learning the coincidence method from Bothe (who received a Nobel Prize for it in 1954), Bruno Rossi, one of the Italian pioneers in the field, set three

Tracks of a pi-meson decaying in a photographic emulsion into an electron and a mu-meson, which then disintegrates into another electron and an invisible neutrino, obtained in 1949 by a group at the University of Bristol under Cecil F. Powell.

counters in a triangular array and deduced the existence of showers of particles produced by stopping the primaries (1932). A new particle, the positive electron (positron), was found among the secondaries in cloud chambers by Carl David Anderson, who worked with Millikan at the California Institute of Technology (Caltech), and by P. M. S. *Blackett and Giuseppe P. S. Occhialini, who used Rossi's electronics. Anderson and Blackett received Nobel Prizes in 1936 (sharing with Hess) and in 1948, respectively.

Cosmic rays became big science when Arthur Holly Compton (Nobel Prize, 1927) undertook to annihilate Millikan's primary photons. Compton designed a method to show that the sea-level intensity of cosmic rays at the poles exceeds that at the equator. If the primaries were charged particles, they would be deflected by the earth's magnetic field more strongly the lower the magnetic latitude. Compton's project mobilized sixty investigators who carried expensive standardized apparatus on eight expeditions. Government agencies used to funding geophysical work and the Carnegie Institution of Washington paid the bills. By 1934 Compton's company had confirmed the latitude effect and silenced the birth cries of infant atoms.

The victory at first appeared to have a heavy cost, however, since *quantum electrodynamics required that the charged primaries, if electrons or protons, be absorbed more quickly by the atmosphere than was compatible with their intensity at the surface. Anderson's cloud chamber soon disclosed the existence of a secondary particle (now called the μ meson or muon) whose

intervention would slow the absorption of cosmic rays. At first identified with the particle Hideki *Yukawa (Nobel Prize, 1949) had postulated as the carrier of nuclear force (now called the π meson or pion), the muon has played an important role in the systematics of the *standard model. It also afforded, through the difference in time between its decay in flight and at rest, the first experimental confirmation of the time dilation required by the theory of *relativity.

The last major contributions of students of cosmic rays to fundamental physics occurred just after World War II. With the help of photographic emulsions developed for wartime use, Cecil Powell (Nobel Prize, 1950) and his colleagues at the University of Bristol (including Cesar *Lattes) caught a pion as it turned into a muon, confirming the growing realization that Anderson's meson was not Yukawa's. In the same year, 1947, two other British physicists, George D. Rochester and Clifford C. Butler, found evidence in cloud chamber tracks of the decay of unknown neutral particles (now called hyperons) into neutrons, protons, and pions. Two years later, Powell, immersed again in emulsions, found a particle (the K meson) that decayed into three mesons.

The data needed to unravel the relations among *elementary particles did not come from cosmic rays, however, but from *accelerators. Already in 1947 the Berkeley synchrotron was making μ and π mesons in greater quantity than cosmic rays furnished. Whereas the synchrotron confirmed and extended the discoveries of mesons made through the study of cosmic rays, cosmic-ray physicists could find the antiproton only by scanning emulsions exposed to the beam of the Berkeley Bevatron, where it was first made and detected in 1955. In compensation, cosmic-ray physicists came to agree that the primary radiation striking the atmosphere consists almost entirely of protons, a substantial sprinkling of alpha particles and other nuclei, and a few electrons.

The invention of artificial *satellites gave cosmic-ray physicists a needed fillip. *Sputnik I* and the U.S. runners-up, *Explorer I* and *Explorer III*, all carried counters to measure cosmic-ray intensity. The *Explorers'* counters stopped working high above the earth. The man in charge of the instrumentation, James A. Van Allen of the University of Iowa, interpreted the silence as evidence that the satellite had passed through a region so full of charged particles that the counters jammed. The region, now known as the Van Allen Belt(s), consists of cosmic-ray and solar particles trapped in the earth's magnetic field. Hess's assertion that the Sun does not contribute to cosmic radiation succumbed to technological advances that replaced the balloons of the years around World War I with the rockets of the Sputnik era.

Bruno Rossi, *Cosmic Rays* (1966). Laurie M. Braun and Lillian Hoddeson, eds., *The Birth of Particle Physics* (1983). Michael W. Friedlander, *Cosmic Rays* (1989).

J. L. HEILBRON

COSMOLOGY. For two millennia the Aristotelian cosmos of rotating spheres carrying the Moon, Sun, planets, and stars around the central earth permeated Western thought. Then in 1543 Nicholas *Copernicus switched the positions of the earth and Sun. To account for the daily motion of the heavens, his scheme had the earth rotating on its axis. Revolutions in science, as in politics, often go further than their initiators anticipate. In Copernicus's system, the outer sphere of the stars was not needed to move them. Human imagination soon distributed the stars throughout a perhaps infinite space.

The earth lost more than its uniqueness in space in the Copernican system. *Galileo's telescopic observations of the Moon's surface emphatically demanded the revolutionary conclusion that the Moon was not a smooth sphere, as Aristotelians had maintained, but resembled the earth, with mountains and valleys. His discovery of four satellites of Jupiter, similar to the earth's single moon, forged another bond between the earth and the other planets.

In late Aristotelian cosmology, solid crystalline spheres carried the planets around the earth and also provided the physical structure of the universe. With his observation of *comets coursing through the solar system, Tycho *Brahe shattered the crystalline spheres. Belief in uniform circular motion, fundamental to both Aristotelian and Copernican cosmology, died early in the seventeenth century. Johannes *Kepler showed that the earth and the other planets all travel around the Sun in elliptical orbits.

Aristotelian physics no longer worked in a Copernican universe. A new explanation of how the planets retrace the same paths forever around the Sun became a central problem of cosmology. Isaac *Newton explained how a force of attraction or gravitation toward the Sun continually draws the planets away from straight-line motion and holds them in Kepler's elliptical orbits. Gravity would collapse the Sun and the fixed stars together unless the stars moved in orbits around the center of a system, as do the planets around the Sun. In 1718 the English astronomer Edmond Halley reported his discovery that three bright stars no longer occupied the positions determined by ancient observations. Formerly fixed, stars were now freed to roam.

Before the observations of William *Herschel (1738–1822), astronomers represented the heavens as the concave surface of a sphere surrounding the observer in the center. Herschel's large *telescopes forced them to treat the heavens as an expanded firmament of three dimensions. Herschel observed that most stars seemed to lie between two parallel planes. He concluded that the Milky Way is the appearance of the stars in the stratum to an observer on the earth.

At the beginning of the twentieth century astronomers reckoned the diameter of our stellar system at a few tens of thou-

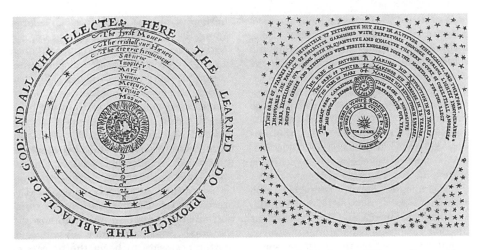

On the left, the ancient earth-centered system as represented in Leonard Digges, *Prognosticon Everlastinge* (1576); on the right, the ultra-modern heliocentric system as represented in the same book by Leonard's son Thomas Digges. Its ultra-modernity lies in its strewing the stars through space rather than, as in Copernicus's original scheme, placing them on the surface of a sphere.

sands of light years across. (A light year, nearly six trillion miles, is the distance light travels in a year in a vacuum.) Also, they generally assumed the Sun to be at or near the center of the universe. By 1920, however, the American astronomer Harlow Shapley had persuaded his peers that the system of stars was ten or even a hundred times larger than previous estimates and that the Sun stood many tens of thousands of light years away from the center. Shapley supposed erroneously that the galactic universe made a single, enormous, all-inclusive unit. In 1925, the American astronomer Edwin *Hubble demonstrated conclusively that spiral nebulae are independent entities far beyond the boundary of our own *galaxy. In 1926 Hubble published the now-standard classification system for the thousands of extragalactic *nebulae then known or within reach of existing telescopes.

At the time, most scientists, including Albert *Einstein, assumed that the universe was static. The Dutch astronomer Willem de Sitter formulated in 1916 a static model of the universe with apparent redshifts in the spectra of nebulae greater for nebulae at greater distances. In 1929 Hubble tested and confirmed de Sitter's predic-

tion. He interpreted the redshifts as Doppler shifts indicative of real velocities of recession. The universe was not static, but expanding. Hubble's velocity-distance relation made possible determinations of greater distance because redshifts can be measured even when individual stars and galaxies cannot be distinguished, out to billions of light years. In 1927 the Belgian priest and astrophysicist Georges Lemaître published a theory of a universe expanding from what he called a "cosmic egg." Initially ignored, it attracted favorable attention in 1931 after Hubble had demonstrated his velocity-distance relation. De Sitter hailed Lemaître's brilliant discovery of the expanding universe and Einstein confirmed that Lemaître's theoretical investigations fit well into the general theory of *relativity.

Advances in nuclear physics helped transform speculations about an expanding universe. Beginning in 1946, the Ukrainian-born American physicist George Gamow set for cosmology the problem of explaining with laws of physics the cosmic distribution of the chemical elements as the result of thermonuclear reactions in an early, extremely hot, dense phase of an expanding universe. The British astronomer Fred *Hoyle in 1950 derisively called Gamow's process the "big bang," and the term stuck. After astronomers agreed that elements heavier than hydrogen and helium could not have been formed during the dense hot origin of the universe, Hoyle explained the later creation of heavy elements in stellar interiors. In opposition to the big bang theory, Hoyle championed steady-state creation, in which the universe expands but does not change in density because of the continuous appearance of new matter. The crucial observation in the cosmological debate came from radio astronomy (see ASTRONOMY, NON-OPTICAL). In 1963 Arno Penzias and Robert Wilson at the Bell Telephone Laboratories measured a faint cosmic background radiation, which was interpreted by Robert Dicke at Princeton University as a remnant of the big bang.

A major conceptual change in cosmology during the 1970s and 1980s focused attention on the question why the universe is the way it is. A young American particle physicist, Alan Guth, proposed in 1979 that important cosmological features can be explained as natural and inevitable consequences of new theories of particle physics. Guth's theory of inflation states that in the first minuscule fraction of a second of the universe's evolution a huge inflation occurred. After that, the inflationary universe theory merges with the standard big-bang theory. If the mass density of the universe exceeds a certain critical value, gravity eventually will reverse the expansion of the universe and reunite everything in a "big crunch." With a density less than critical, the universe will expand forever, resulting in a "big chill." Neither case is indefinitely hospitable to life as we know it. At the critical density, the resulting "flat" universe will continue to expand, but at an ever slower rate.

Why the density of the universe one second after the big bang came very close to the theoretical critical density is called the "flatness problem." The standard big bang theory offers no solution. In the inflationary theory, a brief burst of exponential expansion automatically drives the density, whatever its initial value, incredibly close to the critical density. Another success of inflationary theory involves the "horizon problem." Exchanges of energy cannot be transferred farther than the "horizon distance," the maximum distance that light can have traveled since the beginning of the big bang universe. But the cosmic background radiation is uniform over a space much greater than the horizon distance. The inflationary theory makes the universe far smaller during its initial phase than the standard big bang theory does, smaller than the horizon distance, thus enabling the universe to come to a uniform temperature before the process of inflation switched on. If inflation created a uniform universe, how did the small primordial non-uniformities of matter necessary to begin the process of gravitational clumping come about to seed the evolution of cosmic structure? Answer: inflation could have stretched quantum mechanical fluctuations enormously, and the resulting wrinkles in space-time could have caused large-scale cosmic structures, such as a galaxy or even a cluster of galaxies, to coalesce.

Cosmological theories live or die on the basis of their predictions. When NASA in 1982 approved funding for the Cosmic Background Explorer Satellite, another young American particle physicist, George Smoot, proposed to seek in the cosmic background radiation the tiny differences from uniformity predicted by the inflationary theory. Eventually the experiment involved more than a thousand people, at a cost of over $160 million. After a decade of work, Smoot announced in 1992 what Stephen Hawking called "the scientific discovery of the century, if not of all time." Wrinkles in the fabric of space and time showed that matter was not uniformly distributed, but contained the seeds out of which a complex universe could grow. Furthermore, the fifteen-billion-year-old wrinkles in the background radiation had the size expected.

Astronomers have seen only about 10 percent of the universe; the remaining 90 percent is so far revealed only by its gravitational effects. Weakly interacting massive particles (WIMPs), predicted from theoretical physics, make up most of the missing matter, though they have not been detected in the laboratory. Some of the invisible universe may consist of stars of very low luminosity, such as brown dwarfs, and also primordial black holes. Other forms of dark matter may be topological defects remaining from phase changes in the early universe. The question of large-scale structure and galaxy formation has become central; cosmic strings may provide an answer. Also in question is the expansion of the universe. Many astronomers assumed that either the expansion will proceed indefinitely at a constant rate, or will be slowed and ultimately reversed by gravity. In 1997, however, astronomers observed several gigantic exploding stars or *supernovae at immense distances, seemingly even farther than a constant expansion would have flung them. A runaway universe, its expansion continuously accelerated by some mysterious form of energy, suddenly became a likelihood.

See also ASTRONOMY, NON-OPTICAL; STEADY-STATE UNIVERSE.

Edwin Hubble, *The Realm of the Nebulae* (1936). Richard Berendzen, Richard Hart, and Daniel Seeley, *Man Discovers the Galaxies* (1976). Robert W. Smith, *The Expanding Universe: Astronomy's "Great Debate" 1900–1931* (1982). Michael J. Crowe, *Theories of the World from Antiquity to the Copernican Revolution* (1990). Norriss S. Hetherington, ed., *Encyclopedia of Cosmology: Historical, Philosophical, and Scientific Foundations of Modern Cosmology* (1993). Michael J. Crowe, *Modern Theories of the Universe from Herschel to Hubble* (1994). Norriss S. Hetherington, *Hubble's Cosmology: A Guided Study of Selected Texts* (1996). Helge Kragh, *Cosmology and Controversy: The Historical Development of Two Theories of the Universe* (1996). Alan H. Guth, *The Inflationary Universe: The Quest for a New Theory of Cosmic Origins* (1997).

NORRISS S. HETHERINGTON

CRYSTALLOGRAPHY. Crystals have attracted attention because of their striking and often beautiful forms since early in human history. As an object of systematic study, they have attracted the attention of very different investigators: natural philosophers, mineralogists, chemists, physicists, mathematicians, metallurgists, and biologists. Only in the twentieth century did crystallography become an institutionalized scientific discipline. Perhaps for this reason a comprehensive history of the study of crystals is yet to be written.

In the sixteenth and seventeenth centuries, natural historians thought that crystals, like snowflakes and fossils, bridged the conventional categories of the material world. Like living things, their symmetric form indicated organization. It was even possible to see them growing. Yet they did not seem to be fully alive. Some observers speculated that the variety of crystalline appearance indicated astrological influence, others that it evidenced nature's ability to impose form on matter. Crystals were thought by some to grow from seeds in the earth; by others, from circulating fluids. Mechanical philosophers such as Robert *Boyle and Robert Hooke saw in the regular form of crystals a reflection of the underlying arrangement of corpuscles or atoms. Nicolaus Steno, in his *Prodromus to a Dissertation on Solids Naturally Contained within Solids* (1669), explored possible manners of growth and stated the principle that crystals of the same kind have the same angles between adjacent faces.

In the eighteenth century, Carl Linnaeus proposed that minerals could be classified by counting their faces, a suggestion of limited utility at the time since most minerals do not appear crystalline to the naked eye and since the polarizing microscope was yet to be invented (*see* MINERALOGY AND PETROLOGY). René-Just Haüy, professor of mineralogy at the Muséum d'Histoire Naturelle in Paris, followed earlier work by Romé de l'Isle on the relation of visible crystal forms to the units composing them. In his *Traité de cristallographie* (1822), Haüy suggested that cleaving a crystal divided it into one of six basic polyhedral units, themselves formed of smaller units that might or might not have other polyhedral forms. William Wollaston's invention of a reflecting goniometer that used light rays to measure crystal angles made possible much more precise measurements of crystal forms.

Haüy assumed a univocal relation between crystal form and chemical composition. The chemist Eilhard Mitscherlich

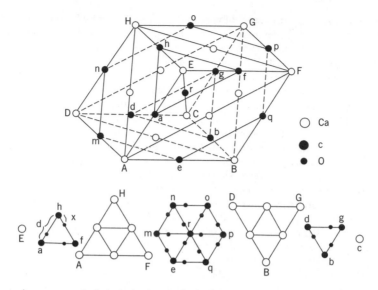

Arrangement of atoms in a crystal of calcium carbonate as determined by x-ray diffraction; the lower diagrams show the arrangement of atoms in special planes.

dissented. He observed that substances with different chemical compositions could crystallize in the same form, a property he called isomorphism. This bothered the young Louis Pasteur who took advantage of the peculiar optical properties of crystals to investigate the matter further. He carefully separated tartrate crystals that differed just slightly in their facial angles. Polarized light passed through a solution of one set of crystals rotated in one direction; when passed through a solution of a second set, in the reverse—indicating, Pasteur argued, crystalline asymmetry at the molecular level.

The optical properties of minerals proved a reliable aid in identifying them. Henry Clifton Sorby's polarizing microscope enabled mineralogists to examine rocks in thin section and identify their previously invisible crystal constituents. This became a basic technique of petrography. In the second half of the nineteenth century, physicists studied the elasticity, density, and electrical properties of minerals. In 1880, Pierre *Curie and his brother Jacques-Paul discovered that pressure exerted at the right point on a crystal produced an electric field, a phe-

nomenon known as piezoelectricity. Pierre later incorporated this effect in an electrometer used in the detection of *radioactivity. Other scientists worked out the mathematical possibilities for crystal structure: Auguste Bravais, professor of physics at the École Polytechnique of Paris, showed in his *Études cristallographiques* (1866) that only fourteen possible arrangements of points in space lattices were possible.

The discovery of *X rays revolutionized crystallography. In 1912 Max von Laue and his group at the University of Munich photographed the diffraction pattern produced by a copper sulphate crystal, showed that X rays passed through a crystal scattered and deflected at regular angles. William Henry Bragg, professor of physics at the University of Leeds, and his son William Lawrence Bragg, then a student at Cambridge, realized that the pattern depended on the atomic structure of the crystal and succeeded in analyzing the crystal structure of the mineral halite (sodium chloride). This in turn led to the invention of the X-ray powder diffractometer. X-ray crystallographers began organizing themselves into formal bodies shortly after

World War I. In 1925, they held an informal meeting in Germany to help restore relations disrupted by the war. At the Royal Institution in London and the University of Manchester, the Braggs trained students from all over the world in the techniques of x-ray crystallography. Societies were founded in Germany in 1929, the United States in 1941, and the United Kingdom in 1943. The *Zeitschrift für Kristallographie* became the leading journal in the field in 1927 when it started accepting papers in French and English as well as in German.

International cooperation in crystallography revived immediately after World War II. The International Union of Crystallography was founded and joined the International Council of Scientific Unions, the intermediary for UNESCO funding. Soon after came the debut of *Acta crystallographica*, quickly to become the premier journal in the field. X-ray crystallography proved crucial to the rapidly expanding field of molecular biology, itself the result of the coalescence of biochemistry and crystal structure analysis. After the discovery that proteins could be crystallized, W. T. Astbury began exploring their structure with X rays. In the late 1950s the double-helix structure of nucleic acid, predicted by Francis Crick and James D. Watson, was confirmed by Maurice Wilkins and Rosalind Franklin; in the same period Max Perutz and John Kendrew determined the structure of hemoglobin and myoglobin.

In the 1960s, x-ray crystallographers began adding computers to their apparatus, increasing the speed and precision of analysis. X-ray crystallography remains the most powerful, accurate tool for determining the structure of single crystals. It is widely used in disparate fields including mineralogy, metallurgy, and biology.

Paul Ewald, ed., *Fifty Years of X-ray Diffraction* (1962). John Burke, *Origins of the Science of Crystals* (1966). Seymour Mauskopf, "Crystals and Compounds: Molecular Structure and Composition in Nineteenth-Century French Science," *Transactions of the American Philosophical Society* (1976). Dan McLachlan Jr., and Jenny P. Glusker, eds., *Crystallography in North America* (1983).

RACHEL LAUDAN

CURIE, Marie Sklodowska (1867–1934), and **Pierre CURIE** (1859–1906), co-discoverers of radium.

This most famous couple in the history of science met in 1894 in Paris where Marie (then Maria) had come from her native Warsaw to study mathematics. With great determination, she came first in her class in physics in 1894 and second in mathematics in 1895.

Pierre was as sunk in his work—teaching physics and studying magnetism—as Maria was in hers. They had other things in common as well. He had been teaching at a new, non-elite technical school (the Ecole Municipale de Physique et Chimie Industrielles) since its foundation in 1882. This attachment and his inherited republicanism gave him the same humanitarian ideals that Maria, who had tried to educate peasants in Poland, not only held but practiced. They also shared a somber love of nature. They married in 1895.

Marie chose for her dissertation the examination of pitchblende, a complex uranium-bearing ore. She hoped that some of its constituents might give off rays of the kind that Henri Becquerel had found in a crystal containing uranium. In conducting her tests, she used several instruments that Pierre had invented. One was a balance that measured weight by electricity. Pierre and his brother Jacques-Paul had discovered its principle, piezoelectricity (the capacity of some crystals to become charged when strained), in 1880. When Jacques-Paul left Paris for a provincial professorship in 1883, Pierre turned from crystallography to magnetism, about which he discovered several notable things: that all substances are diamagnetic (a property overcome in para- and ferromagnetic materials), that magnets become paramagnetic when heated beyond a certain temperature (the Curie point), and that paramagnetism diminishes in strength in inverse proportion to the absolute temperature (Curie's law). Among the instruments he invented during these investigations was a sensitive electroscope with magnetic damping. With this instrument Marie found that pitchblende was more strongly radioactive than the uranium it contained.

Pierre joined her in the laborious task of isolating the source of pitchblende's activity. By July 1898 they had a bismuth sample four hundred times more active than uranium. They called the hypothetical agent it bore "polonium." They then examined the radioactive barium fractions. Radium was the result, their Christmas present to the world. To obtain it in its barium carrier they had processed a few hundred kilograms of pitchblende. That was nothing. Marie turned Pierre's school, at which they did their chemical separations, into a toxic dump. She began with ten tons of pitchblende tailings (the ore minus the uranium). She worked in a large, unheated shed once used for dissections, a cold, smelly, dangerous, eerie place, stirring her cauldrons with a pole as big as herself and filling the endless vessels in which radium was concentrated. Within a year she had a preparation 100,000 times as powerful as uranium. It glowed in the dark. It took three more years to obtain enough moderately pure radium to weigh on Pierre's balance. This singular intellectual and athletic achievement brought Marie not only her doctorate but also a share, with Pierre and Henri Becquerel, of the Nobel Prize in physics for 1903.

The Curies played almost no part in elucidating radioactive lineages or in devising the concept of isotope with which to capture the fact that different radioelements can have identical chemical properties. Their poor showing resulted from fidelity to a peculiarly French notion of scientific theorizing that put them at a disadvantage in competition with Ernest *Rutherford. The Curies believed that scientists should express their ideas in the most general way consistent with the known facts. Only as experiment progressed should options be closed. In contrast, Rutherford followed the British procedure of modeling the phenomena in clear, visualizable images.

Pierre Curie's positivism agreed with the practice of the French establishment that belatedly admitted him to its midst in 1904, as professor at the Sorbonne. Marie Curie held a similar concept of science strengthened by the political activism of her youth. She believed that to throw off the Russian yoke ordinary people would have to learn practical, efficacious, technical, and useful science, with nothing of the speculative or romantic about it.

In 1906 Pierre Curie was killed by a runaway truck. At one stroke Marie lost her husband, best friend, and closest collaborator. The Sorbonne gave her the chair of physics they had created two years before for Pierre. She began her first lecture where he had left off. She described the disintegration theory as the most useful interpretation of the known facts. True to herself and Pierre, she added, "It seems to me useful not to lose sight of the other explanations of radioactivity that can be proposed." In 1911, Stockholm called again, this time with the chemistry prize, for her studies of radium.

Between the announcement and the ceremony, Marie received advice from the chairman of the committee that had proposed her that she should renounce the prize. Stolen letters had revealed that she and Pierre's former student Paul Langevin were lovers. The revelation fueled attacks that linked antifeminism and xenophobia with the defense of the family and old-fashioned morality. The journalistic savaging resulted in no fewer than five duels.

Marie insisted that her prize-winning achievements had nothing to do with her private life. She went to Stockholm. The emotional strain of the year overtaxed even her physical capacity and will power, however, and she never fully recovered her strength. That did not stop her from driving around the battlefields of World War I in one of the ambulances she had equipped with medical x-ray machines; from visiting the United States in a triumphal tour in 1921 to receive a gram of radium from the women of America; or from establishing her Institut du Radium as a premier laboratory for the study of radioactivity and nuclear physics, and the nurturing of women scientists. There in the early 1930s her daughter Irène Joliot-Curie and son-in-law Frédéric Joliot found that bodies could be made artificially radioactive by irradiating them with neutrons. They shared the Nobel Prize in chemistry for 1935. Marie Curie did not live to see this family triumph. She died in 1934 after suffering for some years from the results of her intimate association with radium.

Marie Curie, *Pierre Curie* (1923; reprint 1963). Anna Hurwic, *Pierre Curie* (1995). Susan Quinn, *Marie Curie* (1996).

J. L. HEILBRON

D

DALTON, John (1766–1844), English natural philosopher, proposed the modern atomic theory.

Son of a modest Quaker weaver, Dalton grew up near Kendal, in the Lake District of England. Mentored by a local polymath, Dalton showed early promise in mathematics and natural philosophy, and became a schoolmaster at the age of nineteen. A few years later he was hired at the New College, a "dissenters' academy" in Manchester. He resigned his post in 1800, preferring to support himself by private teaching, which he did for the rest of his life. Dalton was an active and respected member of the flourishing scientific scene in Manchester. He never married.

In 1794, a month after his election to the Manchester Literary and Philosophical Society, Dalton presented his first paper there, an investigation of color-blindness, or "Daltonism," with which he himself was afflicted. The preceding year he had published his first book, *Meteorological Observations and Essays*. These publications reveal a deep and abiding interest in physiology and natural history in addition to a commitment to physics and chemistry. He remained particularly engaged with theories of gases and solubilities, especially the solution of water vapor in air, and of gases in water.

In 1801, Dalton published a new theory of mixed gases applicable to the atmosphere. He supposed that each component in a gaseous mixture exerts pressure independently of the other gases. What later became known as "Dalton's law of partial pressures" suggested that in such mixtures gases interact physically and not chemically, and he applied this thesis to the solution of gases in water. Although it solved some anomalies, the theory created further difficulties. While defending and elaborating his physical theory of gases, Dalton made his greatest contribution to science, his chemical atomic theory.

Influenced by Antoine-Laurent *Lavoisier's new chemistry and by the popular *Newtonianism then prevalent in Britain,

Dalton began to investigate the relative weights of the atoms of the elements. According to Lavoisier's analysis, water was 85 percent oxygen and 15 percent hydrogen by weight. If water consists of a combination of the ultimate particles of oxygen and hydrogen, and if each "compound atom" of water consists of one "simple atom" of each element, then the atom of oxygen must weigh about 5.7 times (85/15) that of hydrogen. Knowing that carbonic oxide was 45 percent carbon and 55 percent oxygen and assuming a "binary" formula for its molecule, the relative weight of a carbon to a hydrogen atom would be about 4.5. The latter weight matched the calculated weight for the carbon atom from the analysis of "olefiant gas" (ethylene), also taken to be binary.

Such atomic and molecular weights appear in an entry in Dalton's notebook of 6 September 1803, and were published two years later without indicating the method of calculation. The first published discussion of Dalton's new atomic theory occurred in the pages of the textbook of his friend, Thomas Thomson, in 1807. Thomson rightly stressed that these calculations could proceed only by assuming molecular formulas for the compounds under study. Dalton always supposed the simplest possible formulas. The finest and most productive new ideas often require arbitrary leaps.

In 1808 and 1810 Dalton published two volumes of *A New System of Chemical Philosophy* in which he developed his atomic theory in detail for the first time. Although much criticized for the simplicity of assumptions it invoked, the book marked an epoch in the history of chemistry. Subsequent elaborators of chemical atomism, such as Jöns Jacob Berzelius, Joseph Louis Gay-Lussac, Amedeo Avogadro, Justus von Liebig, and Jean-Baptiste-André Dumas, began with Dalton's atomistic vision even if they rejected or modified certain details within it.

Frank Greenaway, *John Dalton and the Atom* (1966). Elizabeth Patterson, *John*

Dalton and the Atomic Theory (1970). Arnold Thackray, *John Dalton* (1972). A. J. Rocke, *Chemical Atomism in the Nineteenth Century* (1984).

A. J. ROCKE

DELUC, Jean André (1727–1817), instrument maker, natural historian, savior of Genesis.

Deluc's father, a clock maker, was a leader of the Genevan underclass against the patrician establishment and a friend of Jean Jacques Rousseau, with whom he disagreed about almost everything but the need to curb the governing councils of the city. The elder Deluc was a strict Calvinist, a believer in the literal word of scripture and in a providential God. Jean André and his brother Guillaume Antoine grew up under the strong influence of their father and the counterinfluence of enlightenment ideas as expressed in the writings and life style of Rousseau and the teachings of the advanced natural philosophers of Geneva. Many of these men, though political opponents of the Delucs, shared their notions about God and scripture.

Until he was forty, Jean André dealt largely in textiles and politics. As a businessman, he developed a taste for the sort of arithmetical calculations he would need in his science; as a politician, he joined with his father in remonstrating, in 1762, against the ruling council's condemnation of Rousseau's *Emile* and *Contrat social*. In the susequent struggle, Jean André played a major and successful part as a representative of the bourgeoisie.

The political disturbances undermined Deluc's textile business. In 1773 he sought a place in England, where Genevan gentlemen were welcome as tutors to the aristocracy. Deluc's assets amounted to the ability to speak French (but not English) and a reputation as an exact natural philosopher. He had spent years climbing around the Alps with his brother and making instruments for determining the contours of the earth and the state of the air. Deluc regarded the height of mountains as a material witness to the processes that had formed and shaped them; accurate instruments for determining height therefore could provide information about the creative activities of the God of Geneva.

From his measurements Deluc deduced a famous formula for converting barometric readings to heights and a rule for the decline of the boiling point of water with decrease in air pressure. He also perfected thermometers and hygrometers. In 1768, on the strength of preliminary reports on his hypsometric work, he became a corresponding member of the Académie royale des sciences of Paris. Two years later he dedicated to it his *Recherches sur les modifications de l'atmosphère* (1772), two stout volumes giving a history of barometry, descriptions of his barometers and thermometers, Alpine heights obtained with their aid, and corrections to the standard tables of atmospheric refraction. Just before his arrival in England the Royal Society of London elected Deluc a fellow on the strength of his barometry.

In the late spring of 1773 Deluc's friends among the Genevan expatriates in London managed to present him to George III and Queen Charlotte. Deluc demonstrated his barometer and gave it to the king; the Queen received a hygrometer for regulating the humidity of her greenhouses. In return, after ascertaining that Deluc was a philosopher but not a *philosophe,* Charlotte took him as her tutor, a post he held for 43 years. He advised her on everything from flower arrangements to chambermaids. He occasionally advised the king about politics and acted briefly as George's agent in Berlin during the wars of the French Revolution.

In 1779 Deluc published his second major work, five volumes of *Lettres physiques et morales,* which set the foundations of what, in his introduction to them, Deluc called "geology." According to Deluc, the keys to the accessible past were the nature and distribution of fossils, and the current rates of erosion and sedimentation. From the one he deduced that a sea once covered fossiliferous rocks; from the other, that the length of time since the appearance of the present continents cannot have been long. Deluc invested years in assembling the data to confirm this dating. The geochronometers he devised—the rate of erosion of land, deposit of peat, filling up of river deltas, and blocking of Alpine passages by glaciers—all confirmed the juvenility of the continents. He identified their emergence, and the subsidence of the seabed that revealed the fossils, with the biblical Deluge—not to prove Genesis by geology but to remove all objections to the biblical account based on physics.

Deluc's breadth of vision, his exactness in observation, and his cosmopolitanism earned him a European reputation. Those who admired precision in measurement esteemed him as a leader of the experimental physics that dominated the age. Natural historians followed him in historicizing the earth's development and in refining geochronometers. Traditional believers felt relief at his demonstration that natural science did not threaten scripture. Centrist statesmen admired his resolution of the Genevan crisis of 1768; conservative ones joined in his condemnation of the overturning of traditional values, seigneurial rights, received philosophy, and sound religion by the French Revolution.

Deluc had the misfortune to outlive his reputation. By the time he died in Windsor at the age of ninety, his mixture of Enlightenment ideas with biblical exegesis, and of exact measurement with approximate theory, no longer held much interest for his contemporaries.

Clarissa Campbell-Orr, "Queen Charlotte, 'Scientific Queen,'" in Campbell-Orr. ed., *Queenship in Britain* (2002), 236–266. Roy Porter, *The Making of Geology. Earth Science in Britain, 1660–1815* (1977). Jacques Trembly, ed. *Les savants genevois dans l'Europe intelectuelle du xviie au milieu du xixe siècle* (1987).

J. L. Heilbron

DESCARTES, René (1596–1650), philosopher and mathematician.

Born in La Haye, France, Descartes was educated at the Jesuit College of La Flèche and at the University of Poitiers. In 1618 he journeyed to the Netherlands and Germany, returning in 1622. In the Netherlands he met the natural philosopher Isaac Beeckman, who probably gave him the idea of a universal "mathematics" as the basis of a science of nature. Descartes worked out this idea in Paris between 1625 and

1628; he then worked on analytical geometry and optics, undertaking a general treatise on method (*Regulae ad directionem ingenii*) that was never finished. In 1629 he returned to the Low Countries where he remained until the end of 1649. He died the following year in Stockholm, in the service of Queen Christina.

Descartes was fascinated by the problems of ascertaining natural knowledge. His metaphysics can be seen as an attempt to make a mathematical physics possible while paying tribute to traditional metaphysical and theological concerns like the existence of God and the immateriality of the soul. Later generations regarded his physics, published in its fullest form in his *Principia philosophia* (1644), as a purely hypothetical construction. What lay behind it, however, and what appeared more clearly in Descartes's early work *Le monde* (1633), withdrawn in compliance with the Roman Inquisition's condemnation of Copernicanism in 1633, was the revolutionary idea that nature (as the object of natural science) is not something given but is whatever can be seen as the interpretation of a theoretical model. Descartes expected his approach to provoke a revolution in the general theory of the universe and also in practical disciplines like medicine. He supposed that medicine could be put on a more solid basis by adopting the ideas that animal bodies are machines and that the explanation of any biological function (breathing, sleeping, digesting, etc.) consists in imagining a mechanical device capable of performing it. Descartes linked this idea with the notion that the most fundamental biological process is the circulation of the blood (which, contrary to William Harvey, he explained as a process based on a quality of the blood), because it is self-regulatory and the basis of all other functions. His followers took heuristic advantage of the concept, leaving it to anatomy to discover the actual machine. Descartes's natural philosophy at first spread most rapidly among physicians, especially in the Low Countries, then among philosophers.

Descartes's merits are least contested in mathematics, which fascinated him from his days at La Flèche and which may have provided the model for his method. His mathematical work was based on the idea that algebra, to which he gave a simplified notation still in use, made it possible to bring together problems that in their geometrical formulation appear unrelated. He contributed to optics, demonstrating the theoretical advantages of hyperbolic over spherical lenses in telescopes. He did not manage, however, to develop a controlled, industrial method for producing such lenses. He also gave a demonstration of the law of refraction based on a particulate model of light and an indication of the possibility of mathematical physics in his theory of the *rainbow.

J. F. Scott, *The Scientific Work of René Descartes (1596–1650)* (1952). William R. Shea, *The Magic of Numbers and Motion: The Scientific Career of René Descartes* (1991). Stephen Gaukroger, John Schuster, and John Sutton, eds., *Descartes' Natural Philosophy* (2000). Henk J. M. Bos, *Redefining Geometrical Exactness: Descartes' Transformation of the Early Modern Concept of Construction* (2001).

THEO VERBEEK

DIRAC, Paul Adrien Maurice (1902–1984), British physicist, one of the pioneers of relativistic *quantum physics, *quantum electrodynamics, and *quantum field theory.

All the major developments in quantum field theory in the 1930s and 1940s have as their point of departure some work of Dirac's. Werner *Heisenberg characterized Dirac's postulation of antimatter in 1931 as "the most decisive discovery in connection with the properties or the nature of elementary particles." And all this does not take into account Dirac's famous exposition *The Principles of Quantum Mechanics* (1930), the guide to several generations of physicists.

Dirac was born on 8 August 1902 in Bristol, England, the son of a Swiss father, Charles Adrien Ladislas Dirac, and an English mother, Florence Hannah Holten. Charles Dirac emigrated to England in 1888 and in 1896 obtained a position teaching French in the Merchant Venturers' Technical College (M.V.) at Bristol, Paul's mother's home town. Paul had an older brother and a younger sister. Dirac's father was a strong-willed, dominating personality. He demanded that his children speak to him in French, threatening to punish them for grammatical errors. Dirac traced his legendary taciturnity to this linguistic program. Because the rest of the family could not meet Charles's standards, they ate dinner in the kitchen, while Paul dined with his father—Paul never saw his parents eat a meal together, nor could he recall anyone ever making a social call at the house. Paul's brother Reginald wanted to become a physician, but Charles forced him to study mechanical engineering at Bristol. He obtained a third class degree, accepted a position as a draftsman with an engineering firm, and committed suicide at 24. Thereafter Paul had very little interaction with his father.

Dirac began his secondary education in 1914 at the M.V. His schoolmates recalled him as introverted, reticent, and aloof. He followed Reginald's footsteps to Bristol University and electrical engineering. Although he preferred mathematics, he thought that the only way to earn a living as a mathematician was as a school teacher, a prospect that did not appeal to him. He graduated from Bristol with first class honors in 1921. Unable to find a suitable engineering position owing to the economic recession that then gripped England, Dirac accepted free tuition at Bristol to study mathematics. An 1851 Exhibition Studentship and a grant from the Department of Scientific and Industrial Research led him to Cambridge as a research student in 1923.

At Cambridge, Dirac came under the influence of Ralph Howard Fowler and Arthur Eddington. He attended Fowler's lectures on the old quantum theory and became his research student. Within six months of his arrival, Dirac wrote two papers on statistical mechanics and in May 1924 he submitted his first paper on quantum problems.

Upon first reading, Dirac failed to see the significance of Heisenberg's first paper on quantum mechanics, which he saw in September 1925 in the page proof that Heisenberg had sent Fowler. A week later Dirac realized that the non-commuting quantities that bothered him made up the essence of Heisenberg's new approach. It occurred to him to try to connect the commutator of two non-commuting dynamical variables with their Poisson brackets (a form of the equations of classical mechanics that resembled the commutator). His success convinced him that the new quantum mechanics represented an extension of classical physics rather than, as Heisenberg had argued, a break with it. He stated his position succinctly in the paper he submitted to the *Proceedings of the Royal Society* early in 1926: "Only one basic assumption of classical theory is false... the laws of classical mechanics must be generalized when applied to atomic systems, the generalization being that the commutative law of multiplication, as applied to dynamical variables, is to be replaced by certain quantum conditions."

Dirac's contributions to the development of quantum mechanics were immediately acknowledged as central and by 1927 he had become a member of the core set that judged theoretical advances. In 1930 he was elected a fellow of the Royal Society, and in 1932 Lucasian Professor of Mathematics in Cambridge. In 1933 he and Erwin Schrödinger shared the Nobel Prize in physics for their contributions to the developments of quantum mechanics. In 1937 Dirac married Margit Balasz, the sister of Eugene Wigner, another Nobel Prize–winning mathematical physicist.

During World War II Dirac taught at Cambridge while carrying out research on uranium isotope separation and on atomic bomb design. After the war he was often a visitor at the Institute for Advanced Study in Princeton. In 1969 he accepted a research professorship at Florida State University in Tallahassee, which he held until his death on October 20, 1984.

R. H. Dalitz and R. Peierls, "Paul Adrien Maurice Dirac," in *Biographical Memoirs of Fellows of the Royal Society* 32 (1986): 139–185. H. S. Kragh, *Dirac. A Scientific Biography* (1990).

SILVAN S. SCHWEBER

E

EARTH, AGE OF THE. Before the mid-eighteenth century, few scholars or scientists in the Christian West questioned the adequacy of the chronologies derived from the Mosaic narrative. They believed that the earth was little older than the few thousand years of recorded human history. Beginning in the second half of the eighteenth century, however, investigations of the earth's strata and fossils suggested that the earth's crust had undergone innumerable cycles of formation and decay and that it had supported an ever-changing sequence of living beings long before the first appearance of humans. For geologists such as James Hutton and Charles Lyell, the earth's age seemed too vast for human comprehension or measurement. By 1840, geologists had identified most of the major subdivisions of the stratigraphic column and arranged them in a chronological sequence, but it was a chronology without a measurable scale of duration, a history of the earth without dates.

In 1859, Charles Darwin attempted to determine one geological date by estimating how long it would take to erode a measured thickness of the earth's strata. His conclusion that 300 million years had been required to denude even the relatively recent strata of the Weald, a district of southern England, brought an immediate reaction. In 1860, the geologist John Phillips rebutted with an estimate that the composite thickness of the whole stratigraphic column could be denuded in only 100 million years. Soon afterward, the physicist William *Thomson, Lord Kelvin calculated that 100 million years would be sufficient for the earth to cool from an assumed primordial molten condition to its present temperature. Kelvin's conclusions, based on the widely held hypothesis that the earth had been formed from the scattered particles of a primordial nebula, and supported by the latest theories of thermodynamics, carried great weight during the remainder of the century. Subsequent estimates of rates of erosion and sedimentation, of solar radiation and cooling, and of

the time of formation of the Moon and the oceans converged on his figure of 100 million years.

All this was changed by the discovery in 1903 that radioactive elements constantly emit heat. A year later, Ernest *Rutherford suggested that the ratio of the abundance of radioactive elements to their decay products provide a way to measure the ages of the rocks and minerals containing them. Robert John Strutt and his student, the geologist Arthur Holmes, pursued Rutherford's idea. By 1911, Holmes had used uranium/lead ratios to estimate the ages of several rocks from the ancient Precambrian period. One appeared to be 1,600 million years old. Many geologists were initially skeptical, but by 1930, largely as a result of the work of Holmes, most accepted radioactive dating as the only reliable means to determine the ages of rocks and of the earth itself. The discovery of isotopes in 1913, and the development of the modern mass spectrometer of the 1930s, greatly facilitated radioactive dating. By the late 1940s, the method produced an estimate of between 4,000 and 5,000 million years for the age of the earth. In 1956, the American geochemist Clare Cameron Patterson compared the isotopes of the earth's crust with those of five meteorites. On this basis, he decided that the earth and its meteorites had an age of about 4,550 million years. All subsequent estimates of the age of the earth have tended to confirm Patterson's conclusion.

Joe D. Burchfield, *Lord Kelvin and the Age of the Earth* (1975, 1990). G. Brent Dalrymple, *The Age of the Earth* (1991). C. L. E. Lewis and S. J. Knell, eds. *The Age of the Earth: from 4004 B.C. to A.D. 2002* (2001).

JOE D. BURCHFIELD

EARTH SCIENCE. The discipline of earth science was invented during the 1960s. It replaced *geology as the major institutional framework for studying the earth just as geology had replaced *mineralogy in the early nineteenth century. As with the simultaneous invention of *planetary science, geologists had additional reasons for rethinking the status quo. Geology had lacked focus and exciting new ideas and techniques for the previous half century. Oceanographers and geophysicists (*see* OCEANOGRAPHY; GEOPHYSICS) had new instruments and techniques, particularly in submarine gravity studies, submarine seismology, deep ocean drilling, and paleomagnetism.

The needs of submarine warfare during World War II and the Cold War gave a strong boost to seismology and studies of submarine gravity. Before World War II, the Dutch physicist Felix Andries Vening-Meinesz had developed a pendulum apparatus (now named after him) that made it possible to measure gravity at sea as accurately as on land. Using this apparatus, Vening-Meinesz and others mapped the gravity anomalies of the ocean floors and discovered the areas of downbuckling frequently associated with island arcs. Following the War, Maurice Ewing developed sea-floor seismic equipment, measured velocities in the sediments of the deep ocean, and contributed importantly to the understanding and detection of long range sound transmission in oceans.

In the late 1950s, the *International Geophysical Year (IGY), which included gravity, geomagnetism, oceanography and seismology in its areas of concentration, further encouraged these lines of research. Deep ocean drilling that brought up cores from the sediments on the ocean floor began in the 1960s with the establishment of JOIDES (Joint Oceanographic Institutions for Deep Sea Drilling). At the same time, scientists at the United States Geological Survey, the University of Newcastle, and the Australian National University were racing to produce time scales of global magnetic reversals and theories of the motion of the magnetic north pole.

The promotion of the nascent discipline of earth science owed much to a few pioneering individuals and institutions. Among the individuals, in addition to those already mentioned, were J. Turzo Wilson, who led the Canadian contribution to the IGY and made major contributions to *plate tectonics, and W. H. (Bill) Menard of Scripps Institution of Oceanography, later

Director of the United States Geological Survey. The institutions that saw some of the important early work were spread across the English speaking world: Princeton University, Cambridge University, the University of Toronto, and the Australian National University, together with Lamont Doherty Geological Observatory (founded by Columbia University as a result of Ewing's successes) and Scripps Institution of Oceanography.

By the early 1970s, geophysicists, oceanographers, and geologists had pieced together the theory of plate tectonics. By the end of the decade, the theory had won worldwide acceptance, except in the Soviet Union. The theory vindicated the new approach to the science of the earth. It showed that expensive oceanographic surveys were necessary because the geology of the ocean floor turned out to be unexpectedly different from that of the land surface. It demonstrated that theorizing on a global scale led to the detection of patterns that would never have emerged from the laborious mapping of one square mile of the earth after another. And it showed once and for all that sophisticated instruments and high levels of funding need not be a waste of public money.

Simultaneous developments in other areas supported these conclusions. Geologists showed that the earth had suffered multiple impacts from meteors, asteroids, and other extraterrestrial bodies, ending the century-long convention that extraterrestrial phenomena were irrelevant to the study of the earth (*see* METEORITES). When geologists found evidence supporting the thesis, put forward by Walter Alvarez, that one of these impacts precipitated the widespread extinctions at the Cretaceous–Tertiary boundary, the move to integrate earth and planetary science gained further momentum. The growing environmental movement and the popular belief that in some way all earth systems were interconnected doubtless also helped accelerate the shift to earth science.

During the 1970s and 1980s, established geology departments changed their names to earth science, or earth and ocean science, or earth and planetary science. They hired professors trained in physics and chemistry as well as in geology, and overhauled their curricula. Geological surveys rethought their missions, broadened their scope, and began using new instruments. The movement spread beyond research science. In 1982, a group of scientists founded the History of the Earth Sciences Society, and shortly thereafter began publishing *Earth Sciences History*, now the major journal for the history of geology and earth science. In 1983, the National Earth Science Teachers Association was chartered in the United States and began publishing *The Earth Scientist* to promote the teaching of earth science in the school system. Since the 1970s, instruments have continually improved. Magnetometers for measuring fossil magnetism, which in the 1950s and 1960s were so new and intricate that many scientists believed the instruments created the effects observed by "paleomagicians," became standard, everyday, off-the-shelf tools. Theorizing has continued apace. Plate tectonic theory, which at first swept all before it, has been refined, and to some extent reformulated.

For all the success of earth science, traces of its roots in the formerly distinct areas of geology, geophysics, and oceanography remain. Major societies founded before the disciplinary transition, such as the American Association of Petroleum Geologists, the Geological Society of London, and the American Geophysical Union, still exist and continue to publish journals specializing in their traditional interests. And although histories of plate tectonics and biographies of its creators abound, as yet we have no comprehensive history of the disciplinary change to earth science.

H. W. Menard, *Science: Growth and Change* (1971). Robert Muir Wood, *The Dark Side of the Earth* (1985). H. W. Menard, *Ocean of Truth: A Personal History of Global Tectonics* (1986). Walter Lenz and Margaret Deacon, eds., "Ocean Sciences: Their History and Relation to Man," *Proceedings of the Fourth International Congress on the History of Oceanography, Hamburg, September 1987* (1990).

RACHEL LAUDAN

ECLIPSE. The angular diameters of the Sun and the Moon happen to be nearly the same. Occasionally the two bodies line up and the Sun is eclipsed by the Moon. The first known record of a solar eclipse occurs in a Chinese report of 2136 B.C.; the first recorded British eclipse dates from A.D. 538.

The shadow cast by the Moon only just intersects the surface of Earth, so each total solar eclipse can be observed only over a narrow path. Totality lasts for seven minutes at most. The application of Newtonian gravitational physics to calculate the orbital motion of the Moon enabled Edmond Halley to produce accurate predictions of the path of the solar eclipse of 1715.

A total solar eclipse provides the opportunity to observe the fine features of the Sun. By the middle of the nineteenth century, astronomers had associated the corona and prominences visible during totality with the outer solar atmosphere. Chromospheric flash spectra (*see* SPECTROSCOPY) were taken in 1870, and by the 1930s the highly ionized state of the corona, and its mean temperature of 2,000,000°C, had been established. Study of solar eclipses has also yielded information about matters beyond the Sun. Detailed analysis of eyewitness records of the positions of ancient eclipse paths indicate that the Moon is gradually receding from the Earth, causing the day to lengthen by about two milliseconds per century.

Arthur Eddington used the total solar eclipse of 29 May 1919 to confirm Albert *Einstein's general theory of *relativity. The Sun's gravitational field bent the light beams that passed tangentially to the solar surface by something like the predicted 1.75 seconds of arc.

Lunar eclipses occur when the Moon passes through the earth's shadow cone. Accurate timing of the different phases of a lunar eclipse and knowledge of the angular diameter of the Moon and the earth's radius enabled the Greek astronomer Aristarchus of Samos (c. 310–c. 230 B.C.) to deduce that the Moon lies around 60 terrestrial radii from the earth.

Eclipses elsewhere in the solar system have also been significant in the history of astronomy. The seventeenth-century Danish astronomer Ole Rømer timed the eclipses of Jupiter's moons behind the planet when Earth was at different positions around its orbit. That provided enough information to calculate the velocity of light (*see* LIGHT, SPEED OF). *Galileo, who had discovered the satellites in 1609, proposed that these eclipses, together with optical coincidences between the satellites, could provide the basis of a standard clock, visible in different parts of the earth, for finding differences of longitude. The technique could be used on land but a telescope necessary for determining the moments of coincidence could not be managed at sea. More recently (1985 to 1991) the eclipses of Charon by Pluto have led to accurate calculations of the sizes and colors of these two bodies.

The importance of stellar eclipses was recognized by Edward Pigott and John Goodricke. Observing the variations in brightness of the star Algol (Beta Persei) in the 1780s, they suggested that the alterations might be caused by a planet (half Algol's size) revolving round the star and sometimes eclipsing it. Between 1843 and about 1870 Friedrich Wilhelm August Argelander produced ever more accurate values for variations in brightness throughout stellar eclipses. In 1919 Gustav Müller and Ernst Hartwig produced a catalogue of 131 eclipsing binaries. Joel Stebbins introduced a selenium photometer in 1910 and with it discovered the secondary minimum in Algol's light curve, indicating that its companion was a faint star and not a planet. Analysis of light from eclipsing binaries allows calculation of the sizes and luminosities of the two components. They often show evidence of limb darkening and deviations from sphericity.

Frank Dyson and R. v.d. R. Woolley, *Eclipses of the Sun and Moon* (1937). J. B. Zirker, *Total Eclipses of the Sun* (1995).

DAVID W. HUGHES

EINSTEIN, Albert (1879–1955), theoretical physicist, developer of the special and general theories of relativity.

Einstein transformed and advanced science as only Isaac *Newton and Charles *Darwin had done. His most celebrated contributions were the theories of *relativity: (1) the special theory (SRT), which revised the notions of *space and time, brought together under one view *electric-

ity, *magnetism, and *mechanics, dismissed the concept of the *ether, and revealed as a by-product the equivalence of mass and energy ($E = mc^2$); and the general theory (GRT), which reinterpreted gravitation as the effect of the curvature of space-time and opened the way to the development of a so-far unachieved unified field theory that would also geometrize electromagnetic fields.

From GRT, Einstein deduced the stability of a spatially bounded universe by adding a "cosmological constant" (which he later retracted); gravitational waves; and the imaging of remoter objects using the gravitational field of nearer galaxies. GRT's early successes included the prediction of the degree of deflection of starlight passing close to the sun (observed in 1919 during an *eclipse); the red shift of light traversing a gravitational field; and the explanation of the precession of the planet Mercury. Einstein worked out most of SRT in 1905. The creation of GRT required intense labor from 1907 to 1915/16.

Einstein responded to his successes with expressions of humorous self-derogation. He once said that his greatest gifts were stubbornness and taking seriously the sorts of questions only children ask. His personal behavior and opinions often alarmed his more conventional colleagues. He had bohemian tendencies in dress and demeanor; urged pacifism during World War I and arms control after World War II; decried nationalism and undemocratic, hierarchical rules; supported Zionism (with provision for the Arabs in Palestine); and opposed religious establishments in favor of a cosmic religion in the spirit of Spinoza. Einstein's fame and charisma made him the target of attacks, chiefly by anti-Semites, and also flooded him with adoring or opportunistic appeals. An example of the latter was his acting as intermediary between three of his fellow émigré colleagues, who in August 1939, on the eve of World War II, wanted to alert President Roosevelt to the possibilities of the German war machine's use of nuclear energy. Einstein had been living in Princeton since 1933 and was thought to have the necessary influence.

Einstein discussing atomic energy with the *Pittsburgh Post-Gazette* in December 1934. At the time, four years almost to the day before the discovery of uranium fission, the leading physicists doubted that useful energy could be derived from the atom.

In 1914, Einstein settled in Berlin as professor of physics at the university, among other flattering appointments. The move was the culmination of a long series of events. Born in Ulm, Germany, he had spent his childhood in Munich, where his father, an electrical manufacturer and ultimately unsuccessful entrepreneur, hoped to improve his business. In 1888, Einstein entered the Luitpold Gymnasium, but, disliking its military style of instruction, resorted to self-education. He read geometry, popular science, Kant's *Critique of Pure Reason*, and classic works of literature.

In 1894, Einstein tried, and failed, the entrance examination for the Swiss Polytechnic in Zurich. After two years of further preparation in a secondary school, which repaired his deficiencies in foreign language and brought him to the usual age of entry, he gained admission to the Polytechnic to study to be a schoolteacher. He fell in love with one of his classmates, Mileva Marić, with whom he had a daughter, and after their marriage in 1903, two sons. After a long separation they divorced in 1919, whereupon Einstein married his cousin Elsa Löwenthal.

Failing to find a teaching position or a university assistantship after graduating from the Polytechnic, Einstein moved to Bern in 1904 to become a clerk in the Swiss patent office. His work examining applications for patents on electromagnetic devices may have helped him form ideas used in SRT, described in one of his several groundbreaking publications of 1905 in *Annalen der Physik*. The others—the statistical account of Brownian motion and the postulation of the light quantum—did not have obvious ties to patentable machines but were major theoretical advances. As Einstein's extraordinary talents became known, universities vied for him. He became professor at the University of Prague (1911) and the Swiss Polytechnic (1912) before going to Berlin.

Einstein's great creative work of 1905 stemmed from his preoccupation with fluctuation phenomena. He did not start with a crisis brought on by puzzling new experimental facts, but by stating his dissatisfaction with an asymmetry or lack of generality in a theory that others would dismiss as an "aesthetic blotch." He then proposed several principles, and showed how their consequences removed his original dissatisfaction. All three papers endeavored to bring together and unify apparent opposites, and ended with brief proposals for a series of confirming experiments. Similarly, GRT and the unified field theory arose from his dissatisfaction with the incompleteness of SRT. As he once said, he was driven by his "need to generalize."

In generalizing, he relied on what he called "schemes of thought, the selection of which is, in principle, entirely open to us, and whose qualification can only be judged by the degree to which its use contributes to making the totality of consciousness intelligible." Among his guiding presuppositions, or "themata," were the primacy of formal explanations, unity on a cosmological scale, logical parsimony and necessity, symmetry, simplicity, completeness, continuity, constancy and invariance, and causality. In contrast, the concepts of probabilism and indeterminacy rather than causality and completeness, fundamental in the quantum mechanics of Niels *Bohr and his school, were abhorrent to Einstein.

Of all Einstein's thematic presuppositions, the concept of unity, or, as he described it, the longing to behold the pre-established harmony beyond the harshness and dreariness of everyday life, made him an enduring icon for science.

Paul Arthur Schilpp, ed., *Albert Einstein: Philosopher-Scientist* (1949). Albert Einstein, *Ideas and Opinions* (1954). Abraham Pais, *Subtle Is the Lord...: The Science and the Life of Albert Einstein* (1982). (Various editors), *The Collected Papers of Albert Einstein* (1987-). Gerald Holton, *Thematic Origins of Scientific Thought: Kepler to Einstein*, rev. ed. (1988). Albrecht Fölsing, *Albert Einstein* (1998).

GERALD HOLTON

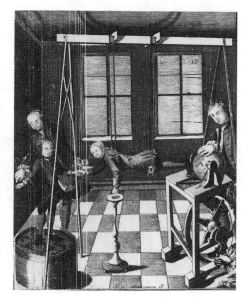

A human capacitor, a standard piece of apparatus in electrostatic experiments around the middle of the eighteenth century. The harness insulates the boy; the electrical machine charges him.

ELECTRICITY was originally the term used to describe the power acquired by certain substances, notably amber, when rubbed, to attract nearby small objects. This power, known to the Greeks, was carefully distinguished by William Gilbert (1600) from magnetism. Seventeenth-century philosophers were unanimous in believing that the rubbing agitated a subtle matter associated with ordinary matter, causing it to be ejected into the surrounding air, where it (or perhaps air rushing in to fill the space left empty) swept up any light objects in the way.

Francis Hauksbee obtained more powerful effects by mounting a glass globe on a spindle and rubbing it as it rotated. He showed that electrification was linked to the emission of light—indeed, in Hauksbee's experiments it seemed that the subtle matter streaming from electrified bodies could be felt, seen, and heard. As elaborated by Jean-Antoine Nollet in the 1740s, the theory envisaged fiery matter streaming from electrified bodies, while other streams flowed in to replace the matter that had left.

In 1731, Stephen Gray discovered that the electrical attracting power could be transmitted over great distances, provided that the conducting line was made of an appropriate material and suitably supported. This led to a distinction between "electrics"—substances electrifiable by friction but poor conductors—and "non-electrics"—substances not electrifiable by friction but good conductors.

In 1746, Petrus van Musschenbroek announced a sensational discovery, the "Leyden experiment," in which a bottle filled with water and electrified by means of a conductor leading from a generating machine into the water delivered a terrible blow when contact was made simultaneously with the conductor dipping into the water and the outer surface of the bottle. "I thought it was all up with me," Musschenbroek reported. Enthusiasts everywhere rushed to repeat the experiment. Nollet delighted the French king by discharging a bottle through a line of monks, making them leap into the air simultaneously. In America, Benjamin *Franklin and his friends amused themselves with electrical party tricks.

Franklin also devised a new theory of electricity that in time became generally accepted. He, too, supposed that ordinary matter was suffused with subtle fluid. For Franklin, however, electrification consisted in the redistribution of this fluid between rubber and rubbed, one finishing up with more than its natural quantity, the other with less. The notion of electrical "charge" thus acquired a meaning, the bodies becoming electrified "plus" and "minus" respectively. Noting the ability of pointed conductors to discharge nearby charged objects, Franklin conceived an experiment, first successfully performed near Paris in May 1752, to demonstrate that thunderclouds were electrified and lightning was an electrical discharge. His conclusion that erecting pointed conductors (*lightning rods) on buildings could protect them from lightning strikes was hailed as a triumph of reason over nature.

Franklin's theory worked well in explaining when a shock would be experienced in a variety of situations. He did not provide a coherent dynamical basis for his theory. In 1759, Franz Ulrich Theodosius *Aepinus published a fully consistent version of Franklin's theory, based explicitly on forces acting at a distance between particles, whether of fluid or of matter. By adding the forces acting in various situations, he gave satisfactory explanations for a wide range of effects, and successfully predicted others. His ideas were taken up by leading investigators of the next genera-

tion such as Alessandro Volta (see GALVANI LUIGI, AND ALESSANDRO VOLTA) and Charles Augustin Coulomb (see CAVENDISH, HENRY, AND CHARLES AUGUSTIN COULOMB), who showed in 1785 that the forces involved obeyed an inverse-square law with respect to distance. On this basis, Simeon-Denis Poisson developed a full mathematical theory of electrostatics in 1812.

To make his theory consistent with elementary observation, Aepinus found it necessary to assume that particles of ordinary matter repelled each other. Many people rejected this idea, supposing instead the existence of two electric fluids that normally neutralized each other but that became separated in electrification. This theory attributed to the second electric fluid the additional repulsive force that Aepinus invoked. Operationally, the one-fluid and two-fluid theories could not be distinguished, and each found adherents.

In the 1790s, controversy erupted between Alessandro Volta and Luigi Galvani over Galvani's experiments in which frogs' legs jerked spasmodically when a conducting circuit was completed between the crural nerve and the leg muscle. Galvani attributed the effect to the discharge of "animal electricity" accumulated in the muscle, which he saw as analogous to a Leyden jar. Volta believed the cause was ordinary electricity, and concluded that dissimilar conductors in contact generate an electromotive force. He built a "pile" comprising pairs of silver and zinc disks separated by pieces of moist cardboard, the electrical force of which he managed to detect with a sensitive electroscope.

Volta's device was the first source of continuous electric current. Its operation was accompanied by chemical dissociations within the moist conductor and in any other conducting solutions forming part of the electrical circuit. These became a focus of research, leading to Humphry Davy's successful isolation of potassium and sodium in 1807 by electrolyzing molten potash and soda, respectively, and eventually to large-scale industrial applications of electrochemical processes.

In 1820, Hans Christian Oersted discovered that a wire carrying an electric current deflects a magnetic needle. André-Marie

Ampère quickly showed that a current-carrying loop or solenoid was equivalent to a magnet, and proposed that all magnetism arose from solenoidal electric currents in molecules of iron (see MAGNETISM). Ampère's discovery led to the construction of electromagnets—iron-cored solenoids carrying ever-larger currents—that produced magnetic effects far more powerful than any previously known. Then, in 1831, Michael *Faraday discovered electromagnetic induction: when an electrical conductor cuts across lines of force, an electromotive force is generated in the conductor. These discoveries underpinned the development, first, of the electric telegraph and, later in the nineteenth century, the electrical power industry, which provided a continuous supply of electrical power for industrial or domestic use from generators in which coils of conducting wire placed between the poles of a powerful electromagnet were rotated by steam or water pressure. In the twentieth century, the increasing availability of such power, and the rapid proliferation of appliances to exploit it, transformed civilized life.

But what is electricity? For most nineteenth-century physicists, sources of electromotive force literally drove a current of the electric fluid (or fluids) around a conducting circuit. In an electrolytic cell, the electrodes acted as poles, attracting the constituent parts of the solute into which they were dipped. For Faraday, however, and later for James Clerk *Maxwell, the energy resided not in electrified bodies but in the medium surrounding them, which was thrown into a state of strain by the presence of a source of electromotive force (see FIELD). If the medium were a conductor, the tension would collapse, only to be immediately restored; the product of this continuous repetition, the "current," was a shock wave propagated down the conductor. Static electric charges represented the ends of lines of unrelieved electric tension.

These ideas were widely influential in late-nineteenth-century physics, even though they left the relationship between electricity and matter far from clear. Following the discovery of the *electron by Joseph John *Thomson in 1897, however, and the recognition that it was a universal constituent of matter, electric currents again came to be conceived in terms of a flow of "subtle fluid"—now clouds of electrons semi-detached from their parent atoms, driven along by an electromotive force. No longer, however, was "charge" defined in terms of accumulations or deficits of this fluid. Rather, charge became a primitive term, a quality attributed to the fundamental constituents of matter that was itself left unexplained. Some kinds of fundamental particles, including electrons, carry negative charge, whatever that might be, whereas others carry an equally mysterious positive charge: in combination, these two kinds of charge negate each other. After Robert Millikan's experiments (1913), most physicists accepted the notion that there is a natural unit of electricity, equal to the charge on the electron. The current *standard model in elementary particle physics, however, assumes the existence of sub-nucleonic constituents ("*quarks") bearing either one-third or two-thirds of the unit charge. Neither these quarks nor their fractional charges have been detected in a free state.

E. T. Whittaker, *A History of the Theories of Aether and Electricity* (2 vols., 1951). René Taton, *History of Science*, trans. A. J. Pomerans, vol. 3, ch. 4, "Electricity and Magnetism (1790–1895)," 178–234; vol. 4, ch. 8, "Electricity and Electronics," 152–202 (4 vols., 1964–1966). L. Pearce Williams, *The Origins of Field Theory* (1966). R. W. Home, *Aepinus's Essay on the Theory of Electricity and Magnetism* (1979). J. L. Heilbron, *Electricity in the 17th and 18th Centuries* (1979). Bruce J. Hunt, *The Maxwellians* (1991).

R. W. HOME

ELECTRICITY, ATMOSPHERIC. See ATMOSPHERIC ELECTICITY.

ELECTROLYSIS. Alessandro Volta's publication of his invention of the battery in 1800 opened an era in science. Learning of Volta's work from Sir Joseph Banks before the publication of Volta's paper, William Nicholson and Anthony Carlisle constructed Voltaic piles of half-crown silver disks, pieces of zinc, and pasteboard soaked in salt water. When they inserted platinum wires from the ends of their pile into a dish of water, hydrogen gas was generated at one wire and oxygen gas at the other. Nicholson announced the results in his journal (*Nicholson's Journal*) before the world learned about the battery from its inventor.

In 1803, Jöns Jacob Berzelius in Stockholm observed that chemical acids formed around the positive pole and chemical bases around the negative pole of a Voltaic pile. Three years later, Humphry Davy summarized his observations of hydrogen, metals, metallic oxides, and alkalies around the negative pole during electrolysis and oxygen or acids around the positive pole. In 1807, Davy succeeded in obtaining the new metals potassium and sodium from electrolysis of their dry fused alkalis.

Both Davy and Berzelius drew the obvious conclusion that electrical attractions and chemical affinities are identical. About 1809, Amedeo Avogadro had devised a scale of oxygenicity or acidity for the affinity of oxygen with other elements. Berzelius transformed Avogadro's concept into a scale of electronegativity of the elements in 1818. The scale supported Berzelius's theory that chemical combination results from the union of atoms containing unequal amounts of positive and negative electrical fluid combining together with the release of heat (caloric fluid). Berzelius suggested that each chemical compound retains a small excess of either negative or positive fluid: acids, with an excess of negative fluid, combine with bases, carrying an excess of positive fluid, to form salts.

Berzelius's dualistic theory enjoyed considerable support until it began to seem irreconcilable with new results in organic chemistry in the 1840s. In the meantime, Michael *Faraday, in collaboration with Whitlock Nicholl, developed the terms "electrode," "electrolysis," and "electrolyte," and with William Whewell, "ion," "cation," "anion," "cathode," and "anode" for the new electrochemical science. Assuming a quantitative relationship between the amount of chemical substance decomposed and the quantity of current that passes through a solution, Faraday devised a "volta-electrometer" to measure the quantity of current. With it, he demonstrated that the quantity of electricity that frees one unit of hydrogen gas liberates

other elementary gases in the amount of their chemical equivalent weights.

As early as 1805, Theodor von Grotthuss contrived a model to explain the appearance of electrolysis products at the electrodes, implying that ions in solution move with equal speeds in different directions. Further work with electrochemical cells suggested that cations move faster than anions, an idea developed in 1854 by Wilhelm Hittorf into a quantitative measure, the "transport number" on fraction of the current carried by a particular ion. In 1857 Rudolf Clausius suggested that some of the molecules in solution exist as ions even before the current is applied. Hittorf and François Raoult accepted the suggestion, which became the underlying principle of Svante Arrhenius's ionic theory.

In his doctoral thesis of 1884, Arrhenius proposed (unclearly) that some "complex molecules" exist in dilute solutions as potential carriers of current before the flow of current begins. By 1887, under the influence of Wilhelm Ostwald and Jacobus Henricus van't Hoff, Arrhenius moved to the truly novel interpretation that a substantial number of charged ions already exist in dilute electrolytes. Ostwald, Van't Hoff, and Arrhenius promulgated this new ionic theory in the new journal *Zeitschrift für physikalische Chemie*, founded in 1887.

Arrhenius's argument that a shifting equilibrium holds between ions and undissociated molecules in strong electrolytes was refuted by the Danish physical chemist Niels J. Bjerrum in 1909 over Arrhenius's persistent objections. A complete and detailed treatment of strong electrolytes, developed by Peter J. W. Debye and Erich Hückel in 1923, and improvements by Lars Onsager in 1926, took into account the Brownian motion of ions surrounded by groups of oppositely charged ions (the ionic atmosphere).

James *Joule made the first attempt to relate the electromotive force of an electrochemical cell to the thermodynamics of chemical reactions in 1840. This problem remained an important theme in thermodynamics, notably in the work of Hermann von *Helmholtz, J. Willard *Gibbs, and Walther Nernst. Merle Randall and Gilbert N. Lewis, who had worked with Ostwald and Nernst, codified the practical application of these studies in their book *Thermodynamics and the Free Energy of Chemical Substances* (1924). The ionist theory and the development of electron theory around 1900 revived interest in electrochemical explanations of the chemical bond, developments to which Lewis decisively contributed, along with Irving Langmuir, between 1916 and 1919.

See also GALVANI AND VOLTA; ION

J. R. Partington, *A History of Chemistry*, vol. 4 (1970). John W. Servos, *Physical Chemistry from Ostwald to Pauling: The Making of a Science in America* (1990). William H. Brock, *The Norton History of Chemistry* (1992).

MARY JO NYE

ELECTROMAGNETISM. The study of electricity and magnetism has alternated between theories that represented the two as manifestations of a single effect and the view that they were separate phenomena, with general inflection points around 1600 (when the two fields were divided) and 1820 (when they reunited).

Ancient and medieval philosophers did not distinguish between the ability of amber to attract objects and the action of the lodestone on iron. In 1600 William Gilbert insisted on a distinction between the two, coining the term *"electricity" to describe the effect of amber and attributing it to many other substances; the lodestone remained for Gilbert the sole source of magnetism, around whose force he constructed an entire cosmology. Gilbert's successors spent the seventeenth century searching for the source of electrical and magnetic phenomena and generally finding it in material emanations, such as René *Descartes's theory of magnetic effluvia that swarmed around iron and accounted for magnetic action.

Over the course of the eighteenth century natural philosophers incorporated the distinct fields of electricity and magnetism within a new quantitative, experimental discipline of physics, aided by new instruments to produce phenomena—electrostatic machines like Francis Hauksbee's spinning glass globe, and the Leyden jar, an early form of capacitor—and to measure them, such as the electroscope, which indicated electrical force by the displacement of threads, straws, or gold leaf. Quantification allowed mathematicians, notably Henry *Cavendish and Charles Augustin Coulomb, to follow Isaac *Newton's example for gravitation and reduce electro- and magnetostatics to distance forces. Philosophers around 1800 explained their measurements in terms of a system of imponderable fluids, with one (or two) weightless, elastic fluid(s) each for electrical and magnetic forces. But despite similar mathematical and conceptual approaches to electricity and magnetism, the phenomena still appeared unrelated. Electricity produced violent action like lightning or sparks, unlike milder magnetic effects, and did not have the same polar behavior.

One of the new instruments, Alessandro Volta's battery, produced a flow of current electricity. It thus enabled the unification, or reunification, of electricity and magnetism, whose interactions stemmed from dynamic electric and magnetic effects. In 1820 Hans Christian Oersted noticed that a current-carrying wire displaced a nearby compass needle. Oersted's response reflected his adherence to *Naturphilosophie, a metaphysical system advanced at the time by German philosophers who sought polar forces underlying various domains of natural phenomena. Oersted thought his experiment gave evidence of such a unifying force and proposed a connection between electricity and magnetism. André-Marie Ampère soon mathematized Oersted's experiment, providing a force law for currents in wires. In 1831 Michael Faraday, who may also have dabbled in Naturphilosophie, explored the reverse effect, the ability of a moving magnet to generate electric currents. Oersted, Ampère, and *Faraday thus established electromagnetism, although all three noted a difference between the linear action of electricity and the circular action of magnetism.

Faraday formulated the concept of *field to explain the action of electric and magnetic forces at a distance. Instead of material, if weightless, fluids carrying the force,

Faraday attributed the source of electromagnetic action to lines of force in the medium between currents and magnets: lines of force ran from one magnetic pole to the opposite pole, and between opposite electrical charges; magnets or conductors experienced pushes or pulls as their motion intersected these lines of force. James Clerk *Maxwell in the 1860s and 1870s systematized the field concept and the interaction of electric and magnetic forces in a set of differential equations, later simplified into four basic equations. Maxwell's theory, which interpreted light as very rapidly oscillating electric and magnetic fields, received experimental confirmation by Heinrich Hertz, who starting in 1887 demonstrated the propagation of electromagnetic waves through space.

Over the course of the nineteenth century experimental work on electromagnetism, especially in Germany, spurred the emergence of precision physics, which in turn promoted the establishment of academic research institutes. Electromagnetism provided one of the two main lines of development of nineteenth-century physics, along with *thermodynamics and the kinetic theory. Whereas thermodynamics derived from the first industrial revolution of the late eighteenth century, specifically the program to describe the working of the steam engine, electromagnetism stimulated the second industrial revolution of the late nineteenth century.

Electromagnetism gave rise to industries around the telegraph, a result of Oersted's discovery of mechanical motion produced by electric current; electric power generation and transmission, with the dynamo deriving from Faraday's work on the transformation of mechanical motion into electric current; and radio, a consequence of Hertz's experiments on the free propagation of electromagnetic radiation. Together with chemistry and the chemical industry, electromagnetism and its applications exemplified the science-based industry of the second industrial revolution. But the electrical industry did not just apply scientific theories developed in isolation from commercial concerns; rather, prominent scientists, such as William *Thomson (later Lord Kelvin) and Oliver Heaviside tackled practical problems, especially in telegraphy, and thus advanced the state of electromagnetic theory.

Since Maxwell's theory established a conception of light as a form of electromagnetic waves, electromagnetism merged with theoretical optics. At the end of the century some physicists sought to extend the domain of electromagnetism to all physical phenomena, including mechanics and gravity, in what was called the electromagnetic worldview. Many electromagnetic theories relied on an all-pervasive *ether as the medium for electromagnetic waves. Albert *Einstein's theory of *relativity at the outset of the twentieth century enshrined the interchangeability of electrical and magnetic forces and banished the ether, leaving electromagnetic waves to propagate through empty space at the speed of light. The concurrent development of quantum theory, which itself arose from the application of thermodynamics to electromagnetic waves (in the form of heat radiation), led to the wave-particle duality, in which electromagnetic radiation at high frequencies may behave either as a wave or a particle. The elaboration of *quantum electrodynamics starting around 1930 provided a quantum theory of electromagnetism and a description of the duality.

Physicists later in the twentieth century came to view electromagnetism as one of four fundamental forces, the others being gravity and the strong and weak nuclear forces. Particle physicists hoped to show that each force was a particular manifestation of a more general force; they thus joined the unified electromagnetism with the weak (in the so-called electroweak force) and then the strong force (in quantum chromodynamics), and sought in vain for final unification with gravity in a so-called *theory of everything.

E. T. Whittaker, *A History of the Theories of Aether and Electricity*, 2 vols. (1951–1953). J. L. Heilbron, *Electricity in the 17th and 18th Centuries: A Study in Early Modern Physics* (1979). Jed Z. Buchwald, *From Maxwell to Microphysics: Aspects of Electromagnetic Theory in the Last Quarter of the Nineteenth Century* (1985). Silvan S. Schweber, *QED and the Men Who Made It: Dyson,*

Feynman, Schwinger, and Tomonoga (1994). Olivier Darrigol, *Electrodynamics from Ampère to Einstein* (2000).

PETER J. WESTWICK

ELECTRON. By 1890, many chemists and physicists believed that atoms of different elements might represent arrangements of a more fundamental unit, a belief that stemmed primarily from the conviction that nature was essentially simple. However, models devised by physicists on the basis of kinetic theory disagreed with the chemists' results of spectral analysis.

Electrical theories of the atom offered a possible solution. These followed from James Clerk *Maxwell's electromagnetic theory and from Michael *Faraday's work on electrolysis. Maxwell proposed that the vibrations of light were not mechanical, as previously thought, but electromagnetic. At the same time, the laws of electrolysis implied that electricity existed in discrete units with a charge equal to that on the hydrogen ion. Might the atom contain such units, whose oscillations would explain the emission of line spectra? In 1891, George Johnstone Stoney named these units of charge "electrons" and attempted to find out how big they were by reconciling the spectroscopic and kinetic data.

Simultaneously, Hendrik Antoon Lorentz and Joseph Larmor were trying to accommodate discrete charges within Maxwell's theory, which was expressed in terms of the continuous *ether. They proposed models of charges as vortices or strain centers in the ether and Larmor adopted the term "electron" for his charge. In 1896, the discovery of the magnetic splitting of spectral lines by Lorentz's student Pieter Zeeman lent support to these theories. For Lorentz and Larmor the electron was embedded within the atom, but played no role in determining its chemical nature. This view changed in the years 1895–1905 following the discovery of *X rays and *radioactivity, and investigations of *cathode rays.

Cathode rays were discovered by Julius Plücker in 1858. They are found when an electric potential is applied across a gas at low pressure, but initially they seemed peripheral to mainstream physics. The discovery of X rays in 1895 revived interest in the cathode rays that caused them. Recognition that cathode rays were negatively charged particles, about 2,000 times lighter than atoms, depended primarily on the work of four men: Philipp Lenard, who showed that cathode rays traveled much farther than expected through gases and that their absorption depended on the molecular weight of the gas; and Emil Wiechert, Walter Kaufmann, and Joseph John *Thomson, who measured the charge to mass ratio of the rays by various means. Of these, Thomson went the furthest theoretically, proposing on 30 April 1897 that cathode rays were subatomic, negatively charged particles from which all atoms were built up: they provided the essential mass of the atom and its chemical constitution. To mark the distinction from electrons, Thomson called the particles "corpuscles."

Thomson's contemporaries at first found his suggestion difficult to accept: it sounded like alchemy. They preferred George Francis FitzGerald's alternative proposal that cathode rays were free Larmor-type "electrons." Thus, the name electron became firmly attached to the particles several years before realization that they were indeed essential constituents of the chemical atom. Chief among the corroborating evidence was work on radioactivity that demonstrated the identity of beta and cathode ray particles and showed that atoms could and did split up and change their chemical nature.

FitzGerald's suggestion ensured that Thomson's cathode-ray particles attracted wide attention by tying them to the attempt by Lorentz, Henri Poincaré, Kaufmann, and others to formulate an entirely electromagnetic theory of matter; an attempt that fostered, but proved incompatible with Einstein's relativity theory. But it was Thomson's concept, as interpreted by Ernest *Rutherford and Niels *Bohr, that made the electron fundamental for new theories of atoms and chemical bonding, and together with *relativity and *quantum physics, to ideas of the nature of matter. Development of these ideas has been very largely the elucidation of the properties of electrons, starting with its charge, first measured by

Thomson's students John S. Townsend and H. A. Wilson (1899 and 1903), but firmly established by Robert A. Millikan's experiments beginning in 1907. Electron spin, proposed by Samuel Goudsmit and George Uhlenbeck in 1925 to complete the explanation of the fine structure of spectra, was the last stage in the development of the old quantum theory; the wave nature of the electron, demonstrated by Clinton Davisson and Lester Germer, and George P. Thomson in 1927, and the relativistic electron, formulated by P. A. M. *Dirac in 1928, became cornerstones of the new quantum mechanics.

*Quantum electrodynamics was developed from the late 1920s by Dirac, Werner *Heisenberg, Ernst Pascual Jordan, and Wolfgang Pauli to describe the interactions between electromagnetic radiation and charged particles such as electrons. It was completed in the early 1950s by Freeman Dyson, Richard *Feynman, Julian Schwinger, and Shin'ichirō *Tomonaga, as "renormalization" became accepted. This provided a way of handling the infinite values for calculated quantities that arose in any attempt to make the electron's (or other particle's) equation of state relativistically invariant.

In the 1950s, accelerated bombardment by electrons suggested that protons and neutrons had a complex structure, the number of elementary particles known had proliferated, and physicists sought further simplification. In 1964, based largely on observed symmetries, George Zweig and Murray Gell-Mann independently proposed that many particles were themselves composed of *"quarks" with charge of one-third or two-thirds that of an electron. The *"Standard Model" of elementary particles synthesizes research suggesting that quarks form a family of six members subject to the "strong interaction" and do not appear independently, being bound together in pairs to form more familiar particles such as the proton and neutron, as described by quantum chronodynamics. A corresponding family of six "leptons" is not sensitive to the strong interaction and has, so far, proved structureless. The electron is the lightest and most stable of the three negatively charged particles that, together

with their associated neutrinos, make up the leptons. Any proof that the electron has a finer structure would undermine the edifice of modern particle physics.

J. L. Heilbron, *Historical Studies in the Theory of Atomic Structure* (1981). Arthur Miller, *Albert Einstein's Special Theory of Relativity: Emergence (1905) and Early Interpretation (1905–1922)* (1981). Jed Buchwald, *From Maxwell to Microphysics* (1985). Silvan S. Schwever, *QED and the Men Who Made It* (1994). Laurie Brown et al., eds., *The Rise of the Standard Model* (1996). Per F. Dahl, *Flash of the Cathode Rays* (1997). Edward Davies and Isobel Falconer, *J. J. Thomson and the Discovery of the Electron* (1997). Jed Buchwald and Andrew Warwick, eds., *Histories of the Electron* (2001).

ISOBEL FALCONER

ELEMENTARY PARTICLES. Modern particle physics began with the end of World War II. Peace and the Cold War ushered in an era of new *accelerators of ever increasing energy and intensity able to produce the particles that populate the subnuclear world. Simultaneously, particle detectors of ever increasing complexity and sensitivity recorded the imprints of high-energy subnuclear collisions. Challenges, opportunities, and resources attracted practitioners: the number of "high energy" physicists worldwide grew from a few hundred after World War II to some 8,000 in the early 1990s.

Developments in 1947 shaped the further evolution of particle physics. Experimental results regarding the decay of mesons observed at sea level presented to the Shelter Island conference led Robert Marshak to suggest that there existed two kinds of mesons. He identified the heavier one, the π meson, with the meson copiously produced in the upper atmosphere in nuclear collisions of *cosmic-ray particles with atmospheric atoms and with the *Yukawa particle responsible for nuclear forces. The lighter one, the μ meson observed at sea level, was the decay product of a π meson and interacted but weakly with matter. A similar suggestion had been made earlier by Shoichi Sakata in Japan. Within a year, Wilson Powell identified tracks showing the decay of a π into a μ meson in a nuclear emulsion sent aloft in a

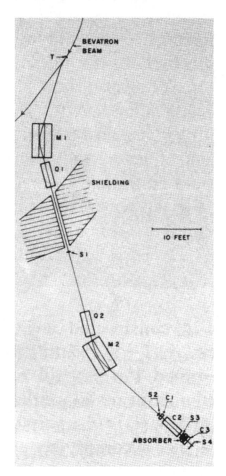

Experimental layout for the detection of the anti-proton: a mixed beam of negative particles from the *accelerator (the Berkeley Bevatron) is bent and focused by the magnets M and Q and analyzed by the counters S and C to distinguish the anti-protons from the much more plentiful pions.

high altitude balloon. During the early 1950s the data obtained from particles produced in accelerators led to the rapid determination of the characteristic properties of the three varieties of π mesons.

The two-meson hypothesis suggested amendments to the list of particles. Some particles ("leptons")—the electron, muon, and neutrino—do not experience the strong nuclear forces. Others ("hadrons")—the neutron, proton, and the π-mesons do interact strongly with one

another. It proved useful to split the hadrons into baryons and mesons. Baryons, of which the proton and neutron are the lightest representatives, have odd-half integer spin and (except for the proton) are unstable, one of the decay products always being a proton. Mesons have integer spin and when free ultimately decay into leptons or photons.

In January 1949, Jack Steinberger gave evidence that the μ-meson decays into an electron and two neutrinos, and shortly thereafter several theorists indicated that the process could be described in the same manner as an ordinary β-decay. Moreover, they pointed out that the coupling constant for this interaction had roughly the same magnitude as the constant in nuclear β-decay. Attempts then were made to extend an idea Oskar Klein had put forth in 1938, that a spin 1 particle, the "W boson," mediated the weak interactions and that the weakness of the β-decay interaction could be explained by making the W mesons sufficiently heavy.

During the first half of the 1950s theoretical attempts to explain pion-nucleon scattering and the nuclear forces were based on *field theoretical models emulating *quantum electrodymics (QED). The success of QED rested on the validity of perturbative expansions in powers of the coupling constant, $e^2/hc = 1/137$. However, for the meson theory of the pion-nucleon interaction the coupling constant had to be large—around 15—to yield nuclear potentials that would bind the deuteron. No one found a valid method to deal with such strong couplings. By the end of the 1950s *quantum field theory (QFT) faced a crisis because of its inability to describe the strong interactions and the impossibility of solving any of the realistic models that had been proposed to explain the dynamics of hadrons. Theorists abandoned efforts to develop a theory of the strong interactions along the model of QED, although Chen Ning Yang and Robert L. Mills advanced a local gauge theory of isotopic spin symmetry in 1954 that proved influential later on. Local gauge invariance, however, implies that the gauge bosons are massless. This is not the case for the pion and thus Yang and Mills's theory was considered an interesting model but not relevant for understanding the strong interactions.

The crisis in theoretical particle physics at the end of the 1950s inspired several responses. It led to the explorations of the generic properties of QFT when only such general principles as causality, the conservation of probability (unitarity), and relativistic invariance figured and no specific assumptions regarding the form of the interactions were made. Geoffrey Chew's *S*-matrix program that rejected QFT and attempted to formulate a theory that made use only of observables was more radical.

Another response to the crisis made symmetry concepts central. First applied to the weak and the electromagnetic interactions of the hadrons, *symmetry considerations were later extended to encompass low-energy strong interactions. Symmetry became one of the fundamental concepts of modern particle physics, used as a classificatory and organizing tool and as a foundational principle to describe dynamics. Interest in field theories, and in particular in gauge theories, revived after theorists had appreciated the notion of spontaneous symmetry breaking (SSB). SSB allows a field theory to have a much richer underlying symmetry than that observed. Usually a symmetry expresses itself in such a way that the vacuum state of the theory is invariant under the symmetry that leaves the description of the dynamics (the Lagrangian) invariant. In the early 1960s Julian Schwinger, Jeffrey Goldstone, Yoichiro Nambu, Steven Weinberg, Abdus *Salam, and others noted that in quantum field theories symmetries could be realized differently: the Lagrangian could be invariant under some symmetry, without the symmetry applying to the ground state of the theory. Such symmetries are called "spontaneously broken" (SBS).

In 1967 Weinberg, and in 1968 Salam, independently proposed a gauge theory of the weak interactions that unified the electromagnetic and the weak interactions and made use of the Higgs mechanism, which generates the masses of the particles associated with the gauge theory. Their model incorporated previous suggestions by Sheldon Glashow (1961) for formulating a gauge theory in which the gauge bosons mediated weak forces. The renormalizabil-

ity of such theories—the existence of consistent algorithms for extracting finite contributions from every order of perturbation theory—was proved by Gerard 't Hooft in his dissertation in 1972. The status of the Glashow-Weinberg-Salam theory changed dramatically in consequence. As Sidney Coleman noted in his article in *Science* describing the award of the Nobel Prize to Glashow, Salam, and Weinberg in 1978, "'t Hooft's kiss transformed Weinberg's frog into an enchanted prince."

As presently described, a common mechanism underlies the strong, weak, and electromagnetic interactions. Each is mediated by the exchange of a spin 1 gauge boson. The gauge bosons of the strong interactions are called gluons, those of the weak interactions, $W^{\pm}$ and Z bosons, and those of electromagnetism, photons. The charges are often called "colors": QED, the paradigmatic gauge theory, works with a single gauge boson, the photon, coupled to a single "color," namely the electric charge. The gauge bosons of the strong interactions carry a three-valued color, those of the weak interactions carry a "two-dimensional" weak color charge. Weak gauge bosons interact with quarks and leptons and some of them, when emitted or absorbed, can transform one kind of quark or lepton into another. When these gauge bosons are exchanged between leptons and quarks they are responsible for the force between them. They can also be emitted as radiation when the quarks or leptons accelerate.

Quantum chromodynamics (QCD) describes the strong interactions between the six quarks: the up and the down, the charmed and the strange, and the top and the bottom. They are usually denoted by u, d, c, s, t, and b. Evidence for the top quark was advanced in the fall of 1994 and confirmed in the spring of 1995. Quarks carry electrical charge and in addition a "three dimensional" strong color charge. QCD is a gauge theory with three colors, eight massless gluons, and color-carrying gauge bosons, six that alter color and two that merely react to it. Gluons do not carry color in the same way as quarks do; they carry a color-anticolor, which enables them to interact with one another.

In QFT the vacuum is a dynamic entity. Within any small volume of space-time the root mean square values of the field strengths (electric and magnetic in QED, color-gluon field in QCD) averaged over the volume do not vanish. Virtual particle-antiparticle pairs are constantly being created, and as demanded by the energy-time uncertainty relations, particle and antiparticle annihilate one another shortly thereafter without traveling very far. These virtual pairs can be polarized in much the same way as molecules in a dielectric solid. Thus in QED the presence of an electric charge e^o polarizes the "vacuum," and the charge that is observed at a large distance differs from e^o and is given by $e = e^o/\varepsilon$, with ε the dielectric constant of the vacuum. The dielectric constant depends on the distance (or equivalently, in a relativistic setting, on energy) and in this way the notion of a "running charge" varying with the distance being probed, or equivalently varying with the energy scale, is introduced. Virtual dielectric screening tends to make the effective charge smaller at large distances. Similarly virtual quarks and leptons tend to screen the color charge they carry.

It turns out however that non-Abelian gauge theories like QCD have the property that virtual gluons "antiscreen" any color charge placed in the vacuum (and in fact overcome the screening due to the quarks). This means that a color charge that is observed to be big at large distances originates in a charge that is weaker at short distances, and in fact vanishingly small as $r \rightarrow 0$. The discovery of the antiscreening in spin 1 non Abelian gauge theories was made independently by 't Hooft in 1972, and by David Politzer and by David Gross and Frank Wilczek in 1973: non-Abelian gauge theories behave at short distances approximately as a free (non-interacting) theory. This behavior, called *asymptotic freedom*, could explain in a natural way the SLAC experiments on deep inelastic scattering of electrons by protons. Some physicists speculate that in non-Abelian gauge theories the complement to asymptotic freedom at short distances is confinement at large distances. This would explain the non-observability of free quarks. In other words, even though the forces among quarks become vanishingly small at short distances, the force between them increases very strongly at large distances. Although to this day confinement has not been proved in a rigorous fashion, non-perturbative calculations point to the correctness of the assumption.

The past two decades have seen many successful explanations of high-energy phenomena using QCD. The detection and identification of the $W^\pm$ and of the Z_0 in 1983 by Carlo Rubbia and coworkers at CERN gave further confirmation. Similarly, the empirical data obtained in lepton and photon deep inelastic scattering, and in the study of jets in high energy collisions, can be accounted for quantitatively by QCD. Furthermore, computer simulations have presented convincing evidence that QCD confines quarks and gluons inside hadrons. Frank Wilczek, one of the important contributors to the field, remarked at a conference in 1992 devoted to an assessment of QCD that it had become mature enough to be placed in its "conceptual universe with appropriate perspective."

L. Hoddeson, L. M. Brown, M. Dresden, and M. Riordan, eds., *The Rise of the Standard Model: Particle Physics in the 1960s and 1970s* (1997). Gerard 't Hooft, *In Search of the Ultimate Building Blocks* (1997).

SILVAN S. SCHWEBER

ENERGETICS. The great unsettled question of late-nineteenth-century physics was the status of the mechanical worldview. For more than two hundred years—from René *Descartes, Christiaan *Huygens, and Isaac *Newton in the seventeenth century to Hermann von *Helmholtz, Heinrich Hertz, and Ludwig *Boltzmann at the end of the nineteenth—physicists had generally sought mechanical explanations for natural phenomena. As the nineteenth century drew to a close, Hertz reaffirmed the classical goal of physical theory: "All physicists agree," he wrote in the preface to his *Principles of Mechanics* (1894), "that the problem of physics consists in tracing the phenomena of nature back to the simple laws of mechanics." But when these words were published, physicists were no longer in general agreement about the nature of their project. Many doubted, and some explicitly denied, that mechanics was the most basic science. Other candidates contended for the honor—thermodynamics and electromagnetic theory, in particular, and several comprehensive alternatives to the mechanical worldview were proposed and vigorously debated throughout the 1890s and early 1900s.

Energetics was one of the alternatives. Tracing its origins to the founders of the law of energy *conservation, especially Robert Mayer (*see* JOULE AND MAYER, and to the thermodynamic writings of Rudolf Clausius, William *Thomson (Lord Kelvin), and Josiah Willard *Gibbs, energetics attempted to unify all of natural science through the concept of energy and by laws describing energy in its various forms. The energeticists believed that scientists should abandon their efforts to understand the natural world in mechanical terms and should give up atomism as well in favor of a new worldview based entirely on relations among quantities of energy.

Energetics as a scientific project of the late 1880s and 1890s took place largely in Germany. (A prominent exception was the work of the French physicist Pierre Duhem.) Its main German proponents were Georg Helm, a Dresden mathematician and physicist, and Wilhelm Ostwald, the professor of physical chemistry at Leipzig. Helm first urged the formulation of a "general energetics" in his *Theory of Energy* (1887), which proposed an "energy principle" (a law more general than the law of energy conservation) as its basis. An essay in 1890 sought to reduce mechanics to energetics by means of this energy principle, and another in 1892 was intended to do the same for electricity and magnetism. In 1894 Helm wrote a book on the energetic development of physical chemistry. These publications elicited an invitation to address the German Association of Scientists and Physicians at their meeting in Lübeck in 1895 on "the current state of energetics."

Ostwald's interest in energy stemmed from his reading, in mid-1886, of Dutch chemist Jacobus van't Hoff's studies in chemical dynamics and from his own efforts, in the late 1880s, to understand the thermodynamic writings of Gibbs, which Ostwald published in German translation in 1892. He was soon converted to the way of "pure energetics," the theory of which he developed in two essays published in 1891 and 1892. He then refined his theory and applied it to a variety of problems in general and physical chemistry in 1893–1894. Always the enthusiast, Ostwald traveled to the 1895 meeting in Lübeck, where he was also in the program, to demonstrate the demise of the mechanical worldview and to promote energetics as its proper replacement.

The heated debate at Lübeck turned out to be a disaster for energetics. The negative reactions of Boltzmann and Max *Planck to the energeticists were taken as definitive by younger physicists such as Arnold Sommerfeld and Albert *Einstein. Helm and Ostwald later replied to these criticisms, only to be rebutted again by Boltzmann (1896–1898). Ostwald published his *History of Electrochemistry* in 1896; Ernst Mach likely hurried his (incomplete) *Theory of Heat* into print in the same year to support the anti-mechanist cause; and Helm, in his history of energetics of 1898, tried to defend his own work. But the damage had been done. Ostwald continued to uphold energetics after 1900, but increasingly as a monistic worldview, not as a scientific project.

The scientific proposals of the energeticists were flawed, but the attention they received undermines the common assertion that the physical scientists of the late nineteenth century were satisfied with the state of their science. The long tradition of mechanical explanation in the natural sciences was coming to an end. The debate over energetics as a viable replacement for the mechanical worldview reflected the difficulties inherent in the mechanical view.

Robert J. Deltete, "Gibbs and the Energeticists," in *No Truth Except in the Details: Essays in Honor of Martin J. Klein* (1995): 135–169. Robert J. Deltete, "Helm and Boltzmann: Energetics at the Lübeck Naturforscherversammlung," *Synthèse* 119 (1999): 45–68. Georg Helm, *The Historical Development of Energetics*, trans. and intro. Robert J. Deltete (2000).

ROBERT J. DELTETE

ENTROPY. Many physicists and chemists quip that the second law of thermodynamics has as many formulations as there are physicists and chemists. Perhaps the most intriguing expression of the law is Ludwig *Boltzmann's paraphrase of Willard *Gibbs: "The impossibility of an uncompensated decrease in entropy seems to be reduced to improbability."

Entropy owes its birth to a paradox first pointed out by William *Thomson in 1847: energy cannot be destroyed or created, yet heat energy loses its capacity to do work (for example, to raise a weight) when it is transferred from a warm body to a cold one. In 1852 he suggested that in processes like heat conduction energy is not lost but becomes "dissipated" or unavailable. Furthermore, the dissipation, according to Thomson, amounts to a general law of nature, expressing the "directionality" of natural processes. The Scottish engineer Macquorn Rankine and Rudolf Clausius proposed a new concept, which represented the same tendency of energy towards dissipation. Initially called "thermodynamic function" by Rankine and "disgregation" by Clausius, in 1865 the latter gave the concept its definitive name, "entropy," after the Greek word for transformation. Every process that takes place in an isolated system increases the system's entropy. Clausius thus formulated the first and second laws of thermodynamics in his statement "The energy of the universe is constant, its entropy tends to a maximum." Hence, all large-scale matter will eventually reach a uniform temperature, there will be no available energy to do work, and the universe will suffer a slow "heat death."

In 1871 James Clerk *Maxwell published a thought-experiment attempting to show that heat need not always flow from a warmer to a colder body. A microscopic agent ("Maxwell's demon," as Thomson latter dubbed it), controlling a diaphragm on a wall separating a hot and a cold gas, could choose to let through only molecules of the cold gas moving faster than the average speed of the molecules of the hot gas. In that way, heat would flow from the cold to the hot gas. This thought-experiment indicated that the "dissipation" of energy was not inherent in nature, but arose from human inability to control microscopic processes. The second law of thermodynamics has only statistical validity—in macroscopic regions entropy *almost* always increases.

Boltzmann attempted to resolve a serious problem pointed out by his colleague Joseph Loschmidt in 1876, and by Thomson two years earlier, that undermined the mechanical interpretation of thermodynamics and of the second law. This law suggests that an asymmetry in times dominates natural processes; the passage of time results in an irreversible change, the increase of entropy. However, if the laws of mechanics govern the constituents of thermodynamic systems, their evolution should be reversible, since the laws of mechanics are the same whether time flows forward or backward: Newton's laws retrodict the moon's position a thousand years ago as readily as they predict its position a thousand years from now. Prima facie, there seems to be no mechanical counterpart to the second law of thermodynamics. In 1877 Boltzmann found a way out of this difficulty by interpreting the second law in the sense of Maxwell's demon. According to Boltzmann's calculus, to each macroscopic state of a system correspond many microstates (particular distributions of energy among the molecules of the system) that Boltzmann considered to be equally probable. Accordingly, the probability of a macroscopic state was determined by the number of microstates corresponding to it. Boltzmann then identified the entropy of a system with a logarithmic function of the probability of its macroscopic state. On that interpretation, the second law asserted that thermodynamic systems have the tendency to evolve toward more probable states. A decrease of entropy was unlikely, but not impossible.

In 1906 Walther Nernst formulated his heat theorem, which stated that if a chemical change took place between pure crystalline solids at absolute zero, there would be no change in entropy. Its more general formulation is accepted as the third law of thermodynamics: the maximum work obtainable from a process can be calculated from the heat evolved at temperatures close to absolute zero. More commonly the third law states that it is impossible to cool a body to absolute zero by any finite process and that at absolute zero all bodies tend to have the same constant entropy, which could be arbitrarily set to zero.

S. G. Brush, *The Kind of Motion We Call Heat: A History of the Kinetic Theory of Gases in the 19th Century*, 2 vols. (1976). Penha Maria Cardoso Dias, "The Conceptual Import of Carnot's Theorem to the Discovery of Entropy," in *Archive for History of Exact Sciences* 49, ed. Penha Maria Cardoso Dias, Simone Pinheiro Pinto, and Deisemar Hollanda Cassiano (1995): 135–161. Robert Locqueneux, *Préhistoire et Histoire de la Thermodynamique Classique* (1996). Crosbie Smith, *The Science of Energy* (1998).

THEODORE ARABATZIS AND
KOSTAS GAVROGLU

**ERROR AND THE PERSONAL EQUA-
TION.** Since Greek times astronomers have recognized that observations were afflicted by errors, that results based on them might only be approximate, and that the quality of data varied. Astronomers in early modern Europe took the first steps toward giving reliable estimates of those errors. Johannes *Kepler, who used Tycho *Brahe's observations to derive the elliptical shape of planetary orbits, was probably the first to construct a correction term that assigned a magnitude to error, and among the first to give a theory of an instrument (the Galilean *telescope) for purposes of improving the accuracy of measurements taken with it.

During the eighteenth century steps were taken toward standardizing the analysis of measurements and understanding the conditions under which different sets of measurements could be combined. Analysts identified two types of errors: constant (affecting the instruments or the conditions of measurement) and accidental (randomly affecting the quality of the measurements themselves). Control over instrumental errors was achieved at first by codifying the behavior and demeanor of the observer, by taking into account the limitations of the human senses (especially vision), by examining how outside sources contaminate experiments, by perfecting the construction of

instruments, and by developing methods for instrument calibration.

The second type of error, the random, relates to classical probability theory. Initially the criteria for the selection of good measurements rested mainly on the notion that the median or the mean of measurements reduced the effect of errors in any of them. In 1756 the mathematician Thomas Simpson countered reports that a single well-taken measurement sufficed by demonstrating the superiority of the mean; his presentation to the Royal Society of London included a discussion of the equal probability of positive and negative errors and an argument that the mean lies closer to the true value than any random measurement. But no consensus existed about the selection or combination of measurements. The first firm parameters of an error theory emerged from the consideration of observations of the Moon's motion, especially its libration; from secular inequalities in the motions of Jupiter and Saturn; and from measurements of the shape of the earth. During the second half of the eighteenth century, Johann Tobias Mayer, Leonhard *Euler, Rudjer J. Boškovié, and Johann Heinrich Lambert developed ad hoc, limited, varied, but effective procedures for combining measurements made under different conditions. In 1774 Pierre-Simon *Laplace deduced a rule for the combination of measurements using probability theory.

The meridian measurements made during the French Revolution to determine the new standard of length, the meter, gave the occasion to devise the first general method for establishing an equilibrium among errors of observation by determining their "center of gravity." This method, the method of least squares, was so employed in 1805 by Adrien Marie Legendre. In 1806 Carl Friedrich *Gauss acknowledged Legendre's work but only to say that he had been using the method for years. A priority dispute ensued. Three years later Gauss published the first rigorous proof of the method of least squares; he demonstrated that if the mean is the most probable value, then the errors of measurement form a bell curve (Gaussian) distribution. The true value (which has the smallest error) lies at the center of the distribution, while the width of the curve determines the precision of the measurement. (Application of the method assumes the absence of constant or systematic errors.) From astronomy the method spread to chemistry, physics, mineralogy, and geodesy. It was also applied to practical projects including the reform of weights and measures, longitude determinations, triangulations, the U.S. Coastal Survey, and cadasters, where it set the boundaries within which dispute could take place. The method of least squares made large-scale projects, like Gauss's magnetic map of the earth, manageable by providing a means to combine and assess data from geographically dispersed locations. Over the course of the nineteenth century, the method dominated error analysis (especially in the German-speaking world) and shaped the development of probability theory.

The power of the method of least squares seemed to eradicate the subjective element in the treatment of measurements. But Gauss and his colleague the astronomer Wilhelm Olbers acknowledged in 1827 that sometimes measurements displayed large deviations from the mean. When was the deviation large enough to justify ignoring a measurement? Gauss could not provide an objective answer and recommended reliance on intuition. In 1852 Benjamin Peirce developed a rigorous method for rejecting outliers.

Deviations of another sort led Friedrich Wilhelm *Bessel to develop the personal equation. In 1823 he noted a constant difference in the measurements taken by the former British Astronomer Royal Nevil Maskelyne and his assistant, which Bessel referred to physiological differences. With the "personal equation" Bessel calculated the average difference between two observers; he could then combine measurements taken by several observers. The personal equation created the factory-like atmosphere of nineteenth-century astronomical observatories where teams of observers were calibrated according to its principles. It was seldom used elsewhere except in psychology. Especially in Wilhelm Wundt's physiological institute at Leipzig, the personal equation became the foundation of a research program in the determination of human reaction times.

The history of error and the personal equation embraces far more than the history of rules and methods. The determination of error is always an estimate; were the true error known, perfectly accurate results could be attained. Because reliable estimates of error generate confidence in results, the history of error theory also sheds light on how trust is established in a scientific community and beyond. In the teaching laboratory, the method of least squares indicated how well student investigators performed an experiment, and thus the level of expertise they had attained. In practical fields like surveying, the introduction of the method aided professionalization. Finally, the method of least squares shaped the moral economy of the sciences by promoting honesty in the execution and reduction of observations. Proper application of the method became a sign of the investigator's integrity.

Mansfield Merriman, *A List of Writings Relating to the Method of Least Squares, with Historical and Critical Notes* (1877). G. E. R. Lloyd, "Observational Error in Later Greek Science," in *Science and Speculation*, ed. Jonathan Barnes, et al. (1982). Stephen Stigler, *The History of Statistics* (1986). Giora Hon, "H. Hertz: 'The Electrostatic and Electromagnetic Properties of Cathode Rays Are Either Nil or Very Feeble,' (1883): A Case-Study of Experimental Error," *Studies in History and Philosophy of Science* 18 (1987): 367–382. Giora Hon, "On Kepler's Awareness of the Problem of Experimental Error," *Annals of Science 44* (1987): 545–591. Giora Hon, "Towards a Typology of Experimental Errors: An Epistemological View," *Studies in History and Philosophy of Science* 20 (1989): 469–504. Gerd Gigerenzer, et al., eds., *The Empire of Chance* (1990).

KATHRYN OLESKO

ETHER, a possibly nonexistent entity invoked from time to time to fill otherwise empty spaces in the world and in natural philosophy. Descended from the Aristotelian quintessence, which occupied the realms through which the planets wandered, and the Stoics' pneuma, which held the world together, ether characterizes theories opposed to atomism, which admits spaces void of matter. Both sorts of theories—plenary and atomistic—enjoyed vigorous revivals during the Scientific Revolution. With the invention of the *barometer and *air pump in the middle decades of the seventeenth century, void and ether became objects of experiment and, in practice, very much the same (no)thing.

The experimental investigation of void began above the mercury in the barometer tube. This space had the property of transmitting light and magnetic virtue, but not sound, and of allowing the free passage of bodies through it. It seemed infinitely compressible or, rather, was so subtle that it could pass right through glass. Were these the properties of a space void of all matter or of one filled with a substance different from ordinary matter? A third way, preferred by René *Descartes, made the special substance and ordinary matter the same thing, except for the size and shape of their constituent parts. Isaac *Newton countered with a solution as hard to grasp as ether itself. In his world system, the planets move through resistanceless spaces replete with the presence of God and perhaps also with springy ethers that mediated gravitational attraction, chemical behavior, *electricity, *magnetism, and the interaction of light and ordinary matter. These ethers, unlike the Stoics' pneuma and Descartes' plenum, admitted voids among their particles.

With the acceptance of the wave theory of light in the nineteenth century, physicists felt obliged to suppose the existence of a subtle, imponderable medium whose undulations constituted the disturbance perceived as light. The first mathematicians to attempt a detailed picture of this "luminiferous ether" modeled it as a mechanical substance with rigidity and inertia. They managed thus to represent most of the properties of light—reflection, refraction, interference, and polarization. As William *Thomson (Lord Kelvin) and the Cambridge mathematician George Gabriel Stokes explained it to other model-makers, the luminiferous ether had to combine the properties of shoemakers' wax, which allows slow bodies to pass through it under steady pressure but shatters when struck a sharp blow, with those of rigid steel, which can support transverse vibrations without suffering permanent distortion.

The ether soon became so familiar that mathematicians assimilated it to ordinary matter, or vice-versa. As latter-day Cartesians, they pictured atoms and molecules as permanent vortex rings in an all pervasive ether (Kelvin, following a hydrodynamical theory of his friend Hermann von *Helmholtz, and Joesph John *Thomson) or as knots or twists in it (Joseph Larmor). This sort of modeling was a specialty of physicists who had passed through the honors course in mathematics at Cambridge, England (the mathematical tripos). Continental physicists, especially the French, regarded it with a mixture of puzzlement and distaste. Nonetheless, James Clerk *Maxwell devised an ether model for the mediation of electrical and magnetic forces that suggested that light was an electromagnetic phenomenon.

Maxwell's theory charged the ether with accounting for electricity and magnetism as well as for light. None of the several models with mechanical properties proposed to effect it succeeded. Hendrik Antoon Lorentz and others then introduced space-filling media that had nonmechanical properties in order to underpin an adequate electrodynamics of bodies moving through the suppositious ether (*relativity). In 1905, Albert *Einstein showed the value of discarding the ether as a substrate and reference frame for electrodynamic phenomena.

The acceptance of relativity theory, however, did not destroy the ether. Einstein himself, in his application of relativity principles to the gravitational theory (1915), supposed that a gravitating body distorts nearby space, and that these distortions determine the trajectory of a passing ponderable body. An entity that can distort its shape, deflect light, and propagate electric and magnetic disturbances can be called a void only by discourtesy. More recently, *quantum electrodynamics has filled the void with a vacuum that undergoes energy fluctuations and acts as a theater for the creation and annihilation of virtual particles. One such fluctuation is said to have given rise to the present universe (*see* Cos-MOLOGY). Physicists appear to need an ether on which to load all the properties of the physical world they cannot otherwise explain. Ether, alias the vacuum, exists. Void is anything but nothing.

E. T. Whittaker, *A History of the Theories of Aether and Electricity*, 2 vols. (1951). E. J. Aiton, *The Vortex Theory of Planetary Motions* (1972).

J. L. HEILBRON

EULER, Leonhard (1707–1783), mathematician, specialist in mechanics. Originally from Basel, Switzerland, Euler was the most productive mathematician of all time. The measure is not the number of papers, for which the current record holder is the Hungarian Paul Erdös, but the number of published pages. Yet productivity was perhaps the least important of Euler's claims to mathematical distinction. One of his great contributions was his clarity, in contrast to French mathematicians of the time, who rarely expressed themselves so lucidly. The polishing that the savants of the previous century carried to extremes was almost wholly abandoned in the prolific eighteenth century.

Eighteenth-century mathematicians often did not cite their sources of insight. Euler appropriated without acknowledgment fundamental ideas of others gleaned from their letters as well as from their works. For example, he borrowed from Jean d'Alembert ideas in fluid dynamics, which d'Alembert had first introduced in his essay on the resistance of fluids, submitted to the Berlin Academy for its prize of 1749. D'Alembert did not win the contest (of which Euler was a judge), but vice did lead to virtue, as Euler expressed limpidly the ideas he pirated.

Euler appropriated freely from other French writers besides d'Alembert. He clarified and developed their ideas well beyond the points to which their originators had brought them. His textbooks incorporate his and their insights. Used for generations, many of these books became classics, for example, *Introductio ad analysin infinitorum* (1748), *Institutiones calculi differentialis* (1755), and *Institutiones calculi integralis* (1768–1770).

Euler's clarity was not the only important factor in establishing his widespread influence. He contributed to every branch of mathematics of his day except probability. He achieved much in the realm of number theory. He arguably founded graph theory and combinatorics when he solved the Königsberg Bridge problem in 1736. These four topics also were among Erdös's specialties. But in addition Euler contributed to ordinary and partial differential equations, the calculus of variations, and differential geometry, which Erdös did not touch.

Moreover, Euler made major contributions to every branch of *mechanics. The motion of mass points, *celestial mechanics, the mechanics of continuous media (mechanics of solids and nonviscous fluids, theories of materials, *hydrodynamics, hydraulics, elasticity theory, the motion of a vibrating string, and rigid-body kinematics and dynamics), ballistics, *acoustics, vibration theory, *optics, and ship theory all received something important from him.

If Beethoven did not need to hear to compose music, Euler did not need to see to create mathematics. He began to go blind in one eye in 1738 and became totally blind thirty years later. This only increased his productivity, since total blindness relieved him of academic chores like proofreading and eliminated unwanted visual distractions. Euler did not miss eyes for another reason; he had a prodigious memory. He could recite the *Iliad* by heart, and he reveled in the most difficult mental computations.

Euler began his academic career in 1727 as a member of the Academy of Sciences of Saint Petersburg. In 1734 he married Katharina Gsell, who also belonged to Saint Petersburg's Swiss colony. She bore him thirteen children, of whom only three sons and two daughters survived early childhood. The xenophobic Russian nobles resented the foreign members of the Academy, and the Russian Orthodox censors did not tolerate new sciences like Copernican astronomy. Furthermore, the Alsatian bureaucrat Johann Schumacher, who presided over the Academy, acted like a despotic rug merchant. Euler had to haggle with him to get his salary raised.

Despite these adversities, Euler cheerfully undertook projects of the Saint Petersburg Academy like Russian mapmaking, which involved painstaking work correcting land maps and reducing astronomical observations. This toilsome and stressful activity took him away from his mathematics and probably affected his sight and health.

In 1741 the new Prussian monarch Frederick II offered Euler a position in the new Berlin Academy. Euler accepted it because of political tensions and heightened hostility towards foreigners in Saint Petersburg. The move did not bring joy. Neither man pleased the other. Frederick was totally uninterested in Euler's mathematics. Euler resented the increasing French influence in the Berlin Academy, whose members, at Frederick's insistence, had to speak and write in French.

Euler came to feel that *belles lettres* had gained too much ground in Berlin at the expense of mathematics and that Frederick, who referred to him as his "one-eyed geometer" or "octopus," was becoming ever less supportive. In 1766 Euler returned to Saint Petersburg, where he remained until he died.

Euler set the trends in research in the mathematical sciences for most of the eighteenth century. The same sort of dominance is not possible for twentieth-century mathematics. The mathematics community of the eighteenth century was small; in the twentieth century it is huge, as is the number of specialized areas of mathematical research. Even Erdös, with his gigantic and pathbreaking output, has not had as much influence on the large-scale development of contemporary mathematics.

Clifford A. Truesdell, "Rational Fluid Mechanics, 1687–1765," in Leonhard Euler, *Opera omnia*, series 2, vol. XII (1954): ix–cxxv. Clifford A. Truesdell, "The Rational Mechanics of Flexible or Elastic Bodies, 1638–1788," in ibid, series 2, vol. XI (1960): 7–435. Thomas L. Hankins, *Jean D'Alembert: Science and the Enlightenment* (1970). Ronald Calinger, "Leonhard Euler: The First St. Petersburg Years (1727–1741)," *Historia Mathematica* 23, no. 2 (1996): 121–166.

JOHN L. GREENBERG

EXPERIMENTAL PHILOSOPHY. "The business of experimental philosophy," says the *Encyclopedia Britannica* in 1771, "is to inquire into, and to investigate the reasons and causes of, the various appearances and phenomena of nature; and to make the truth and probability thereof obvious and evident to the senses, by plain, undeniable, and adequate experiments." Experimental philosophy was thus a method, a practice, and a slogan. It signified to the eighteenth century the means and goal of the suppression of the philosophy of the schools begun in the seventeenth century and, more generally, of any entrenched "system" immune from the challenge of experience. The complacent natural philosophers of the Enlightenment ascribed the method to Francis Bacon, and the first examples of its successful practice to *Galileo and the Accademia del Cimento; next (according to the *philosophes*) the Royal Society of London demonstrated the superiority of the experimental over all other ways of philosophizing, with which the Académie royale des sciences of Paris concurred, once it had rid itself of the system of René *Descartes.

Isaac *Newton's *Opticks* (1704) demonstrated that, and how, the experimental philosopher could move from exact observations and careful manipulations to a new theory of wide application and great importance. Among the few who managed to achieve something similar to the *Opticks* during the eighteenth century were Benjamin *Franklin (*electricity), Henry *Cavendish (electricity), and Antoine-Laurent *Lavoisier (chemistry). Most other "experimental philosophers" repeated demonstrations or found new phenomena that they related to one or another of the dominant natural philosophies. Thus the influential Dutch professors Willem Jacob's Gravesande and Petrus van Musschenbroek, who gave the first sustained university courses in natural philosophy illustrated with experiments, generally followed Newton (*see* NEWTONIANISM); the abbé Jean-Antoine Nollet, who from the 1730s offered courses similar in purpose though wider in coverage outside the universities, generally followed Descartes.

By 1750 professors everywhere were introducing experiments into their courses, particularly in the many small German universities and in the larger Jesuit colleges in France and Italy, in order to help capture the interest (and where possible, the fees) of students. Throughout the second half of the eighteenth century, the German universities combated a shrinking enrollment and the Jesuits faced increasing competition from other teaching orders and a general animosity that led to their suppression in 1773. Meanwhile the niche cut out by Nollet and his predecessors in England—the curators of the Royal Society Francis Hauksbee and John Theophilus Desaguliers—widened to nourish all sorts of lecturers who entertained paying audiences with the instruments of science. For many of these men and most of their auditors, "experimental philosophy" was a misnomer. They showed, looked, and talked, but neither experimented nor philosophized.

Experimental philosophy, as opposed to philosophical entertainment, rested on two principles of unequal weight. The lesser admitted Bacon's criticism of hasty theorizing and his remedies. The greater held that the bigger and more precise the experimental apparatus, the better, even if the size and sophistication served no immediate philosophical purpose. Ever larger electrostatic generators, for example, culminated in the gigantic machine, which could throw a spark a quill thick for two feet or more, built for the Teyler Foundation in Haarlem in the Netherlands (1785). The reason, as given by the experimental philosopher who commissioned it, Martin van Marum: "I took it as certain that, if one could acquire a much greater electrical force than hitherto in use, it could lead to new discoveries." "Twist the lion's tail," Bacon had admonished, and wait for the results. No doubt the method had its successes, including, in electricity alone, the creation of the Leyden jar, the assemblage of the jars into an electric battery, the multiplication of the contact electricity of zinc and silver into the Voltaic pile (see GALVANI, LUIGI, AND ALESSANDRO VOLTA), and the connection of many piles into powerful engines of physical and chemical research.

Other examples of enlargement in the service of philosophy were the long, thin magnetic needles with well-defined poles used by Charles-Augustin Coulomb (see CAVENDISH, HENRY, AND CHARLES-AUGUSTIN COULOMB) to establish the law of magnetic force (1785), Lavoisier's apparatus for his experiments on gases, respiration, and heat (1770s–1780s), and Cavendish's balance for weighing the world in the laboratory (1798). Many significant instrumental improvements, however, did not require enlargement. The English instrument maker Edward Nairne advertised an air pump in 1777 that, he said, reduced the pressure to as little as 1/600 of normal atmospheric (at.) in only six minutes; it did not differ much in size from the standard machine of the time (which gave 1/165 at.) or from 'sGravesande's of the 1720s (at best 1/50 at.).

The most significant instrumental improvements for most experimental philosophers had to do with the microscope and the meteorological instruments. As the microscope became easier to use, Sunday philosophers applied it more readily to increase their wonder at the works of the Creator, especially after Abraham Trembley spied the marvelous self-regenerating freshwater polyp in a drop of dirty ditch water (1740). The measurement of the weather occupied many people wanting to do something useful for themselves and for science. They could measure to many more places of decimals in 1780 than in 1730. At the earlier date, observers did not correct their *barometers for temperature because the correction would have been less than the error of the instrument. After the main cause of error (unwanted air) had been removed, serious philosophers, such as the Genevan Jean André *Deluc, corrected for temperature, capillarity, and the curve of the meniscus, and claimed to be able to read their barometers to a few hundredths of a millimeter.

In 1730 the usually meticulous René-Antoine Ferchault de Réaumur rejected the suggestion that he use brass scales rather than paper ones on his *thermometers; that, he said, would be to carry precision to ludicrous lengths. As late as 1777 the Royal Society found wide variations in the boiling points of their thermometers. A committee chaired by Cavendish established procedures that removed the discrepancies. Meanwhile thermometers marked to a fifth or a tenth of a degree came on the market. Lavoisier's instruments, standardized by methods like those of the Cavendish committee and divided according to the latest technology, could be read reliably to perhaps a hundredth of a degree. The hygrometer, a device for determining humidity by the motion of a piece of animal matter (hair, gut, or ivory) attained similar perfection in the hands of Deluc.

The improvements paid off in several ways. Barometric measurement after suitable corrections could be used to measure heights of steeples and mountains to satisfactory accuracy. The improved thermometers made possible the detailed explorations of heat (see FIRE AND HEAT), especially of the laws of dilation of gases, made by the group of *physiciens* centered on Pierre-Simon *Laplace. And the hygrometer had the honor of helping Deluc to gain the odd post of natural philosopher to the Queen of England. The reasons for his preferment may make an apt symbol for much experimental philosophy in the Age of Reason. Deluc came to England recommended by his grand ideas about the history of the earth, which, as he labored to show, agreed perfectly with the account in Genesis, properly understood. He developed this project with the same keenness for observation and analysis he deployed in perfecting his instruments. Although they had little bearing on his grand philosophy, they propelled him to his royal roost. During negotiations he presented the King with a super-precise barometer for measuring heights during military campaigns and the Queen with the latest hygrometer for determining over-exactly the humidity of her greenhouses.

J. L. Heilbron, *Electricity in the Seventeenth and Eighteenth Centuries: A Study of Early Modern Physics* (1979, 1999). Steven Shapin and Simon Schaffer, *Leviathan and the Air Pump: Hobbes, Boyle and the Experimental Life* (1985). Tore Frängsmyr et al., eds., *The Quantifying Spirit in the Eighteenth Century* (1990). Jan Golinski, *Science as Public Culture: Chemistry and Enlightenment in Britain, 1760–1820* (1992). Thomas Hankins and Robert J. Silverman, *Instruments and the Imagination* (1995). Stephen Gaukroger, *Francis Bacon and the Transformation of Early-Modern Philosophy* (2001).

J. L. HEILBRON

EXTRATERRESTRIAL LIFE. Displaced from the center of Aristotelian cosmology, the earth became one of many planets in the Copernican worldview. Galileo Galilei's telescopic observations of earthlike mountains on our moon, and of moons circling Jupiter, emphasized this displacement. The principle of plenitude, which interpreted any unrealized potential in nature as a restriction of the Creator's power, argued for inhabitants on other worlds. Although there was no evidence of lunar inhabitants, why else, asked the English clergyman John Wilkins in his 1638 *The Discovery of a World in the Moon*, would Providence have furnished the moon with all the conveniences of habitation enjoyed by the earth?

Social critics seized on lunar inhabitants either as members of a perfect society or as exemplars of all earth's vices. This literary convention furnished some defense in attacks against the establishment and helped spread the idea of a plurality of worlds. So did persistent rumors that England intended to colonize the moon.

Life spread beyond the moon in the French astronomer Bernard de Fontenelle's *Entretiens sur la pluralitè des mondes* (1686). During their evening promenades, the conversation of a beautiful marquise and her tutor turned to astronomy. On the second evening they spoke of an inhabited moon, on the third of life on the planets, and by the fifth night they had progressed to the idea of fixed stars as other suns, giving light to their own worlds. More conventional astronomy textbooks repeated these views.

Belief in extraterrestrial life permeated much of eighteenth- and nineteenth-century thought. It allowed an easy attack on Christianity, whose teachings about Adam, Eve, and Christ might appear ridiculous if the earth were not the whole of the habitable creation. Similar concerns regarding the immensity of the universe and the corresponding insignificance of humans appeared in novels.

The gullibility of a public raised on pluralist writings is illustrated by the widespread acceptance of reports in the *New York Sun* in 1835, purportedly from Sir John *Herschel at the Cape of Good Hope, detailing his observations of winged quadrupeds on the moon. *The New York Times* judged these reports to be probable.

Many aspects of the extraterrestrial life debate appeared in Percival Lowell's Martian hypothesis and in reactions to it. Too modest to believe humankind the sole intelligence in the universe, Lowell, a wealthy Boston investor, announced in 1894 his intention to establish an observatory in the Arizona Territory and search for signs of intelligent life on Mars. There was already in America a lively interest in Mars, attributed by cynics to public imbecility and journalistic enterprise. Others hoped that the discovery of intelligent life elsewhere would increase reverence for the Creator. Lowell reported an amazing network of straight lines, which he interpreted as canals, and concluded that Mars was inhabited. Astronomers criticized Lowell for seeing only the evidence that supported his beliefs. Many readers, however, were persuaded by Lowell's literary skill. They also applauded the social arrangements of Mars, as elucidated by Lowell, particularly the abolition of war.

Changes in scientific knowledge in the twentieth century strengthened belief in extraterrestrial life. Larger telescopes expanded the observable universe to millions of galaxies, each containing millions of stars, all rendering it increasingly improbable that our earth alone shelters life. In 1953, Stanley Miller and Harold Urey at the University of Chicago synthesized amino acids, the building blocks of life, from a mixture of methane, ammonia, water, and hydrogen, the supposed ingredients of our primitive earth's atmosphere. Although the late astronomer Fred Hoyle attributed both the origin of life on earth and much of subsequence evolution to showers of microorganisms from space, his was a minority view, though recently resurrected. Most scientists suppose that life occurs inevitably on earthlike planets, of which they estimate that millions exist in the universe. *Star Trek*, the television series, visited some of these planets every week.

Skeptics object that, if intelligent life is inevitable and has had billions of years to evolve and travel through the universe, it should long since have reached our earth. That extraterrestrials are not known argues against their existence. Believers in UFOs (unidentified flying objects) attribute the absence of evidence of extraterrestrials to a government cover-up.

The chemical theory of the origin of life coincided with the space age and did not long remain earthbound. In 1976, in one of the greatest exploratory adventures of the twentieth century, the National Aeronautics and Space Administration (NASA) landed two *Viking* spacecraft on the surface of the planet Mars, at the cost of over one billion dollars. Experiments detected metabolic activity, but probably from chemical rather than biological processes.

Post-*Viking*, interest shifted from micro-organisms to direct communication with interstellar intelligence. Several early radio pioneers, including Guglielmo Marconi in 1920, thought they detected radio signals from Mars. The most comprehensive interstellar communication program was NASA's Search for Extraterrestrial Intelligence (SETI). From a small and inexpensive research and development project during the 1980s, SETI emerged in the early 1990s as a hundred-million-dollar program. A targeted search for radio signals focused on some thousand nearby stars, while a second element of SETI surveyed the entire sky. Ridiculed as "The Great Martian Chase," even after changing its name to "High Resolution Microwave Survey" in a vain attempt to highlight its potential for basic discoveries in astronomy, SETI lost its government funding in 1993. A scaled-back version of the original targeted search has been continued with private funding.

Would intercourse with extraterrestrials be beneficial? Suppose that they were a cancer of purposeless technological exploitation intent on enslaving us, rather than benign philosopher-kings willing to share their wisdom? Even if they should be helpful and benign, superior beings would be menacing. Anthropological studies of primitive societies confident of their place in the universe find them disintegrating upon contact with an advanced society pursuing different values and ways of life.

Steven J. Dick, *Plurality of Worlds: The Extraterrestrial Life Debate from Democritus to Kant* (1982). Michael J. Crowe, *The Extraterrestrial Life Debate 1750–1900: The Idea of a Plurality of Worlds from Kant to Lowell* (1986). Donald Goldsmith and To- bias Owen, *The Search for Life in the University*, 2d Ed. (1992). Walter S. Sullivan, *We Are Not Alone: The Continuing Search for Extraterrestrial Intelligence*, rev. ed. (1993). Steven J. Dick, *The Biological Universe: The Twentieth-Century Extraterrestrial Life Debate and the Limits of Science* (1996). NASA maintains a website on the history of the search for extraterrestrial intelligence, which also links to the official SETI site: http://www.history.nasa.gov/seti.html.

NORRISS S. HETHERINGTON

F

FARADAY, Michael (1791–1867), English natural philosopher and public man of science. Faraday was born on 22 September 1791 in Newington Butts, Surrey. His father was a blacksmith and a member of the Sandemanian Church, to which Faraday had a lifelong commitment. In many ways, Faraday's work can be viewed as his seeking the laws of nature that he believed God had written into the universe at the Creation.

Faraday attended a day school before being apprenticed as a bookbinder, 1805–1812. During his apprenticeship he developed an overwhelming interest in science, which he cultivated by attending various scientific lectures including those by Humphry Davy in the Royal Institution. Faraday was appointed chemical assistant there in 1813, and within a few months he was accompanying Davy on a tour of the Continent. On their return to England in 1815, Faraday was reappointed in the Royal Institution, where he rose to be superintendent of the house (1821), director of the laboratory (1825), and Fullerian professor of chemistry (1833).

Following Hans Christian Ørsted's announcement of electromagnetism, Faraday discovered electromagnetic rotations (1821), the principle behind the electric motor. Ten years later he discovered electromagnetic induction and commenced a remarkable decade of work in which, among other things, he rewrote the theory of electrochemistry (coining in the process words such as electrode, anode, cathode, and ion and establishing his laws of electrolysis). He built the Faraday cage (1836), which showed that measurements of electric charge depended on the electrical state of the observer. This observation led Faraday to develop his theory that electricity was the result of induction between contiguous particles rather than the action of distance forces. In the 1840s he extended his skepticism of scientific theories by arguing against the existence both of chemical atoms and the luminiferous *ether. His skepticism about these theories was strengthened by his discovery of a magneto-optical effect and diamagnetism in 1845 (*see* MAGNETO-OPTICS), and culminated in his establishment of the *field theory of *electromagnetism, which, when mathematized by William *Thomson (Lord Kelvin) and James Clerk *Maxwell, became one of the cornerstones of physics.

Faraday also made a number of purely chemical discoveries: In the 1820s he identified several new carbon chloride compounds, he liquefied gases, and he detected what became known as benzene. In the 1850s he carried out an extensive investigation of colloidal suspensions by passing light through them.

Faraday not only undertook research but was frequently invited to provide practical scientific advice. He helped Davy with the miners' safety lamp in 1815, in the 1820s with the electrochemical protection of the copper bottoms of ships, and, though unsuccessfully, with the improvement of optical glass. In his own right he was scientific adviser to Trinity House (the English and Welsh lighthouse authority) from 1836 to 1865 and oversaw the program to electrify lighthouses. He gave the government scientific advice on wide-ranging problems, including the conservation of pictures, the prevention of explosions in coal mines, the installation of lightning conductors, and the best way of attacking Kronstadt. For twenty years he taught chemistry to the cadets of the Royal Artillery and Royal Engineers at the Royal Military Academy in Woolwich.

Faraday was one of the most popular scientific lecturers of his day. In the mid-1820s he established the Christmas Lectures for children and the Friday Evening Discourses for members of the Royal Institution; both series continue to this day. He used his lectures to criticize passing fashions, for instance, of table-turning in the early 1850s, which offended both his religious beliefs and his sense of scientific propriety.

With his success in scientific research, his value as a government adviser, and his popularity as a lecturer, Faraday became one of the most famous men of the period. He was painted, sculpted, and photographed by leading artists, and was the friend of the scientifically minded Prince Albert. At Albert's suggestion, Queen Victoria gave Faraday in 1858 a "Grace and Favour House" at Hampton Court. Faraday spent an increasing amount of time there. He is buried in the Sandemanian plot in Highgate Cemetery.

David Gooding and Frank A. J. L. James, *Faraday Rediscovered: Essays on the Life and Work of Michael Faraday, 1791–1867* (1985). David Gooding, *Experiment and the Making of Meaning: Human Agency in Scientific Observation and Experiment* (1990). Geoffrey Cantor, David Gooding, and Frank A. J. L. James, *Faraday* (1991). Geoffrey Cantor, *Michael Faraday: Sandemanian and Scientist. A Study of Science and Religion in the Nineteenth Century* (1991). Frank A. J. L. James, *The Correspondence of Michael Faraday*, 4 vols. to date (1991–).

FRANK A. J. L. JAMES

FERMI, Enrico (1901–1954), nuclear physicist, designer of the first nuclear reactor.

Fermi had the widest scope of all the founders of quantum physics. As a theorist, he contributed decisively to quantum mechanics (Fermi–Dirac statistics) and nuclear physics (theory of beta decay). As an experimentalist, he introduced the technique of neutron bombardment to study artificial radioactivity, opening the way to the discovery of nuclear fission. He established a famous school of nuclear physics in Rome, but left fascist Italy because of anti-Jewish legislation. He settled in the United States, where he contributed to the atomic bomb program and served as an influential scientific advisor in postwar American nuclear policy. His name is honored in the unit of length for nuclear dimensions (the fermi, 10^{-13} cm), in the transuranic element of atomic number 100 (fermium), in a class of elementary particles (fermions), and in one of the most important particle physics laboratories in the world (Fermilab, near Chicago).

Born in Rome on 28 November 1901, Fermi studied physics at the University of Pisa, as a fellow of the prestigious Scuola Normale Superiore. Soon after his graduation in 1922, the influential director of the Physics Institute of Rome University, Orso Mario Corbino, a former Minister of National Education and eventually (1923–1924) Minister of National Economy, realized his promise and promoted his career. Corbino sent Fermi to pursue his studies in Göttingen (1923) and Leyden (1924), and in 1926 obtained for him a chair of theoretical physics (the first in Italy) in the institute in Rome. That year Fermi published a seminal paper on the quantization of the monatomic ideal gas, proposing a new quantum statistics for particles with half-integral spin (fermions).

With Corbino's support, Fermi set up a brilliant research group in nuclear physics that included Edoardo Amaldi, Franco Rasetti, Emilio Segrè, Oscar D'Agostino, Bruno Pontecorvo, and Ettore Majorana. The group followed up the then-new phenomenon of artificial *radioactivity by means of neutron bombardment of the elements of the *periodic table. Between March and July 1934, they "discovered" (that is, made and detected) about fifty new radionuclides. A few of these, which they misinterpreted as transuranic elements, turned out to be fission products of uranium, as Otto Hahn and Fritz Strassmann discovered four years later. Also in 1934, Fermi found that neutrons slowed down in passing through light elements and that, when suitably retarded, they became extremely effective in provoking nuclear transmutations. No less important was Fermi's theoretical analysis of nuclear beta decay (1934), which invigorated the study of weak interactions.

The Fermi group itself soon decayed. Most members left Rome and Corbino died in January 1937, thereby depriving Fermi of important institutional support. The fascist racial legislation of 1938 hit Fermi's Jewish wife, Laura Capon, and political boycott added to scientific frustration. Fermi decided to leave Italy. After he collected his Nobel Prize in Stockholm in December 1938 for his neutron work, he and his family sailed to New York. He became a U.S. citizen in 1944.

Initially established at Columbia University, Fermi moved to Chicago in 1942 to work in the Manhattan Project. He led the construction of the first nuclear reactor, which went critical on 2 December, demonstrating the chain reaction and the feasibility of producing plutonium for an atomic bomb. He then collaborated with the Los Alamos teams involved in the construction of the bomb and attended the Trinity Test in Alamogordo, New Mexico, on 16 July 1945. After the war he served on the Atomic Energy Commission's General Advisory Committee, contributing to the definition of American nuclear strategy and research policy. He also spent time in Los Alamos in 1950 to work on the H-bomb program launched by President Harry Truman that January. But Fermi's main base was the University of Chicago, where he inaugurated an important research program in particle physics centered on a new 450 MeV synchrocyclotron. His team studied pion-nucleon interactions (1952–1953), confirming the conservation law of isotopic spin in strong interactions and observing the first pion-nucleon resonance.

In 1952 Fermi was elected president of the American Physical Society. He had to cope with the drama unleashed in the American scientific community by the anticommunist campaign of Senator Joseph McCarthy and the ensuing trial of J. Robert Oppenheimer before the Atomic Energy Commission's Security Board (*see* KURCHATOV AND OPPENHEIMER. He testified in April 1954 in support of the former scientific leader of Los Alamos. In the summer of 1954, Fermi fell ill with stomach cancer. After useless surgery, he died in Chicago on 29 November.

Laura Fermi, *Atoms in the Family* (1954). Emilio Segrè, *Enrico Fermi, Physicist* (1970). Edoardo Amaldi, "Personal Notes on Neutron Work in Rome in the '30s and Post-War European Collaboration in High Energy Physics," in *History of Twentieth Century Physics*, ed. Charles Weiner (1977): 293–351. Gerald Holton, "Fermi's Group and the Recapture of Italy's Place in Physics," in Holton, *The Scientific Imagination. Case Studies* (1978): 155–198. Arturo Russo, "Science and Industry in Italy between the Two World Wars," *Historical Studies in the Physical and Biological Sciences 16* (1986): 281–320.

ARTURO RUSSO

FEYNMAN, Richard (1918–1988). One of the greatest and most original physicists of the twentieth century, Feynman was born in Far Rockaway, New York. His father, a Russian emigré, grew up in Patchogue, Long Island; Feynman's mother came from a well-to-do family, and had attended the Ethical Culture School in New York. After public schools in Far Rockaway, where Feynman had excellent teachers in chemistry and mathematics, he entered the Massachusetts Institute of Technology (MIT) in 1935. Admitted to Princeton University as a graduate student in physics in 1939, he became assistant to the newly arrived John Wheeler, then twenty-seven and full of bold and original ideas. In the spring of 1942 Feynman obtained his Ph.D. and immediately started working on problems related to the development of an atomic bomb. In 1943 he was one of the first physicists to go to Los Alamos. Hans Bethe, the head of the theoretical division, and J. Robert Oppenheimer, the director of the laboratory, quickly recognized Feynman as one of the most valuable members of the theoretical division—versatile, imaginative, ingenious, and energetic. In 1944 he was put in charge of the computations for the theoretical division. He introduced punch card computers to Los Alamos and began a lifelong interest in computing. In 1945 Feynman joined the physics department at Cornell University as an assistant professor. He left in 1951 for the California Institute of Technology, where he remained until his death.

Feynman's first major contribution, the content of his doctoral dissertation and of his article in the *Reviews of Modern Physics* for 1948, was the path-integral formulation of nonrelativistic quantum mechanics, which helped clarify the assumptions that underlay the usual quantum mechanical description of microscopic entities. In Feynman's approach, a particle going from the spatial point x_1 at time t_1 to the spatial point x_2 at time t_2 can take any path, each of which has a certain probability amplitude. The dynamic that results is the outcome of summing over all paths with their respective probability amplitude—a formulation that may well be

Feynman lecturing to a class at Caltech in 1963. The blackboard to the right presents quantum mechanics in P.A.M. *Dirac's "bra" and "ket" formulation.

Feynman's most profound and enduring contribution. It has deepened understanding of quantum mechanics and significantly extended the range of systems that can be quantized. Feynman's path integral has also enriched mathematics and provided new insights into spaces of infinite dimensions.

Feynman was awarded the Nobel Prize for physics in 1965 for his work on *quantum electrodynamics (QED). In 1948, simultaneously with Julian Schwinger and Shin'ichirō *Tomonaga, he showed that the infinite results that plagued QED could be removed by a redefinition of the parameters that describe the mass and charge of the electron, a process called "renormalization." Schwinger and Tomonaga had built on the existing formulation of the theory. Feynman invented a completely new diagrammatic approach that allowed the visualization of space-time processes, clarified concepts, and simplified calculations. Using Feynman's methods it became possible to compute QED processes to amazing precision. The magnetic moments of the electron and muon have been calculated to an accuracy of one part in 10^{12} and found to be in agreement with experiment.

In 1953 Feynman developed a quantum mechanical explanation of liquid helium that justified the earlier phenomenological theories of Lev Landau and Laslo Tisza. Because a helium atom has zero total spin angular momentum, the wave function (the quantum-mechanical formulation) of a large collection of helium atoms is unchanged under the exchange of any two of them. When in this state the system behaves as one unit. Hence helium near absolute zero acts as if it had no viscosity.

In the late 1960s experiments at the Stanford Linear Accelerator indicated an unexpectedly large probability that high-energy electrons underwent large angle scattering in striking protons. Feynman found that he could explain the data by assuming that the proton consisted of small, pointlike entities. He called these subnuclear entities "partons," soon identified with the *quarks of Murray Gell-Mann and George Zweig. Feynman devoted much of his research during the 1980s to studying quarks and their interactions and explaining their confinement inside nucleons and mesons.

Feynman demonstrated his uncanny ability to get to the heart of a problem— whether in physics, applied physics, mathematics, or biology—on prime-time television when he served on the presidential commission that investigated the crash of the *Challenger* space shuttle. He pinpointed the central problem by dropping a rubber O-ring into a glass of ice water to illustrate the cause of the rocket's failure.

In his physics Feynman always stayed close to experiments and showed little interest in theories that could not be tested experimentally. He imparted these views to undergraduate students in his justly famous *Feynman Lectures on Physics* and to graduate students through the widely disseminated notes of his graduate courses. His writings on physics for the interested general public, *The Character of Physical Laws* and *QED*, convey the same message.

Richard Feynman, with R. B. Leighton and M. Sands, *The Feynman Lectures on Physics*, 3 vols. (1963). Richard Feynman, *The Character of Physical Laws* (1965). Richard Feynman, *QED: The Strange Theory of Light and Matter* (1985). James Gleick, *Genius. The Life and Science of Richard Feynman* (1992). Richard Feynman, *The Meaning of It All: Thoughts of a Citizen Scientist* (1998).

SILVAN S. SCHWEBER

FIELD. The field, one of the most important concepts in modern physics, denotes the manner in which magnetic, electrical, and gravitational forces act through space. The field concept alleviated the difficulties many scientists found in assuming the existence of *forces acting at a distance without the intervention of some material entity. Fields thus serve many of the functions of *ether theories, which received their fullest development during the first half of the nineteenth century following the establishment of the wave theory of light by Augustin-Jean *Fresnel.

Michael *Faraday introduced the term "field" into natural philosophy on 7 November 1845, following his discovery of a magneto-optical effect (*see* MAGNETO-OPTICS)

and diamagnetism. He used the term operationally, in analogy to a field of stars seen through a telescope. During the next decade he developed the concept into a powerful explanatory framework for electromagnetic phenomena. This embodied much of his earlier thinking about the nature of magnetic action, especially his use of curved lines of force, which he had employed since the early 1830s to account for phenomena such as electromagnetic induction. It also embodied his opposition to conceptions such as atoms and the ether. As he wrote in 1846, he sought "to dismiss the aether, but not the vibrations."

Initially Faraday's contemporaries ignored his field concept, since it did not have the mathematical precision of action-at-a-distance theories such as André-Marie Ampère's electrodynamics. William *Thomson (Lord Kelvin) reacted with contempt to Faraday's "way of speaking of the phenomena." However, Faraday's field theory had the great merit that in treating electrical events it took account not only of the wire carrying the electric current but also of the insulation of the wire, the surrounding medium, and so on. The notion of field made a good basis for developing a theory of long-distance telegraph signaling, which became a pressing problem in the mid-1850s with the intended construction of the transatlantic cable. Action-at-a-distance theories could not cope with this problem; Thomson, using Faraday's field concept, solved it.

This practical success prompted the adoption of the field concept in Britain. In the hands of Thomson and of James Clerk *Maxwell it became a mathematical theory, much to the bemusement of the nonmathematical Faraday. However, unlike Faraday, who wished to abolish the ether from natural philosophy, both Thomson and Maxwell sought to interpret the field in terms of elaborate mechanical models (involving ethereal vortices) in an endeavor to retain the ether (*see* STANDARD MODEL). This project ultimately failed. Relativity theorists such as Hendrik Antoon Lorentz and Albert *Einstein replaced the ether by the nonmechanical field, a space capable of propagating forces, which is a cornerstone of modern physics.

William Berkson, *Fields of Force: The Development of a World View from Faraday to Einstein* (1974). David Gooding, "Faraday, Thomson, and the Concept of the Magnetic Field," *British Journal for the History of Science* 13 (1980): 91–120. Nancy J. Nersessian, *Faraday to Einstein: Constructing Meaning in Scientific Theories* (1984). Bruce J. Hunt, "Michael Faraday, Cable Telegraphy and the Rise of Field Theory," *History of Technology* 13 (1991): 1–19.

FRANK A. J. L. JAMES

FIRE AND HEAT. The hidden natures of those most obvious of physical phenomena, fire and heat, puzzled natural philosophers until the middle of the nineteenth century. Whether the agent of heat, which could work without glowing, differed from that of fire, to which the ancients had accorded elemental status, became an insistent question toward the end of the eighteenth century.

When seventeenth-century chemists eliminated fire from their list of the five simple principles they believed to constitute all matter, they retained the idea that fire was the most powerful agent available for altering or reducing matter. Their refined distillation procedures enabled them to apply fire with subtle control. Although a few chemists and natural philosophers, including Francis Bacon, regarded fire and heat as an expression of motion of particles of ordinary matter, most believed it to be a special substance. In Newtonian philosophy, this substance was thought to consist of particles that repelled one another but were attracted by ordinary matter. Consequently they spread throughout bodies, reaching an equilibrium in which all bodies in contact with one another contained the same degree of heat as determined by a *thermometer.

Increasingly precise thermometers enabled philosophers to measure even what they did not understand. The mixing of two samples of water at different initial temperatures and the subsequent measurement of the temperature of the combination addressed a question of longstanding interest to physicians: If two items possess the same quality in different degrees, what is the degree of the quality when they are mixed? According to measurements done in 1747–1748 by Georg Wilhelm Richmann, the Saint Petersburg academician later killed attempting Benjamin *Franklin's experiment with *lightning, the mixture's temperature $T = (m_1 T_1 + m_2 T_2)/(m_1 + m_2)$, where the subscripts indicate the temperatures and amounts of the two samples before mixing. The formula intimates the conservation of heat and measures it by the product of mass and temperature.

Daniel Gabriel Fahrenheit gave the first inkling that the mixing business was more interesting than Richmann's formula allowed. Fahrenheit observed that supercooled water when shaken converted to ice while its temperature rose to 32 degrees on his peculiar scale; and also that mercury had a smaller effect in the mixing experiments than an equal weight of water. Joseph Black followed up these observations in the 1750s and concluded that water at the freezing point must contain heat concealed from the thermometer. He confirmed this insight by measuring the time required to melt snow, which he took to be a measure of the amount of heat hidden or latent in ice-cold water. As for the different effectiveness of mercury and water, Black ascribed it to a difference in their capacity for heat. He measured the "specific heat capacity" of a substance by comparing the times required to heat the same weights of water and of the substance through the same interval of temperature (the shorter the time, the smaller the capacity). Black did not publish these measurements, most of which he made in a brewery to secure a steady warmth, nor did he declare his opinion about the true nature of heat.

Johan Carl Wilcke, professor of natural philosophy at the Royal Swedish Academy of Sciences, arrived in 1772 at the concept of latent heat by observing that hot water melted less snow than Richmann's formula called for. Ten years later Wilcke also conceived of specific heat, perhaps independently of Black, whose views had been circulated by his students and colleagues. Wilcke used Richmann's formula to calculate specific heats: if the mixture of a weight W of (say) mercury at temperature T with an equal weight of ice-cold water produces a final temperature Q, and w is the amount of water at temperature T that, if mixed with ice-cold water, also results in the temperature Q, then $Q = (wT + Q \times 0)/(w + W)$, where 0 represents freezing on the temperature scale of Wilcke's countryman Anders Celsius. From the last expression, w/W = specific heat of mercury = $Q/(T-Q)$. Unlike Black, Wilcke did not keep his concept of heat latent. In several publications, he discussed heat as an elastic fluid just like electricity (see IMPONDERABLE).

Natural philosophers quickly accepted the concepts of specific and latent heats, and chemists introduced a new species of combination to explain the presence of bound heat that did not act on a thermometer. Antoine-Laurent *Lavoisier drew on these ideas to define a "matter of fire" that was released in combustion and calcination and gave rise to the liquid and gaseous states. Lavoisier and the mathematician Pierre-Simon *Laplace devised a method in the early 1780s to measure the quantities of fire matter released in physical and chemical processes by the quantity of ice melted when the process took place in a closed environment maintained at the temperature of melting ice. Later they named this apparatus a "calorimeter" and renamed the matter of fire and heat "caloric." The new nomenclature only lightly veiled the descent of their more rigorously defined principle from earlier ideas about the nature of fire.

The question now centered on the relation between the caloric material and the phenomena of fire, light, heat, and (after 1800) radiant heat. Even after Benjamin Thompson, Count Rumford's famous demonstrations in the late 1790s of the improbably huge quantity of heat procurable from a gun barrel by grinding it, most chemists and natural philosophers still believed heat to be material and assimilated it to caloric. Perhaps fire was caloric plus particles of light? With the replacement of the material by the wave theory of *light (see OPTICS and VISION) in the 1820s and 1830s, and the acceptance of the kinetic theory of gases in the 1840s and 1850s (see THERMODYNAMICS), the question resolved itself. Chemists and physicists declared that the matters of fire, heat, and light were all the same in the sense that none of them existed.

Douglas McKie and N. H. De V. Heathcote, *The Discovery of Specific and Latent Heats* (1935). Antoine-Laurent Lavoisier, *Elements of Chemistry* (1965; French orig. 1789). Robert Fox, *The Caloric Theory of Gases: From Lavoisier to Regnault* (1971). J. L. Heilbron, *Weighing Imponderables* (1993).

J. L. HEILBRON

FRANKLIN, Benjamin (1706–1790), natural philosopher, diplomat, and inventor of the lightning rod.

A key figure in the struggle of Britain's American colonists for independence, Franklin served as colonial agent in London (1757–1762, 1765–1775), as a member of the Second Continental Congress, and as an author of the Declaration of Independence. Subsequently American plenipotentiary in Paris, he helped to negotiate the 1783 peace treaty with Britain and, later, to draw up the American constitution.

Franklin's reputation as a natural philosopher preceded his arrival in Europe as a politician. In the 1740s he developed a revolutionary new theory of *electricity that was eventually adopted almost universally. He also devised an experiment to prove the electrical nature of *lightning. Performed successfully in France in 1752, the demonstration caused a sensation, for it offered the prospect of human reason controlling nature's power, of conducting rods projecting above the roofline protecting buildings from lightning strikes.

Born in Boston, Massachusetts, to a tallow-chandler, at the age of 12 Franklin apprenticed with one of his older brothers, a printer. He had little formal schooling but read voraciously and also taught himself to write, modeling his style on Joseph Addison's. In 1723, he ran away to Philadelphia, where in time he built a successful printing business and became a public figure. He published the *Pennsylvania Gazette* and the enormously popular *Poor Richard's Almanack*. He also served as postmaster of Philadelphia from 1737 to 1753. Filled with ingenious ideas, he invented, among other things, the rocking chair, bifocal glasses, and the Pennsylvania fireplace. By 1748, he could retire from business and devote himself to public affairs.

By the mid-1740s, when he took up the study of electricity, Franklin was well-read in experimental natural philosophy, familiar with Robert *Boyle's work, Isaac *Newton's *Opticks*, and the writings of Herman Boerhaave, Willem Jacob 'sGravesande, John Theophilus Desaguliers, and Stephen Hales, among others. Like most of his contemporaries, he attributed electrical effects to the action of a subtle fluid that supposedly pervaded all bodies. However, while others believed that electrification consisted in agitation of this fluid, Franklin saw it as a redistribution. Moreover, for him, the fluid was specific to electricity. When a person rubbed a substance like glass, he said, there was an exchange of fluid between the rubber and the body rubbed, leaving one with a surplus, the other with a deficiency. Hence two kinds of electrical "charge" existed, "plus" and "minus," that could in appropriate circumstances neutralize each other. Later, when Franklin became familiar with Charles-François Dufay's work, he identified these with the two different modes of electrification, "vitreous" and "resinous," that Dufay had discovered some years earlier.

Franklin's way of viewing things enabled him to give a coherent account not only of Dufay's distinction but of a range of electrical phenomena that existing theories had not handled well—phenomena that, in Franklin's terms, involved charging or discharging of bodies. Above all, he managed to explain the most famous electrical phenomenon of the age, the Leyden jar experiment, something existing theories could not do. His observation that pointed conductors drew charge off bodies better than rounded ones did was the key to his idea of the lightning rod. These successes won his ideas many adherents. Other phenomena Franklin dealt with less successfully. Initially, he paid little attention to the traditional starting point of electrical inquiry, the power of rubbed bodies to attract light objects. Later, he suggested that when a body became charged, the additional electric fluid formed an atmosphere around it, and that contact between this atmosphere and other bodies brought about the attraction. However, his discussion of how it did so, and generally of the role such atmospheres might play in electrical attractions and repulsions, left many questions unanswered. In 1759, Franz Ulrich Theodosius *Aepinus brought new clarity to the subject by eliminating the atmospheres and converting Franklin's theory into one based on unexplained forces acting at a distance between particles of electric fluid and particles of ordinary matter.

Although Franklin continued to take an interest in science and to promote the erection of lightning rods on public buildings, his major contributions to electrical understanding were over by 1755. By then, however, he had established a reputation as a natural philosopher that opened doors for his diplomacy in both London and Paris, and enhanced the standing of the American Philosophical Society, which he helped to found and over which in later life he long presided.

Carl Van Doren, *Benjamin Franklin* (1938). I. Bernard Cohen, *Franklin and Newton* (1956). J. L. Heilbron, *Electricity in the 17th and 18th Centuries* (1979). R. W. Home, *Aepinus's Essay on the Theory of Electricity and Magnetism* (1979). I. Bernard Cohen, *Benjamin Franklin's Science* (1990).

R. W. HOME

G

GAIA HYPOTHESIS. Independent scientist James Lovelock published his first book on the Gaia hypothesis in 1979. He gives cell biologist Lynn Margulis credit for copartnering it, and has traced roots of the idea to James *Hutton, Russian geochemist Vladímir Vernadsky, and Swedish chemist Lars Gunnar Sillen. The Gaia hypothesis states that the earth as a planet is a living and evolving organism, maintaining physical and chemical conditions suitable for life. The hypothesis can also be simply stated, "The Earth is homeostatic." Lovelock refers to the science of Gaia as "geophysiology." The Gaia hypothesis (which Lovelock now calls the Gaia theory) should not be confused with the anthropic principle.

Lovelock formulated the Gaia hypothesis by asking, How can we determine from a distance whether or not a planet has life? In comparing the earth to its neighbors, he concluded that it differs in that its atmosphere contains unstable compounds (e.g., O_2, methane) that are present because living organisms modulate the atmosphere. Oxygen content itself is not a prerequisite for life; in fact, oxygen is toxic to many important life forms such as some bacteria.

Before Lovelock's hypothesis, geoscientists had already recognized that the earth has maintained a temperate surface for more than three billion years. They had also described complex interactions among the physical, chemical, and biological systems on Earth that sustain this temperature. For example, Earth has not suffered a runaway greenhouse effect, as has Venus, in part because organisms remove carbon dioxide from the atmosphere, and organic carbon is subsequently stored in the earth's crust. Earth systems have a set of checks and balances. Thus, if more carbon dioxide is emitted from volcanoes, the climate becomes warmer, which in turn favors more weathering and photosynthesis, both of which consume carbon dioxide, driving temperatures back down.

Lovelock named his hypothesis after the Greek goddess of the earth even though he intended the hypothesis to be scientific. Partly because his early treatment included poetic and metaphorical language, some people responded to its possible religious or philosophical meanings, while many scientists were skeptical or dismissive. Now, however, the Gaian concept has entered the mainstream scientific literature. The convening of a Chapman Conference (a focused research symposium) by the American Geophysical Union in 1988 and publication of the proceedings, entitled *Scientists on Gaia*, in 1991 were landmarks in this process.

The Gaia hypothesis continues to cause debate. Is it more than a description of what is and has been? Is it more than metaphorical? Is it testable or falsifiable? Or, as a theory, like *plate tectonics or evolution, can Gaia be accepted by scientists as paradigmatic? Moreover, because it refers to the earth as a living organism, the Gaia hypothesis leads to discussion of the grand question, What is life?

James Lovelock, *The Ages of Gaia. A Biography of Our Living Earth* (1988). Tyler Volk, *Gaia's Body: Toward a Physiology of the Earth* (1998).

JOANNE BOURGEOIS

GALAXY. In 1918 the American astronomer Harlow Shapley argued convincingly that our galaxy is a hundred times larger than previously believed and that the sun is not at its center. Shapley also concluded, erroneously, that the galactic universe made a single, enormous, all-comprehending unit.

A fifteen-year-old crime reporter in a tough Kansas oil town, Shapley discovered a Carnegie library and began to read, and in 1907 he entered the University of Missouri. He earned bachelor's and master's degrees in astronomy at Missouri, and in 1913 a Ph.D. at Princeton University. From there he went to the Mount Wilson Observatory. Using its 60-inch telescope, then the best in the world, Shapley conducted his revolutionary research on the scale of the galaxy. With his quick mind and a complete absence of humility, Shapley followed his own results while ignoring others.

The key to measuring distances was the period-luminosity relation for Cepheid-type variable stars, so named after the constellation Cephus, in which a typical Cepheid star is located. Cepheids are giant stars visible for great distances. The relationship between their periods and luminosities, noted by the American astronomer Henrietta Leavitt in 1908, states that the longer the period, or time, from maximum brightness to minimum and back to maximum, the greater the luminosity of the star. (The observed, or apparent, brightness of a star decreases by the square of its distance from the observer. The absolute magnitude, a measure of brightness, is defined as the magnitude the star would have at a standard distance, approximately 32 light-years.) Shapley collected all the available data on Cepheid stars whose distances and magnitudes he knew, and plotted period against absolute magnitude. Next, he made the reasonable assumption that Cepheids in distant globular clusters resemble the nearby Cepheids upon which he based his period-magnitude curve. Globular clusters, 150 of which exist in our galaxy, are densely packed balls of hundreds of thousands of stars. Shapley observed the periods of Cepheids in globular clusters, read off their presumed absolute magnitudes from his graph of period against magnitude, and compared that absolute magnitude with the observed apparent magnitude. This calculation gave him the distance to the Cepheid variable star and to the globular cluster in which it resided.

Making the assumption that globular clusters constitute a galactic skeleton, Shapley determined the galactic outline, its size, and the place of the solar system within it. He put the Sun far toward one edge of the galactic plane, and made the size of the galaxy far larger than any previous estimate. He concluded that "globular clusters, though extensive and massive structures, are but subordinate items in the immensely greater organization which is dimly outlined by their positions. From the new point of view our galactic universe appears as a single, enormous, all-comprehending unit, the extent and form of which seem to be indicated through the dimensions of the widely extended assemblage of globular clusters." Shapley's big galaxy model argued against the existence of other galaxies. Even the most remote globular cluster should be inside our galaxy. Spiral-shaped *nebulae (concentrations of stars and dust) might lie outside our newly enlarged galaxy, but if so their size would be implausibly large. Shapley made this point in the so-called "Great Debate" before the National Academy of Sciences on 26 April 1920.

In 1921 Shapley left Mount Wilson for the Harvard College Observatory, leaving Edwin *Hubble with the new 100-inch telescope, completed in 1918, to hunt for Cepheids in spiral nebulae. On 19 February 1924, Hubble wrote to Shapley that he had found a Cepheid in the Andromeda Nebula, which (by the period-distance relationship) he reckoned to be roughly a million light years away, far beyond Shapley's limit for our galaxy. Reading Hubble's letter, Shapley remarked to a colleague who happened to be in his office: "Here is the letter that has destroyed my universe." Shapley had enlarged our galaxy a hundred times. Showing that spiral nebulae are galaxies far outside our galaxy, Hubble made the universe larger yet, and shattered Shapley's all-comprehending unity.

Harlow Shapley, *Galaxies*, rev. ed. (1967). Richard Berendzen, Richard Hart, and Daniel Seeley, *Man Discovers the Galaxies* (1976). Robert W. Smith, *The Expanding Universe: Astronomy's "Great Debate" 1900–1931* (1982). Michael J. Crowe, *Modern Theories of the Universe from Herschel to Hubble* (1994). Norriss S. Hetherington, *Hubble's Cosmology: A Guided Study of Selected Texts* (1996).

NORRISS S. HETHERINGTON

GALILEO [GALILEI, Galileo] (1564–1642), natural philosopher and astronomer, discovered the laws of falling bodies and telescopic evidence for Copernicanism.

Galileo was born in Pisa and died near Florence. In 1581 he enrolled at the University of Pisa, studying mostly mathematics but leaving in 1585 without a degree. For several years he did private teaching

The frontispiece to the first collected edition of Galileo's works (1656), which omitted the *Dialogue* that got him into trouble. It shows him presenting his telescope to the muses and pointing to the sun surrounded by the planets. The crown-like band above the upper central planet represents the orbits of Jupiter's moons, whose eclipses Galileo tried to adapt to finding longitude at sea (hence the ship on the left). The cannon indicates the analysis of shell trajectories in the *Discourses on Two New Sciences* (1638).

and independent research. In 1589 he was appointed professor of mathematics at Pisa; from 1592 to 1610, he held a similar position at Padua.

In his Pisan and Paduan years, Galileo studied motion, especially that of falling bodies; opposed Aristotelian physics and followed an Archimedean approach; and pioneered experimentation as a procedure combining empirical observation with quantitative mathematization and conceptual theorizing. By this procedure he discovered an approximation to the law of inertia, the composition of motion into component elements, the laws that in free fall the distance fallen increases as the square of the time elapsed and the velocity acquired is directly proportional to the time, the isochronism of the pendulum, and the parabolic path of projectiles. He did not publish any of these results then, however, and a systematic account appeared only in the *Two New Sciences* (1638).

A main reason for this delay was that in 1609 Galileo became actively involved in astronomy. Until then, although he appreciated the merits of Nicholas *Copernicus's theory, he was more impressed by the evidence against it. He was especially troubled by the astronomical consequences that could not be observed, such as the similarity between terrestrial and heavenly bodies, Venus's phases, and annual stellar parallax. In 1609 he perfected the *telescope enough to make several discoveries, which he immediately published in a little book, *The Sidereal Messenger* (1610): lunar mountains, stars invisible to the naked eye, the stellar composition of the Milky Way and of nebulas, and Jupiter's moons. He became a celebrity. He resigned his professorship, became Philosopher and Chief Mathematician to the grand duke of Tuscany, and moved to Florence. Soon thereafter, he discovered Venus's phases and sunspots.

Galileo judged that his telescopic discoveries strengthened Copernicanism, but not conclusively: there was still astronomical counterevidence (no annual stellar parallax) and the objections drawn from Aristotelian physics had not yet been answered. He conceived a work treating all aspects of the question, but he did not finish it for two decades. It appeared in 1632 as a *Dialogue on the Two Chief World Systems*.

Galileo delayed because he came under attack from conservative philosophers and clergymen, who charged him with heresy for holding a belief contrary to Scripture. They argued that the earth cannot move because many scriptural passages state or imply that it stands still and Scripture cannot err. He defended himself by counterarguing that Scripture is not a scientific authority and does not really favor geostaticism over geokineticism. Responding to a complaint, the Inquisition of the Roman Catholic church launched an investigation, which yielded two results in 1616. Galileo received a private warning forbidding him to hold or defend the earth's motion. Furthermore, the church published a decree declaring the earth's motion to be physically false and contrary to Scripture, condemning attempts to show otherwise, and banning Copernicus's book of 1543 until revised. Issued in 1620, these revisions eliminated suggestions that the earth's motion is physically true and compatible with Scripture and conveyed the impression that Copernicus treated the earth's motion merely as a hypothesis useful for astronomical calculations.

Galileo kept quiet until 1623, when an old admirer became Pope Urban VIII. This encouraged Galileo to write the *Dialogue*, which showed that the pro-Copernican arguments were much stronger than the progeostatic ones. However, his enemies charged that the book defended the earth's motion, something he had been expressly forbidden to do. He was summoned to Rome to stand trial, which began in April 1633. At the first interrogation, Galileo denied that his book defended the earth's motion, claiming instead that it showed the geokinetic arguments to be inconclusive. There followed an out-of-court meeting during which he was persuaded to plead guilty to this charge in exchange for leniency. At the next deposition he admitted having defended the earth's motion but insisted that he did so unintentionally. The trial concluded with a sentence that did not exhibit the promised leniency: the cardinal-inquisitors, instructed by the Pope, found that Galileo's actions made him "vehemently suspected of heresy;" forced him to recite an abjuration, expressing sorrow and cursing his errors; sentenced him to indefinite house arrest; and banned the *Dialogue*.

The interpretation and evaluation of Galileo's work are highly controversial. Nevertheless, his contributions to physics, astronomy, and methodology were epoch-making. Add to them the mythological status into which he was catapulted by the tragic trial, and he may properly be regarded as a martyr as well as a hero of the scientific revolution.

Maurice Clavelin, *The Natural Philosophy of Galileo* (1974). Stillman Drake, *Galileo at Work* (1978). Alexandre Koyré, *Galileo*

Studies (1978). Maurice A. Finocchiaro, *Galileo and the Art of Reasoning* (1980). William A. Wallace, *Galileo and His Sources* (1984). Mario Biagioli, *Galileo Courtier* (1993).

MAURICE A. FINOCCHIARO

GALVANI, Luigi (1737–1797), physician, natural philosopher, and investigator of animal electricity, and **Alessandro VOLTA** (1745–1827), natural philosopher and inventor of the battery.

While belonging to very different cultural milieus, Galvani and Volta shared an interest in both electricity and physiology, not uncommon during the second half of the eighteenth century. After the publication of Galvani's treatise *On the Forces of Electricity in Muscular Motion* (1791), their area of common interest commanded the attention of expert "electricians," physicians, and a wide public throughout Europe, because of their controversy over animal electricity or "galvanism." The controversy magnified their contemporary fame and has since nurtured arguments between physicists (siding with Volta) and biologists (siding with Galvani) as well as historians, many of whom have overemphasized the conflicting traits of their sciences and personalities.

Luigi Galvani was born in Bologna, then part of the Papal States, the son of Domenico, a goldsmith, and Barbara Caterina Foschi. From 1755 he studied medicine and philosophy at the University of Bologna. He also attended the courses in physics, chemistry, and natural history offered by the Bologna Institute of Sciences. He graduated in 1759 and began to practice medicine, including surgery, as he would do throughout his life. In 1762 he started to climb the academic ladder, which raised him to lecturer of practical anatomy at the Institute in 1766.

In 1768 Galvani performed as the anatomist in charge of the "carnival anatomy," an annual celebration during which the body of a recently executed criminal was dissected to the bones in sixteen steps. Besides the professors and students, the event's audience included the local elite. Galvani performed the public anatomy three more times, increasing his fame as well as furthering his career: in 1771 he became a member of the powerful College of Physicians, in 1772 he was made president of the Bologna Academy of Sciences, and from 1775 he occupied the university chair in medicine that had been held by his former teacher who—after Galvani had married Lucia Galeazzi in 1762—was also his father-in-law. Lucia often assisted her husband in his experiments.

Deep religious feelings inform the private papers illustrating Galvani's uneventful but productive life. His religious views supported, if they did not prompt, his decision, late in life, not to swear loyalty to the Cisalpine Republic created in the region with the support of the French, led by General Bonaparte. Galvani died in 1797 deprived of his chair and offices.

Galvani's laboratory notes indicate an interest in electricity from 1780. A likely identity of the electrical and nervous fluids inspired his investigations. By exciting with (static) electricity the spinal medulla of frogs freshly killed, and seeing their legs jump as a consequence, he realized that frogs were the most sensitive detectors of electricity available. He devoted the following seventeen years to exploring the tricky interactions between what he regarded as an electricity internal to animals, in the form of their nervous fluid, and the electricity external to them generated by atmospheric electricity or by instruments like an electrostatic machine, a Leyden jar, or a Franklin square.

Alessandro Volta was born in the commercial town of Como, northern Italy, then part of Austrian Lombardy, a younger son from a family of the lesser nobility. He did not attend university, but as a pupil of the Jesuits developed a lifelong interest in natural philosophy, combining it with a commitment to Enlightenment culture then fashionable among the educated classes and the public administrators of Lombardy. Volta began his international correspondence on scientific and literary topics at the age of eighteen. At twenty-four he published an ambitious treatise "on the attractive force of the electric fire." At thirty he embarked on a career as a civil servant and a teacher in the recently reformed educational institutions of Lombardy. Appointed professor of experimental physics at the University of Pavia in 1778, he held the position until 1820 and managed to build there one of the finest physics laboratories of his time.

Unlike Galvani, who communicated only irregularly with his peers outside Bologna, Volta traveled and corresponded extensively, sharing his enthusiasm for natural philosophy with colleagues in many countries. After his intention to marry an opera singer had brought him close to rupture with his family (and had got him a reprimand from no less a personage than the Emperor of Austria), in 1794 he married the cultured and well-to-do Teresa Peregrini, nineteen years his junior. They had three children, to whose education they attended personally.

By the 1780s Volta had won European fame as an electrician, particularly as an inventor of electrical machines and as a specialist in the chemistry of airs. Before the controversy with Galvani began, Volta had invented the electrophorus, eudiometer, electric pistol, *condensatore* (a device that made weak electricity detectable), and straw electrometer. Volta's contributions to the science of electricity included the notions of tension, capacity, and actuation (an ancestor of electrostatic induction). Thanks to painstaking measurement techniques refined throughout his life, Volta managed to combine these notions into simple quantitative laws. By the 1780s some regarded him as "the Newton of electricity."

Volta was intrigued by Galvani's electrophysiological treatise: it touched on his long-cultivated interest in the Enlightenment theme of the "mind of animals." After repeating and varying Galvani's experiments, Volta rejected a special electricity in animals and developed a theory explaining galvanic phenomena as a consequence of ordinary electricity set in motion by the contact of different metals. The controversy that ensued stirred a crescendo of investigative and theoretical ingenuity on both sides. In 1794 Galvani managed to make his frogs jump without recourse to metals. In 1796 Volta decided to do without frogs as measuring devices. In 1797 Galvani provoked his frogs to jump via a circuit in which only nerves intervened.

Towards the close of 1799, Volta managed to build a mostly metallic device—the first electric battery—that produced a steady flow of electricity while also imitating the electric organs of an animal, the torpedo fish.

Volta's apparatus and its chemical performances made headline news in the European daily press. Bonaparte brought Volta to Paris and rewarded him generously before Europe's cultured elites, then worried by French ambitions. Already in 1794 the Royal Society of London had awarded the Copley Medal to Volta, the first foreigner to achieve the honor. Volta died in 1827 in Como, retaining the title of count conferred on him by Bonaparte, whose star had in the meantime fallen.

Through avenues conspicuously unforeseen by their initiators, Galvani had set in motion what would become electrophysiology, while Volta had laid the foundations of electrochemistry, *electromagnetism, and a new industrial era.

Marcello Pera, *The Ambiguous Frog: The Galvani-Volta Controversy on Animal Electricity* (1992). John L. Heilbron, *Electricity in the 17th and 18th Centuries* (1999). Giuliano Pancaldi, *Volta: Science and Culture in the Age of Enlightenment* (2003).

GIULIANO PANCALDI

GAS DISCHARGE. See CATHODE RAYS AND GAS DISCHARGE.

GAUSS, Carl Friedrich (1777–1855), German mathematician, astronomer, and physicist. Gauss was born in Brunswick; his father was a manual laborer and, before her marriage, his semiliterate mother had worked as a maid. Against the wishes of his father, who wanted Gauss to learn a proper trade, the mathematically precocious boy (he could calculate before he could speak) was encouraged by his teachers to study. At the age of fourteen he was presented to the Duke of Brunswick, Karl Wilhelm Ferdinand, who provided him with financial support until 1806. From 1792 to 1795 Gauss used the ducal subvention to study in the Collegium Carolinum, a preparatory school for university study, where he developed a variety of brilliant mathematical ideas. Between 1795 and 1798 he attended the Uni-

The portrait shows Gauss in middle age in a costume not unusual for a savant of the time.

versity of Göttingen, where he read the mathematical classics and found that many of his discoveries were not new.

Gauss was a gifted linguist, and he did not know at first what path to follow. He decided on mathematics when in 1796 he solved the ancient problem of constructing a regular polygon with seventeen sides using only compasses and a straight edge. In 1798 he returned to Brunswick, where he worked as a private scholar and produced his finest mathematics, which was in number theory. He obtained his doctorate in absentia in 1799, nominally under Johann Friedrich Pfaff at the duchy's University of Helmstedt, after submitting some earlier work on the first of his four proofs of the fundamental theorem of algebra. With publication of his great work, *Disquisitiones arithmeticae* (1801), Gauss became, according to his older contemporary Joseph-Louis Lagrange, the leading mathematician of Europe.

Gauss next turned his attention to astronomy, following up the discovery of the minor planet Ceres by Palermo astronomer Giuseppe Piazzi on New Year's Day 1801. Piazzi soon lost track of his planetoid, but from the few observational data he had provided Gauss determined the orbit and, when Piazzi found Ceres again, it was in the position Gauss predicted. For this extraordinary feat Gauss used an improved orbit theory as well as the method of least squares, which he had previously developed to assist in analyzing geodetic measurements (see ERROR AND THE PERSONAL EQUATION). This work led to his second remarkable book, *Theoria motus corporum coelestium* (1809), which showed him to be one of Europe's finest mathematical astronomers.

With the death of his patron in 1806 following defeat by Napoleon's troops, Gauss had to look for a new source of income. In 1807 he accepted a position at Göttingen University as professor of astronomy and director of the Observatory. By this time he had already produced his main mathematical work and had toyed with non-Euclidean geometry. Much of his time and energy went into equipping and running the Observatory and into teaching. Gauss did not enjoy teaching, yet among his pupils were such outstanding mathematicians as Moritz Benedikt Cantor, Richard Dedekind, and Bernhard Riemann.

Gauss applied his unequalled facility with numbers to various scientific projects, most notably in his geodetic survey of the Kingdom of Hanover, to which Göttingen belonged. From 1818 until 1825, he personally led the triangulation fieldwork and subsequently supervised the data conversion, making more than a million calculations. Among the resulting theoretical advances was his method of overcoming the difficulties of mapping the terrestrial ellipsoid on a sphere (conformal mapping), as described in his *Disquisitiones generales circa superficies curvas* (1828).

In 1828, at the request of Alexander von *Humboldt, Gauss took part in the only scientific convention he ever attended, a meeting of the Gesellschaft deutscher Naturwissenschaftler und Ärzte in Berlin. It inspired him to develop his long-standing interest in earth magnetism. Together with the physicist Wilhelm Eduard Weber, who was professor at Göttingen from 1831 until 1837, he established a magnetic observatory. In 1834 he organized the Magnetic Association, the first major scientific project based on international cooperation. This association, set up to conduct Europe-wide geomagnetic observations, was subsequently expanded by Humboldt into a worldwide network. Gauss summarized the results in a joint publication with Weber, *Allgemeine Theorie des Erdmagnetismus* (1839), followed by an *Atlas* (1840). Gauss also applied himself to the improvement and invention of scientific instruments, developing the heliotrope (a sextant-like instrument that uses reflected sunlight for geodetic measurements), the bifilar magnetometer (in which two silk threads are used for measuring geomagnetic force), and in 1833, together with Weber, the first electrical telegraph.

Gauss's motto was *pauca sed matura*, yet his many (323) *mature* publications were "few" only in relation to the multitude of his ideas. (According to an exact historian of mathematics, Gauss had precisely 404 ideas, 178 of which he discussed in print.) He has been compared to Archimedes and Isaac *Newton, and in the ranking game—which mathematicians play with much greater accord than do other groups of academics—Gauss has been classed as a credible candidate for *princeps mathematicorum*. A preoccupation with his brilliance has deflected many of his biographers from properly historicizing him. The historical conditions under which his talent found a niche, and his work found its initial acceptance, must be sought in the world of German *Kleinstaaterei* (the politics of patchwork-quilt principalities before the 1871 unification of Germany). He remained loyal to the enlightened despots and university officials who supported him. Although not a churchman, Gauss became increasingly preoccupied with the metaphysical issue of immortality and believed staunchly in life after death. He was married twice; the first marriage (1805–1809; one son and one daughter) was happy, the second (1810–1831; two sons and a daughter) less so.

Wolfgang Sartorius von Waltershausen, *Gauss, a Memorial* (1856; trans. 1966). G. Waldo Dunnington, *Carl Friedrich Gauss: Titan of Science* (1955). Kenneth O. May, "Gauss, Carl Friedrich," in *Dictionary of Scientific Biography*, vol. 5 (1972): 298–315.

Karin Reich, *Carl Friedrich Gauss, 1777–1977* (1977). Walter K. Bühler, *Gauss: A Biographical Study* (1981). Horst Michling, *Carl Friedrich Gauss: Aus dem Leben des Princeps Mathematicorum*, 2d ed. (1982).

NICOLAAS A. RUPKE

GEODESY. The librarian of Alexandria, Eratosthenes, called "beta" in antiquity because he was second best at everything, made the first significant measurement of the size of the earth. His technique required the determination of the terrestrial distance between two points on the same meridian and the difference in latitude between them. The figure shows Eratosthenes's stations at noon on the day of measurement, when the sun stood at the zenith in Aswān and a little over seven degrees from it at Alexandria. Since the sun's rays strike the earth almost parallel, the difference in latitude $\Delta\phi$ equals the measurable zenith distance z; the station separation between Alexandria-Aswān is the same fraction of the earth's circumference C as $\Delta\phi$ is of $360°$. Eratosthenes's value of C may have been close to correct (40,000 km [24,800 miles])—the size of his ruler is not known exactly—an extraordinary coincidence since he used gross approximations.

Knowledge of C had a practical value to people planning long trips. The Arabs, who traveled far across deserts, devised several

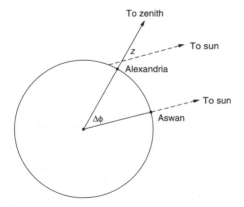

The relation between the difference in latitude between Alexandria and Aswan (Δf) and the distance of the sun from the zenith at Alexandria (z) when it stands overhead at Aswan.

geodetic methods, and the Caliph al-Ma'mūn (reigned 813–833), a great patron of mathematicians, sent out expeditions to deploy them. The principal technique was to take the latitude of a place in a flat region and walk due north or south, measuring the distance, until the latitude had changed by a degree.

Columbus chose the values of the circumference from Greek and Arab sources that made the earth small enough that his project to reach Asia by crossing the great Western ocean might seem feasible. His landing in America revealed the world to be twice the size he had calculated. Renewed interest in geodetic measurement, stimulated by the European voyages of discovery, returned mathematicians to Eratosthenes's technique with the significant difference, however, that thenceforth they would determine distance by trigonometry rather than by walking or guessing.

Although Tycho *Brahe carried out a trigonometric survey of parts of Denmark, the first thorough and accurate geodetic measurement, which became the early-modern exemplar, took place in Holland in 1616. The surveyor, the professor of mathematics at the University of Leyden Willebrord Snel, gave the book in which he published his results the appropriately boastful title, *Eratosthenes batavus* (*The Dutch Eratosthenes*, 1617). He found for C the equivalent of 38,640 km (24,015 miles). He determined the distance between his two stations (around 130 km [80 miles]) by setting out a series of virtual triangles with church steeples and town towers at their vertices, measuring the angles at each vertex, and obtaining the lengths of the sides of the triangles by linking them (by angles taken from the steeples and towers) to base lines laid out by a surveyor's chain.

Snel's technique was first applied on a large scale by Louis XIV's astronomers, who, as members of the new Paris Académie Royale des Sciences, had the chore and challenge of making an exact survey of France. Beginning in the 1660s under the direction of Jean Picard, they measured increasingly long arcs of the meridian through Paris. Picard improved the instrumentation by replacing open sights with lenses and provided a second baseline to check results obtained with the first. He made $C = 40,042$ km (24,886 miles). The effective head of the Paris Observatory, Gian Domenico *Cassini, and then his son and grandson, extended Picard's arc, crisscrossed France with virtual triangles, and delivered the news that the conventional maps of the kingdom ascribed to it lands that in fact lay in the English Channel and the Bay of Biscay. France may have lost more territory to its astronomers than to its neighbors.

The conversion of the Cassinis' measurements to a value for the earth's circumference presented a new problem for geodesy. On a mission for the Académie to the island of Cayenne, Picard found that a pendulum clock went more slowly near the equator than it had done in Paris. Isaac *Newton explained this tropical lethargy with his theory of gravity, which suggested that if the earth had cooled from a spinning fluid ball, its equatorial diameter would exceed the distance between the poles by an amount he could calculate. Consequently geodesists should try to fit their measurements to an ellipse. Cassini's son Jacques reported the first such attempt in his *De la grandeur et de la figure de la terre* (1720). The ellipse that fit his facts had the shape of an American football instead of a Newtonian pumpkin.

Confrontation between the Cartesians, who dominated the Académie, and the slowly growing Newtonian band of younger members came to center on the shape of the earth. In the 1730s the Newtonians, led by Pierre de Maupertuis, convinced the crown to finance a definitive test, in the interest of navigation as well as science. Two expeditions went out, one to Peru, the other, under the direction of Maupertuis, to Lapland. Both braved harsh weather, primitive conditions, uncomprehending natives, and vicious mosquitoes, all to learn whether a degree of the meridian was longer at the pole than at the equator. The outcome "flattened the poles and the Cassinis," according to Newton's popularizer Voltaire, although, in fact, Maupertuis's measurement was so far out that had it been in the opposite direction the expeditions would have found for a pointed earth.

During the second half of the eighteenth century the leading states of Europe undertook trigonometric surveys of their domains, primarily for military and economic purposes, which gave abundant data for fixing the ellipticity of the Newtonian earth. The more the data, the less plausible the ellipsoid. Geodecists assigned the discrepancies to inhomogeneities in the density of the earth and to mountains that drew aside plumb bobs and falsified the verticals necessary for finding latitudes. They came to agree that no fiddling would fit their knowledge to an ellipsoid; the shape of the earth was just that, the earth's shape, a "geoid," defined by Carl Friedrich *Gauss and other leading geodesists as a surface everywhere perpendicular to the direction of gravity.

Beginning in 1787 with the linkage of the meridians through Greenwich and Paris via a series of triangles that crossed the channel, Europe gradually approached unification, trigonometrically speaking. The French furthered unification by running surveys in territories conquered by Napoléon and his generals and by imposing where they could the metric system of weights and measures, itself based on a remeasurement of the Paris meridian from Dunkirk to Perpignan. Introduced during the first years of the French Revolution to replace the welter of provincial units that obstructed commerce and victimized the consumer, the standard meter was defined in theory as one ten-millionth of the distance from the pole to the equator and in practice as the ten-millionth part of ninety times the average length of a degree along the Paris meridian. The meter, its decimal multiples, its progeny of grams and liters, and its imposition on Europe are conspicuous, recurrent reminders of the prestige of geodesy in the eighteenth century, the rationalism of the Enlightenment, and the power of the French Revolution.

After 1800 trigonometrical surveys were pursued outside of Europe notably by the British in India and the new government of the United States. The American Coast and Geodetic Survey became a principal route for the little support the U.S. federal government gave science during the nineteenth century. Warfare, and especially the Cold War, again pushed geodesy to the front in the twentieth century. Plans made from precisely established points played a major part in the artillery duels of World War I and corrected many errors in maps previously regarded as accurate. Development of aerial photography and radio during the war gave geodesy powerful new tools for postwar mapping. Similarly the invention of the laser after World War II enhanced the means, and the creation of intercontinental ballistic missiles increased the importance, of knowing with precision how far one point lies from another on the earth's surface. The emergence of geophysics as a major science during the last fifty years owed much to military interest in the geoid as a firing range.

Hunter Dupree, *Science and the Federal Government* (1957). N. D. Haasbroek, *Gemma Frisius, Tycho Brahe and Snellius and their Triangulations* (1968). Henri Lacombe and Pierre Costabel, eds., *La figure de la terre du xviiie siècle à l'ère spatiale* (1988). J. L. Heilbron, *Weighing Imponderables and Other Quantitative Science around 1800* (1990).

J. L. HEILBRON

GEOGRAPHY has its roots in ancient efforts to describe the surface of the earth, the form and extent of its lands and seas, and all that is observed in them. The peculiarly modern, European art of geography, as it emerged in the Renaissance, combined mathematical delineation and historical description with varying emphases. For Renaissance scholars, geography was both a body of knowledge recovered from the classical past and a newly discovered art of systematically ordering knowledge of the earth. The transmission of the *Geographia* of Claudius Ptolemy (c. 90–168) from Constantinople to Italy and its translation into Latin early in the fifteenth century was central to making the mathematical description of the earth's lands and seas the core of the new art of geography. It gave instructions for arranging geographical knowledge according to a coordinate system of parallels (latitudes) and meridians (longitudes) and for projecting those lines onto flat surfaces. Renaissance geography was closely tied to the production of maps

of the earth's surface and its separate divisions, culminating in the first "atlases" of the Low-country cartographers Gerhard Mercator and Abraham Ortelius in the mid-sixteenth century.

This Renaissance "invention" of geography was sponsored by centralizing states making ever more universal claims to legitimacy. By advertising the usefulness of their art for administering and defending principalities and cities, geographers secured more or less official roles in European states from the sixteenth century. As these states began to explore and trade overseas, geographers used words and pictures to help interpret, manage, and celebrate novel discoveries and new possessions for kings, princes, and churches. Geography also carried all the prestige of humanism and the mathematical arts. Sebastian Münster, author of the *Cosmographica universalis* (Basel, 1544), was typical in advertising geography as the essential study of the active Christian prince. How could anyone understand Homer and Virgil, the campaigns of Caesar and Cato, the acts of God, the wanderings of the Jews, and the missions of the Apostles, without geography?

The Protestant Reformation gave added impetus to official patronage. As reformed and counterreformed princes and cities turned to the foundation of gymnasia and universities, geographical systems assumed a pedagogical and doctrinal function, exhibiting God's wise design of the world. The mathematical basis of geography was progressively refined as a theater for God's providence in the *Systema geographicum* (Hanover, 1611) of the theologian Bartholomäus Keckermann and the *Geographia generalis* (Amsterdam, 1650) of the Dutch physician Bernhard Varenius. Varenius's book was one of the most influential geographical systems of the seventeenth and eighteenth centuries. Isaac *Newton edited the second and third editions despite Varenius's Cartesian leanings. In the work of the Jesuit Athanasius Kircher or the globemaker Vincenzo Coronelli, mathematical techniques imbued geography with a renewed sense of cosmographical mission under the sponsorship of the Roman Catholic Counter-Reformation and royal absolutism.

In the eighteenth century, geography flourished as a descriptive science with a mathematical foundation, often with an appeal to divine design. Enlightenment geography aligned itself programmatically, with the "systematic spirit" and against the "spirit of system," often restricting itself to positive description and tying itself ever more firmly to a technical symbolic language. Although it relied upon the observations of travelers and field surveyors, geography was essentially a product of the cabinet. It expressed its cosmographical ambitions increasingly by attempts to standardize and universalize the language of geographical description—to establish uniform topographical languages and cartographic symbolism (including map scale).

These efforts reached their zenith in Enlightenment France with Cassini's trigonometric charts, the collection of maps for military purposes at the Dépôt de la Guerre (founded 1688), and the closely associated cabinet of J.-B.-B. d'Anville. Geographers such as d'Anville distinguished themselves from lesser workers by their ability to sift and compile geographical information critically and express it in ever more precise and standardized languages, especially its graphic and mathematical expression on a chart. In practice, however, the lines between positive description and philosophical speculation were ill defined. Geographers such as Philippe Buache and Nicolas Desmarest entered into debates over the nature of continents, seas, earthquakes, volcanoes, river basins, and mountain ranges—topics often labeled "physical geography" and "theory of the earth" to distinguish them from "geography" proper.

In Protestant Europe, where geography remained tied to natural philosophy and natural theology, geographers discussed such topics as the causes of the tides, the heat of the tropics, seasonal rains, the decrease of temperature with elevation, and so on. Political geography, human geography, mineral geography, plant geography, and zoogeography emerged as distinct fields, justified principally through their service to natural theology. All exposed the wise hand of Providence by revealing the diversity and distribution of created things, and the balances and economies of nature (language eschewed by French geographers). In Germany and Scotland, at Hamburg, Edinburgh, and Göttingen, in works such as A. F. Büsching's *Neue Erdbeschreibung* (11 vols., 1754–1792) and Christoph Ebeling's *Erdbeschreibung und Geschichte von Amerika* (7 vols., 1793–1816), geography took the form of theologically informed gazetteering aimed at developing industrious, purposeful citizens of the world such as great merchants and administrative officials. At Göttingen, under the auspices of Gottfried Achenwall and A. L. Schlözer, geography was closely allied with statistics and political history (*Staatenkunde*). Immanuel Kant, who spent most of his forty years at Königsberg lecturing on "physical geography" (including physical anthropology), promoted geography's "extensive utility," for arranging our knowledge and "sociable conversation." Johann Reinhold Forster and his son Georg Forster, who together accompanied James Cook on his second expedition around the world from 1772 to 1776, similarly used geographical exploration to demonstrate the providential design of the world and to make its audience into active witnesses and instruments of providence. In his German translations of contemporary voyages and in his briefer, essayistic works, the nomadic younger Forster decisively focused geography on active exploration and the gradual emergence of underlying dynamic laws and processes.

Toward the end of the eighteenth century, the natural-historical concern with systematic collection, classification, and exposition gave way to philosophical concern with dynamic processes and physical causes. Taking inspiration from the Forsters' "philanthropic" approach to understanding the reciprocal influences of nature and human civilization, the Prussian mining official and naturalist Alexander von *Humboldt sought physical and historical laws in systematic geographical investigations of everything from climate to language and art. Humboldt shunned the term "geography" and referred to his science as "physics of the earth." Humboldt put the systematic spatial analysis of phenomena usually left to physicists and natural historians—rocks, terrestrial magnetism, atmospheric temperature and chemistry, plant and animal species, anything open to calibrated and standardized perception—at the center of geography. He also emphasized the aesthetic and emotional responses of the traveler. Although Humboldt omitted God from his unfinished *Kosmos* (1845–1862, subtitled "Sketch of a Physical Description of the World"), his dual revelation of the lawfulness of nature and the progress and limits of human knowledge of nature was implicitly congenial to a recognition of divine design.

Humboldt's contemporary Carl Ritter presented geography as a comparative study of the world's "terrestrial units" that supplied the key to a developmental understanding of the history of civilization, expressly overseen by divine providence. As delivered for three decades at the newly founded Berlin University and at the Berlin Military Academy and in his massive but unfinished *Erdkunde* (1817), Ritter's demonstration that civilization had migrated from East to West influenced generations of statesmen, soldiers, and scholars. Among them was Arnold Guyot, whose Protestant geographical theodicy found a home at Princeton University after 1848. In Britain, Mary Somerville discovered a similar providence at work in geography: both objectively, in the interaction of physical forces over the surface of the globe to produce geographical laws (*Physical Geography*, 1848), and subjectively, in the progressive interaction of the different branches of the physical sciences in geographical science (*On the Connexion of the Physical Sciences*, 1834).

Geographical societies began emerging in the late eighteenth century, imbued with a civilizing mission and informed by this sense of divine lawfulness The Association for Promoting the Discovery of the Interior Parts of Africa, founded in London in 1788, was followed by groups in Paris (1821), Berlin (1828), London (1830), Saint Petersburg (1845), and New York (1851). These metropolitan societies brought together army and navy officials, statesmen, and scholars to promote exploration and publish maps and accounts of voyages. Between 1870 and 1890, the number of geographical societies in Europe quadrupled to over eighty. In 1875, these societies began convening quadrennial International Congresses of the Geographical Sciences. They also encouraged the proliferation of popular geographical magazines and literature. Taking their cue from Germany, where geography was established in natural science faculties, and where academic geography and geographical societies had close ties, geographical societies urged that geography be taught in universities and secondary schools. In 1871, after the Prussian victory over France, the Paris society and the ministry of education established a Committee on the Teaching of Geography and a number of professorships at French universities. In the 1880s, the Royal Geographical Society of London succeeded in establishing chairs of geography at Oxford and Cambridge.

In the universities, geography was dominated by the prevalent enthusiasm for evolutionary theories. Friedrich Ratzel, who trained with Ernst Haeckel in Jena before settling at Munich, Halford Mackinder at Oxford, Alfred Hettner at Tübingen, and William Morris Davis at Harvard all argued that the geographical diversity of humans resulted from the variation of physical conditions over time and space. The Russian exile Peter Kropotkin and the Scottish social critic Patrick Geddes stressed the independent laws of organisms in interpreting responses to inorganic conditions. Paul Vidal de la Blache and Otto Schlüter insisted that human modes of social life (lifestyles, *Kulturformen*, genres de vie) shaped the landscape.

Regardless of their understanding of evolution, academic geographers took the region or the landscape as the fundamental unit of analysis. For all their sometimes bitter disagreement, Carl Sauer at Berkeley and Richard Hartshorne at the University of Wisconsin agreed that regional description lay at the heart of scientific geography. Between the 1950s and the 1970s, geographers enthusiastic about the so-called "quantitative revolution" rejected Hartshorne's strictures, and argued that by using quantification and statistics, geographers could discern the laws of social dynamics. More recently, geographers have analyzed the subjective experience of space and produced "behavioral geographies" and "mental maps."

The very ubiquity of geographical knowledge made it difficult to define the field. Hartshorne's *The Nature of Geography* (1939) drew up a historical lineage and philosophical foundation for the discipline that established geography as a descriptive science of regional differentiation. It remains an influential interpretation of the history and philosophy of geography. Other geographers have adopted philosophies from phenomenology and logical positivism to Marxism, structuralism, and postmodernism in the search for philosophical foundations. In the 1990s, the emphasis began to shift to studying the institutional, political, and social contexts, interests, and languages of earlier geographers. The bread and butter of geography, though, has remained the training of school teachers and the preparation of regional descriptions applicable to a variety of policy needs.

See also CARTOGRAPHY; GEODESY

International Geographical Union, *Geographers: Biobibliographical Studies* (1977–), 3 vols. so far. Margarita J. Bowen, *Empiricism and Geographical Thought from Bacon to Humboldt* (1981). David Stoddart, ed., *Geography, Ideology, and Social Concern* (1981). David Livingstone, *The Geographical Tradition* (1992). Morag Bell, Robin Butlin, and Michael Heffernan, eds., *Geography and Imperialism, 1820–1940* (1995). Anne Godlewska, *Geography Unbound: French Geographic Science from Cassini to Humboldt* (1999).

MICHAEL DETTELBACH

GEOLOGY. The word geology as a general term for the study of the earth was popularized in the late eighteenth century by the Swiss naturalists Jean André *Deluc, who made his career in England, and Horace Bénédict de Saussure, famous for his voyages in the Alps. Abraham Gottlob *Werner disliked the term as being too suggestive of theorizing, and promoted the alternative "geognosy." But when a group of Englishmen decided in 1807 to found a society for the study of the earth that eschewed the practical, utilitarian goals of the Continental mining schools, they called

it the Geological Society of London. Within two or three decades, geology had its own specialists, societies, textbooks, and journals, the Geological Society's *Transactions* being the first. Worldwide, the word geology (or its cognates) rapidly became the preferred term for the study of the earth.

Geologists generally divided their specialty into two parts: historical geology, which reconstructed earth history using *stratigraphy and *paleontology, and physical geology, which investigated the earth's structure and causal processes. The latter consisted of *mineralogy, diminished from the overarching category for the study of the earth to a mere subdiscipline, petrology, and structural geology.

The successes of historical geology were manifest in the outlines of the stratigraphic column as worked out by the mid-nineteenth century. Field work with a hammer and a map was the preferred way of investigating, and the geological map an ingenious way of summarizing the results. Georges Cuvier and Alexandre Brongniart produced a pioneering map of the Paris area in 1812. The first national geological map, of England, was prepared by William Smith in 1815; the Geological Society's map followed in 1819. Although these maps were expensive and time-consuming to produce, state governments set up geological surveys to prepare them because of their economic importance for the extractive industries and agriculture. The French began fieldwork for a national geological map in 1825. The British Geological Survey was founded in 1835, followed by the Canadian (1842), the Irish (1845), and the Indian (1854). In Europe, the Austro-Hungarian Empire and Spain created surveys in 1849, Sweden and Norway in 1858, Switzerland in 1859, Prussia and Italy in 1873, and Russia in 1882. The American states founded their surveys early: Massachusetts and Tennessee in 1831 and Maryland in 1833, but a national survey was not created until 1879.

Though the work of mineralogists and petrologists attracted less public attention than that of the historical geologists, their field rapidly moved ahead in the nineteenth century. And although they too carried out field studies, they retained their ties to chemistry and the laboratory.

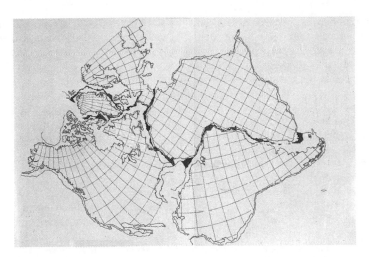

The Atlantic continents snuggled up at 500-fathoms in the influential diagram by Edward C. Bullard et al. The dark patches indicate gaps and overlaps.

Historical and physical geology supposedly worked in concert, but the results obtained in the one frequently held little interest for the other. Nonetheless, geologists sought ways of presenting a coherent picture of the earth. One possible strategy, tried by Charles *Lyell in his *Principles of Geology* (1830–1833), made the study of present geological processes a necessary key to understanding earth history. Although rejected by most geologists, it provoked a useful methodological discussion.

A second strategy looked for an overarching causal process that could elucidate the details of the stratigraphic record. For most nineteenth-century geologists, the preferred hypothesis was that the earth had cooled and contracted. The nebular hypothesis proposed by Pierre Simon *Laplace and William *Herschel, and the work on rates of cooling by Jean Baptiste Fourier, supported their model. In 1831, the French geologist Élie de Beaumont suggested that as the earth had cooled from a molten body, the crust at intervals had buckled under the strain, throwing up mountain chains and exterminating whole genera in the great floods that coursed down their sides. Variants of this theory, and criticisms of it, flourished for the rest of the century and culminated in the four-volume *Face of the Earth* (1883–1904) by

Eduard Suess, Professor of Geology at the University of Vienna. According to Suess, the molten center of the earth had once been covered with a thin, solid crust. As the earth cooled, portions of the crust collapsed, creating ocean basins. Later, the remaining higher areas became unstable, and collapsed in turn, forming new ocean basins and leaving former ocean beds exposed as new continents.

It was a triumphant moment for geology. Two great geologists summed up the history of their science during the nineteenth century. Sir Archibald Geikie, head of the Geological Survey of Great Britain, published his *Founders of Geology* in 1905. He celebrated the achievements of British geologists, especially the Scots James Hutton and Charles Lyell. He arranged his history around battles between Neptunists and Plutonists (*see* NEPTUNISM) and Uniformitarians and Catastrophists (*see* UNIFORMITARIANISM AND CATASTROPHISM). Karl Zittel, a professor at the University of Munich and a renowned paleontologist, published a *History of Geology and Paleontology* in 1901. He gave more weight to mineralogy, petrology, and theories of mountain elevation, and praised Suess as having achieved "almost general recognition for the contraction theory."

This celebration of geology's progress

was soon to seem inappropriate. The cooling earth with its foundering continents did not survive as a synthesis for more than a decade. The discovery of *radioactivity revealed a heat source within the earth that counteracted the cooling from some original molten state. The discovery of *isostasy made it highly improbable that continents could have foundered. Detailed studies of the Alps made it clear that simple up-down forces acting on the earth's crust could not explain the tens or even hundreds of miles of foreshortening revealed by their folded strata.

The resultant theoretical vacuum led to a proliferation of alternatives in the 1910s and 1920s. Some scientists, such as Harold Jeffreys in England and Hans Stille in Germany, attempted to revamp contraction theory in light of the criticisms. Others preferred more radical alternatives: the planetismal hypothesis advanced by Thomas Chamberlin, the radiogenic by John Joly, and the theory of continental drift of Alfred Wegener. None of these theories succeeded in garnering enough evidential support to win widespread acceptance.

For the next fifty years, geologists hunkered down and continued their map-making and surveying. Mineralogy and petrology made important advances. Underwater exploration revealed interesting gravity anomaly patterns around island arcs (see OCEANOGRAPHY). Geologists found new work in the oil industry, which joined geological surveys and mining as the main sources of employment outside the universities. The American Association of Petroleum Geologists, founded in 1917, had become the world's largest professional geological society by 2000, with over 30,000 members in more than 100 countries. By this time, however, geology was no longer the umbrella discipline for the study of the earth, but just one branch of the *earth sciences.

Mott Greene, *Geology in the Nineteenth Century* (1982). Rachel Laudan, *From Mineralogy to Geology* (1987). Gabriel Gohau, *History of Geology* (1990). David Oldroyd, *Thinking About the Earth* (1996).

RACHEL LAUDAN

GEOPHYSICS. Until the late nineteenth century no institutionalized discipline of geophysics, the physics of the earth, existed. Most geophysical efforts responded to specific national interests, relying heavily on government funding.

In the eighteenth century, although many natural philosophers still produced speculative cosmogonies based on Newtonian or Cartesian physics, quantitative, empirical geophysical work gradually displaced speculation. In particular, physicists and cartographers alike sought to determine the figure of the earth (see GEODESY).

In the nineteenth and early twentieth centuries, researchers turned to problems of *terrestrial magnetism, regional variations in the earth's gravitational field (gravity anomalies), and *seismology in addition to geodetics. In the 1830s, scientists in Germany and England including Alexander von *Humboldt, Carl Friedrich *Gauss, Wilhelm Weber, and Edward Sabine proposed and initiated far-flung surveys of magnetic declination and dip. In the 1850s, surveyors for the Great Trignometrical Survey of India, established early in the century, found that their plumb bobs were not deflected by the gravitational force of the Himalayas to the extent they had predicted. This led to theories of *isostasy that preoccupied physicists and geologists in the first half of the twentieth century. In the last decades of the nineteenth century, the invention of reasonably accurate and compact seismographs spurred the development of seismology into a quantitative discipline.

By the early twentieth century, geophysicists clustered in their own societies and institutes. In 1880 John Milne had founded the Seismological Society of Japan, and the Japanese quickly joined the Germans and the Americans as leaders in geophysics. The geophysical laboratory of the Carnegie Institution of Washington, founded in 1902, became a major center for American geophysics. The American Geophysical Union, established in 1919 as an affiliate of the National Academy of Sciences by the National Research Council of the United States, quickly assumed world leadership in geophysics.

World War I and World War II provided new applications (and research funding) for geophysics—seismology, geomagnetism, and submarine cartography, for instance, aided submarine warfare. The history of the military's role in the development of geophysics has yet to be written as many of the relevant documents have only recently been declassified. The *International Geophysical Year that began in 1957 gave a further boost to geophysics. The U.S. government established the National Aeronautical and Space Administration (NASA) in 1958 and the National Oceanic and Atmospheric Administration (NOAA) in 1970.

The prestige of geophysics rose with spectacular discoveries such as *cosmic rays, the core and mantle of the earth, sea floor spreading, and El Niño. Applied geophysics boomed as geophysicists turned their attention to problems such as global warming, ozone depletion, water supply and quality, and earthquake prediction. At the end of the twentieth century, the largest international organization of geophysicists was the American Geophysical Union, with over 35,000 members in more than 100 countries. Although geophysics retained a separate identity, by the late twentieth century it was much more tightly linked to geology and oceanography than ever before. Through most of its history, geophysics was practiced by physicists chiefly interested in problems arising from physics itself. Geology, by contrast, was from its start in the early nineteenth century the province of professional geologists. Their primary goal was to tell the history of the earth. Although they were interested in geological causes, they inferred these from current geological processes and past geological monuments, not from physical first principles. Insofar as geologists and geophysicists interacted, the encounters were often tense, as in the debate between William *Thomson, Lord Kelvin, and the geological community over the age of the earth (see EARTH, AGE OF THE). Oceanography had its origins in the mapping of the oceans and in exploration rather than in the physical sciences. With the creation of the overarching discipline of *earth science and the broad acceptance of the theory of *plate tectonics in the 1960s and 1970s, geophysicists, geologists, and oceanographers began working on related problems using the same techniques and implements. Simultaneously the creation of the related umbrella category of *planetary science meant that geophysicists, geologists, and oceanographers extended the scope of their investigations beyond the earth itself to the domain formerly dominated by astronomers, the solar system. Though traces of the independent historical trajectories of these disciplines remain, their merging in earth and planetary sciences, although not always easy, has led to rapid scientific advances.

American Geophysical Union, *History of Geophysics* (1984, ongoing). Robert Muir Wood, *The Dark Side of the Earth* (1985). Gregory Good, ed., *The Earth, the Heavens and the Carnegie Institution of Washington* (1994). John Leonard Greenberg, *The Problem of the Earth's Shape from Newton to Clairaut* (1995).

RACHEL LAUDAN

GIBBS, Josiah Willard (1839–1903), mathematical physicist.

The only son of the Professor of Sacred Literature at Yale College, Gibbs seldom ventured beyond New Haven. He suffered constantly from ill health and after the death of his parents continued to live in the family house with his two sisters and one brother-in-law. His ordered existence centered on Yale, as a student (1854–1863), tutor (1863–1866), and professor (of mathematical physics, 1871–1903). Gibbs did his graduate work at Yale's advanced engineering school, from which he emerged as one of the very first Ph.Ds in the United States. His thesis on gear design scarcely suggested the direction of the researches in fundamental physics that made his reputation. In 1866, Gibbs tore himself away from New Haven for a three-year stay in Europe, during which he attended courses in mathematics and experimental physics at the Collège de France in Paris and the universities of Berlin and Heidelberg, and read the works of European mathematicians extensively. Gibbs left no repository of personal papers for historians trying to understand the enigma of the mind behind the papers and the man behind the work.

After his return to the family house, Gibbs taught and worked gratis at Yale until 1879, when the university gave him a salary to counter an offer that Johns Hopkins had made on the basis of Gibbs's growing reputation in Europe. He had developed this reputation by staying away from meetings and sending copies of his work to European and American physicists. He was elected to the major physical and mathematical societies of Europe, but did not achieve the same reputation in the United States, where empiricism dominated the sciences. He seems to have thought deeply for a long time before committing his ideas to paper, publishing only a few, crucial works. Yet, after some years working on his pivotal and complex thermodynamics paper of 1879, he judged it to be too verbose.

Gibbs's research interests encompassed engineering, *mechanics, *thermodynamics, the electromagnetic theory of light, vector analysis, and statistical mechanics. He always worked from general principles, basing his mechanics on Hamilton's equations, his thermodynamics on the first and second laws, and his statistical mechanics on ensembles of particles that obeyed Hamiltonian dynamics. He explored the implications of these basic laws at the most general level, moving from the simplest to the most general case. He sought physically significant, rather than mathematically elegant, results, and avoided speculations about molecules.

Gibbs transformed thermodynamics. Whereas physicists agreed in their understanding of the law of conservation of energy, they held three different interpretations of the second law, and many notions of the relationship between energy and entropy. In 1873, Gibbs restated both laws of thermodynamics distinctly, established *entropy as a function of state; investigated the thermodynamic properties of a homogeneous substance, and developed new thermodynamic diagrams to display these properties; investigated the solid, liquid, and gaseous states of a substance; showed the conditions for their coexistence; and identified the critical points. By 1878 he had developed a theory of chemical mixtures and arrived at his phase rule, which set both the conditions under which a particular chemical mixture was stable, and if unstable, the direction of change to reach a new equilibrium.

James Clerk *Maxwell first drew chemists' attention to the significance of Gibbs's work, which was then exploited by the energeticists of the 1880s and 1890s (see *ENERGETICS). In turn, reading Maxwell led Gibbs to develop vector analysis, based on William Rowan Hamilton's quarternions and William Kingdom Clifford's mechanics. As a result, he drew the ire of the Edinburgh professor of physics Peter Guthrie Tait, a fiery defender of British priority in science, who insisted on keeping quarternions pure and undefiled. Gibbs countered by demonstrating the use of vectors in astronomy and *electromagnetism.

In his last publication, Gibbs in 1903 created a statistical mechanics more general than those of Maxwell or Ludwig *Boltzmann. Initially he considered the behavior of ensembles of mechanical systems obeying Hamilton's mechanics and sharing the same total energy but having different positions and velocities. He moved from these "microcanonical ensembles" to "canonical ensembles," whose only common quantity was the same number of particles, and demonstrated that such ensembles behaved as thermodynamic systems. His "grand canonical system" was an ensemble of systems with different numbers of particles.

Gibbs was well aware of the major problems within his system, particularly its incorporation of the equipartition theorem (that the average energy for every degree of freedom of motion is equal) and the ergodic hypothesis (that all the systems in an ensemble passes through every possible configuration consistent with its total energy). He was also dissatisfied with his understanding of irreversibility and did not see his work as a description of real gases.

Gibbs's approach ended in paradox. The entropy of the Grand Ensemble depended on whether the particles in the systems were distinguishable or indistinguishable from those in other systems of the ensemble. For Gibbs, this was a technical problem; for his critics, it addressed the properties of real molecules.

Gibbs's paradox was not so much solved as bypassed by the development of quantum physics. Max *Planck and Albert *Einstein postulated statistical distributions of energy for black-body radiation (photons) and the photoelectric effect (electrons), respectively, avoiding the conundrum of deriving them from mechanical principles.

Josiah Willard Gibbs, *Elementary Principles of Statistical Mechanics* (1903). Josiah Willard Gibbs, *The Scientific Papers of J. Willard Gibbs*, Henry Andrews Bumstead and Ralph Gibbs Van Name, eds., 2 vols. (1906). Lynde Phelps Wheeler, *Josiah Willard Gibbs: The History of a Great Mind* (1952). Martin J. Klein, *The Physics of J. Willard Gibbs in His Time*, in *Proceedings of the Gibbs Symposium, Yale 1989*, D. G. Caldi and G. D. Mostow, eds. (1990).

ELIZABETH GARBER

GLACIOLOGY is the discipline that examines how glaciers and ice sheets behave. It studies their origin and accumulation, deformation and movement, sublimation and melting, as well as how glacial ice interacts with climate. A subdiscipline of *geophysics, glaciology also has strong ties to the atmospheric sciences, particularly climatology, and to glacial geology, which analyzes the history and geological effects of glaciers, glaciation, and *ice ages. The study of mountain glaciers and polar ice caps has required mountaineering skills and other high-risk tactics, so the history of glacier exploration and examination is also a history of adventure and, in some cases, tragedy.

The first International Polar Year (1881–1883) was largely responsible for the first studies of high-latitude glaciers and ice sheets. Expeditions to Greenland directed by Fridtjof Nansen in the 1890s and by Alfred Wegener in the early 1900s surveyed the ice sheet. Robert Scott's Antarctic expedition of 1901–1904 conducted the first studies of the southern ice sheet. The renowned valley glaciers of southeast Alaska were first examined by Grove Karl Gilbert while on the Harriman Alaska expedition in 1899.

In the early twentieth century, R. M. Deeley and P. H. Parr made the first suc-

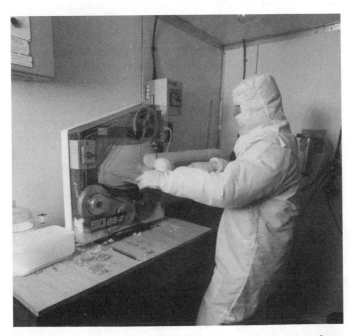

A glaciologist shown sawing through a section of an ice core from Antarctica. He works in a refrigerated clean room in Cambridge, England, to minimize melting and contamination.

cessful mathematical models of glaciers as viscous fluids. The first professional society for glaciology was founded in the 1930. In the 1950s—recognized within the field as the beginning of its modern era—John Nye led in putting glaciology on a sound physical footing. In the same decade the International Geophysical Year (IGY) (1957–1958) sparked further studies of high-latitude phenomena.

Ice sheets preserve one of the finest records of climate change over the last hundred thousand years or more. Among the notable discoveries of recent decades is that the ice record of atmospheric carbon dioxide content shows a marked positive correlation with fluctuations in global temperature. Fears of global warming have impelled scientists also to examine short-term behavior of these large ice sheets, which hold much of the world's water and hence have an important influence on global changes in sea level. Programs are in place for drilling and examining cores from the Greenland and Antarctic ice sheets. Other glaciologists are studying the rapid surging and retreat of ice shelves and of glaciers, the basal boundary condition of glaciers, and the interaction of glaciers with their substrates. The study of ice on other planetary bodies, notably on Jupiter's moons Europa and Ganymede, on Saturn's icy satellites, and on Triton, is an emerging field. Icy outer crusts on these bodies exhibit phenomena uncommon to Earth—impact cratering preserved in ice, and ice volcanism—eruption of liquid (probably water) through the icy crust.

"The Histories of the International Polar Years and the Inception and Development of the International Geophysical Year," *Annals of the IGY*, vol. 1 (1959). G. K. C. Clarke, "A Short History of Scientific Investigations on Glaciers," *Journal of Glaciology*, special issue commemorating the fiftieth anniversary of the International Glaciological Society (1987): 4–24.

JOANNE BOURGEOIS

H

HALLEY, Edmond (1656-1742), astronomer, eponym of a comet.

Born into a prosperous London family engaged in business, Halley attended Queens College, Oxford. He made astronomical observations there and with John Flamsteed at the Royal Observatory, then under construction. In 1676, his third year at Oxford, Halley published three papers in the *Philosophical Transactions of the Royal Society,* on sunspots, on an occultation of Mars by the Moon, and on determining a planet's *orbit around the Sun.

In October 1676 Halley sailed for St. Helena, the southern-most territory then under English rule and a three-month passage from London. For a year Halley measured positions of stars; he also observed eclipses and a transit of Mercury across the face of the Sun. His father paid for his instruments. At the request of King Charles II, the East India Company provided for his transportation on one of their ships and for his maintenance on the island. Although tormented by clouds and rain, Halley produced a chart of stars of the southern hemisphere, which he dedicated to the king. It contained a newly depicted constellation, *Robur Carolinum (Charles's Oak).* Soon after Halley's return, the Royal Society elected him a fellow, at the early age of 22.

In August 1684, Halley asked Isaac *Newton what sort of planetary orbit an inverse square force of attraction would produce. Newton answered that it would be an ellipse; indeed, he had already calculated it. But he could not find his calculation. In November, Newton provided Halley with new calculations showing not only that the orbit would be an ellipse, but also incorporating *Kepler's second and third laws of planetary motion. Halley arranged for the Society to publish Newton's work in a vastly expanded version, the world-famous *Principia mathematica philosophiae naturalis* (1687). Halley collected material, edited the book, saw it through the printers, corrected proofs, and bore all costs.

In 1695 Halley agreed to calculate *comet orbits for a second edition of the *Principia.* He soon realized that the comets of 1531, 1607, and 1682 had similar orbits; he deduced that they were the same comet turning around the Sun in a pronounced elliptical orbit. Halley died before the predicted return of his comet in 1758. He also died before the transits of Venus of 1761 and 1769, for the observation of which he prepared detailed instructions. Following them his successors would be able to deduce the size of the solar system.

In the 1690s Halley also discovered, from his study of solar and lunar eclipse observations by an Arab astronomer in the ninth century, that the Moon's mean motion had accelerated. Later, in 1718, Halley announced that a comparison of stellar positions in his day with those measured by Hipparchus revealed that in the course of 1800 years several of the supposedly fixed stars had altered their places relative to other stars; they had motions of their own.

Halley found time to undertake the attempted salvage of gold from a sunken ship, which involved improving the diving-bell; to pinpoint the time and place of Julius Caesar's invasion of Britain from classical sources, astronomical dating of a full moon, and a knowledge of tides; to develop mortality tables showing the distribution of deaths by age group; to estimate the acreage of England and Wales; to study trade winds; to serve two years as Deputy Comptroller of the Mint at Chester; to estimate the maximum thickness of an atom; to experiment with linking the scale of a thermometer to the boiling point of alcohol; to study the relationship between the pressure and velocity of a fluid; to survey harbors on the Dalmatian coast for potential use by the British fleet in the anticipated war of the Spanish succession; and to serve for several years as captain (an extremely rare appointment for a civilian) of his Majesty's ship *Paramore* carrying out a study of tides in the Channel and of magnetic variation in the Atlantic. During his second voyage, Halley discovered and took possession in his Majesty's name of a small, uninhabited volcanic island in the South

Atlantic. A more significant outcome of his voyage was a world map exhibiting lines of equal magnetic declination – an attempt at a method for finding longitude at sea.

Halley was appointed Savilian Professor of geometry at Oxford in 1704, and produced editions of several classics of ancient Greek geometry. He was elected Secretary of the Royal Society in 1713; he resigned the position in 1720 when he succeeded John Flamsteed as Astronomer Royal, though he remained active in the Society. Halley carried out astronomical observations until a few months before his death, at Greenwich, in 1742.

Angus Armitage, *Edmond Halley* (1966). Alan Cook, *Edmond Halley: Charting the Heavens and the Seas* (1998).

NORRISS S. HETHERINGTON

HAWKING, Stephen (1942–), and **Carl SAGAN** (1934–1996), icons of science who starred on television and wrote popular books, but whose scientific reputations are very different.

Carl Sagan's thirteen-part television series *Cosmos* (1980) has been seen by an estimated 400 million people. *Time* magazine featured on its cover the entrancing visionary with the strangely halting yet melodic voice. Sagan was "the Prince of Popularizers," "the Showman of Science," and "the Nation's Scientific Mentor to the Masses." Many of his peers, however, never accepted Sagan as a serious scientist. He made connections and identified goals, but had a short attention span and often failed to follow through on details. Abrasive, arrogant, and egomaniacal, Sagan made few friends and many enemies. Envy no doubt also played a part.

Sagan began as an undergraduate at the University of Chicago in 1951; he finished a controversial doctoral dissertation, including sections on possible life on the Moon and a greenhouse model for the atmosphere of Venus, in 1960. His showmanship and chutzpah were evident early. A campus science lecture series he organized on the creation of life in the universe had several distinguished faculty and one young graduate student as lecturers.

Sagan won a postdoctoral fellowship to the University of California at Berkeley that year and went on to a junior appointment at Harvard University in 1963. His book *Intelligent Life in the Universe* (1966), co-authored with the Russian scientist I. S. Shklovskii, gave Sagan a taste of fame. But Harvard was not impressed, and Sagan, denied tenure, left for Cornell University. There he developed into an admired superstar, flitting like a butterfly from one scientific topic to another while becoming the preeminent voice of American space science and a national celebrity. For the 1972 *Pioneer 10* spacecraft, Sagan and his wife designed a plaque, a potential message to other civilizations, showing the location of Earth and naked male and female figures (*see* EXTRATERRESTRIAL LIFE). His book *The Cosmic Connection*, on the origin of life, extraterrestrial life, and space travel, came out a year later.

In 1985 Sagan published a science fiction novel, *Contact*. The heroine's successful attempt to communicate with alien intelligences helped win in the real world a temporary reprieve from congressional budgetcutters for the SETI (Search for Extraterrestrial Intelligence) program. The novel was later made into a movie.

Sagan used his fame to champion the "nuclear winter" hypothesis: that a nuclear exchange might trigger urban fires, cloud the earth's atmosphere, block sunlight, lower temperatures to freezing, prevent crops from growing, and destroy civilization. This hypothesis remains contested. Its conclusion that a nuclear war is not winnable does not seem to have had much effect on nuclear strategy.

Sagan was rejected for membership in the National Academy of Sciences (NAS) in 1992, but later the NAS awarded him its Public Welfare Medal for his popularization of science. In 1994, he was diagnosed with myelodysplasia, a blood disorder leading to acute leukemia. A bone-marrow transplant prolonged his life until 20 December 1996.

In 1962 Stephen Hawking was at Cambridge studying for a doctorate in *cosmology. He had only worked about an hour a day as an Oxford undergraduate, physics and mathematics problems having presented little challenge to the young genius, and his laziness caused him to flounder at

Cambridge. That year he was diagnosed with amyotrophic lateral sclerosis (motor neuron disease, also known as Lou Gehrig's disease). Thought and memory are not affected, but paralysis progresses. The disease rapidly worsened, then slowed. Now he realized that there were many worthwhile things he could do if reprieved. He completed his dissertation in 1965 and obtained a research position at Gonville and Caius College.

Einstein's general *relativity theory predicted superdense neutron stars, black holes, and points of infinite density in the universe, but their existence seemed highly improbable. That changed with the discoveries of the *pulsar and quasar in the 1960s. Astronomers soon ascribed the power of quasars to the rotational energy of gigantic *black holes and identified pulsars as rotating neutron stars.

Meanwhile, Hawking, in collaboration with Roger Penrose, a mathematician at the University of London, was developing a new mathematical technique for analyzing the relation of points in space-time, and proved mathematically that *space and time would have a beginning in a Big Bang and end in black holes. Undertaken as an esoteric mathematical study, Hawking's work was suddenly transformed into a major contribution to the most exciting scientific topic of the decade. Just a few years out of school, Hawking already had the reputation of a new Einstein.

In the early 1970s, Hawking, unable physically to manipulate mathematical equations on paper, combined relativity theory and *quantum physics to produce a new understanding of black holes. He became a fellow of the Royal Society in 1974 at the unusually early age of thirty-two. Scientific audiences gathered to see the sticklike figure hunched in a wheelchair mumbling almost incomprehensibly. In 1985, Hawking caught pneumonia and had a tracheostomy. Since then he has used a computer and a speech synthesizer to communicate.

Public fame intensified after Hawking was profiled on a BBC television program in 1983. In need of money for his children's schooling, his own nursing, and his family's support after his death, Hawking obtained a large advance for a popular book. Millions of copies, in twenty different languages, of *A Brief History of Time: From the Big Bang to Black Holes*, have been purchased, if not read, since 1988. The film version appeared three years later, followed by a six-part television series, *Stephen Hawking's Universe*, in 1997.

Implications drawn from Einstein's general theory of relativity won fame and respect for Hawking after the discovery of quasars and pulsars demanded explanations that his work provided. In his person, Hawking is an inspiration for the physically handicapped. Sagan inspired searches for extraterrestrial life. It is perhaps possible that early fame and the duties of a popularizer short-circuited Sagan's potential for fundamental scientific advance and the type of respect that Hawking earned.

Michael White and John Gribbin, *Stephen Hawking: A Life In Science* (1992). Stephen Hawking, *Black Holes and Baby Universes and Other Essays* (1993). Keay Davidson, *Carl Sagan: A Life* (1999). William Poundstone, *Carl Sagan: A Life in the Cosmos* (1999). Web Sites: http://www. Hawking.org. uk; http://www.sciam.com/explorations/ 010697sagan/ 010697explorations.html.

NORRISS S. HETHERINGTON

HEISENBERG, Werner (1901–1976), and **Wolfgang PAULI** (1900–1958), theoretical physicists.

Heisenberg and Pauli were key members of the small group of theorists who invented quantum mechanics, the physics of atomic and molecular processes, during the 1920s. Pauli was called "the conscience of physics" for his mastery of theoretical physics and his critical assessment of ongoing work. His review articles and books on *relativity theory, statistical mechanics, and *quantum physics are masterpieces of physical insight. Heisenberg also worked on hydrodynamics. His decisions to remain in Germany after Hitler's rise to power and to work on applications of nuclear fission during the war still evoke controversy.

Heisenberg was quiet and friendly, at once retiring and daring, virtuous and athletic. Pauli to the contrary was outspoken and aggressive, systematic yet inclined to mysticism and risqué nightlife. Like most quantum physicists of the 1920s, Heisenberg and Pauli were products of the European upper-middle-class cultural elite. Both of their fathers were professors: Pauli's, of colloid chemistry at the University of Vienna; Heisenberg's, of Byzantine philology at the University of Munich. Both sons attended outstanding humanistic gymnasia (high schools), which emphasized classical languages and literature. Both were attracted to physics by the excitement attending Albert *Einstein's theory of *relativity, and both earned doctorates in theoretical physics with Arnold Sommerfeld in Munich. Pauli arrived at the University of Munich in 1918, two years ahead of Heisenberg. Heisenberg's early encounter with Pauli, then in his last semester, helped to turn Heisenberg toward atomic theory. Their semester together marked the beginning of the lifelong collaboration that encouraged their most significant contributions to quantum physics. Their voluminous technical correspondence offers a rich resource for the history of twentieth-century physics.

Pauli and Heisenberg each pursued postdoctoral work in quantum physics with Max Born in Göttingen and Niels *Bohr in Copenhagen. Although still in their twenties, both assumed full professorships in theoretical physics in 1928: Heisenberg at the University of Leipzig, Pauli at the Swiss Federal Institute of Technology in Zurich. There they trained students from the world over until well after the outbreak of war. In 1942 Heisenberg transferred to the Kaiser Wilhelm Institute for Physics in Berlin, where he headed German fission research, and a year later assumed a professorship at the University of Berlin. Fearing a German invasion of Switzerland, Pauli moved to the Institute for Advanced Study in Princeton. He returned to Zurich after the war, while Heisenberg remained head of his institute, renamed for Max *Planck, in Göttingen and Munich. He was heavily involved in West German science and cultural policy.

Heisenberg and Pauli entered their profession at a time of great ferment. New data and analyses indicated the inadequacy of the planetary quantum model of the atom developed earlier by Bohr and Sommerfeld. At the same time, turmoil surrounding the German defeat in World War I, hyperinflation, and a boycott of German science made professional advancement difficult for young physicists. Special scholarships and foreign philanthropy helped maintain the financial stability of German physics, while social disengagement and intense work by its practitioners fostered rapid developments.

Together with other physicists—P. A. M. *Dirac, Ernst Pascual Jordan, Hendrik Kramers, Alfred Landé, and Erwin *Schrödinger—and their mentors—Bohr, Born, and Sommerfeld—Heisenberg and Pauli probed the boundaries of the quantum theory in its applications to spectroscopy, atomic models, and the interactions of light with atoms. Pauli's emphasis on empirical data and his constant criticism of Heisenberg's inconsistencies helped push Heisenberg to achieve the first breakthrough to quantum mechanics in 1925. Heisenberg's often radical approach helped to encourage Pauli's well-known "exclusion principle" in the same year, a principle limiting the number of electrons in each quantum state of atoms and molecules. After further correspondence with Pauli in 1926 and 1927, Heisenberg, then in Copenhagen, proposed his well-known principle of uncertainty, or indeterminacy, one of the pillars of the current Copenhagen Interpretation of quantum mechanics (see COMPLEMENTARITY AND UNCERTAINTY).

The relativistic formulation of quantum mechanics and its extension to the electromagnetic field during the late 1920s formed the background to the only joint Heisenberg-Pauli publication, a two-part paper in 1929 and 1930 presenting a general quantum theory of fields. It provided the foundation for subsequent work. But their paper also confirmed the existence of infinite results in the calculation of observed finite quantities, a problem that plagued field theories for years. Two other collaborations were influential but unpublished at the time: a report for the canceled 1939 Solvay Congress, and a proposed unified theory from which Pauli resigned shortly before his death.

New data raised additional challenges for quantum mechanics during the 1930s. Among these were the discovery of the neutron, the energy distribution of electrons emitted by neutrons in beta decay, and the mysterious showers of particles created when a cosmic ray smashes into matter. Correspondence, conference meetings, and collaborations between their institutes enabled Heisenberg, Pauli, and their coworkers to achieve fundamental new results. Heisenberg presented the first neutron-proton model of the nucleus in 1932. A year later Pauli proposed a solution to the energy puzzle in beta decay by postulating the existence of what became the neutrino, an elementary particle later associated with the weak force, one of the four forces of nature. Utilizing new data on cosmic rays and their interactions with matter, Heisenberg and Pauli produced influential studies of another of the four forces, the proposed strong force, and its elementary particle, the meson, which challenged the adequacy of then-current quantum mechanics.

As *accelerators replaced cosmic rays after the war as sources of elementary-particle data, Heisenberg and Pauli remained at the forefront of work on high-energy interactions and the search for a unified theory of the four forces (or fields). Nevertheless, they were skeptical that the infinities of field theory could be effectively eliminated through mere subtraction of the offending terms, a process known as "renormalization" that is now widely accepted. They argued instead for a new transformation in *quantum physics.

Max Jammer, *The Conceptual Development of Quantum Mechanics* (1966). Charles P. Enz, *No Time to be Brief: A Scientific Biography of Wolfgang Pauli* (2002). Wolfgang Pauli, *Wissenschaftlicher Briefwechsel mit Bohr, Einstein, Heisenberg u. a.*, eds. Armin Hermann, Karl von Meyenn, Victor Weisskopf (1979–). Abraham Pais, *Inward Bound: Of Matter and Forces in the Physical World* (1986). David C. Cassidy, *Uncertainty: The Life and Science of Werner Heisenberg* (1992).

DAVID C. CASSIDY

HELMHOLTZ, Hermann von (1821–1894), physiologist, physicist, philosopher, and statesman of science; and **Heinrich HERTZ** (1857–1894), physicist.

Hermann Ludwig Ferdinand von Helmholtz came from a lower-middle-class family that stressed education and culture. As a youth, he wanted to become a physicist, but financial considerations forced him instead to become a medical doctor. In 1838, he enrolled in the Friedrich-Wilhelms-Institut in Berlin, the Prussian military's medical-training institution.

While serving as a Prussian military doctor in Potsdam (1843–1848), Helmholtz published articles on heat and muscle physiology. In 1847, he also published his epoch-making memoir, *On the Conservation of Force,* which provided the conceptually clearest and most general presentation to that time of what became known as the law of conservation of energy (*see* CONSERVATION PRINCIPLES).

Helmholtz left military service for academic life in 1848. He taught for a year at the Art Academy and served as an assistant at the Anatomical Museum in Berlin. In 1849, he became an associate professor of physiology at the University of Königsberg. He soon announced a major invention, the ophthalmoscope, and a major discovery: nervous impulses are propagated at a finite, measurable velocity, hence a short time intervenes between human thought and bodily reaction. He soon became a full professor (1851). Before leaving Königsberg for Bonn in 1855, Helmholtz contributed important studies of color, human perception, electric currents within the human body, and the eye's accommodation.

Helmholtz remained at Bonn for only three years, during which time he published the first part of his magisterial *Handbook of Physiological Optics* (1856). He left for Heidelberg in 1858 and there completed the second (1860) and third (1867) parts of the *Handbook,* which synthesized the empirical results of physiological optics and propounded Helmholtz's own theory of perception. During the same period, he revolutionized musicology with his *On the Sensations of Tone as a Physiological Basis for the Theory of Music* (1863). Along the way, he independently discovered non-Euclidean geometry and gave popular lectures on science to lay and professional audiences. These last were published as *Popular Scientific Lectures* (1865, 1867, 1876).

At Heidelberg Helmholtz's intellectual interests had shifted from physiology to physics, especially hydrodynamics, acoustics, electric currents, and electrodynamics. With the founding of the German empire in 1871, Helmholtz was called to Berlin as professor of physics. Now widely perceived as Germany's foremost man of science, he was received into the Prussian nobility. From 1887 to 1894, he served as the founding president of the Imperial Institute of Physics and Technology in Berlin, an institute devoted to pure and applied physics, the setting of physical standards, and testing.

Helmholtz's most gifted student, Heinrich Hertz, came from a well-situated and highly cultured family. Intending to become an engineer, Hertz studied at the Dresden Polytechnic and the Munich Technische Hochschule, but in 1877 he switched to physics and the University of Munich. A year later, he transferred to Berlin to study with Helmholtz.

In the 1860s and 1870s, Helmholtz was much concerned with evaluating competing theories of electrodynamics. Under Helmholtz's direction, Hertz first worked on a topic on electric currents. When Helmholtz suggested in 1879 that Hertz test experimentally the assumptions underlying James Clerk *Maxwell's theory of *electromagnetism, Hertz demurred, in part because he thought it too difficult. Instead, he completed a doctoral dissertation on electromagnetic induction.

During the next three years (1880–1883), Hertz served as Helmholtz's assistant. In 1883, he left Berlin for a lectureship at the University of Kiel. There he produced a penetrating analysis of Maxwell's electrodynamic equations, showing their essential validity even as he doubted Maxwell's contiguous-action interpretation of them.

In 1885, Hertz became professor of physics at the Karlsruhe Polytechnic. During the next four years, he confronted the problem that Helmholtz had posed in 1879: He tested Maxwell's theory, demonstrating experimentally the existence of electromagnetic waves propagated at finite velocity in air. (During the course of his work, he also discovered, in 1887, the photoelectric effect.) Helmholtz arranged for the quick publication of Hertz's results. These years marked his greatest achievements in physics.

In 1889, Hertz moved to the University of Bonn. He further analyzed Maxwell's theory, now adopting Maxwell's contiguous-action interpretation of his equations, which he (and others) put into their canonical form. He also took up the new problem of the electrodynamics of moving bodies. He had clarified Maxwell's theory and thus fulfilled Helmholtz's program of clarifying electrodynamics, while simultaneously achieving world fame for himself.

During the 1880s Helmholtz had sought to reformulate mechanics and to provide a mechanical version of the second law of *thermodynamics. Hertz now undertook to develop this program. In *The Principles of Mechanics* (1894), Hertz eliminated the concept of force and developed an ether-based physics. The *Principles* represented one of the last statements of the mechanical view of nature. The book was received with much skepticism, even by Helmholtz, who wrote a preface to it.

Helmholtz sought to unify physics, if not all the sciences, and indeed hoped for the ultimate unification of all culture. Hertz, by contrast, worked within the physics that Helmholtz had outlined. Together they cleared the ground in electrodynamics and mechanics, and so paved the way for Max *Planck, Albert *Einstein, and others at the turn of the century. Furthermore, Hertz's results in electrodynamics proved as seminal for technology as for physical theory, for they set the stage for revolution in wireless communication.

Leo Koenigsberger, *Hermann von Helmholtz* (1906; reprint 1965). David Cahan, ed., *Hermann von Helmholtz and the Foundations of Nineteenth-Century Science* (1993). Jed Z. Buchwald, *The Creation of Scientific Effects: Heinrich Hertz and Electric Waves* (1994). Davis Baird, R. I. G. Hughes, and Alfred Nordmann, eds., *Heinrich Hertz: Classical Physicist, Modern Philosopher* (1998).

DAVID CAHAN

HERSCHEL FAMILY: William (1738–1822), discoverer of Uranus, builder of giant telescopes, and observer of nebulae; **Caroline HERSCHEL** (1750-1848), first notable woman astronomer; and **John HERSCHEL** (1792-1871), embodiment of the ideal of universal knowledge.

Frederick William Herschel, born in Hanover, joined his father in the Hanoverian military band. After a disastrous encounter with a French army, William left for London, in 1757. Finding London crowded with musicians, William tried his luck in the country. There followed a brief stint as bandmaster of a militia regiment and a succession of free-lance engagements: composing, directing, and playing at Sunderland, Newcastle, Edinburgh, and Pontefract. He served as director of public concerts in Leeds for four years, and as organist, briefly at Halifax and then, beginning in 1766 in Bath.

Music directed William's interest to harmony and to Robert Smith's *Harmonies, or the Philosophy of Musical Sounds*, which in turn led to telescopes via Smith's other book, *System of Optics*. Soon William was cutting back on musical pupils to devote more time to constructing telescopes and observing the heavens.

Observing with a *telescope of his own construction on 13 March 1781, Herschel noticed an object visibly larger in appearance than nearby stars. He recorded in his journal the discovery of "either a nebulous star or perhaps a comet." *Comets usually are detected approaching the Earth, apparently increasing in size, and William perceived the object to be growing larger. Actually it was receding from the Earth; preconception affected perception. Astronomers calculated an orbit, revealing the object's planetary nature, and William was famous. He named his discovery "Georgium Sidus" after the king. George III reciprocated with an annual pension of £200, enough to allow William to devote all his time to astronomy. He also profited from selling telescopes he made, and in 1788 he married a wealthy widow.

William's sister Caroline Lucretia Herschel received an annual pension, too, of £50. She had moved to England in 1772 and sung in performances conducted by William. When he took up astronomy, she joined him in all-night observing, polished mirrors, and made calculations. Observing with her own small telescope after William's marriage, she discovered eight comets. After his death she compiled a catalog of his observations of clusters and nebulae.

William's increasingly powerful telescopes disclosed two moons around Uranus in 1787 and Saturn's sixth and seventh moons in 1789, and hundreds of double stars; it also resolved fuzzy nebulae into clusters of stars. In 1783 he completed a telescope with a 20-foot focal length and an 18.7-inch mirror, and in 1789 a monstrous telescope with a 4-foot mirror (weighing nearly a ton) at the end of a sheet-iron tube nearly 40 feet long. George III provided construction grants totaling £4,000, plus £200 a year for upkeep. The cumbersome telescope was soon abandoned.

William believed that the Milky Way is composed of many millions of stars, and that the Earth is embedded among them. He proposed to locate the solar system in the stratum by counting numbers of stars in different directions. Astronomers then had little interest in the stars, which they regarded primarily as a backdrop against which to record the motions of planets and comets. Nor were William's giant reflecting telescopes reliable in use, and his necessary working assumption that stars were equal in brightness was invalid. He reached farther into space than anyone had before and began to outline the structure of our galaxy, but his speculative cosmology did not attract disciples.

William's only child, John Frederick William Herschel, attended St. John's College, Cambridge. In 1813 he scored highest on the university examination in mathematics and was elected a Fellow of the Royal Society on the basis of a brilliant paper on mathematics published in the Society's *Philosophical Transactions*. He helped found the Analytical Society, which in a few years reformed British mathematics, replacing Isaac Newton's awkward calculus of fluxions with modern infinitesimal calculus. He began reading law in 1814 while also studying mineralogy. In 1815 he narrowly missed election to the Chair of Chemistry at Cambridge. In 1816 he received an MA degree and a fellowship at St. John's, and gave up law. Later in the year he resigned his university career to assist his ailing father with astronomical observations.

Experiments in *optics led John Herschel to important discoveries in polarization. Concurrently, his observations of double stars won awards in London and Paris. And he published papers on *electricity and *magnetism. He was Secretary of the Royal Society from 1824 to 1827 and then President of the Astronomical Society. He turned away feelers regarding the Lucasian Professorship at Cambridge (earlier held by Newton) and a Chair of Mathematics at London University. From 1825 to 1833 John used his father's 20-foot telescope to revise his father's catalog of nebulae and clusters, and also prepared a new catalog of 5,075 double stars, 3,347 of them discovered by himself. Further, he found time to write his *Preliminary Discourse on the Study of Natural Philosophy*, an introduction to the nature and method of science plus surveys of its various branches, published in 1830, and a *Treatise on Astronomy* for general readers, published in 1833.

From 1834 to 1838 John was at the Cape of Good Hope extending to southern skies with his father's 20-foot telescope his father's systematic survey of the heavens. John refused offers of financial support from both the Admiralty and the Royal Society, relying instead on family wealth. He did not want to become a national scientific figure, but hoped to enjoy peace in which to work out the results of his researches.

Back in England, John wrote up the results of his four years at the Cape, served another term as president of the (now Royal) Astronomical Society (1839–1841), but declined to be considered for election to the presidency of the Royal Society. Nor did he accept the Savilian Professorship of Astronomy at Oxford or become Member of Parliament from the University of Cambridge. In 1847 he moved to an isolated country house in Kent. Experimenting there in *photography and photochemistry, he used his knowledge of chemistry to devise a superior method of fixing images. His revised *Treatise on Astronomy*, now *Outlines of Astronomy* running some 700 pages, went through 12 editions between

1849 and 1873. Then he decided to take up public office, and was appointed Master of the Mint (a position once held by Newton) in 1850. He carried out the drastic reform mandated by Parliament, but failed in his effort to change British currency to the decimal system. In bad health, he retired in 1856.

On his death, John was hailed as the greatest scientist of the century and interred near Newton in Westminster Abbey. More recent appraisal rates the father higher for his creative imagination, and credits the son with stronger analytical powers.

Angus Armitage, *William Herschel* (1963). Michael Hoskin, *William Herschel and the Construction of the Heavens* (1963). Günther Buttmann, *The Shadow of the Telescope: A Biography of John Herschel* (1970). Michael Hoskin, *Caroline Herschel's Autobiographies* (2003).

NORRISS S. HETHERINGTON

HERTZ, Heinrich. See HELMHOLTZ, HERMANN VON, AND HEINRICH HERTZ.

HIGH-ENERGY PHYSICS. The field of high-energy physics emerged after World War II out of research in *nuclear physics and *cosmic ray physics. Its name referred to the energies of the nuclear and subnuclear particles that physicists sought to study, and hence the field was also sometimes called elementary particle physics.

Physicists in the first decades of the twentieth century used particles emitted by *radioactivity as a probe of the atomic nucleus. But many of the charged alpha and beta particles emitted in natural radioactive decay had insufficient energy to overcome the electrical barrier of the nucleus, and exist in insufficient quantities to provoke enough reactions for convenient study. Starting in the 1920s nuclear physicists sought to increase the energy of the particles by accelerating them. Early *accelerators passed electrically charged particles through a single large voltage drop (as in the Cockcroft-Walton and Van de Graaff accelerators) or multiple, smaller drops provided by high-frequency oscillators (as in the cyclotron and linear accelerator, or linac) to kick particles to higher speeds.

Technical refinements and industrial-size apparatus allowed particle accelerators to produce ever higher energies through the 1930s, beyond ten million electron volts.

In the meantime, starting in about 1930, physicists began to exploit a natural source of high-energy particles in cosmic rays. A thriving field of cosmic ray research focused on sorting the nature of cosmic radiation as well as new phenomena and new particles, including the positron (detected in 1932 by Carl David A. Anderson) and a mysterious new particle with mass midway between that of the electron and proton, which was hence called the meson. Whereas nuclear physics in the 1930s emphasized the particle accelerator, cosmic ray research relied on particle detectors, especially *cloud chambers and Geiger counters.

World War II interrupted the pursuit of high-energy particles but provided new resources, both technical and political, to exploit after the war. Technical resources included advances in microwave electronics and new designs for accelerators, both of which allowed physicists to push into energies of billions of electron volts. By about 1950 physicists could perceive some separation between nuclear physics and a new subfield of particle physics, which became known as high-energy physics.

New political resources stemmed from the contributions of physicists to military technology during World War II and their continued mobilization during the Cold War. National governments supported high-energy physics as a way to train new scientists and engineers who might then work on problems of national interest; as a means to keep talented scientists on tap in national laboratories in case of military emergency; and as a hedge against scientific discoveries that might have military or industrial applications. High-energy physics also provided a surrogate arena for international competition, with American, Soviet, and European labs jockeying for the claim to high-energy hegemony. High-energy physicists accepted and encouraged such justifications of state support, in exchange for the opportunity to pursue interesting and challenging scientific problems at higher energies.

Postwar high-energy physics combined detectors and accelerators in laboratories that typified a new type of big science, characterized by large and expensive equipment, state support, cooperative team research, collaboration of scientists and engineers, and industrial-style management. American laboratories led the postwar development of big-science accelerator programs, notably at the University of California at Berkeley and Brookhaven in New York, although the Soviet Union and a new European lab at CERN soon offered strong competition. The United States in the 1960s would choose to centralize its largest accelerators in a new lab at Fermilab near Chicago, which would thereafter challenge CERN for the claim to the highest energies.

High-energy physics had theoretical and experimental components. Experimentalists pushed toward higher energies, lured by the possibility of producing previously unseen particles with masses equivalent to the energy of particle collisions. In 1955 Berkeley scientists achieved the exemplary discovery of the antiproton using the first generation of postwar accelerators. Higher energies required ever larger magnets and vacuum chambers and bigger budgets, all of which constrained the ambitions of accelerator builders. The invention of strong focusing in 1952 provided a way around the problem of magnets and vacuum volumes, both of which contributed to costs. The principle used an alternating gradient in the magnetic field to compress the beam of particles, and quickly led to proposals from Brookhaven, where the idea originated, and CERN, whose design problems had inspired the Brookhaven work, for machines in the range of 25 billion electron volts.

Physicists still needed ways to study the output of accelerators, and particle detectors grew in importance and size in the 1950s and beyond. They fell into two general categories. Image detectors, such as cloud and bubble chambers and nuclear emulsions, produced a snapshot of the tracks left by individual particles at a particular time and place. Logic devices, such as electronic counters, accumulated statistical counts of large numbers of particles. By the 1970s physicists were combining the two types in new, even larger detectors.

Experiment led theory in high-energy physics through the 1950s, as theorists struggled to explain new resonances and particles emerging from accelerators. To make sense of the proliferating particles theorists classified them according to such properties as strangeness and isospin; then, in the 1960s, theorists developed the *quark model of matter to resimplify the concept of fundamental particles and began to seek a unified theory of the four fundamental forces that would incorporate the quark model. By the 1970s theory was driving experiment, as particle physicists directed their research to find particles predicted by the new theories. Their efforts paid off in the detection of the J/psi and W and Z particles, weak neutral currents, and several of the postulated quarks, thus confirming the so-called *standard model of the theorists and encouraging speculation about a grand unified theory, or *theory of everything.

The spiraling energies available in accelerators changed the definition of high energy. Accelerators that had pushed back the high-energy frontier in the late 1940s were considered medium-energy machines by the mid-1960s; succeeding generations of devices might last only a decade or so at the cutting edge. Theorists, in the meantime, extended their equations far beyond energies available in the laboratory or in nature, and particle physics began to merge with *cosmology; the phenomena postulated by theorists existed only in the milliseconds after the Big Bang and disappeared as the universe expanded and cooled.

Scientific and popular attention to the quest of particle physicists made theirs the most glamorous field of physics in particular, if not science in general, in the second half of the twentieth century. High-energy physicists claimed to engage in the most fundamental research, from which all other science was derived, since they dealt with the elementary constituents of matter. Physicists in other fields, such as *solid-state physics, questioned the pretensions of particle physicists and sought to divert some of their substantial government

funding. The perceived fundamental nature of high-energy physics alone could not ensure continued state support, especially after the end of the Cold War. American physicists in the 1980s proposed a massive new accelerator, the Superconducting Super Collider (SSC), to pursue the Higgs boson and other exotic phenomena predicted by theory in the neighborhood of 40 trillion volts. In 1993 the U.S. Congress voted to end the multi-billion dollar project, unconvinced that the scientific results justified the social investment.

Laurie Brown and Lillian Hoddeson, eds., *The Birth of Particle Physics* (1983). Andrew Pickering, *Constructing Quarks: A Sociological History of Particle Physics* (1984). Laurie Brown, Lillian Hoddeson, and Max Dresden, eds., *Pions to Quarks: Particle Physics in the 1950s* (1989). Peter Galison, *Image and Logic: A Material Culture of Microphysics* (1997). Lillian Hoddeson, Laurie Brown, Max Dresden, and Michael Riordan, eds., *The Rise of the Standard Model: Particle Physics in the 1960s and 1970s* (1997).

PETER J. WESTWICK

HOMEOSTASIS. The American physiologist Walter Cannon coined the word "homeostasis" in 1926 to designate the coordinated physiological reactions that maintain steady states in the body. He believed that a special term was necessary to differentiate the complex arrangements in living beings, involving the integrated coordination of a wide range of organs, from the relatively simple physico-chemical closed systems in which a balance of forces maintains an equilibrium. "Changes in the surroundings," Cannon wrote, "excite reactions in [the open system that constitutes a living being], or affect it directly, so that internal disturbances of the system are produced. Such disturbances are normally kept within narrow limits, because automatic adjustments within the system are brought into action, and thereby wide oscillations are prevented and the internal conditions are held fairly constant."

Cannon illustrated his concept by describing a variety of mechanisms that maintain constant conditions in the fluid matrix, or "internal environment," of higher animals. These included materials such as glucose and oxygen in the blood, as well as the fluid matrix's temperature, osmotic pressure, and hydrogen-ion concentration. The knowledge of the mechanisms that Cannon showed to be involved in these reactions came largely from his own previous experiments on the role of the autonomic nervous system and the adrenal secretions. The close association he established between homeostatic mechanisms and the preservation of conditions in the internal environment, however, derived in large part from the inspiration of the nineteenth-century French physiologist Claude Bernard, whom Cannon acknowledged as the first to give a "more precise analysis" to general ideas about the stability of organisms. Cannon quoted particularly Bernard's "pregnant sentence" that "It is the fixity of the 'milieu interieur' which is the condition of free and independent life."

Originally a physiological principle, homeostasis took on broader meanings after World War II, with the recognition of the similarity of homeostatic mechanisms to feedback controls in servo-mechanisms. Biologists applied the concept at all levels—cellular, organ system, individual, and social systems. The maintenance of steady concentrations of the intermediates of a metabolic pathway despite a constant flux of matter and energy through the pathway became an example of homeostatic regulation. Cannon himself had asked in 1932 in his popular book *The Wisdom of the Body*, in an epilogue entitled *Relations of Biological and Social Homeostasis,* whether it might not "be useful to examine other forms of organization—industrial, domestic, or social—in the light of the organization of the body?" His suggestion has been followed, and homeostasis in its widest sense now means the "maintenance of a dynamically stable state within a system by means of internal regulatory processes that counteract external disturbances of the equilibrium."

Walter Cannon, "Organization for Physiological Homeostasis," *Physiological Reviews,* 9 (1929): 399–427. Claude Bernard, *Lectures on the Phenomena of Life Common to Animals and Plants* (1974).

FREDERIC LAWRENCE HOLMES

HOOKE, Robert (1635–1702), natural philosopher and inventor. Hooke was born in Freshwater, Isle of Wight, in 1635 to John Hooke, a minister. He showed a talent for invention as a young child, making sundials, a wooden clock, a model man-of-war that could sail and fire its guns, and other mechanical toys. Sent to London in 1648, he was taken into the home of Richard Busby, master of Westminster School, where he devoured classical languages and mathematics (the first six books of Euclid in a week) and continued his career of invention.

Hooke matriculated at Oxford in 1653 as a chorister and by 1655 had found his way into the brilliant circle there that was fashioning the new mechanical philosophy.

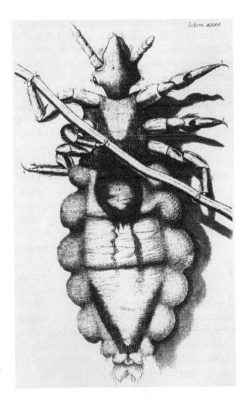

Hooke's louse as drawn by him from observations through his microscope. From Hooke's *Micrographia* (1665), which contains, besides many other beautifully executed drawings of microscopic objects, a description of the interference phenomenon known as Newton's rings.

He became assistant to Robert *Boyle, who perceived Hooke's genius for devising simple yet conclusive experiments. Hooke built an air pump for Boyle around 1657, and collaborated in Boyle's investigations with the pump and related instruments. The experiments confirmed what we know as Boyle's law. Hooke's diary and manuscripts mention many techniques and devices for flying. He invented important improvements to the clock, including spring-controlled escapements, and probably the spiral spring for watch balances and the anchor escapement for pendulum clocks. His work with springs led by 1678 to his law of elasticity (Hooke's law), "*Ut tensio sic vis*" or "Force is proportional to extension."

With the restoration of the English monarchy in 1660, the Oxford group moved to London and joined with others to found the Royal Society. In 1662, the Society appointed Hooke curator of experiments, charging him to furnish "three or four considerable experiments" for each weekly meeting. This seemingly impossible goal unleashed in him a flurry of creativity. From drawings for the Society of microscopical observations came the *Micrographia* (1665), one of the great works of the century, filled with images that still excite admiration. The *Micrographia* included speculations on light, which Hooke regarded as a rapid vibration of small amplitude. He discovered and explained the interference phenomena occurring between two lenses, later known as "Newton's rings." He introduced the idea of a wave front, explained its deflection at the boundary between media, and described diffraction. He invented telescopic sights, crosshairs, and a clock-driven telescope and the conical pendulum that drove it, and carried out experiments showing how the Moon's features might have been caused by impacting bodies or volcanic action. The universal joint is also his invention.

Hooke has been called the first scientific meteorologist, organizing cooperative weather observations and designing a standard *thermometer, graduated from a single fixed point, which modern analysis has proved remarkably accurate. He also invented a hygrometer, a wind gauge, numerous types of barometers, and a weather clock that recorded the readings of all of them. Hooke had an abiding interest in the earth's history, clearly recognizing the organic origin of fossils and suggesting that earthquakes, crustal movements, and other changes had cast marine fossils high upon dry land.

Hooke was also an accomplished architect. Appointed surveyor to the City of London after the Great Fire of 1666, he resurveyed most of the City; his buildings—the Royal College of Physicians and Bedlam Hospital among them—compare in quality to those of Christopher Wren, his close friend and collaborator.

Many of Hooke's inventions involved him in controversy. As always, mathematicians and natural philosophers pursued closely related problems, and betrayal was common. One of Hooke's backers probably provided Christiaan *Huygens with information on the spiral spring, which Huygens then claimed to have invented; considerable evidence also points to treachery on the part of Henry Oldenburg, secretary of the Royal Society, against Hooke and others. Hooke felt most bitterly toward Isaac *Newton. Hooke proposed that the planets' orbital motions are compounded of a straight-line motion along the tangent and a centripetal gravitational motion, and he postulated the inverse-square behavior of gravity. In 1679 he communicated these ideas to Newton. Newton, nursing a hatred against Hooke from an earlier controversy over their theories of light, proved these hypotheses mathematically, but refused to acknowledge Hooke in the *Principia*. Hooke, for his part, believed that Newton had stolen his ideas about gravity.

The historiographical tradition makes Hooke a difficult, jealous, and ungenerous man. The primary sources belie this portrayal. His diary shows him pursuing an active social life, always at coffeehouses and taverns with his many friends, enjoying deep and loving relationships. Historians have also faulted Hooke for taking on much but finishing little. On the contrary, he certainly completed his many scientific instruments, mechanical devices, surveys, and buildings. The truth seems to be that he had a practical rather than a theoretical mind, and that this has hurt his reputation at the hands of a historiography that values theory over practice.

Michael Hunter and Simon Schaffer, eds., *Robert Hooke: New Studies* (1989). Ellen Tan Drake, *Restless Genius. Robert Hooke and His Early Thoughts* (1996). Steven Inwood, *The Man Who Knew Too Much: The Strange and Inventive Life of Robert Hooke* (2002). Lisa Jardine, *The Curious Life of Robert Hooke* (2003).

THEODORE S. FELDMAN

HOYLE, Fred (1915-2001), and **Martin RYLE** (1918-1984), Cambridge astronomers and bitter rivals.

Hoyle, born in Yorkshire, the son of a wool merchant, won a scholarship to Emmanuel College, Cambridge, to study mathematics. As a graduate student he took up mathematical problems in astronomy. He was elected to a fellowship at St. John's College in 1939. During the war he worked for the Admiralty on radar.

Ryle was born in Brighton, the son of a physician who became Regius Professor of Physic at Cambridge in 1935, Physician Extraordinary to the King in 1936, and Chair of Social Medicine at Oxford in 1943. Ryle's uncle was the philosopher Gilbert Ryle. Ryle studied physics in Oxford and after graduating in 1939 joined a radio research team at the Cavendish Laboratory, Cambridge. During the war he developed radar for the Royal Air Force.

After the war Hoyle returned to Cambridge. He calculated the statistical distribution of nuclei resulting from nuclear reactions in a star and in 1953 predicted an energy level in carbon-12 essential for the synthesis of carbon from helium. The culmination of this work was the famous "B^{2}FH" paper of 1957 on nucleosynthesis, co-authored with Geoffrey and Margaret Burbidge and William Fowler. Fowler alone received the Nobel Prize for this work, in 1983. By then Hoyle was attributing the origin of life on Earth to an infall of organic matter from space, and influenza epidemics to viruses carried in meteor streams. Perhaps the Royal Swedish Academy had not wished to associate itself with these ideas. In 1997 the Academy did award Hoyle its prestigious Crafoord Prize, recognizing outstanding basic research in fields not covered by the Nobel Prize, for pioneering contributions to the study of nuclear processes in stars and stellar evolution.

In his autobiography, Hoyle wrote that, although the scientist must go against accepted scientific opinion to achieve anything really worthwhile in research, he or she needs fine judgment to avoid becoming a crackpot. His judgment may have faltered in later years, not only about the origin of life on Earth, but also in his continuing objection to big bang theory and his increasingly desperate *ad hoc* modifications of steady-state theory (*see* COSMOLOGY).

Hoyle first proposed steady-state theory in the late 1940s, with Hermann Bondi and Tommy Gold. The three had worked together on radar during the war. Their steady-state universe expanded but did not change in density because of the continuous creation of new matter. The most serious challenge to steady state theory came from Ryle. Back at Cambridge after the war, he had adapted surplus radar equipment to study radio emission from the Sun. Soon he discovered other radio sources. In 1951 Ryle believed that these sources were within our *galaxy, and hence of no cosmological interest; but over the next few years he concluded that most of his radio sources were extragalactic and that his observations could be used to test cosmological models. His survey of almost 2,000 radio sources completed in 1955 seemed to contradict steady-state theory.

Hoyle felt that Ryle was motivated not by a quest for the truth, but by a desire to destroy steady-state theory. Ryle felt that even if he never produced measurements with sufficient accuracy to refute any cosmological theory, at least the threat of doing so might discourage theoretical cosmologists, who would otherwise irresponsibly postulate theories with no chance of being disproved.

Refutation of cosmological theory required more precise location of radio objects, and this seemingly required impracticably large aerials. Instead, Ryle developed aperture synthesis, moving small aerial elements about to occupy successively the whole of the effective aperture of a much larger hypothetical aerial. This work earned him the Nobel Prize in physics in 1974, following many lesser but prestigious awards from scientific societies and academies. He had already been elected to a Fellowship at Trinity College in 1949, appointed to a new Chair of Radio Astronomy in 1959, knighted in 1966, and named Astronomer Royal in 1972.

In 1958 Hoyle became Plumian Professor of Astronomy at Cambridge, where he founded the Institute of Theoretical Astronomy in 1966. He served as vice president of the Royal Society from 1969 to 1971, as president of the Royal Astronomical Society from 1971 to 1973, and as a member of the Science Research Council from 1967 to 1972. He helped plan the 150-inch Anglo-Australian Observatory and became chairman of its board in 1973.

A bureaucratic squabble that included Ryle led to Hoyle's resignation in 1972, ending 39 years at Cambridge. He retired to a remote part of England's Lake District, and later to the south coast. Although he held visiting appointments at the California Institute of Technology and Cornell University, Hoyle had fewer contacts with observational astronomers after leaving Cambridge, probably to the detriment of his scientific work. He became more inclined to deny direct evidence he had not seen, allowing his fertile imagination great play. He explored topics as disparate as Stonehenge, panspermia, Darwinism, paleontology, and viruses from space.

Hoyle wrote many popular books, beginning with *The Nature of the Universe* in 1950, based on a series of popular radio broadcasts. In them Hoyle introduced the term "big bang" to discredit the rival theory, and the name stuck. His classic sci-fi novel *The Black Cloud* was published in 1957, and *A for Andromeda* was written in 1962 as a television play. *A Different Approach to Cosmology: From a Static Universe through the Big Bang towards Reality* appeared in 2000.

W. T. Sullivan, *The Early Years of Radio Astronomy: Reflections Fifty years after Jansky's Discovery* (1984). Fred Hoyle, *Home is Where the Wind Blows: Chapters from a Cosmologist's Life* (1994). Helge Kragh, *Cosmology and Controversy: The Historical Development of Two Theories of the Universe* (1996).

NORRISS S. HETHERINGTON

HUBBLE, Edwin (1889–1953), astronomer, showed that spiral nebulae are independent island universes beyond our galaxy and that the universe is expanding.

Hubble was born in Marshfield, Missouri. Attending the University of Chicago on an academic scholarship, he played on the undefeated 1909 basketball team and won letters in track. In 1910 he went to Oxford University as a Rhodes scholar. There he studied law, participated vigorously in sports, and traveled on the Continent. Back in the United States in 1913, Hubble taught high school physics and Spanish, and coached basketball. He joined the Kentucky bar, but never practiced law. In 1914 Hubble returned to the University of Chicago and its Yerkes Observatory. When the United States declared war on Germany in April 1917, Hubble rushed through his doctoral dissertation, took his final examination, and reported to the army three days later. He served in France, reaching the rank of major. Discharged in 1919, Hubble joined the Mount Wilson Observatory, just east of Los Angeles.

Hubble's doctoral dissertation was a photographic investigation of *nebulae, cloudy patches of light in the sky, of which little then was known. A few are nearby agglomerations of gas and dust, but most are distant clusters of stars. About 17,000 small, faint nebulous objects resolvable into groupings of stars had been cataloged, and some 150,000 were within reach of existing telescopes. A classification system was needed, and Hubble developed the now-standard one for galaxies during the 1920s.

Harlow Shapley's determination of distances to globular clusters had suggested by 1918 that the universe was a single, enormous, all-comprehending unit. Soon Hubble destroyed Shapley's big *galaxy model. Using the new 100-inch telescope at Mount Wilson, Hubble discovered Cepheid variable stars in several spiral-shaped nebulae (*see* GALAXY). Centuries of speculation over the possible existence of island universes similar to our galaxy came to an abrupt end in 1925 when Hubble announced his discovery of Cepheid variables in spiral nebulae and distance calculations placing them decisively beyond the boundaries of our galaxy. This multiplied the known size of the universe by about a factor of ten.

Early in the twentieth century most scientists, including Albert *Einstein, assumed that the universe is static. The Dutch astronomer Willem de Sitter formulated in 1916 a static model of the universe with apparent (but not real) redshifts in the spectra of nebulae greater for nebulae at greater distances. Normally, a shift of a spectral line toward the red end of the spectrum indicates that the object is moving away from the observer. In 1929 Hubble tested de Sitter's prediction. He had distances to forty-six extragalactic nebulae and redshifts measured by Milton Humason at Mount Wilson under Hubble's direction. The redshifts were greater at greater distances. De Sitter's static model of the universe, however, was abandoned and the redshifts generally interpreted as real Doppler shifts indicative of real velocities of recession. The empirical relationship between redshift and distance (actually, between redshift and size or brightness, unambiguous indications of distance) became generally understood as a correlation between velocity and distance. Hubble's famous velocity-distance relation showed that more distant nebulae are receding from us at greater speeds, that the universe is not static but expanding.

The velocity-distance relation made possible determinations of greater distances because redshifts could be measured even when individual stars and galaxies were too faint to distinguish. Hubble increased the size of the known universe by yet another factor of ten.

Hubble traveled to Europe several times to give lectures and receive important honors. He read widely in the history of science and was a trustee of the Huntington Library. He also partied in nearby Hollywood with movie stars, including Charlie Chaplin and Greta Garbo, and with famous writers, including Aldous Huxley, Christopher Isherwood, and Anita Loos. He fished in Scotland and the Colorado Rockies.

World War II interrupted Hubble's work on cosmology. He served as chief of ballistics and director of the Supersonic Wind Tunnels Laboratory at the Army Proving Grounds in Maryland, and was awarded the Medal of Merit for his wartime work. Hubble died in San Marino, California, of a heart attack soon after completion of the giant 200-inch telescope on Palomar Mountain, south of Los Angeles, and too soon for conclusive answers from the research program he had planned for the new telescope.

Edwin Hubble, *The Realm of the Nebulae* (1936). Richard Berendzen, Richard Hart, and Daniel Seeley, *Man Discovers the Galaxies* (1976). Robert Jastrow, *God and the Astronomers* (1978). Robert Smith, *The Expanding Universe: Astronomy's "Great Debate" 1900–1931* (1982). Gale E. Christianson, *Edwin Hubble: Mariner of the Nebulae* (1995). Norriss S. Hetherington, *Hubble's Cosmology: A Guided Study of Selected Texts* (1996).

NORRISS S. HETHERINGTON

HUMBOLDT, Alexander von (1769–1859), naturalist, geographer, explorer, scientific popularizer.

Born in Berlin the son of a Prussian officer and his Huguenot wife, Alexander von Humboldt and his brother Wilhelm became by 1810 internationally known figures in science, education, and administration. Initially educated at home, Humboldt's formal education encapsulated older traditions in cameralism (the management of the state's natural, financial, and population resources) and exposed him to novel changes in administration and management, scientific and philosophical thinking, and observational practices. Between 1787 and 1792 he studied at the Universities of Frankfurt-an-der-Oder and Göttingen, the Hamburg Commercial Academy, and the Freiberg Academy of Mines. He also studied engineering in Berlin.

He published initially on volcanism (1790), muscle and nerve stimulation (1790), plants in relationship to the environment (1793), and galvanism (1797). Later he took photometric measurements of the southern stars and explained why sound is amplified at night (the Humboldt effect). He spent his early professional years as an administrator or traveler. His best known scientific expedition was to Latin America with the French botanist Aimé-Jacques-Alexandre Bonpland. In 1826 the Prussian king called him back to Berlin to be privy councilor and court tutor to the crown prince; he also gave a popular university course on physical *geography. Humboldt helped to organize the first international scientific conference in Berlin in 1828. Thereafter, he divided his time between Paris and Potsdam until his death in 1859.

Humboldt was an anti-Kantian who rejected innate ideas. Antoine-Laurent *Lavoisier's *Elements of Chemistry* (1789) convinced him to favor knowledge based on precise measurements, not imprecise metaphysical constructions. He eventually learned to use nearly every measuring instrument available at the time. By the 1790s he had committed himself to a "terrestrial physics" that substituted precise measurements and detailed natural drawings for verbal description and symbolic representations. The geographer Georg Forster in 1790 urged him to assemble observations into a meaningful whole; Jena philosophers in 1794 suggested he view nature aesthetically. With the assistance of leading French scientists from 1805 to 1826 Humboldt reduced his Latin American measurements, obtained estimates of their errors, and interpreted the results of his travels.

His terrestrial physics is expressed best by his *Physical Profile of the Andes and Surrounding Country* (1807), a chart of twenty-one columns presenting a grand combination of measurement-based results and natural observations. The "Physical Profile"—a representation of Mount Chimborazo, which Humboldt had climbed to within 365 meters (1,200 ft) of its peak— was to accompany his *Essay on the Geography of Plants* (1805) that argued for the dependency of plant life on geographical location. He was the first to use isotherms and isobars to depict average thermometric and barometric conditions. Averages represented to Humboldt lawful expressions of nature's equilibrium, the balance of all forces. His integration of his results into a tabular and pictorial whole was a monu-

mental achievement. He tried to apply his method of averages to the geographical distribution in plants (1817). Humboldt's precision never captured all observational errors, however, and his averages, sometimes difficult to define and use, were often secured only by acts of faith. Others adopted the overall contours of *Humboldtian science—its focus on averages and the integration of nature's features—with some success.

Humboldt's *Views of Nature* (1808, 1826, 1849), partly based on results reported to him by others, popularized his idea that order could be achieved by finding the equilibrium points of nature's forces. His magisterial multivolume *Cosmos* (1845–1862) was an attempt to create for the entire universe the conceptual and historical unity he had already achieved for Latin America. A history beginning with the universe's primeval origins and ending with the creation of the physical landscape and life forms, *Cosmos* wove together Humboldt's prodigious correspondence with leading scientists of his day. Although often maligned by experts, *Cosmos* secured Humboldt's reputation as Germany's first widely read popularizer of science.

Humboldt's views of nature and politics intersected. Civilizations were to him morally developed to the degree that they overcame nature's challenges. Committed to a constitutional monarchy after the French Revolution, Humboldt looked upon civil disturbances as geological faults: pressures seeking equilibrium. He considered silver mines in the New World of value to European countries even though he was critical of slavery and political domination. Believing average conditions to be politically desirable, he advocated to colonial powers that they adopt his scientific observational and measuring methods as tools for the efficient management of local resources. Yet the European revolutions of 1848–1852 shocked him. Ironically, in the reactionary 1850s *Cosmos* contributed to the perception that Humboldt stood for impiety, republicanism, and revolution.

Douglas Botting, *Humboldt and the Cosmos*, 2d ed. (1973). Margarita Bowen, *Empiricism and Geographical Thought: From Francis Bacon to Alexander von Humboldt*

(1981). Kurt-Reinhold Biermann, *Alexander von Humboldt*, 3d ed. (1987). Alexander von Humboldt, *Cosmos: A Sketch of the Physical Description of the Universe*, 2 vols. (1996). Alexander von Humboldt, *Personal Narrative* (1996). Alexander von Humboldt, *Island of Cuba: A Political Essay* (2001).

KATHRYN OLESKO

HUMBOLDTIAN SCIENCE. Historians use the term "Humboldtian science" to describe a type of scientific practice during the nineteenth century that resembled the work of Alexander von *Humboldt, whether or not it resulted from Humboldt's direct influence. Susan Faye Cannon coined the term in 1978 to signify a scientific style that conducted observations with the latest instruments, corrected measurements for errors, and linked these to mathematical laws; constructed maps of isolines connecting points with the same average values; identified large, even global, units of investigation; and used nature rather than the laboratory as a site of investigation. The term, as applied to nineteenth-century science, has since acquired other connotations, including connecting different types of large-scale phenomena, demonstrating their interdependencies, seeking a universal science of nature, and using large-scale international organizational structures to execute local readings as part of a global effort.

Scholars since Cannon have been careful to differentiate what Humboldt did from Humboldtian science. Humboldt deliberately avoided speculation and description as was found in natural history and natural philosophy. The major theme of his master science of terrestrial physics was the equilibrium of the earth's forces. For Humboldt isolines represented not merely average values, but also a natural aesthetic order (as they did in patterns of maximum areas of concentration in the regional distribution of plants) and even political stability. This aesthetic sensibility rarely appears in Humboldtian science despite its importance to Humboldt, who in many ways incorporated the Romantic emphasis on aesthetics, the imagination, and the picturesque in image and word. Nature was to Humboldt not only the as-

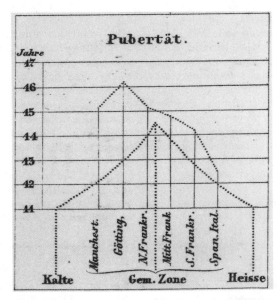

Onset of sexual activity as a function of latitude. The primitive inhabitants of hot and cold climates reach puberty early; the refined citizens of the university town of Göttingen follow five years later.

semblage of averages and the balance of forces, but also an aesthetic composition. The language of his travel books projected strange worlds in living color, a literary quality with great public appeal. He also wanted to retain in nature study the morally didactic qualities that eighteenth-century aesthetic theory had valued. In this sense, Humboldt's science, in contrast to Humboldtian science, was not merely knowledge to be learned, but to be lived. Hence some of the social values associated with the "European tour" in the nineteenth century came from their association with Humboldt's science.

Humboldtian science, as it has been used, also does not include two diametrically opposed directions in which Humboldt's work was taken: popular science and disciplinary specialization. The aesthetic image of the unity and balance of nature's forces particularly appealed to the public. The theme of harmony in Humboldt's work acquired political, social, and religious connotations, and helped to make nature study a part of liberal culture in Germany. The image of nature's order became an antidote to social and political disarray after the European revolutions of 1848. Immediately following Humboldt's death in 1859, Humboldt Associations, and later Humboldt Festivals, were established throughout the German states. They promoted interest in nature through public participation in nature walks and specimen collecting. More focused disciplinary uses of Humboldt's work appeared after industrialization exuded pollution, damaged forests, and in other ways highlighted the interconnectedness of environmental conditions. August Grisebach published in 1872 the first comprehensive classification of the earth's vegetation according to climatic conditions in which he adopted Humboldt's techniques in plant geography, especially Humboldt's notion of "social plants" (plant species in a regional environment forming special communities, such as heaths, savannahs, and bogs, that excluded the germination of other species).

Humboldtian science differed markedly from the institutionalized, disciplinary-based sciences that took shape in the nineteenth century. Although Humboldtian science shared the methodological rigor of nineteenth-century scholarship (*Wissenschaft*), it eschewed the intellectual specialization of, and sharp divisions between, scientific disciplines. For example, Humboldt and his Berlin colleague the physicist Heinrich Wilhelm Dove took meteorological measurements, but Dove's meteorology never reached out to other scientific disciplines like botany and never included naturalistic drawings in color. Dove also conducted some of his experiments in a laboratory; a Humboldtian scientist worked only in nature. The laboratory scientist posed detailed, particular questions in a contrived environment where certain variables could be held constant. Practitioners of Humboldtian science viewed nature as an ensemble, an organic whole whose interrelatedness could be captured in a geography broadly defined and keenly sensitive to large-scale issues. British natural philosophers especially viewed Humboldtian science as a rigorous counterbalance to the specialization and professionalization then shaping the content and practice of science.

Despite its marginal institutional position, Humboldtian science had a powerful effect on certain areas of science in the nineteenth century, especially geomagnetism, as well as on the development of certain scientific communities. Historians have found Humboldtian science more in evidence in Anglo-Saxon regions than elsewhere. Humboldtian science appeared most frequently in Victorian Britain with its vast empire over which scientists could collect data on the scale advocated by Humboldt. The best example of Humboldtian science is the British Magnetic Crusade, an effort dedicated to measuring the magnetic features of the earth in the British empire and beyond. Humboldt himself had proposed to the Royal Society of London that its members undertake global geomagnetic observations; in the 1830s the British government donated over £100,000 to the enterprise. Directed by Humphrey Lloyd of Trinity College, Dublin, the Magnetic Crusade consisted of a chain of fixed magnetic observatories with standardized instruments whose measurements—magnetic declination and the horizontal and vertical components of magnetic intensity—were sent to England where Edward Sabine reduced them. Methods of observation were printed on instruction sheets and distributed; adherence to the rules of observation was upheld by naval officers who conducted many of the observations. Through these observations Sabine could correlate deviations in Earth's magnetic phenomena to the action of sunspots and demonstrate the eleven-year cycle in sunspots.

The British viewed Humboldt as writing within their tradition of providential design in nature, and were annoyed that he did not specify the design's spiritual agency. Yet they believed that Humboldt's measuring methods were suited for shaping character and tempering laziness in young men. Humboldtian science defined the work of two leading British scientific organizations, the Royal Society of London during the 1820s and the British Association for the Advancement of Science in the 1830s. British imperial activity also bore the mark of Humboldtian science. Charles Darwin carried a copy of Humboldt's *Personal Narrative* aboard the voyage of HMS *Beagle* from 1831 to 1836. Humboldt's aesthetic image of nature recurs in the paragraphs on the tangled bank near the end of Darwin's *Origin of Species* (1859). In Australia magnetic observations at the Rossbank Magnetic Observatory between 1840 and 1854 and astronomical ones at Melbourne's Flagstaff Observatory (founded 1857) were further examples of Humboldtian science. The Rossbank data became Australia's first project in physics, inspiring a younger generation, including Georg Neumayer, the German scientist who founded the Flagstaff Observatory. In characteristic Humboldtian fashion, Neumayer gathered data from remote locations for map construction by means of the telegraph. British colonial administrators believed Humboldtian science was a part of their "civilizing mission." Even British imperial literature captured its importance. Rudyard Kipling's *Kim* (1901) not only immortalized the empire's penchant for gathering data on natural and human activity (Colonel Creighton was an ethnographer, Kim a surveyor), but also took place during the period of political unrest when natural

and social data about India became a form of surveillance deployed to hold the empire together.

Elsewhere, Humboldt directly influenced the Berlin geographer Carl Ritter, but he, like other Humboldt-inspired German scientists, directed no large-scale projects. When Humboldt proposed to the British that they inaugurate global geomagnetic observations, he did not know that Carl Friedrich *Gauss had already published a mathematical theory of magnetic intensity in 1833. Working with the physicist Wilhelm Weber, Gauss nonetheless made the Magnetic Observatory at the University of Göttingen (then a part of the British empire) the center of the British Magnetic Crusade. Gauss and Weber designed the instruments for the project; all chronometers were calculated, and all observations were reduced, according to Göttingen mean time. Neither *Gauss nor Weber, however, held up Humboldt's work as a model; both continued to work within the framework of laboratory- or observatory-based disciplinary sciences.

Many other scientists, even among the British, felt that Humboldt's science led to writing in outmoded traditions of travel literature and that the integration it claimed was rarely achieved. In the context of discipline-building and professionalization in Germany, Humboldt's science came to be viewed as amateurish, a throwback to an earlier era. Humboldtian science of a sort did promote scientific internationalism, but to what degree is still disputed. Although a concept constructed by historians rather than by historical actors to explain their own actions, Humboldtian science seems to capture the sciences of empire.

John Cawood, "Terrestrial Magnetism and the Development of International Collaboration in the Early Nineteenth Century," *Annals of Science* 34 (1977): 551–587. Susan Faye Cannon, *Science in Culture: The Early Victorian Period* (1978). Jack Morrell and Arnold Thackray, *Gentlemen of Science* (1981). Malcolm Nicolson, "Alexander von Humboldt, Humboldtian Science and the Origins of the Study of Vegetation," *History of Science* 25 (1987): 167–194. Mary Pratt, *Imperial Eyes: Travel Writing and Transculturation* (1992). W. H. Brock, "Humboldt and the British: A Note on the Character of

British Science," *Annals of Science* 50 (1993): 365–372. R. J. Home, "Humboldtian Science Revisited: An Australian Case Study," *History of Science* 33 (1995): 1–22. Michael Dettelbach, "Humboldtian Science", in *Cultures of Natural History*, eds. N. Jardine, J. A. Secord, and E. C. Sparry (1996): 287–304. David Philip Miller and Peter Hans Reill, eds., *Visions of Empire* (1996).

KATHRYN OLESKO

HUTTON, James (1726–1797), Scottish natural philosopher.

James Hutton was born in Edinburgh of merchant stock. He went to Edinburgh University, where he attended courses on mathematics and natural philosophy given by the famous Newtonian exponent Colin Maclaurin. Hutton then dithered about his future, trying law and medicine, spending four years in Paris, and receiving a medical degree from the University of Leiden in 1749 before returning to Scotland to improve a small farm inherited from his father. In 1768, he moved to Edinburgh, settled with his unmarried sisters, and threw himself into the brilliant intellectual life of the city at a time now described as the Scottish Enlightenment. Apart from a tour of England and some excursions in Scotland, he spent the rest of his life in Edinburgh.

Hutton numbered the chemist Joseph Black, the political economist Adam Smith, the philosopher and historian Adam Ferguson, the inventor James Watt, and the mathematician (and later disciple) John Playfair among his friends. Stimulated by their company, he refined his intellectual commitments. A deist in religion, he believed that a powerful and benevolent deity governed the universe, and dismissed the Biblical miracles as fables. A Newtonian in natural philosophy, Hutton took the laws of mechanics as the basis of a system of the earth—when supplemented by Black's principles of heat and chemistry.

In 1783 an informal philosophical society to which Hutton belonged was reconstituted as the Royal Society of Edinburgh. Two years later, he outlined his theory of the earth for the members. "This globe of the earth," he declared, "is evidently made

for man.... It is a habitable world; and on its fitness for this purpose, our sense of mission in its formation must depend." So that the globe would remain indefinitely habitable, the deity had designed a cycle of continuous decay and renewal. Soil, washed into the ocean by rain and rivers, sank to the bottom, where it consolidated into rock and was elevated by heat to form new habitable land. Hutton took Black's theory of specific and latent heat for granted. Solar heat in latent form consolidated rocks; as specific heat, it elevated them.

In 1788, a version of Hutton's presentation of 1785 to the Royal Society appeared in the Society's *Transactions*. Most of the British and Continental natural philosophers, chemists, and mineralogists who read it found it incomprehensible. Many argued that the soil, far from constantly disappearing, was a mantle that protected the earth. Although they agreed that fossils in calcareous rocks indicated that these strata had once been underwater, they ridiculed the idea that rocks had been consolidated and elevated by heat. Everyone knew that heat caused limestone to disintegrate to quicklime. Hutton's readers were equally baffled by his claim that granite had consolidated from a melt. Melts of materials similar to granite (the sand that made glass for example) cooled to glasses, not to crystalline rocks.

To counter these criticisms, Hutton set about writing his two-volume *Theory of the Earth* (1795). Simultaneously, he worked on other books: *Dissertations on ... Natural Philosophy*, mainly *meteorology, chemistry, and matter theory (1792), *Philosophy of Light, Heat and Fire* and *Investigation of the Principles of Knowledge and the Progress of Reason* (both 1794), and a 1,000 page manuscript on the principles of agriculture unpublished at his death.

Hutton's Scottish admirers, notably John Playfair and James Hall, developed his ideas (*see* NEPTUNISM AND PLUTONISM). Charles Lyell made Hutton the hero of the influential introduction to his *Principles of Geology* (1830) and Archibald Geikie, in his *Founders of Geology* (1901), treated Hutton as a founding father of the discipline. Hutton thus came to be seen as the founder of modern *geology, a Plutonist, a uniformitarian, and a hardheaded, no-nonsense scientist reporting what he saw in the field, untouched by religious pre-conceptions. In the mid-twentieth century, historians of science have tried to put Hutton back into his intellectual context in the Scottish Enlightenment.

See also UNIFORMITARIANISM.

E. B. Bailey, *James Hutton: The Founder of Modern Geology* (1967). Roy Porter, *The Making of Geology* (1977). Rachel Laudan, *From Mineralogy to Geology* (1987). G. Y. Craig and J. H. Hull, eds., *James Hutton* (1999).

RACHEL LAUDAN

HUYGENS, Christiaan (1629–1695), mathematician, astronomer, and natural philosopher.

Huygens was the second son of Constantijn Huygens, an important secretary of the independent Dutch States and a clever humanist. Christiaan's mother died when he was eight years old, but his father took great care with his sons' education. Interested in everything new, even optics and mechanics, Constantijn became a good friend of René *Descartes, who had moved to Holland in 1629. Christiaan was thus one of the first important natural philosophers to begin with Descartes's ideas rather than Aristotle's. Huygens learned Cartesian geometry privately from Frans Van Shooten before attending the University of Leyden. Although he accepted the new natural philosophy and geometry, he nevertheless studied carefully the treatises of Archimedes and *Galileo.

At the age of eighteen, Huygens was already corresponding with Marin Mersenne about the main problems in geometry and mechanics then under discussion in Paris. He traveled to London with his father and to Paris several times until he was employed officially as a foundation member of the Paris Royal Academy of Sciences in 1666. The main reason for his call to Paris was his work on instruments—*telescopes and lenses, whose power he demonstrated by resolving the usual blurred image of Saturn into a spherical body surrounded by a ring, and the pendulum clock, the first reliable timekeeper for astronomical and navigational use.

Huygens remained a French academician for fifteen years, pursuing his own work and collaborating with Gian Domenico *Cassini, the effective head of the then-new Observatory of Paris. Huygens's work covered most of the questions of interest to mathematicians and natural philosophers of his time: quadratures and rectifications of curves, practical optics and theoretical models for light, motion and space, gravity, and all mechanical devices related to clocks, telescopes, *air pumps, boats, coaches, windmills, and so on. He returned to Holland in 1681 for health reasons. He remained there after his recovery because an increasing intolerance toward Protestants in Paris made his return impossible. After the death of his father in 1687, Christiaan became a recluse in the old family country home of Voorburg, where he died.

Huygens's most famous publication, *Horologium oscillatorium* (1660), goes far beyond the practice and theory of clocks. It contains his main results about mechanics: free fall, the compound pendulum, and centrifugal force. Some geometrical results, about evolutes and involutes for instance, come in when necessary for physics. But he omitted a very important mechanical problem, although he found its solution in 1656: the demonstration of the classical rules for elastic collision, which he deduced from Galilean invariance (the principle that the velocity of a body measured by an observer in motion is the difference between the body's velocity and that of the observer, with respect to a third object supposed to be at rest). The invariance of physical laws with regard to any uniform motion, as determined by observers in uniform motion among themselves, such as that illustrated by Galileo's famous example of a falling weight in a moving boat, led him to the idea of an indefinite empty space where motion can only be relative. He held this idea strongly but confidentially, as he debated it only with Gottfried Leibniz.

Huygens's second most important book, the *Traité de la lumière* (1690), is the source of his reputation among modern physicists. He was the first to describe mathematically the propagation of light as a wave of motion and to deduce from it a construction for the laws of refraction that can still be found in schoolbooks.

Less famous than Galileo, Descartes, Isaac *Newton, or Leibniz, Huygens nevertheless belongs among them for his exceptional mathematical and experimental skills and, more importantly, his deep intuitions about motion and light. He developed the best ideas of Cartesian mechanism with the help of Galilean principles and classical geometry. This syncretism did not result from conflict but from Huygens's personal sense of intelligibility. But since his convictions were intuitive and scarcely changed from his youth to his death, he had insuperable difficulties in arguing and publishing them. Thus, though his contemporaries admired his technical results and manual dexterity, his methods and principles remained almost unknown.

H. J. M. Bos et al., *Studies on Christiaan Huygens* (1980). Joella G. Yoder, *Unrolling Time: Christiaan Huygens and the Mathematization of Nature* (1988). Fokko Jan Dijksterhuis, *Lenses and Waves: Christiaan Huygens and the Mathematical Science of Optics in the Seventeenth Century* (1999).

CHRISTIANE VILAIN

HYDRODYNAMICS AND HYDRAULICS. The tremendous growth of hydraulic construction during the Roman Empire and the medieval period led to important innovations, such as aqueducts and waterwheels, but added little to Archimedian concepts of fluid equilibrium and motion. Leonardo da Vinci brought late-medieval mechanics and the emerging experimental trend to bear on these questions. His insights into the pressure-head relation, eddy formation, flux conservation, and open-channel dynamics probably aided *Galileo's disciple Evangelista Torricelli, who in 1644 established the proportionality between the efflux velocity and the square root of the water head. Torricelli also explained the principle of the Florentine barometric tube by the balance between the weight of supported mercury and atmospheric pressure.

In seventeenth-century France, Blaise Pascal formulated the law of isotropic pressure, and persuaded his brother-in-law to verify the altitude-dependence of barometric pressure, which in Pascal's view excluded the Aristotelian *horror vacui.* Hydrostatics thus reached maturity. Fluid motion still challenged the new mechanical philosophy. In his *Principia mathematica* of 1687, Isaac *Newton discussed fluid resistance in order to show, contra Aristotle and René *Descartes, that matter could not fill interplanetary space. By theory and experiment, he established that the resistance of a fluid to motion through it was proportional to the cross section of the moving object, to the fluid density, and to the squared velocity.

Newton's reasoning used the balance between the momentum lost by the object and that acquired by the fluid and a drastic simplification of the flow pattern. In contrast, the Swiss geometer Daniel Bernoulli based his *Hydrodynamica* of 1738 on Leibnizian *vis viva* (kinetic energy) conservation, thus obtaining the relation between wall pressure, velocity, and height (Bernoulli's law). His word "hydrodynamica" expressed the synthesis between conceptions of hydrostatics and hydraulics. Only after suitable extensions of Newtonian dynamics and differential calculus did "hydrodynamics" come to mean a general theory of fluid motion. In 1744 and 1752, Jean Le Rond d'Alembert published treatises in which he applied his general principle of dynamics to fluid motion and established the paradoxical lack of resistance to the motion of a solid through a perfect fluid. Probably motivated by this breakthrough, in 1755 the Swiss geometer Leonhard *Euler obtained the partial-differential equation for the motion of a perfect fluid by equating the forces acting on a fluid element to the product of its acceleration and mass. He also showed how to derive Bernoulli's law from this equation. In his *Mécanique analitique* of 1788, Joseph Louis Lagrange solved Euler's equation for simple cases of two-dimensional fluid motion and proved a few important theorems.

D'Alembert's paradox and the nonlinear structure of Euler's equation deterred geometers and engineers from applying the new fluid mechanics to concrete problems. The French masters of late eighteenth-century hydraulics, Jean-Charles Borda, Charles Bossut, and Pierre-Louis-Georges Du Buat, combined experiment, global balance of momentum or *vis viva*, and physical intuition. Borda and Du Buat corrected Newton's misconceptions about fluid resistance; Borda completed Bernoulli's ideas on efflux and on head loss in a suddenly expanding pipe; Bossut and Du Buat established the proportionality between the loss of head in a long pipe or channel and the squared velocity; Du Buat formulated the general condition of permanent (constant velocity) flow as the balance between wall friction, pressure gradient, and weight.

The scope and accuracy of this semi-empirical approach grew considerably in the nineteenth century with the work of French, British, and German engineers. In the 1830s, the increased interest in canal building and river navigability led Jean-Baptiste Bélanger and Gaspard Gustave de Coriolis to compute the backwater caused by weirs. Between 1850 and 1870, Henri-Philibert-Gaspard Darcy and Henri-Émile Bazin made extensive measurements of flow in pipes and channels. In Britain, John Scott Russell, William Rankine, and William Froude studied how wave formation, streamlining, and a vessel's skin friction affected ship resistance. Most influential were Froude's model-towing experiments and his formulation of the laws that relate small-scale data to true-scale resistance.

Fundamental hydrodynamics also progressed. The mathematicians Simeon-Denis Poisson, Augustin-Louis Cauchy, George Biddell Airy, George Gabriel Stokes, William *Thomson (Lord Kelvin), Joseph Boussinesq, and John William Strutt (Lord Rayleigh) provided solutions of Euler's equation for ocean waves, ship waves, canal waves, and solitary waves. In 1858, while studying the aerial motion in organ pipes, Hermann von *Helmholtz discovered that rotational motion in a perfect, incompressible fluid obeyed remarkably simple conservation laws. Ten years later, Thomson exploited the resultant steadiness of annular vortices to represent atoms of matter. Meanwhile, Helmholtz and Rayleigh argued that the formation of

highly unstable vortex sheets (thin layers of uniformly rotating fluid) behind solid obstacles provided a solution to d'Alembert's paradox.

In 1822, the French engineer-mathematician Claude-Louis Navier inserted a viscosity-dependent term in Euler's equation. As Stokes demonstrated in his memoir on pendulums of 1850, the equation correctly described the regular flows observed for small characteristic lengths and velocities (e.g., the diameter of the bulb and its velocity in the pendulum case), but it seemed useless for the irregular flows observed in hydraulic cases. In the 1840s, Navier's disciple Adhémar Barré de Saint-Venant suggested that a variable effective viscosity (viscosity depending on local agitation) could be used to describe the average large-scale motion in pipes and channels. Saint-Venant's protégé Boussinesq successfully implemented this approach in the 1870s.

Whereas the French separately studied the two kinds of flow—laminar and turbulent, in Thomson's parlance—in 1883 the British engineer Osborne Reynolds studied the transition between them. He found it to occur very suddenly (as had been observed by Gotthilf Hagen in 1839) for a given value of the number LUD/η, where L and U are the characteristic length and velocity of the flow, D the fluid's density, and η the fluid's viscosity. By astute experiments and by analogy with the kinetic theory of gases, Reynolds shed light on the implied instability. Thomson and Rayleigh then pioneered the mathematical study of this question.

Despite the practical orientation of some of its theorists, nineteenth-century hydrodynamics failed to meet hydraulic and other engineering needs. It did produce, however, some of the key concepts that permitted the success of applied fluid mechanics in the twentieth century. Helmholtz's theorems on vortex motion and his concept of surfaces of discontinuity (or vortex sheets) served Ludwig Prandtl's and Theodore von Kármán's theories of fluid resistance, Frederick Lanchester's, Martin William Kutta's, and Nikolai Joukowski's theories of the airfoil, and Vilhelm Bjerknes's theory of meteorological fronts. Reynolds's and Boussinesq's theories inspired Geoffrey Taylor's, Johannes Burgers's, Prandtl's, and Kármán's statistical approaches to turbulence.

The newer fluid mechanics bridged fields as diverse as hydraulics, marine architecture, *meteorology, and aeronautics. Large laboratories were built to combine model measurements, theoretical analysis, and technical forecast. Prandtl's Göttingen institute set the trend early in the twentieth century; similar institutes were created in other industrializing countries. The United States and the Soviet Union became leaders in theoretical and applied fluid mechanics.

René Dugas, *Histoire de la mécanique* (1950). Hunter Rouse and Simon Ince, *History of Hydraulics* (1957). G. A. Tokaty, *A History and Philosophy of Fluid Mechanics* (1971). Paul Hanle, *Bringing Aerodynamics to America* (1982). Thomas Wright, *Ship Hydrodynamics,* 1770–1880 (Ph.D. diss., London Science Museum, 1983).

OLIVIER DARRIGOL

I

ICE AGE. The *Etudes sur les glaciers* (1840) by the Swiss botanist and geologist Louis *Agassiz opened the eyes of geologists to the possibility that a great ice age had occurred in the recent geological past. The evidence for the hypothesis was already well known. In northwestern Europe, the location of almost all early geological research, a thick layer of boulder clay covered the bedrock, huge stones (erratics) turned up far from their mother strata, bare rocks showed long, parallel scratches, and the remains of earlier beaches appeared well above existing sea levels. Until Agassiz, though, these manifestations had been put down to a vast flood. Since most naturalists still believed in the relative youth of the earth, they relied on written testimony as much as on field evidence. An ancient flood figured prominently in Greek authors such as Ovid as well as in Genesis. Moreover, from the early nineteenth century, most geologists held that the earth had cooled from an originally hot state, making it difficult to contemplate a past ice age (*see* GEOLOGY).

Louis Agassiz drew on work by earlier Swiss geologists to reinterpret the phenomena and argue that most of Europe had been covered by an ice sheet. By the early 1860s, even the most reluctant geologists, such as Charles *Lyell, had grudgingly come to agree. What caused the ice age, Agassiz never explained. In the nineteenth century, a Scottish autodidact, James Croll, set forth the most satisfactory attempt in his book, *Climate and Time* (1875). He suggested that as the shape of the earth's orbit slowly changed as a result of gravitational interactions with other planets, the variations in ellipticity caused occasional ice ages. One of the consequences of his theory was that glacial conditions in one hemisphere would be opposed by interglacials in the other hemisphere.

Scientists soon invoked ice ages to solve other problems. They explained the well-known changes in sea level around the Baltic by the locking up of sea water as ice or by the depression of the crust under the weight of the ice. They suggested that humans had reached North America by crossing from Eurasia on a bridge of ice. They told the story of much of human pre-history as a series of adaptations to life on the southern edge of the Eurasian ice cap.

By the 1920s, using fieldwork by the United States Geological Survey and studies on gravel river terraces by German geologists, scientists had decided that Agassiz's great ice age had in fact consisted of four different stages of advance and retreat. Since these stages occurred in the southern hemisphere as well as in the northern, geologists dropped Croll's theory. Following suggestions that radiation was the decisive cause of glaciation, Milutin Milankovich, a Serbian mathematician, calculated the radiation received in the two hemispheres at various times in the last half million years. At the time, these calculations did not seem to offer the needed support for the four-stage theory.

Following World War II, scientists abandoned the four stage theory as inconsistent with their findings from deep-sea cores. Correlations between the ages of the cores, their paleomagnetism, and their temperature at the time of formation (using oxygen isotopic ratios) revealed a more complicated story that seemed to fit better with information about radiation cycles. As regards the cause of ice ages, though, the verdict is still out.

John Imbrie and Katherine Palmer Imbrie, *Ice Ages: Solving the Mystery* (1979).

RACHEL LAUDAN

IDEAL GAS. An ideal gas consists of a vast number of elastic spheres in rapid rectilinear motion with no forces acting between them. Alternatively, it is a fluid that satisfies the relation $pv = nRT$ between its pressure p, volume v, and temperature T; R is the "gas constant," and n the molar amount of substance, or the number of gas volumes containing Avogadro's number of molecules, in the gas sample. The ideal gas law rests on Boyle's law, which Robert *Boyle was not the first to state, but the first to publish. In 1661 he heard of a quantitative relation between volume and pressure of an enclosed sample of air proposed by Richard Towneley and Henry Power. Boyle's assistant Robert *Hooke confirmed that he had made observations consistent with the statement that the product of the pressure and volume of air is a constant.

In 1718 the Swiss mathematician Daniel Bernoulli correlated air pressure with particle motion on the assumption that pressure arose from the impact of confined particles against the walls of the container. Bernoulli derived the relation $p = nmv^2/3s^3$, where n is the number of molecules, m the mass of a single particle, v the mean velocity of the molecules, and s the length of a side of the cubic container. Since s^3 represents volume, the relation can be restated as Boyle's law with $nmv^2/3$ expressed as a constant. (Later $mv^2/2$ was identified with temperature.) In the early 1800s publications of Joseph Louis Gay-Lussac on the expansion of gases, along with unpublished work by Jacques-Alexandre-César Charles, established the linear relationship between the volume and temperature of gases.

Working within the tradition of molecular physics, Rudolf Clausius in 1857 deduced that the average speed of gas molecules is several hundred meters per second. He also worked out an equation for the ratio of specific heat at constant pressure to specific heat at constant volume for a gas particle. Clausius, James Clerk *Maxwell, and Ludwig *Boltzmann further developed the kinetic theory of gases using statistical and probabilistic laws (see THERMODYNAMICS AND STATISTICAL MECHANICS).

Nineteenth-century researchers found that all gases deviate from the ideal gas laws at high pressures and low temperatures. The Irish physical chemist Thomas Andrews particularly investigated the conditions under which a gas can be liquefied by the application of pressure in his studies of the critical temperature above which the gas cannot be liquefied. In 1873 the Dutch physicist Johannes Diderik van der Waals modified the familiar ideal gas law to take into account the weak attractive forces between molecules of a gas and the size of the molecules. His equation, $(p + a/V^2)(V - b) = nRT$, accounted for reduced pressure owing to attractive force (the term a/V^2), and for the excluded volume of the container owing to the finite size of the molecules (b). The intermolecular attractive forces or "Van der Waal forces" received an explanation in Fritz London's application of quantum mechanics to the problem of dispersion forces in 1931.

Stephen G. Brush, *The Kind of Motion We Call Heat* (1976). Keith J. Laidler, *The World of Physical Chemistry* (1993).

MARY JO NYE

IMPONDERABLES. Around 1800 physical science enjoyed a fleeting unification under a scheme that developed from study of the phenomena of heat (see FIRE AND HEAT) and *electricity. The discovery of the conduction of electricity down damp threads by Stephen Gray in 1729 prompted the assimilation of the agent of electrical attraction to water running through a pipe, an analogy strengthened by Benjamin *Franklin's comparison (1751) of the machines used for generating electricity (globes or cylinders of glass spun against the hand) to pumps, and Leyden jars (condensers) to reservoirs. For those who accepted Robert Symmer's version of Franklin's theory (1759), which made negative charge as real as positive, electricity was served by two fluids, which, since ordinary bodies appeared to weigh no more when electrified than when not, were taken to be imponderable.

In order to explain the most evident electrical phenomena, natural philosophers ascribed repulsive and attractive forces to the droplets of the electrical fluids: repulsive between droplets of the same fluid, attractive between those of different fluids and between the fluid(s) and ordinary matter. To account for the differences in the degree of electrification or tension exhibited by insulated conductors of different shapes and sizes electrified in the same way by the same machine, philosophers ascribed a pressure to the electrical fluid(s) and specific electrical capacities to the conductors. Johan Carl Wilcke and Alessandro Volta developed these concepts in the 1770s (see GALVANI, LUIGI, AND ALESSANDRO VOLTA).

Most chemists and natural philosophers of the eighteenth century traced the action of heat to a special substance, which, like electricity, was understood to be an expansive fluid because of its spontaneous "flow" from hot to cold bodies. Also like electricity, its parts were taken to be self-repellent in order to explain the expansion of bodies when heated. Within the Newtonian philosophy the self-repellency of heat arose from a repulsive force acting between the particles of the heat fluid. With the discoveries of latent and specific heats by Wilcke and Joseph Black in the 1770s and 1780s, the parallels between the heat fluid and the agent of electricity broadened: latency could be regarded as a bonding between heat and matter; specificity indicated an analogy between temperature and heat capacity, on the one hand, and tension and electrical capacity, on the other.

The standard representations of magnetism and visible light easily fit the imponderable model. By analogy to electricity, *magnetism came to be regarded as a distance force arising from magnetic fluid(s) whose particles obeyed the same rules of attraction and repulsion that regulated the traffic of the electrical fluids. The main distinction—that nothing comparable to conductors of electricity existed for magnetism—was regarded as a question of degree not kind. The magnetic fluid(s) stayed in magnetic substances as electrical fluid(s) did in strong insulators. Franz *Aepinus worked out these parallels in detail in 1759. As for *light, its particulate nature was assured by the widely held optical theories of Isaac *Newton, which endowed light particles with short-range forces by which they interacted with matter to produce the phenomena of reflection, refraction, and inflection (diffraction). The capstone of the arch of imponderables was the discovery by William *Herschel in 1800 of radiant heat beyond the red end of the visible spectrum. Infrared light connected heat and ordinary light and, via the analogies between heat and electricity, light with magnetism. More speculative philosophers added fire, flame, phlogiston and to the generally accepted five (or seven) imponderables.

The scheme, which functioned as a *standard model for physical science around 1800, had two important assets.

For one, it immediately explained the existence of the phenomena it covered by the mere presence of the relevant agent. For another, it lent itself to the fashion of science of the time, quantification. In 1785, Charles Augustin Coulomb (see CAVENDISH, HENRY, AND CHARLES AUGUSTIN COULOMB) established to the satisfaction of the members of the Paris Académie des Sciences (and few others) that the interfluid forces in electricity and magnetism declined, as did the force of gravity, with the square of the distance between interacting elements. Pierre-Simon *Laplace and his school pursued the program of quantifying the distance forces supposed to act between elements of the heat fluid (which they called caloric) and between light particles and matter. Laplace and Jean-Baptiste Biot managed to give detailed accounts of refraction, both single and double; polarization; and other optical phenomena in these terms. By taking literally the concept of heat as a conserved fluid, Laplace created a brilliant theory of adiabatic processes that resolved the long-standing and scandalous discrepancy between theoretical treatments and measurements of the speed of sound in air (see ACOUSTICS AND HEARING). Although it did not appeal to the notion of distance forces, the adiabatic theory encouraged belief in the existence of caloric.

A serious fault with the scheme of imponderables, apart from its multiplication of weightless fluids at a time when chemistry was learning to live strictly by the balance, was the ontological independence of the several fluids. The unification the scheme brought rested on parallel treatment of diverse phenomena, not on connections among their agents. This weakness was partially overcome by the linking of electricity and magnetism beginning with the discovery of the action of a current-carrying wire on a magnet by Hans Christian Oersted in 1819. But the replacement of the particulate by the undulatory theory of light in the early nineteenth century, and the annihilation of caloric by the kinetic theory of heat in its middle decades, destroyed the old standard model. A new synthesis seemed imminent and immanent in James Clerk *Maxwell's unification of electricity, magnetism, light, and radiant heat, and the program pursued around 1900 to reduce ponderable matter to *electromagnetism. That program failed, leaving the *electron, the electric current, and the flow of heat as residues and reminders of the first standard model in physics.

Tore Frängsmyr et al., *The Quantifying Spirit in the 18th Century* (1990). J. L. Heilbron, *Weighing Imponderables and Other Quantitative Science around 1800* (1993).

J. L. HEILBRON

INSTRUMENTS AND INSTRUMENT MAKING.

Although the concept of a scientific instrument may seem clear, the historian is bound to find it problematic, not least because the term did not come into common use until the second half of the nineteenth century. The present usage, in its application to the past, comes partly from a projection of current scientific practice onto earlier activities and partly from choices made by collectors, curators, and dealers as they constructed a specialist interest in material culture that would guide their activities in museums and salesrooms.

An important function of contemporary scientific instruments is to investigate the natural world, to discover new truths about nature. But until the beginning of the seventeenth century instruments had no such role. In the terminology of the time, some "mathematical instruments" played an integral part in certain mathematical arts or sciences, but had no place in the science that dealt with causes and explanations in the material world, *natural philosophy. Practitioners used these to solve problems and produce practical results, such as casting a horoscope, telling the time, finding the latitude, or drawing a map.

Astronomy provides the earliest record of the use of mathematical instruments and through astronomy many of the technical characteristics of mathematical instruments developed. Astronomical instruments fixed in observatories were generally made by bringing together local resources such as woodwork, metalwork, and masonry, whereas personal portable instruments like astrolabes and sundials came from mathematical instrument makers and specialist workshops in recognized centers

A trade card from the mathematical instrument maker Thomas Wright (1718). The card gives pride of place to the orrery, a clockwork planetarium, shown under the globe (upper center), armillary ring (left), and theodolite (right).

of production. One of the earliest makers on whom we have much biographical information was Jean Fusoris, university educated and a church canon, who set up a workshop in Paris and produced astrolabes and clocks in the early fifteenth century.

Other early makers or founders of workshops were leading mathematicians and astronomers. Johannes Regiomontanus established a workshop in Nuremberg in the 1470s; several surviving instruments are attributed to him. Scholarly mathematicians from the sixteenth century with a strong commitment to the development of instrumentation include Peter Apian in Ingolstadt and Gemma Reines (called Frisius) in Louvain. Gemma worked in association first with Gerardus Mercator and later with his nephew Gualterus Arsenius to produce a great many instruments and a variety of new designs. An international trade developed. The Louvain workshop benefited from an understanding with Christophe Plantin whose printing house in Antwerp was used for their ordering and distribution. This arrangement represented a wider association between instrument production and other mathematical commerce, such as mapmaking and book publishing.

The sixteenth century saw a remarkable development in instrumentation as part of a general flourishing of practical mathematics. The development of navigation and commerce and changes in the conduct of warfare inspired new ways of harnessing geometry to more effective action. This activity tended to take place at court or in the city rather than in the university. Mathematics occupied an inferior position in the university hierarchy to medicine, law, and natural philosophy, but at court it could be used as a tool for political and territorial advance, and instruments were used for persuasion as well as action. The appearance of many surviving instruments from this period indicate that their role was partly rhetorical.

Among many new designs from the sixteenth century were different types of sundials, quadrants, and nocturnals for finding the time, universal astrolabes, theodolites, and other surveying instruments, the cross-staff and backstaff for astronomical navigation, and the sector for a wide range of calculations. The publication of many books on instrumentation and the spread of centers of production, notably to Florence, Venice, Nuremburg, Augsburg, Ingolstadt, Louvain, Paris, Antwerp, London, and Prague, accompanied and supported the development in the range of designs.

Because these mathematical instruments did not engage with natural philosophy, they did not have to respect the received account of the natural world. Terrestrial globes rotated on polar axes in advance of the publication of the Copernican theory simply for convenience. Different projections of the celestial sphere could be used on the two sides of a single astrolabe according to their intended applications, a freedom that reflected the variety of projections used in contemporary cartography. In both fields, convenience and efficiency were the criteria of success, not fidelity to nature. At the same time, the status enjoyed by practical mathematicians in nonacademic contexts for work, and their freedom from the disciplinary restrictions of the university, gave them a relative confidence and autonomy that would eventually facilitate the application of their practices to the reform of a demoralized natural philosophy.

The application of instruments to discovering the truths of nature began in the late sixteenth and early seventeenth centuries, most significantly in the use of the *telescope in astronomy. Instrumentation had not been applied to the natural philosophy of the heavens. *Galileo insisted that his telescopic discoveries from 1609 onwards gave evidence for the Copernican cosmology and so thrust the telescope into the forefront of a dispute in natural philosophy, where its reliability as a tool of discovery would be a critical issue.

The telescope and the microscope created a new domain of instrumentation separate from the established trade of mathematical instruments. A different category of artisan, the more able and enterprising among the spectacle makers, produced the new optical instruments. Like the telescope, the microscope first arose in a commercial rather than a learned context. It was an optical toy with no agenda for use in natural philosophy until the mid-seventeenth century. Then an increasing interest in explaining natural phenomena through the interaction of invisible particles acting as tiny machines made the microscope a likely arbitrator of the claims of the *mechanical philosophy. Through the development of the microscope and telescope a new trade was born. By the late seventeenth century, although they included spectacles among their stock, some specialists had become "optical instrument makers."

The natural philosophers involved themselves closely in this development. Johannes *Kepler and René *Descartes had been concerned with the true form of an aplanatic lens, one that did not suffer from spherical aberration. Christopher Wren designed an unsuccessful machine for grinding hyperbolic lenses to remove the defect, while other early fellows of the Royal Society ground telescope objectives, as did Christiaan *Huygens in the Netherlands. In the case of microscopes, Robert *Hooke associated with the London makers and frequented their workshops, while in Delft Antoni van Leeuwenhoek made his own extraordinary microscopes with their single, tiny, spherical lenses. The best optical glass came from Italy, where Eustachio Divini in Rome and Giuseppe Campani in Bologna led the field. Italian natural philosophers, such as Gian Domenico *Cassini for the telescope or Marcello Malpighi for the microscope, could rely on the products of the best commercial workshops. Leeuwenhoek had to make everything himself.

By the late seventeenth century, the commercial trade in optical instruments was particularly vigorous in London, where visitors to the shops of Christopher Cox, John Yarwell, or Richard Reeves might expect to buy a fine telescope or a microscope equivalent to the one illustrated in Hooke's *Micrographia* (1665). But the early promise of microscopes as arbitrator of philosophy proved hollow. If fleas and other tiny things were as complicated as they appeared to be, the fundamental mechanical corpuscles lay far beyond the instrument's reach. Microscopy declined in natural philosophy. Nonetheless Hooke's astonishing illustrations had made their mark: through much of the eighteenth century a widespread interest in natural history would supply the makers with a ready clientele for microscopes.

A third category of instrument with which the natural philosophers had an even stronger engagement than with optical instruments emerged in the later seventeenth century. These "instruments of natural philosophy," unlike the mathematical and optical ones, had no location of their own within the trade. Natural philosophers themselves designed the instruments and contracted assembly to artisans. Philosophical instruments included air pumps and electrical machines, which, unlike the passive telescopes and microscopes for observing, intervened and interfered with nature. They literally implemented the collaborative, public, and institutionalized experimental philosophy practiced in the Royal Society of London and other societies that cultivated natural knowledge. Experimental demonstrations were to be performed in public before witnesses and repeated at will: they created a need for instruments of natural philosophy.

Practical applications continued to drive improvements in mathematical instruments even while optical and philosophical instruments began to capture attention. Edmund Gunter in Gresham College, London devised a quadrant for telling the time and performing other astronomical calculations, a sector for navigating by the Mercator chart, and a rule for achieving the same end with logarithmic scales. The ubiquity and longevity of the "sliding Gunter" or logarithmic slide rule testify to the sophistication of mathematical instrumentation in the early-modern period.

As instruments changed over the seventeenth century so did their provenance and markets. London grew into a major center for instrument making and, in the eighteenth century, dominated the trade. Makers in London could belong to any guild or company—they could nominally be grocers, or haberdashers, or fishmongers—and the companies did not restrict the production methods, designs, or materials. This freedom, which contrasted with the centralized and regimented situation in Paris, suited a trade that needed to combine disparate materials, adopt new designs, adapt working practices, and merge artisanal skills. New configurations in the trade began to emerge, as certain London makers at the turn of the century, notably Edmund Culpeper and John Rowley, traded across the traditional boundaries by offering both mathematical and optical instruments.

During the eighteenth century the most ambitious makers dealt in "mathematical, optical and natural philosophical instruments." Demonstration apparatus became fully commercial under the stimulus of subscription courses in experimental natural philosophy, such as those given by Francis Hauksbee in London and the abbé

Jean-Antoine Nollet in Paris. Makers offered books (often written by themselves, in the cases of George Adams or Benjamin Martin), demonstrations, and courses of lectures in addition to instruments. Shops presented their wares within the context of the regular trade in luxury goods, intriguing foreign visitors. The growth of material consumption within the middle classes benefited the makers and natural philosophy had a fashionable following, encouraged by entertaining lecture courses or domestic demonstration from itinerant lecturers. The formation of instrument collections spread from institutions, universities, and the aristocracy to the homes of the bourgeoisie—a development encouraged by entrepreneurial traders in a buoyant market.

The rise of a consumer market directed the production of instruments toward the elegant, such as barometers, globes, and orreries, and the spectacular, such as air pumps and electrical machines. Telescopes, particularly the Gregorian reflector, and microscopes multiplied in the same context, and their designs reflected their intended station in a library or a drawing room. The solar microscope was developed to project large images of microscopic subjects onto a wall to entertain a group. At the same time, however, London manufacturers consolidated a leading position in the most exacting part of the trade: measuring instruments for astronomers, navigators, and surveyors.

A succession of outstanding makers of precision instruments in eighteenth-century London, beginning with George Graham and continuing through Jonathan and Jeremiah Sisson, John Bird, Jesse Ramsden, and John and Edward Troughton, raised the status of makers among the community of mathematicians and natural philosophers to an unprecedented level. These makers produced observatory instruments for fundamental measurement in astronomy—at first mural quadrants, transit instrument, and zenith and equatorial sectors, and later meridian circles—and sextants, theodolites, and other precision measuring instruments for everyday professional use. Their work was complemented by that of several outstanding optical instrument makers, John and

Peter Dollond for lenses and James Short for telescope mirrors.

While individual skills must figure in the explanation of this development, certain institutional factors also played an important part—commissions from the Royal Observatory and later the Ordnance Survey, and the liberality of the Royal Society, which elected the leading makers as fellows, awarded them medals, and published their papers in the *Philosophical Transactions*. The activities of the Board of Longitude charged with administering the longitude prize established in 1714 were also influential. One of the contending methods, that of lunar distances, demanded exact and robust instruments. The board publicized methods it rewarded, for example, Bird's prescriptions for making quadrants.

At the end of the eighteenth century makers began to move away from a concentration on handwork in small workshops. Jesse Ramsden employed some fifty men in his premises in Piccadilly: the Board of Longitude published his description of the dividing engine he built for the mechanical graduation of scales on sextants and other instruments. Hand division had previously been the most prized skill in the precision trade; it was now mechanized and, comparatively speaking, deskilled. Subcontracting, buying in parts, even buying whole instruments and adding the retailer's name, became common in the eighteenth century as the trade grew and its organization became more complex. These trends accelerated during the nineteenth century.

The acceleration proved costly to the London workshops. They lost their dominance, partly through complacency, partly through loss of status in the community of learning, partly through the vigorous rise of other centers of innovation. The downgrading of manual skill in the social changes brought by the industrial revolution may have been a factor in the loss of status, but the makers as individuals never regained their positions of respect. Even in astronomy, scientists with mechanical flair like George Biddell Airy designed the major instruments and commissioned components from different makers, adopting the division of labor from contemporary industry. Germany and France became increasingly competitive.

Munich was the center of the resurgent German industry. The able makers there included Georg von Reichenbach, Joseph Liebherr, Joseph von Utzschneider, Joseph Fraunhofer, Traugott Lebrecht Ertel, Georg Merz and Carl August, Ritter von Steinheil. They focused on precision instrumentation, which they pursued in partnerships of opticians and mechanicians. Thus they benefited from research-based improvements in glass quality as well as from innovative designs in structures and mountings. The workshop of the Repsold family formed another center in Hamburg, while Karl Philipp Heinrich Pistor and Johann August Daniel Oertling were active in Berlin from 1813 and 1826, respectively. Observatories, other than British ones, equipped in the nineteenth century usually had German instruments, whereas the many eighteenth-century foundations, including French ones, had been supplied from London.

The French Revolution swept away the old restrictive practices of the guilds and put in place reforms, such as the metric system of weights and measures, that would create work for instrument makers and encourage innovation. Étienne Lenoir, Jean Nicolas Fortin, and François-Antoine Jecker seized the new opportunities, supplied standards of length, weight, and capacity throughout the country, and met the renewed demand for portable instruments from mathematical professionals. Prominent and successful workshops in nineteenth-century France included Gambey, Lerebours and Secretan, Gautier, Morin, and others. The French developed a particular expertise in physical optics, led by Jules Duboscq and later by the aptly named Jean-Baptiste-François Soleil.

From 1851 onwards international exhibitions furthered the international character of the instrument trade. The exhibitions' reports give a good indication of the relative strengths of the contributing nations. The British were taken aback at the success of their rivals at the Great Exhibition in London (1851). Microscopes, largely through the introduction of the achromatic objective, had again become serious tools of scientific research, and here London makers continued to shine; but elsewhere they had lost much ground. By the end of the century, the German workshops of Zeiss and Leitz had seized the lead in microscopy.

By 1900 the "scientific instrument" in the modern sense of the term had arrived. As makers sought to respond flexibly to rapid innovation, the old characterizations and distinctions became irrelevant. *Spectroscopy opened up a vast new area of instrumentation for chemistry and astronomy. Industrial instruments greatly expanded the market open to instrument entrepreneurs, while the coming of the electrical industry and the spread of power supply, of the electric telegraph, and then of radio, opened up a large field for collaboration between scientists and manufacturers. Techniques of detection and measurement had not only to work on the laboratory bench, they had to be standardized and made sufficiently robust to travel successfully to distant stations.

The twentieth century was characterized by ever larger manufacturing units, close liaison with research laboratories in universities, in institutions, or in-house, a bewildering array of new techniques and instruments, and the growing irrelevance of regional contexts other than as economic determinants. This flexibility has been particularly marked as electronic technologies have increasingly displaced mechanical ones. Instrument making has become difficult to isolate from science itself. That may always have been the case. The ubiquity of instrumentation in today's science makes the relation obvious.

A further feature of the twentieth century was the rise of collecting, both institutional and private. Museums now have large collections of instruments—transferred to them from societies, universities, manufacturers, industries, collectors, dealers, and salesrooms. Historians have not realized the potential of this resource fully. Instruments have been integral to the story of science. Although material evidence may be more intractable and awkward to use than written sources, historians of science can scarcely afford to neglect it.

See also STANDARDIZATION.

Maurice Daumas, *Scientific Instruments of the Seventeenth and Eighteenth Centuries and Their Makers*, trans. Mary Holbrook (1972). Jim Bennett, *The Divided Circle: A History of Instruments for Astronomy, Navigation and Surveying* (1987). Anthony Turner, *Early Scientific Instruments, Europe 1400–1800* (1987). Gerard L'E. Turner, *Nineteenth-Century Scientific Instruments* (1983). Robert Bud and Deborah Jean Warner, eds., *Instruments of Science: An Historical Encyclopedia* (1998).

JIM BENNETT

INTERNATIONAL GEOPHYSICAL YEAR. The International Geophysical Year (IGY) was an ambitious international scientific project that ran from July 1957 to December 1958. In 1950 American geophysicists had proposed a Third International Polar Year that would make significant advances on the earlier ones of 1882–1883 and 1932–1933 by using rocketry, information processing, and other instrumentation developed during World War II. The project quickly widened to geophysics as a whole. Sanctioned by the International Council of Scientific Unions—the parent body of international scientific organizations—and implemented by national committees in participating countries, it involved some eight thousand scientists from about sixty different nations. It was timed to coincide with the twenty-fifth anniversary of the Second International Polar Year and with a peak in the sunspot cycle.

The IGY Special Committee decided to concentrate on the topics most likely to benefit from a global approach: aurora and airglow, cosmic rays, geomagnetism, glaciology, gravity, ionospheric physics, longitude and latitude determinations, *meteorology, *oceanography, *seismology, and solar activity. The project produced an unparalleled database as well as a number of major discoveries. Oceanographers confirmed the existence of a continuous worldwide system of submarine midocean ridges (actually huge mountain chains), one piece of evidence that contributed to *plate tectonics in the mid-1970s. Satellites launched by the United States detected belts of radiation around the earth (named the Van Allen belts) and the influx of charged solar particles believed to be responsible for the auroras (see COSMIC RAYS). Scientists in Antarctica determined the size and shape of the land mass underlying the ice and discovered a jet stream circling the continent.

The IGY was a prime example of "big science." Governments contributed major funding. Scientists participated in much larger numbers than had been normal in peacetime. International cooperation led to the invention and dissemination of new, intricate, and very expensive instruments. Scientists became more involved in and gained new standing in politics and international law. IGY's success encouraged the United States to commit to Skylab; it led to more intense Antarctic exploration; and it prompted further international cooperation in the International Year of the Quiet Sun (1964–1965), the International Hydrological Decade (1965–1975), and the International Decade of Ocean Exploration (1970–1980).

Walter Sullivan, *Assault on the Unknown* (1961). J. Tuzo Wilson, *The Year of the New Moons* (1961). Karl Hufbauer, *Exploring the Sun* (1991). G. E. Fogg, *A History of Antarctic Science* (1992).

RACHEL LAUDAN

ION. William Whewell gave the world the English term "scientist" at a meeting of the British Association for the Advancement of Science in 1833. In that same year Whewell and Michael *Faraday collaborated on the creation of a new vocabulary to describe the results of Faraday's investigations in electrochemistry. In 1834 Faraday published an article in the *Philosophical Transactions* of the Royal Society of London in which he systematically introduced a new vocabulary for electrochemistry, including the terms "ion," "anion," and "cation" ("cathion").

Like Michael Faraday and Humphry Davy in London, Jöns Jacob Berzelius in Stockholm used the battery to decompose compound substances by electrical current, depositing the separated components at the electrical poles or electrodes. Around 1812 Berzelius came to identify the force of chemical affinity with the force of electrical attraction, assuming that charged atoms or groups of atoms are held together in a molecule by slight opposite electrical charges or forces. This theory came to be called a dualistic or electrochemical theory of chemical composition and decomposition. Berzelius's electrochemical theory was gradually undermined by the recognition that molecules such as hydrogen contain like atoms, as well as by the discovery that (negatively charged) chlorine atoms can substitute for (positively charged) hydrogen atoms in hydrocarbons. Berzelius's molecular constituents anticipated ions joined in polar chemical bonds.

Svante Arrhenius's proposal in 1887 that molecules of electrolytes break up into charged ions in dilute solution, whether or not electric current is present, furthered the conviction that ions exist and participate in many chemical and physical processes. Scientists studied ionized particles with increased fervor in the late nineteenth century following investigations of the nature of *cathode rays and canal rays in the 1880s and 1890s, both of which consist of charged particles.

*X rays and *radioactivity became detectable by their capacity to produce charged ions and (in the case of alpha and beta particles) by their identification as charged ions. The improvement of electrometers allowed the detection of small amounts of ionization, as did the invention in the late 1920s of the automatic electrical counter dubbed the "Geiger counter" after its invention by Hans Geiger and Walther Müller. *Cosmic radiation also could be detected through electrical effects and ionized particles could be tracked as vapor trails in the *cloud chamber invented by C. T. R. Wilson in 1899.

The electrochemical hypothesis that "ions" exist fleetingly within activated chemical molecules gained adherents in the early 1900s when the electron theory of valence was being developed by Joseph John *Thomson, Walther Kossel, Gilbert N. Lewis, and Irving Langmuir. Lewis, and then Langmuir, identified the electron as the essential constituent of polar (ionic) and nonpolar (covalent) chemical valence bonds between 1916 and 1919. In the 1920s and 1930s Arthur Lapworth, Robert Robinson, and especially Christopher Ingold

revolutionized the understanding of reaction mechanisms in organic chemistry by applying theories of electrons and ions to aromatic and aliphatic compounds. The history of modern chemistry is inseparable from the history of ions.

See also ATOMIC STRUCTURE; ELECTROLYSIS; ELECTRON.

William H. Brock, *The Norton History of Chemistry* (1992). Elisabeth Crawford, *Arrhenius: From Ionic Theory to the Greenhouse Effect* (1996).

MARY JO NYE

IONOSPHERE. Beginning with the discovery in the mid-eighteenth century that air can be electrically charged, some investigators speculated about electrical conditions at higher elevations. Erik Pontoppidan suggested in 1752 that the aurora borealis is electrical. Jean-Baptiste Biot and Joseph Louis Gay-Lussac carried electrometers in a balloon ascent in 1804 that reached an elevation of several kilometers, far short of the ionosphere. By the 1830s, scientists such as John *Herschel had concluded that diurnal variation of geomagnetism depended on changes in the electric state of the atmosphere.

Balfour Stewart first suggested the existence of electrical currents in a high, conductive layer of the atmosphere in 1884 to explain geomagnetic variations. Arthur Schuster in 1889 supported this supposition with an application of the spherical harmonic analysis of Carl Friedrich *Gauss. Studies of atmospheric conductivity and ionization by Schuster, Hans Linns, Franz Exner, and Svante Arrhenius at this time indicated that gas ions give rise to electrical processes in the atmosphere. In 1902, Arthur E. Kennelly in the United States and Oliver Heaviside in England explained Guglielmo Marconi's trans-Atlantic radio transmission of 1901 as arising from reflection from this layer.

Although many scientists still doubted the existence of an ionized layer, W. H. Eccles termed it the Heaviside layer in 1912 and soon others called it the Kennelly-Heaviside layer. By the 1920s researchers determined that there must exist several ionized layers. Merle A. Tuve and Gregory

Breit in the United States and Edward Victor Appleton in England used radio waves purposefully to probe these conductive layers. Tuve and Breit determined a height of 230 km (140 miles) for one layer (the Appleton layer, now the F-layer), and Appleton 90 km (55 miles) for another (the Kennelly-Heaviside layer, now the E-layer). Appleton's sustained investigation of physical processes in these layers earned him the Nobel Prize for physics in 1947.

In 1926, specialists meeting at conferences and in committees named this region the ionosphere, in analogy with the troposphere and stratosphere, two terms then gaining currency. Robert Watson-Watt first used the name in print that year, and Hans Plendl next used its German cognate in 1931. The term became widely accepted during the Second International Polar Year of the 1930s. E. O. Hulburt and Sydney Chapman proposed the first detailed theories of the ionosphere at this time. Extreme solar ultraviolet radiation was found to cause much of the ionization.

Detailed examination of accumulated geomagnetic observatory data by Chapman in the 1940s indicated the existence of a tightly defined eastward electric current along the magnetic equator. Chapman called this current the Equatorial Electrojet. The electrical conductivity of the ionosphere came under study by sounding rockets beginning in the late 1940s in the United States and in the 1950s in Europe. Rocket-borne electrical research, along with investigations of cosmic rays and high-altitude chemical composition, greatly multiplied during the International Geophysical Year. The Van Allen Radiation Belts were discovered in 1958, and Thomas Gold termed the region of near space around Earth the magnetosphere in 1959.

Rockets and satellites carrying magnetometers, mass spectrometers, and electrostatic probes greatly increased knowledge of the ionosphere from the 1960s on. The vertical distributions of oxygen, ozone, helium, and hydrogen ions have been studied, as have electron densities. Likewise, investigators used these new capabilities to study the ionospheric effects of solar and interplanetary phenomena.

See also COSMIC RAYS.

C. S. Gillmor, "The History of the Term 'Ionosphere'," *Nature* 262 (1976): 347–348. Marco Ciardi, "Atmosphere, Structure of," in *Sciences of the Earth: An Encyclopedia of Events, People, and Phenomena*, ed. Gregory A. Good (1998): 47–50.

GREGORY A. GOOD

ISOSTASY the idea that different parts of the earth's crust are in gravitational balance and by implication floating on a semifluid substrate, had its origins in the mid-nineteenth century in mapping and geodetics. It played a crucial role in debates about earth tectonics in the first half of the twentieth century.

In the 1850s, surveyors working for the Great Trigonometrical Survey of India ran into a problem. They knew that the gravitational force of mountains should deflect their plumb bobs. The Himalayas, though, deflected their bobs less than anticipated. For the surveyors, this underperformance created cartographic problems. For scientists, it invited speculations about the structure of the earth. George Biddell Airy, the British Astronomer Royal, proposed that high mountains have low-density crustal roots extending to greater depth than the surrounding crust. John Henry Pratt, a mathematician as well as Archdeacon of Calcutta, proposed instead that the density of crustal columns varies inversely with their height. Formally these suggestions were equivalent. In either case, at some finite depth the load on the substrate would be everywhere the same.

Geologists were intrigued. The American geologist Clarence Dutton named the phenomenon isostasy and speculated about its tectonic consequences. European geologists saw in it an explanation for the changing levels of land and sea in Scandinavia. They proposed that during the *Ice Age the weight of the ice sheets had depressed land that was now rebounding. Further developments in *geophysics and *geodesy changed the status of isostasy from an exciting idea to a widely accepted truth. Geophysicists, notably the Finn Weikko Heiskanen, explored the implications of Airy's mechanism. Geodesists preferred Pratt's mechanism for convenience. Here the leader was John Hayford, Inspector of Geodetic Work and head of the U.S. Coast and Geodetic Survey. Geodesists knew only too well that determinations of latitude and longitude made by triangulation frequently differed from those made by astronomical observations. They attributed these discrepancies to deflections of plumb bobs. Hayford undertook a heroic series of calculations using Pratt's formulation of isostasy. He showed that plumb-bob deflections varied systematically. Boundaries between land and sea caused much larger variations than local topography. These results in hand, Hayford announced in 1909 a widely acclaimed new model for the figure of the earth. Furthermore he confirmed what some geologists had long suspected, namely that material of the continents was less dense than that of the ocean floors.

Geologists quickly realized that isostasy menaced the theory of the cooling contracting earth that had underpinned mainstream thinking about tectonics. If continents were lighter than ocean floors, then the proposal that continents foundered to form new ocean basins, put forward by Eduard Suess in his magisterial *Face of the Earth* (1883–1904), could not be true. Geologists and physicists scrambled to produce alternative tectonic theories. None of them, though, succeeded in dealing with the problems caused by isostasy. Continental drift, for example, seemed impossible: how could less dense continents move through more dense ocean floors? Only in the 1960s, when earth scientists developed the theory of *plate tectonics, was a satisfactory alternative found that sidestepped the problems of isostasy. Placing the definitive boundaries between plates of equivalent thickness, not between continents and oceans, made horizontal plate movement possible.

Mott Greene, *Geology in the Nineteenth Century* (1982). Gabriel Gohau, *History of Geology* (1990). David Oldroyd, *Thinking About the Earth* (1996). Naomi Oreskes, *The Rejection of Continental Drift* (1999).

RACHEL LAUDAN

J

JOULE, James (1818–1889), scientific brewer, and **Robert MAYER** (1814–1878), physician, natural philosophers commonly known as "co-discoverers" of the principle of energy conservation.

Joule was educated at home and by the natural philosopher John *Dalton. As the son of the wealthiest brewing family in Manchester, England, he had the opportunity to choose his profession freely. But he actively participated in the brewing business while planning to become a natural philosopher. The brewery and Manchester industry in general made an ideal environment for studying the most current problems in science and technology. The new forces of *electricity and *magnetism then enjoyed the attention of most people interested in natural philosophy. Joule's first major research project was his investigation into the electric motor as a possible alternative to the steam engine. His observations of conversion processes displayed by the engine led him to estimate its duty (efficiency) and to ponder the ontological status of the forces involved.

Expressing phenomena numerically had become a habit of Joule's when he worked in the brewing world. Producing beer under a tax regime taught him to work by, and to trust in, numbers: every minute process of brewing was controlled and measured by the tax authorities, and expressed numerically. Joule successfully applied this moral economy and its related skills to natural philosophical research. He hoped that by quantifying his electrical research he could gain deeper insights into the relationship between electricity and heat. He succeeded in deriving a quantitative law of heat production by a voltaic current: heat is proportional to the product of the resistance and the square of the intensity of the current ("Joule heat"). Joule hoped to reduce chemistry to absolute measures and elucidate the concept of latent heat (*see* FIRE AND HEAT).

Joule's experiments to determine the mechanical equivalent of heat (the ratio of the mechanical work performed to the heat produced) were part of a series of investigations. The experiments drew on the thermometric skills he had acquired in the brewery, and invoked a close collaboration with the local instrument maker and natural philosopher John Benjamin Dancer. Joule and Dancer produced the most precise working mercury *thermometer available at the time. Their contemporaries still used air thermometers, although the constancy of the expansion coefficient for all gases had been doubted. Joule recognized the need for a more sensitive mercury thermometer in heat research.

In the manuscript of his paper "On the mechanical equivalent of heat," Joule discussed his "paddle wheel" experiments (in which stirring water raises its temperature) and concluded that friction consisted in the conversion of mechanical force into heat. He based his claim on precision. But the scientific community (represented by the Royal Society, to which he submitted the paper) doubted the scientific brewer's claim. His conclusion about the conversion of stirring into heat does not appear in the printed paper. Joule was characterized as a "gentleman specialist" for having established the mechanical equivalent of heat through exact measurement, but his related reflections on the dynamical nature of heat and its significance for *thermodynamics carried little weight before he began his collaboration with William *Thomson. Thomson made "Joule's constant" (the ratio of mechanical work to heat) the building block of the science of energy. As a result of his new status as the master of precision measurement, Joule became a core member of the British Association for the Advancement of Science's program to establish absolute units based on the new science of energy.

Mayer, the youngest son of an apothecary from Heilbronn, Germany, devoted his free time to chemical and physical experiments, trained in the medical faculty of the University of Tübingen, and passed his medical examination in 1838 with a dissertation, *Über das Santonin*. He visited Paris in 1839–1840 and in 1840–1841 worked as the doctor aboard a Dutch vessel sailing to the East Indies. In 1842 he returned to Heilbronn. There he investigated the mechanics of heat and developed an overarching theory of force. Like Joule, Mayer was an outsider to the rising community of physical science. Through his practice as a physician, his upbringing in an apothecary household, and his childhood play with mechanisms, Mayer gathered that neither a physical machine nor the human body could generate motion out of nothing: thus, production is proportional to consumption.

He made the curious observation that the blood drawn from Europeans in Java differed in color from blood taken in Europe. He ascribed the difference to a disparity in the heat requirements of the body in the two regions. He wrote: "For the maintenance of a uniform temperature of the human body the heat *produced* in it must necessarily stand in a quantitative relationship to its heat *loss*, also to the temperature of the surrounding medium." His subsequent scientific contributions focused on challenging the then-dominant materialistic notion of forces. For Mayer, force was distinct from matter, indestructible, transformable or mutable, and immaterial, but converted in invariable quantitative relations from one form to another. He insisted on the mechanical equivalence (not equivalent) of heat, which could be measured as a constant relationship between motion (expressed as mechanical work) and heat. Although he urged that physical sciences be grounded in exact measurements, he based his own claim of equivalency on disputed gas expansion coefficients. For those like Joule, who favored precision measurement, Mayer's

knowledge claim—implying a strong anti-materialism—was mere speculation. Mayer managed neither to persuade his contemporaries (such as the editor of *Annalen der Physik und Chemie*, Johann Christian Poggendorff, who rejected his major article of 1841), nor to forge an alliance with Justus von Liebig and Hermann von *Helmholtz, respectively the dominant chemist and physicist of Germany at the time. Mayer suffered immensely both mentally and physically from being marginalized in the newly professionalized scientific community.

Joule and Mayer both participated actively in the rising culture of the exact sciences, which they wished to see anchored in quantitative measurement. Although neither of them coined or used a phrase equivalent to "conservation of energy," their investigations of the mechanics of heat became seeds for the nineteenth-century science of energy. However, theirs is not a case of simultaneous discovery. They did not announce the same thing at the same time, and they did not consciously seek to uncover a new principle hidden in nature.

Donald S. Cardwell and James Joule, *A Biography* (1989). Kenneth L. Caneva, *Robert Mayer and the Conservation of Energy* (1993). H. Otto Sibum, "Reworking the Mechanical Value of Heat. Instruments of Precision and Gestures of Accuracy in Early Victorian England," *Studies in History and Philosophy of Science* 26, 1 (1995): 73–106. H. Otto Sibum, "Les gestes de la mesure. Joule, les pratiques de la brasserie et la science," *Annales: Histoire, Science Sociale* 4–5 (1998): 745–774. Crosbie Smith, *The Science of Energy. A Cultural History of Energy Physics in Victorian Britain* (1998).

H. OTTO SIBUM

K

KAPITSA, Pyotr (1894–1984), one of the most famous physicists in the Soviet Union, a pioneering investigator of superconductivity.

Born on the island of Kronshtadt near St. Petersburg to Leonid and Olga Stebnitskii Kapitsa, in 1919 Pyotr Kapitsa graduated from the Petrograd Polytechnic Institute where he studied under A. F. Ioffe. For two years Kapitsa taught in this same institute and then transferred to Cambridge University in England, where he received a Ph.D. in 1923. During the 1920s and early 1930s Kapitsa remained in Cambridge, where he enjoyed the full support of Ernest *Rutherford. With financial support procured by Rutherford, Kapitsa built the Mond Laboratory within the Cavendish Laboratory at Cambridge as a center for the study of very high magnetic fields. Nobody expected Kapitsa to return to Russia, especially at a time when Joseph Stalin's ruthless policies were becoming more and more apparent. While he was visiting friends in Russia in 1934, however, the police detained Kapitsa on Stalin's orders and refused to permit him to return to England.

Kapitsa complained that he did not have the resources in Moscow that he had had in Cambridge. Stalin decided to give Kapitsa everything that he wanted materially. The Soviet government purchased the equipment in the Mond Laboratory and brought it to Moscow. Named the Institute of Physical Problems and located in a small woods near the Moscow River, Kapitsa's new institute could pursue delicate experiments under quiet conditions. Stalin also gave Kapitsa a Cambridge-style home in Moscow next to his institute. Although at first depressed about his captivity, Kapitsa soon responded scientifically to the favored conditions Moscow offered. During the first three years of his detainment, at the peak of Stalin's purges, Kapitsa accomplished the most important work of his scientific career, research on the superconductivity of liquid helium at temperatures near absolute zero. In 1976 he received a Nobel Prize for this work.

Kapitsa married twice, first Nadezhda Tschernovsvitova and then Anna Krylova, by whom he had two sons. He received many honors, both in the Soviet Union and abroad, and held a U.S. patent on a turbine for the production of liquid air.

Kapitsa retained a remarkable independence of spirit. When the secret police arrested several other prominent Soviet physicists, including Lev Landau and Vladimir Fok, Kapitsa played an important role in gaining their release. During World War II Stalin asked him to work on the atomic bomb project. Kapitsa refused, not because of objections to the bomb itself but because he could not tolerate Lavrenty Beria, the head of the Soviet secret police who was also in charge of the bomb project. As a result of this refusal, Kapitsa was removed from the directorship of his institute and kept under house arrest at his country dacha, where he attempted to continue his experimental work. After the death of Stalin in 1953 Kapitsa returned to his former position.

During the 1960s and 1970s Kapitsa provided places in his institute for young physicists to do research and also for artists to display paintings that did not follow official cultural policies. When the Soviet Academy of Sciences tried to expel the dissident physicist Andrei Sakharov, Kapitsa and others rose to his defense and the effort failed. When Kapitsa died in 1984, he was the only member of the presidium of the Soviet Academy of Sciences not also a member of the Communist party.

Only after Kapitsa's death did even his close friends learn that all during the Soviet period he wrote letters of protest and complaint to government leaders. The archives contain dozens of letters to Stalin, Vyacheslav Molotov, Georgy Malenkov, and Nikita Khrushchev in which Kapitsa criticized Soviet policies toward science, excoriated the secret police for arresting scientists, and practically challenged the police to detain him. We now know that Beria intended to arrest Kapitsa but that

Stalin intervened, saying, "I'll take care of him personally. Don't you touch him." Stalin might have been motivated by respect for Kapitsa's opinions and appreciation that Kapitsa never made his protests public.

Peter Kapitsa, *Experiment, Theory, Practice* (1980). Lawrence Badash, ed., *Kapitza, Rutherford, and the Kremlin* (1985). "Peter Kapitsa: The Scientist Who Talked Back to Stalin," *Bulletin of Atomic Scientists*, vol. 46, no. 3 (April 1990): 26–33.

LOREN R. GRAHAM

KEPLER, Johannes (1571–1630), astronomer and mathematician.

Kepler was born on 27 December 1571 in the imperial free city of Weil der Stadt, where his grandfather was mayor. His frequently absent father was a ne'er-do-well and sometime mercenary. Kepler's education in the fine school system of the Duchy of Württemberg culminated in a scholarship to study theology at the University of Tübingen. He passed his B.A. by examination in 1588, proceeding to Tübingen in 1589. The theologian Jacob Heerbrand and the astronomer Michael Mästlin, both adherents of Philipp Melanchthon in their view of the natural world, were among his teachers. After receiving his M.A. in 1591, Kepler continued advanced theological studies until he was assigned to teach at the Protestant school in Graz, Styria. He held the position of mathematics teacher and district mathematician from 1594 until the expulsion of the Protestants from Graz in 1600. He then became one of Tycho *Brahe's assistants in Prague. Upon Tycho's death in 1601, Kepler became imperial mathematician to Holy Roman Emperor Rudolf II, a post he retained under emperors Matthias and Ferdinand II until his death. He moved to Linz in 1612 to become mathematician to the estates of upper Austria. When civil war and the expulsion of Protestants made his position untenable, he became mathematician to Albrecht Wallenstein in Sagan in 1628. He died in Regensburg.

Kepler's work was distinguished most of all by his fervent, realist defense of Nicholas *Copernicus's heliocentric system, and by his insistence that astronomy

unite its physical (cosmological) and mathematical (technical) parts. Kepler's conception of physical reasoning was broadly Aristotelian. It included the notion of formal cause deriving from Kepler's belief in God's providential design. Thus his first publication, the *Mysterium cosmographicum* (1596), argued for Copernicus from the physical premise that God had constructed the solar system using the perfect Platonic solids as archetypes, determining the planets' unique commensurable spacing through the inscribing and circumscribing of their spheres around the solids.

When his work under Tycho Brahe shunted him into detailed work on planetary theory, Kepler pursued a controversial physical approach, analyzing Mars's motion in terms of a force coming from the sun that moved the planets around their orbits. In this way he was led to the first two of what are now called his laws of planetary motion—that the orbits of the planets are elliptical with the sun at one focus, and that the planet-sun radius sweeps out equal distances in equal times. He published the long justificatory narrative of this research in his *Astronomia nova* (1609).

Kepler returned to cosmological writing in his *Harmonice mundi* (1619), taking up especially the relationship of the planets' distances and periods. Almost accidentally, he discovered the third law of planetary motion—that the ratio of the square of a planet's period to the cube of its average distance was constant for all of the planets. This finding fit only awkwardly within the book's broader argument that God had painstakingly arranged the planet's distances and periods so that their angular motions viewed from the sun would embody all musical harmonies.

Kepler inherited from Tycho the prestigious responsibility for the *Tabulae Rudolphinae* (1627). These were astronomical tables based on Tycho's unparalleled observations using Keplerian theory. The tables—the first ever using logarithms—were abstruse by contemporary standards, and their immense superiority was only made manifest by Kepler's prediction of the first observed transit of Mercury in 1631. In an effort to smooth the acceptance of his unorthodox astronomy, Kepler also published the *Epitome astronomiae Copernicanae* (3 vols., 1618–1621), a detailed exposition of Keplerian astronomy in textbook form.

In addition to his astronomical work, Kepler made important contributions to optics and mathematics. His *Astronomia pars optica* (1604) was the foundational work of seventeenth-century optics. It included the inverse-square diminution of light and the first recognizably modern description of the formation of the retinal image. His *Dioptrice* (1611), the first theoretical analysis of the *telescope, included an improved telescope design. In mathematics, his *Nova stereometria doliorum vinariorum* (1615), concerning the volumes of wine casks, was a pioneering work of precalculus, and his *Chilias logarithmorum* (1624), an important early work on logarithms. He also published chronological works and balanced defenses of astrology.

Kepler's mathematical genius and his important conviction that astronomical theory must be derived from physics places him without doubt among the greatest early modern applied mathematicians, although the technical difficulty of his works and an unwarranted reputation for mysticism impeded his recognition by historians.

Max Caspar, *Kepler*, trans.C. Doris Hellman (1959; reprinted 1993). Alexandre Koyré, *The Astronomical Revolution*, trans. R. E. W. Maddison (1973; reprinted 1992). Bruce Stephenson, *Kepler's Physical Astronomy* (1987). Owen Gingerich, *The Eye of Heaven: Ptolemy, Copernicus, Kepler* (1993). Bruce Stephenson, *The Music of the Spheres: Kepler's Harmonic Astronomy* (1994). James R. Voelkel, *The Composition of Kepler's Astronomia nova* (2001).

JAMES R. VOELKEL

KURCHATOV, Igor Vasilyevich (1903–1960), Soviet nuclear physicist; and **J. Robert OPPENHEIMER** (1904–1967), American theoretical physicist, director of the Los Alamos laboratory and government advisor.

Oppenheimer was born to a well-to-do family in New York. Nominally Jewish, Oppenheimer obtained his early education at the assimilationist Ethical Culture School. He entered Harvard in 1922 and majored in chemistry, but studied widely in physics, mathematics, philosophy, literature, and languages. His undergraduate work under Percy Bridgman convinced him to switch to physics, in which he displayed remarkable analytical ability but little experimental dexterity. To pursue physics he went to the Cavendish Laboratory at Cambridge, where he stuck to theory, mastered quantum mechanics, and overcame his emotional insecurities. His first papers from Cambridge in 1926 won him an invitation to study under Max Born at the University of Göttingen, where he obtained his Ph.D. in 1927 with a dissertation on the quantum theory of continuous spectra. The sixteen papers Oppenheimer published by 1929 marked him as a rising theoretical physicist.

After several stints as a postdoctoral researcher in Europe, Oppenheimer accepted joint appointments at the California Institute of Technology (Caltech) and the University of California at Berkeley. His research extended into *quantum electrodynamics, *cosmic rays, *nuclear physics, and *astrophysics, including the first theoretical suggestion of *black holes. He presided over the main American school of theoretical physics in the 1930s, prolific in publications and students. Although he could be arrogant and impatient, Oppenheimer exuded aesthetic style, wit, and charisma, and students adopted his mannerisms, if not his tendency for sloppy calculation. The depression and the rise of fascism in the 1930s steered him to left-wing politics, although he shed formal political affiliations as World War II approached.

Kurchatov was born in 1903 in the southern Urals; his family moved to the Crimea in 1912, where Kurchatov attended the Gymnasium through the hardship of the first world war and then studied physics at the local state university. Unlike many Soviet physicists of his generation, and unlike Oppenheimer, Kurchatov did not study in Europe. In 1925 a former classmate recommended him to A. F. Ioffe, the director of the Leningrad Physical Technical Institute. Ioffe's institute, known as the Fiztekh, was a leading center for physics in the Soviet Union. At the Fiztekh Kurchatov worked on dielectrics and pioneered the study of ferro-electricity. He also demonstrated abundant reservoirs of energy, enthusiasm, and confidence, and, unlike Oppenheimer, dexterity in the lab.

In 1932 Kurchatov shifted to the emerging field of nuclear physics and Ioffe named him, at the age of twenty-nine, to lead the Fiztekh's nuclear physics program. Kurchatov helped make the Fiztekh a center for Soviet nuclear physics. His leadership and administrative abilities won him the nickname "the General." Although his father suffered through internal political exile, Kurchatov survived the Stalinist terror of the 1930s.

Kurchatov organized the first Soviet study of nuclear fission in 1939, which concluded that a self-sustaining chain reaction was possible. A world away, in Berkeley, Oppenheimer came to similar conclusions. A program to pursue an atomic bomb gradually coalesced in the United States, and in 1942 Oppenheimer joined it as head of research on fast neutrons, which was then underway at several sites. When the program came under the direction of the U.S. Army that summer, Oppenheimer suggested to General Leslie Groves that the design of a bomb be centralized in a new laboratory, in order to overcome the effects of secrecy and distance on communication. Groves accepted Oppenheimer's advice as well as his suggestion of a site, a remote mesa called Los Alamos in New Mexico, and named Oppenheimer to direct the lab. It was a surprising choice—first, since Oppenheimer, a theorist, would be leading a largely experimental program, and second, because he lacked a Nobel Prize and hence the stature of other possible candidates. Still, it was an inspired one, as Oppenheimer added his charisma and talent for identifying fruitful lines of research to the sense of purpose of the enterprise. In June 1945 Los Alamos exploded the first nuclear weapon in a test code-named Trinity; although Oppenheimer and some scientists who witnessed the test professed unease at the implications of their work, Oppenheimer and other project leaders concurred in the use of atomic bombs in combat against Japan.

Kurchatov's laboratory helped lead the exploration of fission in the Soviet Union, including the important discovery in 1940, to which Kurchatov contributed, of spontaneous fission. Germany's invasion of the Soviet Union diverted scientists to more immediate problems, and Kurchatov worked on the defense of ships against magnetic mines. After Soviet espionage in 1942 brought news of Britain's pursuit of an atomic bomb, the Soviet government revived fission research and selected Kurchatov as scientific director of the project. Like Oppenheimer, Kurchatov lacked prestige within the physics community—a nomination to the Soviet Academy of Sciences in 1938 had failed to win him election—but he made up for it with charisma and leadership.

Unlike the United States, the Soviets pursued only a small bomb project during the war. Continuous spy reports from Britain and the United States guided the Soviet work. At the end of the war Kurchatov's group was still a long way from an atomic weapon, but the awesome effects of the American bombs on Japan jolted Stalin and he immediately launched a crash project. Kurchatov and his colleagues did not display any moral scruples about atomic weapons, as did Oppenheimer and some Americans after Trinity and Hiroshima; they viewed atomic weapons at the time as necessary for the defense of their homeland and worked on them with patriotic dedication. Kurchatov continued to direct the work, displaying remarkable ability to navigate the treacherous political waters swirling around Stalin. In 1949 the Soviets tested their first atomic bomb, based on intelligence from the United States but also reflecting Soviet talent and commitment. Kurchatov was named a Hero of Socialist Labor after the test. The Soviets followed that with a test of a thermonuclear device in 1953 based on original work in Kurchatov's project. The Soviet thermonuclear test affected Kurchatov in the same way that Trinity and Hiroshima affected Oppenheimer. The stressful preparations for the test itself took a toll and then a tour of the destruction at ground zero brought home the power of the weapon. Kurchatov afterward increasingly turned to development of fission and fusion reactors for peaceful atomic energy.

Oppenheimer would also leave weapons work, although involuntarily and without the honors bestowed on Kurchatov. At the end of the war Oppenheimer resigned from Los Alamos and accepted appointment as director of the Institute for Advanced Study at Princeton, which during his tenure became a thriving center for theoretical physics. He continued to exercise substantial influence on nuclear policy, however, as chair of the General Advisory Committee to the Atomic Energy Commission from 1947 to 1952, during which time the United States accumulated a stockpile of nuclear weapons. Oppenheimer led the committee's opposition to development of the hydrogen bomb, not only on moral grounds but also because fusion did not appear technically feasible, and because a crash program would divert scarce resources from new fission weapons. President Harry Truman instead heeded the advice of the Air Force and hawkish anticommunists and approved a crash program for the H-bomb. Oppenheimer's opponents in the H-bomb debate would lead an effort a few years later to revoke his security clearance, based as much on his advice on the H-bomb as on his mendacity in reporting contacts with communists during the war. The hearings on the matter in 1954 divided the scientific community and ended with a judgment that Oppenheimer constituted a security risk. After a dozen years as a leader of the American nuclear program, Oppenheimer found himself cut off from inside circles of nuclear policy. He continued to direct the Institute at Princeton until 1966.

Alice Kimball Smith and Charles Weiner, eds., *Robert Oppenheimer: Letters and Recollections* (1980). Richard Rhodes, *The Making of the Atomic Bomb* (1986). David Holloway, *Stalin and the Bomb: The Soviet Union and Atomic Energy 1939–1956* (1994). Richard Rhodes, *Dark Sun: The Making of the Hydrogen Bomb* (1995). S. S. Schweber, *In the Shadow of the Bomb: Oppenheimer, Bethe, and the Moral Responsibility of the Scientist* (2000).

PETER J. WESTWICK

L

LAPLACE, Pierre-Simon de (1749–1827), mathematician, astronomer, and physicist.

Laplace transformed the study of mathematical astronomy with his five-volume *Traité de mécanique céleste* (1799–1825). Using calculus and differential equations, he expressed *Newton's *Principia* in its modern language, solved many of the issues left open by his predecessor, and demonstrated the stability of the solar system. At the time of his death, almost exactly a century after Newton's, Laplace had won every honor available. He was dean of the Académie des Sciences, decorated with titles and medals, a member of the Académie Française, and a marquis serving in the Chamber of Peers.

Laplace began life as the second of three sons of an average Norman farmer who became mayor of Beaumont-en-Auge. Destined for the clergy, he attended a local Benedictine school and then the nearby University of Caen, where he found his vocation as a mathematically adept natural philosopher. As a student, he took a special interest in current discussions over the validity of the Newtonian system of the world. During his stay at Caen, he rejected traditional Catholic theology and remained opposed to it for the rest of his life.

At the age of twenty, Laplace set out for Paris with a recommendation to Jean Le Rond d'Alembert, who sponsored him for a post as a teacher of mathematics at the École Militaire in Paris. The young instructor submitted a battery of original papers on differential equations and astronomy to the Paris Academy, where he gained election in 1773. He rose rapidly in its ranks. Laplace participated actively in the academy's transformation during the French Revolution, when it became part of the Institut de France, and he assumed a preponderant role in all its affairs. During the Revolution, he taught mathematics for a few months at the École Normale de l'an III and served on the board of directors of the École Polytechnique. Laplace was a central figure in the creation of the French Bureau of Longitude, whose activities he dominated for thirty years. Although he served briefly as minister of interior under Bonaparte and as vice chancellor of the emperor's Senate, his political activities were unremarkable.

Laplace's philosophical conviction in favor of determinism led him to explore probability theory and to generate its classical synthesis in the *Théorie analytique des probabilités* (1812). His most notable innovation was the establishment of Bayesian inductive probabilities for calculating the likelihood that a particular explanatory hypothesis is the cause for a set of known phenomena. He applied such a calculation in proposing the nebular hypothesis for the origins of the solar system that held sway for much of the nineteenth century.

Starting in 1781, Laplace collaborated with Antoine-Laurent *Lavoisier on several problems, notably the behavior and nature of heat, for which he invented an ice calorimeter to measure heat transfers, and the systematization of the concepts and nomenclature of chemistry. Lavoisier inspired their colleague Claude Louis Berthollet to write an *Essai de statique chimique* (1803) explaining chemical phenomena as consequences of Newtonian laws of attraction. These were the laws used by Laplace to explain the phenomena of capillarity, optical refraction, the cohesion of solids, and tidal motions. He expected that all aspects of the inorganic world could be elucidated in terms of Newton's laws.

With his neighbor Berthollet, Laplace organized a gathering of young scientists in 1806 named the Société d'Arcueil to encourage them to pursue research in their Newtonian style. Many of these disciples became the leaders of French science in the next generation. Jean-Baptiste Biot and Simeon-Denis Poisson furthered Laplace's program productively, but others, like François Arago, Étienne Louis Malus, and Joseph Louis Gay-Lussac, who attended the Arcueil meetings, struck out on their own.

Laplace wrote two popular works without equations that have been translated and reprinted often. During the Terror, he had time for an *Exposition du système du monde* (1796), which contains a chapter on the history of astronomy since antiquity. In 1814, he published his *Essai philosophique sur les probabilités*, developed from the introduction to his earlier mathematical treatise. It was here that he inserted his famous assertion about the determinism of the universe.

Maurice P. Crosland, *The Society of Arcueil* (1967). Roger Hahn, *Laplace as a Newtonian Scientist* (1967). Charles C. Gillispie, *Pierre-Simon Laplace, 1749–1827. A Life in Exact Science* (1997).

ROGER HAHN

LATTES, Cesar (b. 1924), physicist, co-discoverer of the pi meson, and **Jose LEITE LOPES** (b. 1918), physicist, pioneer of "science for development," who predicted the Z° boson.

Brazilians place Cesar Lattes and Jose Leite Lopes among the builders of the twentieth century. Born in Recife, Leite Lopes studied at the Recife Chemistry School until he shifted to physics and specialized at the University of São Paulo (USP) in 1944. In 1946 he completed his Ph.D. at Princeton, working with Wolfgang Pauli (*see* HEISNBERG AND PAULI) and writing articles on meson theory and electron radiation. Leite Lopes became professor of theoretical physics in Rio de Janeiro, a career that became unsafe after the military came to power in 1964. The military suspected him for his views on development and forbade him from appearing in public. Leite Lopes accepted a position at the Orsay Faculty of Sciences, Paris, returning to Brazil in 1967. In April 1969, he and other researchers who favored an independent Brazilian nuclear program were banned from university activities. In September 1969, Leite Lopes again left Brazil, this time for Carnegie-Mellon University, Pittsburgh, and thence the Université Louis Pasteur, Strasbourg. He returned to Brazil in the 1980s.

In 1949 Leite Lopes cofounded the Centro Brasileiro de Pesquisas Físicas (CBPF), the Brazilian Center for Physical Research, which he directed from 1960 to 1964. He coordinated the center's theoretical work, while Lattes oversaw its experimental research. Until 1954, CBPF was a cornerstone in the efforts of Brazilian President Getúlio Vargas and National Research Council President Alvaro Alberto to apply nuclear science to development. A combination of American and Brazilian conservative forces stopped the plan. With Juan José Giambiagi (Argentina), Marcos Moshinsky (Mexico), and Lattes, Leite Lopes founded the Escola Latinoamericana de Física (LASP, the Latin American School for Physics) in 1959. LASP has been fundamental in improving the quality of Latin American research.

In 1958, Leite Lopes published his important paper on the unification of electromagnetic and weak forces, in which he predicted the boson Z°, a subatomic particle with integral spin and mass 91 GeV. It is the carrier of the weak nuclear force. Confirmation of the prediction at CERN by a team led by Carlo Rubbia and Simon Van der Meer brought them the Nobel Prize. For his part Leite Lopes received the 1993 Mexican Prize for Science. It is widely believed in Latin America that he should have shared the Nobel Prize with Rubia and Van der Meer.

Leite Lopes originated much of the accepted wisdom on science for development. His *Ciência e Desenvolvimento* (*Science and Development*, 1964) shaped a generation of researchers and practitioners. It represented science as a liberating force, a necessary component of development, sovereignty, and democracy. A liberated Latin America required investment in science and in new generations of scientific workers. France recognized Leite Lopes for his contributions to science and science policy with the Academic Palm and the National Order of Merit (1989). In 1999 he received the UNESCO Science Prize for his application of science to development.

Lattes, born in Curitiba, completed the physics course at USP in 1943, working with Gleb Wataghin and G. P. S. Occhialini on *cosmic rays. After the war, he joined Occhialini and Cecil Powell, who were trying to detect in photographic emulsions particles produced by Powell's cyclotron at the University of Bristol. Lattes suggested that they expose the plates to cosmic rays. Occhialini took plates prepared by Lattes to the Pyrenees, and Lattes exposed some himself at the Chacaltaya meteorological station in the Bolivian Andes. Lattes's addition of boron to the emulsions made possible the detection of the pi meson (predicted by Hideki *Yukawa in 1934) and demonstrated its decay into a mu meson and an electron. Powell received the Nobel Prize of 1950 for the discovery, an award that Latin Americans feel should have gone to Lattes. A letter from Niels *Bohr, to be opened in 2012, supposedly explains Lattes's exclusion.

In 1948 Lattes, still interested in the production of mesons by *accelerators, visited the Radiation Laboratory (now the Lawrence Berkeley Laboratory) of the University of California at Berkeley. Working with Eugene Gardner, he discovered that the 470-centimeter (184-inch) cyclotron had been producing mesons, which Gardner had not been able to detect, for over a year. Within two weeks, Gardner and Lattes could announce their results. With prestige unprecedented for a Brazilian physicist, Lattes returned to cofound CBPF and the Brazilian National Research Council. His work transformed the Chacaltaya station into an international center for cosmic-ray research. Lattes also organized high-energy research at USP and, in 1962, at Campinas University (UNICAMP). Today, UNICAMP's Gleb Wataghin Physics Institute, which ranks among Latin America's best, has a worldwide reputation.

Lattes was elected to the Brazilian Academy of Sciences and to the physics societies of Brazil, the United States, Germany, Italy, and Japan, among others. He received Brazil's Einstein Prize and Santos Dumont Medal. The Organization of American States and the Third World Academy of Sciences also honored him. Hundreds of streets and libraries have been named after him, a popular recognition of his efforts for human development.

Lattes and Leite Lopes represent the best in Latin American science and serve as role models for scientists. Both have contributed greatly to science for development and participated in the making of institu-tions in Brazil and Latin America. The fact that others received Nobel Prizes based in part on their work epitomizes a central issue in development: knowledge producers from developing countries are rarely accepted on equal terms by their European and North American counterparts. Both Lattes and Leite Lopes have pointed out that, as long as this continues, scientists will be as much to blame for underdevelopment as financial institutions and power grabbers.

Jose Leite Lopes, *Ciência e Desenvolvimento* (1964). Jose Leite Lopes and Michel Paty, eds., *Quantum Mechanics, Half a Century Later* (1977). Simon Schwartzman, *Formação da Comunidade Cientifica no Brasil* (1979). Regis Cabral, *The Brazilian Nuclear Debate, 1945–1955* (1994). Ana Maria Ribeiro De Andrade, *Físicos, Mésons e Política: A Dinâmica da Ciência na Sociedade* (1998). A. M. R. Andrade, *Cesar Lattes, a Descoberta do Méson Pi e Outras Histórias* (1999).

REGIS CABRAL

LAVOISIER, Antoine-Laurent (1743–1794), founder of modern chemistry, discoverer of oxygen's role in chemical reactions, and public servant.

Born in Paris to Jean-Antoine Lavoisier, solicitor to the Parlement of Paris, and to Émilie Punctis, the daughter of a wealthy attorney, Lavoisier was raised, after the early death of his mother, by a maiden aunt. His family life was warm and affectionate. He enrolled in 1754 at the Collège Mazarin, where he received an outstanding education that included both classical and literary training, science, and mathematics. Although he finished his education in the Faculty of Law in 1763, he quickly acquired such a consuming interest in the natural sciences that he never practiced law. With Jean-Étienne Guettard, who was planning a geological and mineralogical atlas of France, Lavoisier undertook several extensive expeditions, mapping, examining rocks, and observing the ordering of strata. In describing these formations, Lavoisier introduced the innovative practice of drawing vertical cross sections to represent the stratigraphical order. Recognizing the close relationship between field

An experiment in Lavoisier's laboratory drawn by his wife (who sits recording the results). The apparatus, more elaborate than typical for the time, collects gases respired by Lavoisier's seated assistant for later analysis.

*mineralogy and the chemical analysis of minerals, Lavoisier attended the popular chemical lectures of Guillaume-François Rouelle, and set up a laboratory in his own home.

Talented and ambitious, Lavoisier undertook various projects that he hoped would win him enough recognition that, with the help of his father's connections, would gain him election to the Academy of Sciences. He attained that goal in 1768 after presenting a paper on an improved hydrometer with which he claimed he would reform the analysis of mineral water. To secure an assured income, Lavoisier at the same time joined the Ferme Générale, a private consortium that collected taxes and customs duties for the French government. Although this position, whose responsibilities he took seriously, caused him to spend much time away from Paris on inspection trips, it also enabled him to afford the expensive instruments and apparatus that his innovative programs of experimental investigation required. In 1771, he married Marie-Anne-Pierrette Paulze.

During 1772, Lavoisier became interested in a problem then attracting attention among French chemists: the processes by which air is fixed in or released from other bodies. In the fall of that year, he discovered that phosphorus and sulfur gain weight when they are burned and showed that a lead ore reduced in the presence of charcoal released an air. Convinced that this last discovery was "one of the most interesting...since the time of Stahl," he decided to read everything published on airs from the time of Stephen Hales in 1728 and Joseph Black in 1756 through the identification of several new "airs" by Joseph Priestley. Lavoisier judged that this work had set the stage for a "revolution" in physics and in chemistry.

His own early experiments in the field were only marginally successful. His first publicly announced theory, that Black's "fixed air" was absorbed in combustion and in the calcination of metals and released in the reduction of metallic calces, collapsed during his effort to support it experimentally. In the process, however, he devised novel techniques for measuring the quantities of materials—the airs as well as solids and liquids—that took part in a chemical "operation," and to reason about composition in the terms that have become known as his "balance sheet" method. His growing mastery of these techniques and his persistence in overcoming the many pitfalls arising in their use, rather than his advocacy of the generally accepted principle that matter is neither created nor destroyed in chemical changes, most distinguished Lavoisier from his contemporaries. These talents gave him the critical foundation for theoretical reasoning that enabled him to surpass other chemists of his time and impose his views upon them.

By 1777, aided in part by Priestley's discovery that the air released by a mercury calx reduced without charcoal is "better" than common air, Lavoisier had identified the substance that takes part in the changes he had been studying as a particular portion of the atmosphere that he called at first "eminently respirable air," because only this portion supported the life of animals and humans. Because he had also found that several common acids contain the "base" of this air, he renamed the base the "acidifying principle," or "oxygen." By this time, he could also conclude that "fixed air" was a combination of oxygen and "charcoal." Thereupon, Lavoisier offered a new "general theory of combustion," which he claimed could explain the phenomena better and more simply than could Stahl's phlogiston theory.

It required ten more years, the discovery of the decomposition of water, and an extensive campaign to convert the leading chemists of Europe to adopt the new theory. The climax of the campaign came in 1787 when Lavoisier and his French followers, Guyton de Morveau, Claude Louis Berthollet, and Antoine Fourcroy, proposed a reform of the chemical nomenclature that expressed rationally the composition of compound bodies and embodied the new theories into the language of chemistry.

Despite his monumental achievements in chemistry, Lavoisier's work in that field constituted only one of his many activities. In addition to his duties in the Tax Farm, Lavoisier took on important administrative functions in the Academy of Sciences, developed an experimental farm at a country estate he had purchased outside Paris, made improvements in the manufacture of gunpowder, wrote important papers on economics, and became deeply involved in reformist political and administrative roles during the early stages of the French Revolution. As a member of the Temporary Commission on Weights and Measures from 1791 to 1793, he played an important part in the planning for the metric system. He devised plans for the reform of public instruction and the finances of the revolutionary government. In his efforts to deal rationally with the turbulent events of the Revolution, however, Lavoisier repeatedly underestimated the power of public passion and political manipulation, and despite his generally progressive views and impulses, he was guillotined, along with twenty-eight other Tax Farmers, as an enemy of the people at the height of the Reign of Terror in 1794.

Henry Guerlac, *Antoine-Laurent Lavoisier: Chemist and Revolutionary* (1975). Bernadette Bensaude-Vincent, *Lavoisier: Mémoires d'une revolution* (1993). Arthur Donovan, *Antoine Lavoisier: Science, Administration, Revolution* (1993). Jean-Pierre Poirier, *Lavoisier: Chemist, Biologist, Economist* (1996).

FREDERIC LAWRENCE HOLMES

LAWRENCE, ERNEST O. See BLACKETT, PATRICK, AND ERNEST O. LAWRENCE.

LEE, T. D. (b. 1926); **C. S. WU** (1912–1997); and **C. N. YANG** (b. 1922), Chinese-American physicists.

Chien-Shiung Wu was born in Liuhe, a small town near Shanghai, China, months after the downfall of the Qing (Manchu) dynasty. Girls had few educational prospects in China at the time, but Wu's father, an educator, encouraged her to study in Shanghai with a prominent scholar; her evident intellect led her to the National Central University in Nanjing for undergraduate work in physics. She moved to the United States in 1936 and earned her Ph.D. in physics at the University of California at Berkeley in 1940, with a dissertation on the fission products of uranium. She spent the war in various teaching positions and then in work on radiation detectors at Columbia University for the Manhattan Project. She stayed at Columbia after the war, in the end remaining for thirty-five years. Her postwar research focused on beta decay, the spontaneous, radioactive disintegration of an atom through emission of an electron. Her precise experimental technique confirmed Enrico Fermi's theory of beta decay and established her as an expert on the phenomenon.

T'sung-Dao Lee was born in Shanghai and came of age during the war with Japan. Lee studied at the National Chekiang University and then the National Southwest Associated University, a combination of Tsing Hua University and other leading Chinese schools that relocated to the interior away from the combat. After graduating in 1946 he came to the United States to study with Fermi at Chicago. Fermi steered him towards *astrophysics, and Lee's Ph.D. thesis of 1950 examined white dwarf stars. After stints at Berkeley and Princeton, he won appointment at Columbia in 1953 and stayed there for the rest of his career. He thus came into contact with Wu, a connection of consequence for physics. His research dealt with the theory of turbulence, statistical mechanics, and quantum field theory.

Chen Ning Yang grew up around the Tsing Hua University in prewar Beijing, where his father, a doctor of mathematics from Chicago, was a professor. Yang also attended Southwest Associated University, a few years before Lee. After graduate work at Tsing Hua he won a fellowship for further study in the United States and arrived there in the fall of 1945. Like Lee he chose Chicago, and obtained his Ph.D. there in 1949, studying with Edward Teller, Fermi, and Samuel Allison; his dissertation, directed by Teller, applied group theory to nuclear reactions. He then went to the Institute of Advanced Study in Princeton, where he would stay for seventeen years, working in quantum field theory and theoretical particle physics. In 1954, while spending a year at Brookhaven National Laboratory, he developed with R. L. Mills a nonlinear gauge field theory, later known as the Yang-Mills theory.

In the 1950s physicists faced a proliferation of new particles produced by high-energy *accelerators. Theoretical physicists struggled to compose a taxonomy for the denizens of the particle zoo, based on mass, charge, lifetime, spin, and other observed properties. Lee and Yang had started to collaborate in 1951 at Princeton on the theory of *elementary particles. Starting in 1955 they tackled the so-called θ - τ (theta-tau) puzzle: experimental evidence conflicted over whether the theta and tau particles were the same animal or two different ones. To solve the puzzle, Lee and Yang in 1956 proposed that parity, or right-left symmetry in space, is not conserved in weak interactions—that is, that two weak interactions that are mirror images but otherwise identical do not behave the same way. Theta and tau were in effect the same particle in different parity states, one right-handed and the other left-handed. Lee and Yang's idea challenged current theories of beta decay in particular and fundamental assumptions about the *symmetry of natural laws in general.

Lee and Yang proposed several experiments to test the hypothesis, including ones for beta decay, a prominent example of a weak interaction. Wu, the beta-decay expert, discussed experimental approaches with Lee at Columbia and became convinced that some were feasible. She suggested an experiment using cobalt-60, cooled to a fraction of a degree above absolute zero. Wu collaborated with several scientists at the National Bureau of Standards to take advantage of their expertise in cryogenics (see COLD AND CRYONGENICS). The experiment demonstrated that in beta decay the cobalt atoms gave off electrons in one direction more than the other, against the requirements of symmetry. The confirmation of parity nonconservation led to challenges to other long-held principles of symmetry, especially with respect to electric charge and time. It also won Lee and Yang, but not Wu, the Nobel Prize in physics for 1957, the quickness of the reward for their work indicating the perceived importance of symmetry violation for physics.

Lee and Yang were the first Nobel laureates of Chinese descent, and thus a source of ethnic pride. Wu, Lee, and Yang were part of the large Chinese-American scientific diaspora in the second half of the twentieth century. The diplomatic rapprochement between the United States and China in the early 1970s renewed scientific exchange between the two countries, and Lee and Yang played leading roles in the exchange. Their international stature as scientists also gave them a voice in Chinese science policy. Their advice sometimes diverged: Lee advocated vigorous government support of basic research in China, including expensive accelerators, as a way for China to stay at the cutting edge of science; Yang, although he supported basic research, opposed high-energy physics in China as a waste of money given the backward state of the economy. Lee and Yang would also provide strong responses to the Tiananmen Square massacre in 1989, both arguing for continued exchange with Chinese colleagues instead of sanctions, though Yang again favored economic development before political reform. Although Wu did not engage in foreign policy to the same degree as Lee and Yang, she enjoyed prominent status in the American community: she was elected president of the American Physical Society in 1975, the first Chinese and the first woman to hold the post, and won the Wolf Prize the following year.

Laurie M. Brown et al., eds., *Pions to Quarks: Particle Physics in the 1950s* (1985). Robert Novick, ed., *Thirty Years since Parity Nonconservation: A Symposium for T. D. Lee* (1988). C.S. Liu and S.-T. Yau, eds., *Chen Ning Yang: A Great Physicist of the Twentieth Century* (1995). Zuoyue Wang, "U.S.-China Scientific Exchange: A Case Study of State-Sponsored Scientific Internationalism in the Cold War and Beyond," *Historical Studies in the Physical and Biological Sciences* 30 (1999): 249–277.

PETER J. WESTWICK

LIGHTNING. Until the seventeenth century most ideas about lightning in the Western world were based on Aristotle's *Meteorologica*, which ascribes lightning and thunder to a dry, warm exhalation "forcibly ejected" from one cloud onto another. The impact produced the sound. As for the light, it came from a wind burning "with a fine and gentle fire."

In spite of his espousal of the mechanical philosophy, René *Descartes's account in his *Meteorology* (1637) resembled Aristotle's. Thunder occurs when a high cloud condenses and falls against a lower one. When the air between these clouds contains "very fine, highly inflammable exhalations," it usually produces "a light flame which is instantaneously dissipated." Others in the seventeenth century attributed lightning to combustion of "sulphurous vapors." In the early eighteenth century Francis Hauksbee and William Wall conjectured that lightning was related to "electric fire." As late as 1746 John Freke argued that both *electricity and lightning arose from "a great quantity of the elementary fire driven together."

This echo of Aristotle underlay Benjamin *Franklin's famous identification of lightning and static electric discharge. Franklin suggested in 1750 that an iron rod extending 9 meters (30 ft) above a church steeple could be used to collect lightning (*see* LIGHTNING CONDUCTOR), and two years later Thomas François Dalibard successfully performed the experiment in France. Franklin soon after tried his famous kite experiment and survived, though Georg Wilhelm Richmann died in 1753 when he personally conducted lightning from rod to ground. Franklin concluded that clouds are electrified, the bottom usually negatively, while John Canton discovered that clouds may be either positive or negative. In the 1770s Giambattista Beccaria investigated the frequency of lightning strikes.

Little more was learned until the twentieth century when photography, spectroscopes, and electromagnetic devices provided new tools of investigation. In the 1930s, T. E. Allibon and J. M. Meek concluded that sparks move from cloud to ground, ending in terminal "brushes," increasing gradually in length until the atmosphere is ionized. Meek found that a "pilot" discharge travels from the negative cloud bottom, followed by other discharges. A positive current follows the completed discharge path to the cloud.

Many scientists tried to explain how clouds become charged. G. C. Simpson hypothesized in 1927 that as raindrops separate the droplets take opposite charges. About the same time C. T. R. Wilson, Julius Elster, and Hans Geitel suggested that the charge separation intensifies as induction polarizes drops in a cloud. Bigger drops, falling, pick up electrons, making the bottom of the cloud increasingly negative. Others proposed that up-drafting air carries charged ions. In 1960 C. B. Moore and Bernard Vonnegut cast doubt on the idea that the ionization of droplets must occur to produce lightning since electrical fields in and around clouds can precede precipitation.

Although lightning scientists still use ground-based detection networks, in the second half of the twentieth century they have relied increasingly on high-altitude airplanes, sounding rockets, satellites, and the space shuttle. In 1990 this technique resulted in the dramatic discovery of "red sprites"—lightning that goes from cloud tops to the ionosphere—and other stratospheric discharges.

Park Benjamin, *A History of Electricity (The Intellectual Rise in Electricity) from Antiquity to the Days of Benjamin Franklin* (1898). Peter E. Vermiester, *The Lightning Book* (1972).

GREGORY A. GOOD

LIGHTNING CONDUCTOR. The similarity between electrical discharge and *lightning occurred to several minds after the invention of the electrical machine and the Leyden jar in the late 1740s made impressive sparks and shocks available in the laboratory. Even with this modest equipment, Jean-Antoine Nollet made a line of insulated monks jump by using them to short-circuit a condenser. Nollet was the first to list the analogies between electricity and lightning (in 1748), and another Frenchman, Jacques de Romas, probably was the first to propose to bring atmospheric charge to earth via a kite (1750 or 1751). Benjamin *Franklin, with his characteristic boldness and optimism, first proposed an experiment to show that lightning could produce some of the effects of artificial electricity and offered measures to protect buildings from lightning strokes.

The experiment, first performed in France in 1752 under the instructions of the Comte Georges-Louis de Buffon, who expected, rightly, that a positive outcome would undermine the position of Nollet as Europe's leading electrician, employed a tall pointed insulated iron rod exposed to the elements. During a storm a dispensable old soldier approached his knuckle to the rod, drew a spark, and tried to confirm the theory. It worked. The soldier survived because the rod detected fluctuations in the electrical state of the lower atmosphere. Had he received a lightning stroke, the veteran would not have reported his success. The first and last natural philosopher who succeeded with Franklin's experiment was Georg Wilhelm Richmann.

After Richmann's electrocution in 1753, Franklinist electricians, particularly Giambattista Beccaria and Alessandro Volta, monitored atmospheric electricity remotely (*see* GALVANI AND VOLTA). Others developed the art of protecting buildings from thunderbolts. Deployment progressed slowly partly because Franklin's theories

A method, not recommended, for protection against a lightning stroke–the parapluie-paratonnerre ('umbrella-lightning rod') invented by Jacques Barbeu Dubourg, Benjamin Franklin's French translator.

suggested that a sharply pointed high metal pole, perfectly grounded, could despoil a cloud of its electricity silently as well as channel the stroke if it came. There arose a squabble over whether pointed rods might entice lightning that otherwise would strike somewhere else. The last significant holdouts, Nollet (died 1770) and Benjamin Wilson (died 1788), rejected as silly and arrogant the claim that puny man could disarm a thundercloud. Nollet recommended sitting in a grounded metal cage. Wilson preferred conductors that ended in balls or knobs so as not to attract the attention of passing clouds. If a stroke came anyway, the blunt rod would be at least as effective as the pointed one in disposing of the lightning. Nollet's cage, though inconvenient, would have worked. Wilson's obtuse ends would have been no better than Franklin's points.

During the nineteenth century, many public and private buildings were armed with lightning rods installed so as to connect all the metal parts to the same grounded system. Failures in protection could almost always be traced to imperfect grounding. Thus electrical science saved property and lives, especially of people who rang church bells in an effort to break up storm clouds, and made its first significant contribution to technology.

J. L. Heilbron, *Electricity in the 17th and 18th Centuries* (1979, 2000). I. Bernard Cohen, "Prejudice against the Introduction of Lightning Rods," Franklin Institute, *Journal* 253 (1952): 393–440.

J. L. HEILBRON

LIGHT, SPEED OF. Several investigators tried to measure the speed of light in the seventeenth and eighteenth centuries. Galileo, or perhaps his followers, flashed light to an assistant, but the speed was too great to measure. In 1675 the Danish astronomer Ole Rømer noticed that intervals between eclipses of Jupiter's moons are less when Jupiter and the Earth approach each other; he correctly attributed the phenomenon to the time it takes light from Jupiter to reach the Earth. Using contemporary estimates of satellite periods and distances, Rømer calculated a velocity of 214,000 kilometers per second (k/s). James

Bradley's discovery of the *aberration of starlight provided a second means of estimating the speed of light, since the aberration angle depends upon the ratio of the speeds of the observer and of light. If extended to the speed of light, Bradley's calculations would have produced about 264,000 k/s.

In the nineteenth century, French physicists Armand-Hippolyte-Louis Fizeau and Jean Bernard Léon Foucault made terrestrial measurements of the speed of light by passing a beam through the gaps in a rapidly spinning toothed wheel. They obtained values of 315,300 and 298,000 k/s. The American physicist Albert A. Michelson improved on their experiments by measuring the interference fringes produced by a light beam split up in his "interferometer" and made to traverse slightly different paths through it. That raised the light velocity to 299,910 k/s (1879). Next Michelson attempted to detect the expected change in the speed of light caused by its motion through the hypothetical luminiferous *ether. The surprising null result of the 1887 ether-drift experiment suggested that the speed of light is constant, independent of the speed of the emitting body. This result became a crucial element in the pedagogy, and perhaps in the discovery, of Einstein's *relativity theory. Subsequent measurements of the speed of light over precisely measured distances and timed electrically ended in the spurious agreement of several sets of pre-1941 measures at 299,776 k/s. Holding Michelson in awe, followers ended their searches for flaws in their experimental apparatus when their measures agreed with his. New technologies later increased the accepted speed by nearly 17 k/s, more than four times Michelson's purported 4 k/s margin of error, to 299,792.5 k/s. Newer technology, including radiotelemetry, also improved the determination of planetary distances and the astronomical unit.

Dorothy Michelson Livingstone, *The Master of Light: A Biography of Albert A. Michelson* (1973). Albert Van Helden, *Measuring the Universe: Cosmic Dimensions from Aristarchus to Halley* (1985).

NORRISS S. HETHERINGTON

LOMONOSOV, Mikhail Vasilievich (1711–1765), Russia's first eminent natural philosopher, who still occupies an important symbolic place in the Russian world of scholarship. Moscow University and a number of other institutions and academic prizes are named after him.

Lomonosov was born in the small village of Mishaninskaia near the far north of European Russia, on the White Sea. Although legally peasants, Lomonosov and his family enjoyed a freedom not known to the serfs on the estates of central Russia. Lomonosov's father, an active merchant, owned several fishing and cargo ships, and his mother was the daughter of a deacon. As a child Lomonosov learned to read and write, in both Russian and Church Slavonic.

Desiring further education, Lomonosov in 1730 applied for admittance to the Slavic-Greek-Latin Academy in Moscow, the best higher educational institution in Russia at that time, although devoted to theology and the preparation of clergy. Since peasants could not attend the academy, Lomonosov concealed his origins; he claimed to be the son of a priest, which his knowledge of Church Slavonic supported. By the time his superiors learned the truth, Lomonosov had so impressed them by catching up with his classmates (most of whom were younger than he) in Latin and surpassing them in other subjects that they permitted him to stay.

In 1735 the Imperial Academy of Sciences in St. Petersburg, then just ten years old, requested monasteries and ecclesiastical academies to send students to its fledgling university to study under foreign academicians. Lomonosov was chosen. In St. Petersburg he learned mathematics and physics. A year later the academy sent him to Western Europe to study chemistry and mining. For almost five years Lomonosov studied natural science at the universities of Marburg and Freiburg. The professors who made the deepest impression upon him were Christian Wolff at Marburg and Johann Friedrich Henkel at Freiburg.

After returning to St. Petersburg in 1741 with his German wife, Elizabeth Zilch, Lomonosov was made an adjunct of the Academy of Sciences in physics and later professor of chemistry. In 1748 he opened the first scientific chemical laboratory in Russia, equipped with balances and other equipment similar to what he had seen in Europe. Later he became head of the geographical department of the Academy of Sciences.

Lomonosov's scientific activity can be divided into three phases. From 1740 to 1748 he concerned himself with speculative physics, particularly the corpuscular philosophy, the nature of heat and cold, and the elasticity of air. He compiled a syllabus in physics and in 1746 delivered the first public lecture on the subject ever presented in the Russian language. Most of his scientific writings, however, were in Latin. From 1748 to 1757, after the construction of his chemical laboratory, Lomonosov worked on the characteristics of saltpeter, the nature of chemical affinity, the production of glass and mosaics, the freezing of liquids, and the nature of mixed bodies. From 1757 to his death in 1765 Lomonosov devoted his time to scientific administration, exploration, mining, metallurgy, and navigation. Throughout his professional life he wrote poetry and promoted the Russian language and Russian history.

Lomonosov's most significant work in natural philosophy was his extension of the corpuscular or *mechanical philosophy common in the seventeenth and early eighteenth centuries to a wide variety of phenomena. He liked to describe nature in concrete pictures and mechanical models, and to reason by analogy from them. This approach, applied literally and speculatively, sometimes led Lomonosov to concepts that seem prescient to modern readers, such as the lowest possible temperature occurring as the state in which all particles are motionless. However, much Soviet literature on Lomonosov contains exaggerated claims about his achievements. No truly definitive study of his scientific work exists.

Lomonosov symbolized emerging Russian scholarship, talented but still only partially developed. He had brilliant ideas but lacked discipline and scattered his efforts over too broad a front. Nonetheless, he was an unprecedented phenomenon, a native Russian champion of science and learning who, in important areas, stood at the forefront of knowledge. He would serve as a model for young Russian scientists for generations.

Henry M. Leicester, ed., *Mikhail Vasil'evich Lomonosov and His Corpuscular Theory* (1970). Galina E. Pavlova and Aleksandr S. Federov, *Mikhail Vasil'evich Lomonosov: His Life and Work*, trans. Arthur Aksenov (1984).

LOREN R. GRAHAM

LOPES, Jose Leite. See LATTES, CESAR, AND JOSE LEITE LOPES.

LOW-TEMPERATURE PHYSICS. The field of low-temperature physics, or cryogenics, emerged in the late nineteenth century. In 1877, two researchers succeeded in liquefying oxygen independently within days of each other. Raoul-Pierre Pictet, a Swiss physicist, and Louis Paul Cailletet, a French mining engineer, both cooled oxygen gas under pressure, then rapidly expanded the volume to condense the gas. Cailletet soon reproduced the feat with nitrogen. In 1895, Carl von Linde in Germany and William Hampson in England developed a method for industrial-scale production of liquid air, which aided subsequent cryogenics research. The three leading low-temperature laboratories—at the University of Leyden (under Heike Kamerlingh Onnes), the University in Cracow (Karol Olszewski), and the Royal Institution in London (James Dewar)—engaged in a race to lower temperatures and the liquefaction of hydrogen and, after its discovery on Earth in 1895, helium. In 1898 Dewar won the race for hydrogen using a variation on Linde's technique to reach about 20° Kelvin. In 1908 Kamerlingh Onnes used liquid hydrogen to cool helium enough to condense it at low pressure, at around 5° K. The Leyden laboratory thereafter enjoyed a fifteen-year monopoly in the production of liquid helium.

The early history of cryogenics illustrates the proliferation of academic research laboratories in the late nineteenth century and their move into fields requiring such relatively expensive apparatus. The work involved chemists, engineers,

Heike Kamerlingh Onnes, professor of physics at the University of Leyden, seated in front of his elaborate equipment for low-temperature work. He and his colleagues succeeded in liquefying helium and in discovering superconductivity in 1908 and 1911, respectively.

skilled glassblowers, and instrument makers as well as physicists. Kamerlingh Onnes's success with liquid helium owed as much to his ability to form a team with physical, mechanical, and chemical expertise as to his high standards of experimental precision and keen grasp of thermodynamic and electromagnetic theory.

Cryogenics demonstrates the interpenetration of industry and science in the second industrial revolution—in particular, the budding refrigeration industry, which had emerged as a rival to natural ice in the late nineteenth century, especially for brewing lager beer and shipping meat to Europe from Argentina and New Zealand. Industrial uses of liquid air and its components, such as liquid oxygen for oxyacetylene blowtorches, spurred the formation of firms such as Linde Air (founded by von Linde), British Oxygen Company, and L'Air Liquide. Dewar's development of a silvered, double-walled flask with an intervening vacuum led quickly to a commercial market in thermos bottles (although Dewar failed to patent his flask). Kamerlingh Onnes had ties to Dutch industry. The industrial relevance of low-temperature research induced national governments to sponsor it in their standards laboratories such as the U.S. National Bureau of Standards and the German Physikalisch-Technische Reichsanstalt (PTR).

Academic physicists instigated their own low-temperature programs. The elucidation of specific heats at low temperatures, explored by Dewar and extended by Walther Nernst and F. A. Lindemann, provided important evidence for the fledgling quantum theory before World War I and helped establish low-temperature research as a fruitful field of physics. Pyotr *Kapitsa's investigation of magnetic effects at low temperatures in Cambridge, which he continued in Moscow, and William Giauque's work on magnetic cooling and specific heats in Berkeley illustrate the spread of low-temperature physics in the 1930s. Cryogenic techniques would find application after World War II in the production of the first thermonuclear fusion weapons, in rocket propellants, and in bubble chambers for particle physics experiments. The volatile liquids and high pressures of cryogenics required elaborate safety precautions, although they did not always prevent catastrophic explosions.

Two important lines of research emerged from peculiar phenomena observed at low temperatures. In 1911, Kamerlingh Onnes and his collaborators found that electrical resistance in mercury suddenly vanished at 4° K. This "superconductivity," as Kamerlingh Onnes called it, puzzled theorists for decades. In 1933, Walther Meissner and Robert Ochsenfeld at the PTR in Berlin found that magnetic induction as well as electrical resistance disappears in a superconductor, and hence showed that superconductivity included more than its name indicated. In 1957,

John Bardeen, Leon Cooper, and John Schrieffer of the University of Illinois produced a satisfactory microscopic explanation of superconductivity, based on quantum mechanical coupling of electrons with opposite spin.

Superconductivity promised spectacular technological applications in low-loss electrical power transmission and high-power superconducting electromagnets. But known materials that exhibited superconductivity proved too fragile for power transmission; and, like heat, high magnetic fields, as in electromagnets, destroy the superconductive state. Hopes for new technologies rekindled with the discovery of so-called type II superconductivity in 1961, which persevered in the presence of magnetic fields, and then flared anew with the announcement in the mid-1980s of a class of ceramic materials that stayed superconductive at temperatures up to 100° K.

The second peculiar phenomenon that stemmed from observations by Kamerlingh Onnes was a drop of density in liquid helium below about 2° K. With the spread of low-temperature physics, unexpected results with liquid helium began to accumulate, suggesting that it existed in two different states: normal helium I and low-temperature helium II. In 1937 and 1938 several physicists established that the viscosity as well as the density of helium II seemed to vanish, and that it could form a thin film that swept up the sides of vessels; Kapitza termed the effect "superfluidity." In 1955 Richard *Feynman arrived at a microscopic theory of superfluidity based on quantization of vortices in the fluid. The theories of both superconductivity and superfluidity rely on interactions among individual particles or atoms; and Bose-Einstein statistics instead of Fermi-Dirac statistics govern the behavior. In other words, superconductivity and superfluidity are macroscopic quantum phenomena; hence their novelty and interest, and the difficulty in accommodating them within physical theory.

See also COLD AND CRYONICS.

Kurt Mendelssohn, *The Quest for Absolute Zero* (1977). Kostas Gavroglu and Yorgos

Goudaroulis, *Methodological Aspects of the Development of Low Temperature Physics 1881–1956* (1989)). Per F. Dahl, *Superconductivity: Its Historical Roots and Development from Mercury to the Ceramic Oxides* (1992). Ralph G. Scurlock, ed., *History and Origins of Cryogenics* (1992).

PETER J. WESTWICK

LYELL, Charles (1797–1875), originator of the doctrine of uniformitarianism in *geology.

Lyell came from a Scottish family with a long tradition of serving in the English navy. Although he was educated and spent most of his life in England, he remained attached to the intellectual traditions of the Scottish Enlightenment. His father intended him to be a barrister. While studying at Oxford, he developed problems with his eyesight that made reading difficult and the prospect of a legal career unappealing. The geological lectures of William Buckland fascinated Lyell, and his ambitions began to shift. In 1819 he joined the Geological Society of London. He took several field trips to the Continent, some of them with another budding geologist, Robert Murchison. Sicily, in particular, impressed him, with its evidence of major geological changes in the recent past.

In the 1820s, Lyell read Jean Baptiste de Lamarck's *Philosophie Zoologique* (1809) and John Playfair's *Illustrations of the Huttonian Theory of the Earth* (1802). Lamarck's suggestion that one species could transmute into another appalled Lyell as speculative and irreligious. Playfair's exposition of Hutton's theory of an indefinitely habitable earth, on the other hand, attracted Lyell, a deist like his fellow Scots James Hutton and John Playfair. Lyell thought that the deity would have approved of Hutton's plan. Playfair's use of the vera causa method also attracted him. Lyell began planning an ambitious book to bring Huttonianism up to date, establish the principles of reasoning in geology, and use them to discredit the theories of the transmutation of species and the cooling, contracting earth. He also hoped that the book would make his reputation and bring in enough money so that he could live comfortably and help his unmarried sisters.

In 1830, Lyell published the first of the three volumes of his *Principles of Geology*. He introduced the volume with a cautionary history to make plain the perils of an ill-chosen methodology, described the presently observable geological causes that would have been adequate to produce past geological effects, and proposed a theory of climate change based on alternating positions of land and sea that made the supposition of a cooling earth unnecessary. In the second volume (1832), he tackled the vexing question of the fossil record, maintaining that although individual species had become extinct and been replaced by newly created ones, the same broad classes had always existed. In the third volume (1833), he responded to his critics and introduced a new classification of the Tertiary strata.

The *Principles* succeeded brilliantly. It was a master work, rigorous in its argument yet accessible to a wide, educated public. It won respectful (though deeply critical) reviews from George Poulet Scrope and William Whewell. Lyell spent much of the rest of his life revising, reworking, and abstracting the *Principles*, which went into a twelfth edition after his death. In 1832 he married Mary Horner, who acted as his secretary and became expert in the fossil shells so important to Lyell's classification of the Tertiary. In 1831–1832, he taught geology for a brief period at King's College London but did not enjoy it and never again took a university position. From 1834 to 1836, he was president of the Geological Society. He and Charles Darwin started a lifelong friendship when Darwin returned from the *Beagle* voyage in 1836.

In 1841–1842, the Lyells traveled in the United States, where Charles lectured at the Boston Atheneum. They found much to admire, and returned to the United States in 1845–1846 and 1852–1854. Lyell's *Travels in North America* (1845) became a small classic. In 1848 he was knighted. Soon he began worrying once again about whether one species could have evolved into another. In 1863, in his *The Antiquity of Man*, he published a broad range of evidence for man's great age and descent from the lower animals, but left his readers to draw their own conclusions. In 1864 he announced at the Royal Society that Darwin's argument had finally persuaded him, and between 1865 and 1868 he rewrote the *Principles* for the tenth edition to reflect his change in belief.

Leonard Wilson, *Charles Lyell, The Years to 1841* (1972). Charles Lyell, *Principles of Geology*, ed. Martin Rudwick (3 vols., 1991). Leonard Wilson, *Lyell in America: Transatlantic Geology, 1841–1853* (1998). Derek J. Blundell and Andrew C. Scott, eds., *Lyell: The Past is Key to the Present* (1998).

RACHEL LAUDAN

M

MAGNETISM. William Gilbert's *De magnete* (1600) was the starting point for the study of magnetism during the scientific revolution. For Gilbert, magnets were characterized by their north and south "poles," the opposite ends of the axis along which freely suspended magnets aligned themselves with the earth. Magnets brought together with like poles facing fled each other, but if unlike poles were facing, they came together. The earth itself, Gilbert concluded, was a large magnet. He described magnetic attraction in animistic terms as coition, two magnets coming together in mutual harmony.

Magnetism challenged seventeenth-century mechanists who proclaimed that all natural powers arose from matter and motion alone. *Descartes's success in providing a mechanistic explanation within the framework of his overall world picture was a triumph for the new philosophy. He envisaged distinctively shaped corpuscles of subtle matter passing through a magnet along channels suitable to receive them, then returning through the external air to form a loop of matter in motion. Such channels were peculiar to iron. The patterns formed by iron filings around a magnet showed the streamlines of flow of the subtle matter through the air. Pressure from the streaming matter caused other, nearby magnets to align themselves as they did. While later authors often altered details, Descartes's depiction was widely accepted for more than a century. Even *Newton, who tried measuring the force between two magnets, adopted it.

In 1759, Franz Ulrich Theodosius *Aepinus proposed a new theory based on forces acting at a distance, analogous to his improved version of Benjamin *Franklin's theory of *electricity. Magnetization, he suggested, involved not currents of subtle matter but redistributions within samples of iron of a subtle fluid specific to magnetism, which left one part of a sample with a surplus of fluid and another with a deficit. These regions corresponded to the poles. Particles of fluid repelled each other but were attracted to particles of iron, which likewise repelled each other; other kinds of matter did not act on the fluid. Magnetic poles were therefore centers of force, not entry or exit points for streams of subtle matter. By adding the forces acting in various situations, Aepinus accounted semiquantitatively for the known phenomena and predicted new effects, particularly in improving compass needles, that were quickly confirmed.

A modified form of Aepinus's theory became widely adopted following Charles Augustin *Coulomb's advocacy in the 1780s. Unhappy with Aepinus's notion that particles of iron repelled each other, Coulomb invoked a second subtle fluid as the carrier of this force. He also developed a molecular theory of magnetization, restricting the redistributions Aepinus discussed to individual molecules; the alignment of polarized molecules of iron made macroscopic magnets. In 1785 Coulomb reported that the forces between magnetic poles followed an inverse-square law. This opened the way to Simeon Denis Poisson's fully mathematized version of Coulomb's theory.

From Gilbert on, magnetism and electricity had been regarded as unconnected (if analogous) phenomena. However, in 1820 Hans Christian Ørsted discovered that a wire carrying an electric current affected a compass needle. André-Marie Ampère quickly showed that a current-carrying loop or solenoid was equivalent to a magnet. He argued from this that the magnetic fluids were a fiction and that magnetization derived from tiny solenoidal electric currents in molecules of iron. In the manner of *Laplace, he showed how the observed forces between current-carrying wires could be compounded from elementary forces exerted by electrical charges in motion. Meanwhile, electromagnets—ironcored solenoids carrying ever-larger currents—generated magnetic effects far more powerful that any previously obtained. In 1831, Michael *Faraday discovered the reverse of Ørsted's effect. Electromagnetic induction became the basis of the electrical power industry, power being generated by rotating coils of wire in a magnetic field.

Dissatisfied with theories based on action at a distance, Faraday shifted attention to the *field surrounding a magnet. He regarded magnetic lines of force not as mere geometrical constructs showing the direction of a compass needle at any point, but as lines of strain in space. Poles as centers of force did not exist—matter interacted with lines of force, conducting them well, as in iron, or poorly, as in diamagnets such as bismuth, the anomalous magnetic behavior of which was another of Faraday's discoveries. James Clerk *Maxwell started with Faraday's ideas when in the 1860s he developed his dynamical theory of the electromagnetic field, which represented magnetic lines of force as lines of rotational strain in the *ether.

In the twentieth century, the equations of Maxwell's theory survived the abandonment of the ether that had initially given them physical meaning, and continued to provide the basis for understanding magnetic interactions. Meanwhile, modern theories of atomic structure gave Ampère's ideas on the origins of magnetism new currency, the motions of electrons in atoms constituting precisely the kind of elementary current loops he envisaged. New and much more powerful permanently magnetic materials—ferrites and, from the 1970s, various rare-earth alloys—became indispensable to much of late twentieth-century technology. In the 1970s, magnets exploiting the phenomenon of superconductivity at low temperatures (see LOW-TEMPERATURE SCIENCE) also became widely available; these provided stronger and more homogeneous fields that found widespread application in new forms of laboratory apparatus.

E. T. Whittaker, *A History of Theories of the Aether and Electricity* (2 vols., 1951). René Taton, *History of Science*, trans. A. J. Pomerans (4 vols., 1964–1966), vol. 3, 178–234; vol. 4, 142–151. L. Pearce Williams, *The Origins of Field Theory* (1966). R. W. Home, *Aepinus's Essay on the Theory of Electricity and Magnetism* (1979).

R. W. HOME

MAGNETO-OPTICS. The effects of magnetic fields on light have had an importance for physical theory far beyond their significance in nature since their first detection in 1845. That occurred because William *Thomson, Lord Kelvin, inferred from Michael *Faraday's ideas about electricity (see FIELD) and his own ideas about light that glass stressed by an electric field should rotate the plane of polarized light passing through it. Faraday looked, found nothing, and substituted a magnetic field, which worked. Success prompted a characteristic leap: Faraday inferred that magnetism could be concentrated, as it had been in the experiment, by materials other than ferromagnetics. This insight, the first spin-off from a magneto-optical effect, led him to the discoveries of para- and diamagnetism.

In 1862 Faraday sought to change the spectrum of a light source by placing it in a magnetic field. No luck. But the fact that he had tried encouraged a repetition some thirty years later. Pieter Zeeman had just obtained his doctorate from the University of Leyden with a prize-winning thesis on a second magneto-optic effect, the change in polarization of plane-polarized light reflected from an electromagnet. In 1896 Zeeman saw the bright yellow lines of sodium broaden when he placed their source in a magnetic field. He brought this news to his professor, Hendrik Antoon Lorentz. From his own model for the emission of light, in which the radiator is an "ion" of unknown charge e and mass m, Lorentz predicted that the broadened lines should be resolvable into triplets polarized in certain ways and separated by a distance proportional to e/m. Zeeman confirmed the prediction and deduced the value of e/m. It came out close to the number that Joseph John *Thomson had found for the ratio of charge to mass of *cathode-ray particles. The "Zeeman effect" played a major part in the establishment of the *electron as a building block of matter. Zeeman and Lorentz shared the Nobel Prize for physics in 1902.

Most Zeeman patterns differ from Lorentz's triplet. Explaining quartets, quintets, and so on proved too much for both classical and early quantum theories of *atomic structure and spectral emission.

After World War I demobilizing physicists found the "anomalous Zeeman effect," which most of them had ignored, high among the outstanding problems of atomic physics as described in Arnold Sommerfeld's comprehensive survey, *Atombau and Spektrallinien* (1919).

Lorentz's theory derived the magnitude of the line splitting from g, the ratio of the magnetic moment to the orbital angular momentum of the radiating electron. Alfred Landé, a young Jewish theorist at the University of Frankfurt am Main, managed to refer the refractory splitting to an anomalous value of g and to accepted rules for quantum transitions. Attempts to derive the anomalous g from an atomic model failed until the introduction of electron spin in 1925 by Samuel Goudsmit and George Uhlenbeck, young Dutch physicists who, like Landé, made their careers in the United States. Meanwhile, analysis of the systematics of quantum transitions in the anomalous effect helped to guide Wolfgang Pauli to the most striking of all the discoveries prompted by magneto-optical phenomena: the "Exclusion Principle," which ascribes four quantum numbers to each electron in an atom and prohibits any two electrons from having the same values for all their quantum numbers. Pauli's fourth quantum number was soon associated with Goudsmit and Uhlenbeck's spin. Particles that, like electrons, are exclusive, divide the universe with particles that, like photons, are gregarious. Both sorts have revealed much when under the influence of a magnetic field.

Paul Forman, "Alfred Landé and the Anomalous Zeeman Effect," *Historical Studies in the Physical Sciences* 2 (1970): 153–261. J. L. Heilbron, "The Origins of the Exclusion Principle," *Historical Studies in the Physical Sciences* 13 (1982): 261–310. Theodore Arabatzis, "The Discovery of the Zeeman Effect," *Studies in the History and Philosophy of Science* 23 (1992): 365–388.

J. L. HEILBRON

MASS EXTINCTIONS. See METEORITES.

MASS SPECTROGRAPH. The mass spectrograph is an electromagnetic instrument for separating ions on the basis of their charge to mass ratio (e/m), and hence for studying their mass and chemical nature. In 1912 Joseph John *Thomson and his assistant Francis Aston, analyzing positive rays (ions that stream through a hole in the cathode of a gas-discharge tube), discovered an ion closely associated with that of neon, atomic mass 20, but corresponding to mass 22. For several years identification of this ion as a compound, a new element, or an isotope of neon remained uncertain, but from 1913 Frederick Soddy actively promoted it as evidence of isotopes in nonradioactive substances.

The subsequent deveopment of the mass spectrograph is intimately connected with Soddy's concept of isotopes and the Rutherford-Bohr atom (*see* ATOMIC STRUCTURE). The first instruments, invented by Arthur Dempster in Chicago in 1918 and Aston in Cambridge in 1919, both attempted to separate isotopes unambiguosuly using variations on Thomson's poitive-ray apparatus. But it was Aston, working in the Cavendish Laboratory at Cambridge under Ernest *Rutherford, whose name became linked with the mass spectrograph.

The two instruments relied on different focusing techniques and produced different types of results. In Aston's mass spectrograph perpendicular electric and magnetic fields focused ions with different masses at different points on a photographic plate. Aston identified and measured the atomic weights of a large number of isotopes; established the "whole number rule" for atomic weights (that isotopic masses are integral multiples of that of hydrogen, then known only as a single isotope of mass 1); and, with his second instrument of 1925, measured deviations from this rule, the "packing fraction" of nuclei (a measure of the mass equivalent of the energy binding the constituents of a nucleus together). He received the Nobel Prize for chemistry in 1922 for his work. Mass spectrographs are used extensively for accurate atomic weight determination.

Dempster's instrument, which Aston refused to call a "mass spectrograph" because it provided a momentum rather than mass spectrum, established a tradition of "mass spectrometers." In this design, a magnetic field perpendicular to the plane of a beam of ions causes them to move in a circle whose radius depends on the mass and velocity of the particles. In half a turn around a narrow vacuum chamber, ions of the same mass and velocity can be caught in a cup, the others having ended in the walls of the chamber. By varying the accelerating potential, Dempster selected different ions and measured their relative abundance. Although he failed to distinguish unambiguously between the hypothetical isotopes of magnesium and chlorine, Dempster opened the way for accurate abundance determinations and subsequently discovered many isotopes.

From the mid-1920s on several different types of mass spectrograph have been developed to meet the needs of spectroscopists, atomic-weight chemists, and specialists in radioactivity. The most important of these instruments were those of Kenneth Bainbridge (1933), who provided the first experimental proof of the Einstein mass-energy relationship, and Alfred Neir, who introduced a 60 degree (rather than 180 degree) analyzer in 1940. In the 1930s and 1940s physicists invented other electromagnetic means of separating ions: time-of-flight, radio-frequency, and cyclotron-resonance instruments in particular. By this time isotopes had become fundamental to many physical sciences, commercial instruments were available, and the earlier distinction between mass spectrographs and spectrometers had been lost as mass spectrometry became established as a central technique in an era of increasing reliance on instrumentation. Much of the enriched uranium for the nuclear bomb that destroyed Hiroshima passed through a cascade of mass spectrographs called calutrons.

Francis Aston, *Mass Spectra and Isotopes* (1933). H. E. Duckworth, R. C. Barber, and V. S. Venkatasubriamanian, eds., *Mass Spectroscopy*, 2d ed. (1986). J. A. Hughes, *The Radioactivists: Community, Controversy and the Use of Nuclear Physics* (Ph.D. diss., Cambridge University, 1994).

ISOBEL FALCONER

MATHEMATIZATION AND QUANTIFICATION. Early modern mathematics drew on a tradition dating to antiquity and included all practical subjects involving extensive calculation: *astronomy, navigation, *mechanics, civil and military engineering, surveying and *cartography, geometrical *optics, *acoustics, perspective, and music. The field had low status, as the pejorative use of "mechanic" suggests.

*Natural philosophy directed the eye of wisdom towards the natural, as distinguished from the divine, realm. Synonymous with "physics" in the early modern period, the field included subjects that today would be classified with biology, *geology, and psychology, as well as chemistry and *physics. From the beginnings of the Scientific Revolution well into the eighteenth century, natural philosophy remained a literary discipline, as nonquantitative as it was broad.

From the beginnings of the Scientific Revolution, some natural philosophers, notably *Galileo, *Descartes, *Huygens, and *Newton, aimed to bring their discipline under the sway of mathematics. Mathematicians working in established mathematical disciplines sometimes appropriated parts of natural philosophy. For example, over the course of the sixteenth and seventeenth centuries astronomy took over from natural philosophy the concept of the cause of planetary motion, and mathematized it. Newton's triumph applied calculus to that cause: a force that, inherent in infinitesimal parts of matter, draws the planets infinitesimally into their orbits during infinitesimal moments of time (*see* FORCE; GRAVITATION).

Experimental physics resisted quantification until the late eighteenth century (*see* EXPERIMENTAL PHILOSOPHY). Even Newton, attacking *electricity, employed a kind of pictorial explanation often labeled "Cartesian" and popularized by successive generations of experimental natural philosophers (*see* MECHANICAL PHILOSOPHY). Newton's followers—Francis Hauksbee and Petrus van Musschenbroek among them—missed Newton's application of infinitesimals and groped after a mathematical law of magnetism based on the gross characteristics of the magnet. The problem lay partly in the subject matter: astronomers had studied the comparatively simple regularities of the heavenly bodies over millennia; phenomena from *magnetism to *meteorology presented a greater challenge to mathematization. The orientation of natural philosophy contributed to the challenge: the discipline aimed at edification and entertainment rather than measurement and calculation. Nor were scientific (or, as they were then called, philosophical) *instruments suitable for quantitative investigations. Mathematical instruments, intended for astronomy, surveying, engineering, and the other parts of mathematics, were not very accurate. Philosophical instruments—aids to natural philosophy such as the *thermometer and *barometer—offered at best a qualitative indication, despite the numbers penned on their paper scales. Indications such as "blood heat" on thermometer scales reflect the anthropomorphic coloring of early modern measurement. Measures were also local: barometer readings taken in local inches, for example, frustrated attempts to coordinate meteorological observations across Europe.

Around 1760 a quantifying spirit swept over natural philosophy. A typical early result was Joseph Black's discovery of the latent heat of evaporation (see FIRE AND HEAT). Black calculated the rate at which a can of water, set on a stove with the fire "pretty regular," heated from room temperature to boiling; this quantity, multiplied by the time taken to boil off the water, gave the "degrees of heat…contained in [its] vapor." The mathematics, no higher than and perhaps borrowed from bookkeeping, characterized other early successes as well. Benjamin *Franklin applied similar reasoning to quantities of charge collected by a person drawing electricity from an electrostatic generator, as did Antoine-Laurent *Lavoisier to the quantities of oxygen fixed during combustion.

Mathematical instruments played a defining role in quantification. In the age of the emergence of the national state, governments and their armies and navies required accurate navigational and cartographic information; gentleman farmers draining fens and businessmen financing canals needed good surveys; engineers wanted precisely machined parts for the engines of the Industrial Revolution. The mathematical instrument trade responded, developing sophisticated precision instruments as it moved from a craft-based to a protocapitalist form of organization. Philosophical instruments required for survey like the thermometer and barometer benefited, but natural philosophers, increasingly involved in research, also demanded precision in instruments like the electrometer that had no practical application. Equally important, they developed a methodology of exact measurement, standardizing instrument scales, correcting for disturbing factors, and repeating extensive series of readings.

In the last third of the eighteenth and the first years of the nineteenth century, enlightened governments organized precise measurement in many fields, including cartography, meteorology, national population census, natural resources survey, and the calculation of mathematical constants. The tabular format was perfected for the presentation of large amounts of information. The quantifying spirit penetrated even to fields like natural history, where the Linnaean method applied the simplest part of arithmetic, counting, to the classification of species into higher taxa. Calculation came to be regarded as the highest form of intellectual activity. Large quantities of reliable data provided the basis for later fruitful theorizing in the nineteenth century; the late Enlightenment was not given to synthesis of insight from information.

In natural philosophy, nevertheless, quantification yielded important exemplars in *electricity, *magnetism, and the theory of heat. The military engineer Charles Augustin Coulomb (see CAVENDISH, HENRY, AND CHARLES-AUGUSTIN COULOMB) applied the techniques of engineering to the construction of his famous magnetic torsion balance, with which he demonstrated the inverse-square law of magnetic and electric forces. Coulomb represents an important late-eighteenth-century trend: the cross-fertilization of engineering and experimental physics. Through the crossover of personnel and with the increasing availability of mathematical training at institutions like the École du Génie at Mézières and the École Polytechnique, mathematicians and natural philosophers, and their respective fields, became steadily less distinguishable. The process was one of the strongest threads in the quantification of natural philosophy.

J. L. Heilbron, *Electricity in the 17th and 18th Centuries* (1979). Theodore S. Feldman, "Applied Mathematics and the Quantification of Experimental Physics: The Example of Barometric Hypsometry," *Historical Studies in the Physical Sciences* 15 (1985): 127–197. Tore Frängsmyr, J. L. Heilbron, and Robin Rider, eds., *The Quantifying Spirit in the Eighteenth Century* (1990). J. L. Heilbron, *Weighing Imponderables and Other Quantitative Science around 1800* (1993).

THEODORE S. FELDMAN

MATTER. During the Renaissance the Aristotelian conception of matter came under fire from several sides, notably in the sixteenth century by the physician Paracelsus, who defended an account of bodies as composed of three material elements (salt, sulfur, and mercury).

In the seventeenth century, Paracelsian chemistry gave way to the chemical theories of Jean Baptiste van Helmont, who attributed active powers of *sympathy and antipathy to matter. By far the dominant conception, however, was that of the *mechanical philosophy. In his *Discourse on Method* (1637), René *Descartes argued for the fundamental distinction between matter and mind. He made matter pure extension; all apparent qualities of bodies, such as color and texture, are merely appearances that result from the motions of the particles of bodies (corpuscles) impinging upon our nerves and exciting sensations in us.

In the next generation, Robert *Boyle defended an experimental version of corpuscularianism heavily influenced by his interest in chemistry. He agreed with Descartes that the diversity of bodies arose from the configuration of one universal matter. In his *Origin of forms and qualities* (1666), however, he argued that some configurations (primary concretions) remained relatively permanent because he found that they could not be broken down by chemical analysis. John Locke, in his *Essay concerning human understanding* (1690), elaborated upon the distinction between the primary qualities of matter, which refer to such quantitative aspects as size and shape, and the secondary qualities of matter, which refer to the powers that bodies have because of their particular configurations to produce sensations in human observers. But, Locke added, we can only know matter through the powers of these particular configurations and thus the general idea of a "substance" as that which possesses powers is a mere name signifying an unknown support for the qualities we experience.

Isaac *Newton went a long way toward eliminating the concept of matter entirely. According to his *Principia mathematica* (1687), matter is what resists change of motion and causes change of motion in other bodies. He distinguished between material objects, which have weight because of mass, and forces, which measure the interaction of material objects. An object's response to force depends upon its quantity of matter (mass). By focusing upon mass as a quantifiable aspect of matter, Newton could calculate and predict the motions of bodies. This understanding of matter differed significantly from the earlier conception of matter as substance.

After Newton, investigation into the underlying substrate became an empirical question. In the late eighteenth and nineteenth centuries, Joseph Priestley, Joseph Black, Antoine-Laurent *Lavoisier, Humphry Davy, and John *Dalton pursued the idea that chemical as well as physical phenomena could be explained by assuming that all material substances possessed mass and were composed of atoms. In the early twentieth century advocates of electromagnetic conceptions of basic physical phenomena questioned the need to postulate the existence of any underlying material substratum. In their approach, which had a brief vogue, matter became a label for objects studied by classical mechanics and no longer represented a scientific explanatory category.

Stephen Toulmin and June Goodfield, *The Architecture of Matter* (1963). P. M. Heimann and J. E. McGuire, "Newtonian Forces and Lockean Powers: Concepts of Matter in Eighteenth-Century Thought,"

Historical Studies in the Physical Sciences 3 (1971): 233–306. Ernan McMullin, *Newton on Matter and Activity* (1978). Ernan McMullin, ed., *The Concept of Matter in Modern Philosophy* (1978). P. M. Harman, *Energy, Force, and Matter* (1982).

ROSE-MARY SARGENT

MAXWELL, James Clerk (1831–1879), physicist, creator of the electromagnetic theory of light and the statistical theory of gases.

Maxwell was born in Edinburgh, his father John Clerk having taken the name Maxwell as heir to estates in Galloway in Scotland. The young man's abiding concern with philosophical principles was established while a student at Edinburgh University (1847–1850). Writing substantial papers on the geometry of rolling curves and the theory of elastic solids, he became interested in color vision. In the 1850s, building on work by Thomas Young, Hermann von *Helmholtz, and Hermann Grassmann, he used red, green, and blue primaries to form color combinations in experiments with tinted papers and on the mixture of spectral colors. Awarded the Royal Society's Rumford medal in 1860, he projected the first trichromatic color photograph in May 1861.

Admitted to Peterhouse and then Trinity College, Cambridge, Maxwell became a pupil of the mathematics coach William Hopkins, graduating second wrangler in 1854. He became a fellow of Trinity in 1855, professor of natural philosophy at Marischal College, Aberdeen, in 1856, the same at King's College London (1860–1865), and professor of experimental physics at Cambridge and director of the Cavendish Laboratory (1871). Building upon his own work on establishing a standard unit of electrical resistance, Maxwell directed experiments on precision measurements in electricity and edited Henry Cavendish's *Electrical Researches*, published shortly before his death from cancer in November 1879.

Maxwell's most sustained achievement was in formulating the theory of the electromagnetic *field, developing work by Michael *Faraday and William *Thomson (Lord Kelvin). Guided by Thomson, he illustrated Faraday's "lines of force" by the analogy of streamlines in a fluid, establishing a geometry of field relations. Seeking physical foundations, he proceeded in a deliberately hypothetical style, imagining a mechanical *ether model where rotating vortices representing magnetism were separated by particles whose motion represented the flow of an electric current. Maxwell obtained an unexpected result: the close agreement between the velocity of transverse waves in an electromagnetic ether and the measured velocity of light. His electromagnetic theory of light, first proposed in 1862, unified *optics and *electromagnetism. He subsequently discarded the ether model, placing emphasis on the transmission of energy in the field, and stated equations of the electromagnetic field that later became codified as the four "Maxwell equations." Maxwell expounded his theory in his *Treatise on Electricity and Magnetism* (1873), deploying vectors, integral theorems, topology, and analytical dynamics, a style that joined geometry to dynamics and freed physical quantities from representation by a mechanical model. Heinrich Hertz's production of electromagnetic waves in 1887 led to the general acceptance of Maxwell's field theory (*see* HELMHOLTZ AND HERTZ).

In his 1857 Adams Prize essay at Cambridge on the stability of the motion of Saturn's rings, Maxwell had concluded that the system consists of concentric rings of particles. Alerted to problems of describing particle collisions, in spring 1859 he noticed a paper by Rudolf Clausius on the kinetic theory of gases, and was intrigued by his use of a probabilistic argument to calculate the motions of gas molecules. Maxwell introduced a statistical function, identical in form to the distribution formula in the theory of errors, to calculate the distribution of velocities among molecules. He applied this model to obtain results for gaseous diffusion, viscosity, and thermal conductivity. The theory was provisional and Clausius was able to point out some of its deficiencies; and Maxwell's own experimental study of the viscosity of gases, by observing the decay in the torsional oscillation of discs, led him to reconstruct his argument. In his mature paper of 1867, he revised his theory of gas molecules in a form consonant with his findings on viscosity. He provided a rigorous derivation of the distribution law, a formulation soon enlarged by Ludwig *Boltzmann in seeking statistical foundations for thermodynamics.

In a famous argument, and in characteristic style, Maxwell ingeniously expanded his reasoning. According to the second law of thermodynamics recently developed by Clausius and Thomson, heat flows from hot to cold bodies. But because of the statistical distribution of molecular velocities in a gas, there will be fluctuations in the flow. The intervention of Maxwell's "finite being" (termed "demon" by Thomson) would be needed to select molecules so as to make the fluctuations detectable; but they constantly take place at the molecular level. The second law of thermodynamics, Maxwell concluded, is a statistical law, which applies to systems of molecules, not to individuals.

Lewis Campbell and William Garnett, *The Life Of James Clerk Maxwell* (1882; Repr. 1969). C. W. F. Everitt, *James Clerk Maxwell: Physicist and Natural Philosopher* (1975). Stephen G. Brush, *The Kind of Motion We Call Heat: A History of the Kinetic Theory of Gases in the Nineteenth Century* (1976). P. M. Harman, *Energy, Force, and Matter: The Conceptual Development of Nineteenth-Century Physics* (1982). P. M. Harman, ed., *The Scientific Letters and Papers of James Clerk Maxwell*, 3 vols. (1990–2002). Daniel M. Siegel, *Innovation in Maxwell's Electromagnetic Theory: Molecular Vortices, Displacement Current and Light* (1991). P. M. Harman, *The Natural Philosophy of James Clerk Maxwell* (1998). Olivier Darrigol, *Electrodynamics from Ampère to Einstein* (2000).

P. M. HARMAN

MECHANICAL PHILOSOPHY. Many seventeenth-century natural philosophers sought to explain all physical properties and processes in terms of the motion of the least parts of matter of which physical bodies are composed. They usually referred to these least parts as corpuscles so as not to confuse the mechanical position with the type of ancient atomism that Pierre Gassendi had tried to revive early

in the century. Although the mechanical philosophers (or corpuscularians) agreed in rejecting Aristotelian philosophy and most of the mystical elements associated with Renaissance naturalism, they divided over the positive formulation of their position. "Mechanical philosophy" is a cover term for a continuum of positions from a pure kinetic theory of motion to a robust matter theory. These variations can be seen in the works of René *Descartes, Francis Bacon, *Galileo, Robert *Boyle, and Isaac *Newton.

Descartes maintained in his *Discourse on Method* (1637) that matter is pure extension, from which it followed that all motion must result from direct contact. All physical processes, therefore, were to be explained by the laws of motion that the least parts of matter obey. Only the human soul escaped mechanical explanation. The world was a vast machine made up of smaller machines (including human and animal bodies) consisting of inert particles moved by physical necessity. Although Descartes located the origin of motion in God, his principles of inertia and the plenum allowed him to describe a deterministic system where, on impact, motion is transferred but not destroyed.

Cartesianism dominated corpuscularism throughout the seventeenth and into the eighteenth centuries, but it had several strong competitors. In his *Novum organum* (1620), Bacon advocated explanations in terms of the motion of matter. In his investigation of heat, for example, he concluded that bodies feel warm when the particles of matter that compose them move rapidly. Unlike the later Cartesians, however, Bacon insisted that experimental and observational techniques had to be developed to discover the true nature of the particles responsible for such qualities. Galileo's *Assayer* (1623) offered a similar account of motion as the cause of heat and his *Discourse on the Two New Sciences* (1638) presented detailed experimental studies of mechanical subjects. Both Bacon and Galileo brought the practices of craftsmen and mechanics to bear on natural philosophical issues. In this tradition, the mechanical philosophy helped to elevate the intellectual, social, and economic status of the technical arts.

Many natural philosophers in England in the generation after Bacon took his works as their model. Boyle followed Bacon's experimental program and believed in the Baconian ideal of useful knowledge. Writing in the 1660s, Boyle was also influenced by the works of Descartes, Gassendi, and Galileo, and introduced the term "mechanical philosophy" in 1674 to refer to all explanations of physical phenomena in terms of matter and motion. Unlike philosophers before him, however, Boyle tried to use chemical analysis to turn the mechanical philosophy into an experimentally based theory of matter. He also elaborated upon the distinction, first introduced by Galileo, of the primary and secondary qualities of bodies. In *The Origin of Forms and Qualities* (1666), Boyle maintained that quantifiable properties, such as size and shape, are primary because all material bodies possess them. Other qualities, such as color or texture, arise in us in consequence of the particular configurations of corpuscles in the bodies that we see or touch.

Newtonian or classical mechanics is often taken as the paradigm of mechanical explanation. Yet in his *Principia mathematica* (1687), Newton upset the mechanical philosophy by introducing the concept of force. Unlike Descartes and Boyle, for whom force amounted to the pressure of one body on another, Newton's force was the measure of the change in motion of a moving body. Thus he added a third element to the original principles of matter and motion. At first mechanical philosophers, especially Cartesians, rejected Newtonian forces as a throwback to *sympathies and antipathies. But his scheme gradually gained acceptance as mathematicians succeeded in deploying gravitational force to ever finer phenomena and other examples of distance forces turned up in *electricity and *magnetism.

Following Newton's achievements, mechanical conceptions came to be applied to all areas of learning, not always with advantage. Ernst Mayr, for example, in *The Growth of Biological Thought* (1982), argued that reliance upon mechanics advanced physical sciences but led to the neglect of the biological ones. This assessment can of course be extended to the human sciences as well. Sociologists and psychologists in the nineteenth and twentieth centuries often attempted to find deterministic laws covering the behavior of groups and of individuals.

At a more global level, some have argued that ecological disaster can be attributed, at least in part, to mechanistic ideas of nature. In *The Death of Nature* (1980), Carolyn Merchant put forward the still controversial thesis that with the mechanical philosophy scientific inquiry became a masculine activity imposed upon a passive, feminine nature. According to Merchant, this attitude set the stage for, and ultimately justified, the "rape" of nature.

E. J. Dijksterhuis, *The Mechanization of the World Picture* (1961; 1989). Marie Boas Hall, *Robert Boyle on Natural Philosophy* (1965). Richard S. Westfall, *The Construction of Modern Science* (1977). Peter Alexander, *Ideas, Qualities and Corpuscles: Locke and Boyle on the External World* (1985). Daniel Garber, *Descartes' Metaphysical Physics* (1992). Michael Hunter, ed., *Robert Boyle Reconsidered* (1994).

ROSE-MARY SARGENT

MECHANICS. Motion, forces, and machines have been the subject of two distinct sciences: the science of motion, and mechanics. In the Aristotelian tradition, the science of motion belonged to physics, or natural philosophy, the science of natural bodies "insofar as they are natural." Aristotle opposed natural motion (a falling rock) to violent motion (a weight raised by a pulley). Violent motions and the machines that created them were the concern of the mechanical arts—acting *against* nature for practical ends—and of mechanics, or the science of weights. The division of mechanics into manual or practical, and rational or theoretical, goes back to Pappus of Alexandria in the early fourth century A.D. and was current thirteen hundred years later.

The Renaissance inherited many important achievements in rational mechanics, including the parallelogram rule for the composition and resolution of motions, early forms of the principles of virtual work and of virtual velocities, demonstrations of the law of the straight and angular lever, an embryonic notion of moment (torque), and determination of centers of gravity. Equally important was the understanding that geometrical demonstration was essential to mechanical theory. This legacy underpinned the evolution of mechanics in the early modern period. Simon Stevin and *Galileo simplified Archimedes' proof of the law of the lever; Christiaan *Huygens devised a more rigorous proof (1693). Galileo used the angular lever and the notion of moment to determine equilibrium conditions on the inclined plane (*Le meccaniche*, c. 1593).

The parallelogram rule for motions uses geometrical displacements straightforwardly. Not so the corresponding rule for forces, because it is not obvious what their "composition" and "resolution" mean. Following the work of Stevin and others, Pierre Varignon based his theory of equilibrium on the rule for forces (*Projet de la nouvelle mécanique*, 1687). Isaac *Newton recognized its indispensability, and cannily prepared for his own proof of the rule in the *Philosophiae naturalis principia mathematica* (1687) by stipulating in his second Law of Motion that every change in motion takes place "along the straight line in which the force is impressed." This ensured *by definition* the geometrical equivalence of the rules for forces and motions. As for the principle of work, Newton claimed that it depends on the equality of action and reaction (Law III). In René *Descartes's formulation of the principle of mechanical work, the same force that raises one hundred pounds through two feet will raise two hundred pounds through one foot (and so on). This he used effectively in short treatises on machines he sent in 1637 to Marin Mersenne and Constantijn Huygens.

These examples belong to "rational mechanics" as understood by Pappus. They remind us that the major figures, better known for their contributions to the science of motion, also contributed to mechanics, though their thinking in these areas revealed the conceptual fluidity characteristic of pivotal transformations in the development of science. Two contrasting signs of this fluidity were the creative coupling of principles from both sciences, and

indecisiveness about the relations between mechanics and physics. Galileo corroborated his law of free fall through experiments on the inclined plane, and the equilibrium conditions on the inclined plane played key roles in the formal demonstration of the law in the Third Day (or part) of his *Discourses concerning Two New Sciences* (1638). In the fourth part, Galileo used the law of fall and the composition rule to demonstrate the parabolic path of projectiles, a result he had obtained and confirmed experimentally around 1608. Apart from certain medieval innovations in the geometrization of natural motion, natural philosophy had not been mathematical, whereas mechanics had never been anything else. Galileo decisively blurred that dichotomy by showing that natural motions could be given mathematical descriptions in accord with experiment and mechanical principles. Still, he would have agreed that *Le meccaniche* belonged in a different disciplinary pigeonhole from the third and fourth part of his *Discourses*.

In *Principia philosophiae* (1644), Descartes set out his pioneering three "Laws of Nature," according to which a bodily state persists until forced to change by external causes, a moving body endeavors to move always in a straight line, and exchanges of motion (size × speed) between colliding bodies are determined by the contests between their forces of persistence and by Descartes's conservation law. Descartes claimed that his laws and the collision theory derived from them could explain the whole physical universe, including machines. Yet there is nothing on mechanics in his *Principia*, nothing that explains the work principle of 1637. Christiaan Huygens's masterly solution to the problem of center of oscillation (*Horologium oscillatorium*, 1673) required an insight from Galileo's *Discourses* (third part) enunciated as a principle by Evangelista Torricelli (1644): a system of heavy bodies cannot move of its own accord unless the common center of gravity descends. Torricelli's principle also played a crucial role in Christiaan Huygens's collision theory, out of which tumbled the result that in perfectly elastic collisions the quantity mv^2 remains constant. Huygens regarded this result as a notable corollary of his collision rules; to Gottfried Wilhelm von Leibniz it suggested the universal conservation of *vis viva*, a force measured by mv^2, to add to the already known conservation of "directed motion." This confirmation of force as a metaphysical reality led Leibniz to the creation in 1691 of a new science of force, which he baptized *dynamics*. It led in turn to a protracted argument in the early decades of the eighteenth century about whether motive force should be mv, the Cartesian and Newtonian measure, or mv^2, the Leibnizian measure.

Newton transformed Descartes's "Laws of Nature" into three "Axioms" or "Laws of Motion" according to which a body persists in its state of rest or straight-line motion (Law I), the force impressed on a body is proportional to its change of motion (Law II), and action and reaction are equal and opposite (Law III). From these laws Newton demonstrated the mutual dependence of his inverse-square law of universal gravitation and Johannes *Kepler's first and third laws of planetary motion—the second law being a consequence of inertial motion under any central force. The basic problem—to determine the central force given the planet's deviations from inertial motion summed as an *orbit, and conversely—was quite foreign to traditional mechanics. Although in the *Principia* Newton links Law III to the work principle and proves the parallelogram rule for forces (which is "abundantly confirmed from mechanics"), the *Principia* is not a treatise on mechanics, but on "the mathematical principles of natural philosophy."

However, Newton's *Principia* was a major exercise in rational mechanics in a new sense that emerged in the work of Isaac Barrow and John Wallis. Barrow had argued (1664–1666) that geometrical theorems apply to all of physics, so that the principles of mechanics and of physics become identical. For Wallis, mechanics was "the part of geometry that deals with motion, and investigates, apodictically and using geometrical reasoning, the force with which such and such a motion takes place" (*Mechanica*, 1670–1671). Similarly, for Newton rational mechanics was "the science, set out in exact propositions and demonstrations, of the motions that result from any forces whatever and of the forces that are required for any motions whatever" (*Principia*, "Preface"), and natural philosophy was basically the problem of "finding the forces of nature from the phenomena of motions and then to demonstrate other phenomena from these forces." The same ideas were to inform Leonhard *Euler's *Mechanica* (1736), significantly subtitled *The Science of Motion Expounded Analytically*.

"Rational" or "theoretical" mechanics in the older sense should be distinguished from the post-Newtonian sense of "rational mechanics," which comprised dynamics and statics. Mechanics in its golden age (the eighteenth century) was not merely a set of variations on the principles and methods of Newton's *Principia*. Among additional ingredients were the concepts and symbolism of Leibnizian differential and integral calculus, which became standard in treatises on analytical mechanics of the period, and the new mathematics, particularly the calculus of variations, which made it possible to formulate new principles that solved new problems.

To take some notable examples, a general theory of rigid-body motion became a desideratum following the work on centers of oscillation of Huygens, Jakob Bernoulli (1703), and Jean d'Alembert (*Traité de dynamique*, 1743), and the researches on lunar libration and equinoctial precession by Newton (*Principia*), d'Alembert, and Joseph Louis Lagrange. Here the principal figure was Euler (memoirs of 1750, 1758), whose *Theoria motus corporum solidorum seu rigidorum* (1760) provided a general theory of rigid-body motion. Euler's researches depended on a "new principle of mechanics" (1750), his recasting of Newton's Law II in the analytic form $mdv_{x,y,z} \propto f_{x,y,z} \, dt$. Euler introduced moment of inertia (1749) and principal axes of rotation (suggested by the rolling of ships about three orthogonal axes), which he applied in his theory of the spinning top, an exceptionally difficult problem that had not even been recognized as a problem since the early seventeenth century.

Suppose in a system of bodies in mutual constraint the motion applied to each body

a_i resolves into the motion actually acquired v_i and another motion V_i. That is the same as if the v_i and V_i had initially been communicated together, so that the system would have been in equilibrium had the V_i alone been present. That is "d'Alembert's Principle," the centerpiece of his *Traité de dynamique*, which allowed the methodological reduction of dynamics to statics. Lagrange reformulated the principle and coupled it with the principle of virtual velocities to obtain the first formulation of what became "Lagrange's Equations" (*Mécanique analytique*, 1788). Lagrange showed that the conservation of linear and angular momentum, and of *vis viva*, and the principle of least action follow from his equations, rather than being foundational principles in their own right.

Pierre de Maupertuis's principle of least action had sounded a new note. Reflecting on the controversy of the 1660s over Fermat's least-time optical principle, Maupertuis argued (1744) that in all bodily changes, the "action" (Σ mass $\times$ speed $\times$ distance) is the least possible, a principle that for Maupertuis and Euler—though not for d'Alembert and Lagrange—pointed to the governance of all things by a Supreme Being. (Σ signifies a sum over all particles in the system under consideration.) Given the principle of virtual velocities, Σ force $\times$ increment of distance $= 0$, which means, by the principles of the integral calculus, that the integral of this sum is a maximum or minimum. Euler developed the least action principle clearly and rigorously for a single particle (1744), and, in a memoir on lunar libration (1763), Lagrange extended Euler's result to an arbitrary system of bodies and derived a general procedure for solving dynamical problems.

The nineteenth century saw new departures in the application of principles established in the preceding two centuries. The relativity of motion, a central theme since the work of Galileo and Huygens, had received further study in the work of Alexis-Claude Clairaut, who asked how a system of moving bodies would behave if the system moved along noninertial curves (1742). Gaspard Gustave de Coriolis showed (1835) that the Newtonian laws of motion apply in a rotating reference frame

if the equations of motion include a "Coriolis acceleration" in a plane perpendicular to the axis of rotation, a kinematic acceleration which Coriolis interpreted as an extra force (the "Coriolis force") and which became important in ballistics and meteorology. An important step in the formalization of mechanics was Heinrich Hertz's attempt to remove inconsistencies arising from the assumption within traditional classical mechanics that forces are ontologically prior to the motions they cause. He treated forces as "sleeping partners" in a formalized mechanics that depended on the operationally understood notions of time, space, and mass, and when necessary on linkages to hidden masses with hidden motions with respect to hidden coordinates (*Die Prinzipien der Mechanik*, 1894).

Maupertuis's variational principle enjoyed an improved mathematical treatment by William Rowan Hamilton (1834, 1835), whose transformation of Lagrange's equations was modified and generalized by Carl Gustav Jacobi in the form now known as the Hamilton-Jacobi Equation (1837). In turn, the Hamilton-Jacobi Equation found fruitful application in the establishment of the quantum mechanics of Louis de Broglie (1923) and Erwin *Schrödinger (1926).

See also HELMHOLTZ, HERMANN VON, AND HEINRICH HERTZ; RELATIVITY.

René Dugas, *Mechanics in the Seventeenth Century* (1958). Clifford Truesdell, *Essays in the History of Mechanics* (1968). Stillman Drake and I. E. Drabkin, eds., *Mechanics in Sixteenth-century Italy. Selections from Tartaglia, Benedetti, Guido Ubaldo, and Galileo* (1969). Richard S. Westfall, *Force in Newton's Physics: The Science of Dynamics in the Seventeenth Century* (1971). Ernst Mach, *The Science of Mechanics: A Critical and Historical Account of its Development* (1974). Pierre Duhem, *The Evolution of Mechanics* (1980). René Dugas, *A History of Mechanics* (1988). H. J. M. Bos, "Mathematics and Rational Mechanics," in *The Ferment of Knowledge: Studies in the Historiography of Eighteenth-Century Science*, ed. G. S. Rousseau and Roy Porter (1980): 327–355. Ivor Grattan-Guinness, "The Varieties of Mechanics by 1800," *Historia Mathematica* 17 (1990): 313–338. Alan Gabbey, "Newton's Mathematical Principles of Natural Philosophy: a Treatise on 'Mechanics'?" in *The Investigation of Difficult Things: Essays on Newton and the History of the Exact Sciences*, ed. P. M. Harman and Alan Shapiro (1992): 305–322.

ALAN GABBEY

MEITNER, Lise (1878–1968), atomic physicist, co-discoverer of nuclear fission.

Lise Meitner was born in Vienna, the third of eight children of Philipp (a lawyer) and Hedwig (née Skovran) Meitner. Lise grew up in a family that was intellectual and socially progressive. Both parents were of Jewish origin, but the religion played no part in the children's upbringing and all were baptized as adults.

Meitner entered the University of Vienna in 1901, four years after the university first admitted women; earned a doctorate in physics in 1906; and worked in Berlin from 1907 until her forced emigration from Germany in 1938. Primarily an experimentalist, Meitner made important contributions to *radioactivity, *nuclear physics, and the discovery of nuclear fission. Her most formative teachers, however, were the theoretical physicists Ludwig *Boltzmann and Max *Planck, and her lifelong interest in theory informed and guided her work.

In Berlin, Meitner began an interdisciplinary collaboration with Otto Hahn, a chemist about her age. Together they identified several new radioactive species, discovered and developed the physical separation method known as radioactive recoil, and were the first to use photographic methods for studying magnetic beta spectra. Their radiochemical expertise permitted them to search for a rare "missing" element, and in 1918 they reported the discovery of a long-lived isotope of element 91, protactinium (Pa).

Meitner's career was a series of firsts for the inclusion of women into German science. She held no position whatsoever until Planck appointed her his assistant in 1912. In 1913 she was given a position and salary comparable to Hahn's in the Kaiser-Wilhelm Institute for Chemistry (KWI). After volunteering as an X-ray nurse in the Austrian army during World War I, she returned to the KWI to head her own physics section in 1917, acquired the title of profes-

sor in 1920, and served as an adjunct professor at the University of Berlin from 1926 until she was dismissed by the Nazi regime in 1933. After fleeing Germany in July 1938, she held positions at the Nobel Institute for Experimental Physics and the Royal Institute of Technology in Stockholm. She was recognized by many international scientific societies and repeatedly nominated for a Nobel Prize.

In the 1920s Meitner achieved exceptional prominence for her pioneering studies of the nucleus. Using magnetic beta-gamma spectra, she was the first to describe radiationless orbital electron transitions in 1923, an effect now named for Pierre Auger, and in 1924 she proved that gamma radiation follows particle emission in radioactive decay. Convinced that quantization must extend to the nucleus, she was dismayed when Charles D. Ellis proved the existence of the continuous primary beta spectrum in 1927. After Meitner confirmed the result in 1929, Wolfgang Pauli (*see* HEISENBERG AND PAULI) proposed a new nuclear particle, the neutrino, which was quickly incorporated into nuclear theory. Always close to theory, Meitner measured the Compton scattering of high-energy gamma radiation, verifying the formula of Oskar Klein and Yoshio Nishina that was based on the relativistic electron theory of P. A. M. *Dirac. In 1932, Meitner and her coworkers were the first to observe positrons from a noncosmic source and to observe the formation of electron-positron pairs in a cloud chamber.

In 1934 Meitner recruited Hahn and the chemist Fritz Strassmann for the "uranium project" that culminated in the discovery of nuclear fission. Believing that they were synthesizing artificial elements beyond uranium, the team spent four years disentangling a complex mixture of radioactive species, nearly all of which were later found to be smaller nuclei produced by fission. Only in December 1938, when barium was identified among the uranium products, was it recognized that the uranium nucleus had split. Although Meitner was in Stockholm during the final experiments, she collaborated closely with Hahn through correspondence. She and her nephew, Otto Robert Frisch, provided the first theoretical interpretation of fission and calculated the energy released in the process. Their interpretation was regarded as seminal, but Meitner's contribution to the discovery itself was obscured and Hahn alone received the 1944 Nobel Prize in chemistry. With recent historical correctives, the discovery is now more generally understood as the result of an interdisciplinary collaboration in which Meitner and nuclear physics played an essential role.

Meitner was invited to join the Manhattan Project in Los Alamos in 1943, but she was unwilling to work on a nuclear weapon and remained in Sweden. She retired in 1954 and in 1960 moved to England, to be near Frisch and his family in Cambridge.

Lise Meitner, "Looking Back," *Bulletin of the Atomic Scientists* 20, no. 11 (1964): 2–7. O. R. Frisch, "Lise Meitner 1878–1968," *Biographical Memoirs of the Fellows of the Royal Society of London,* 16 (1970): 405–420. O. R. Frisch, *What Little I Remember* (1979). Ruth Lewin Sime, *Lise Meitner: A Life in Physics* (1996). Elisabeth Crawford, Ruth Lewin Sime, and Mark Walker, "A Postwar Tale of Nobel Injustice," *Physics Today* 50, no. 9 (1997): 26–32. Ruth Lewin Sime, "Lise Meitner and the Discovery of Nuclear Fission," *Scientific American* 298, no. 1 (1998): 80–85.

RUTH LEWIN SIME

METEORITES, pieces of rock from space that reach the surface of the Earth. While in space, they are meteoroids. The luminous phenomena of their fiery passages through the Earth's atmosphere are called meteors.

Not until the beginning of the nineteenth century did astronomers accept the old conjecture that stones from space rained on the earth. In 1802 Charles Howard, an English chemist, reported that several meteorites all contained nickel alloyed with iron, a form almost never found on Earth, and thus they "might be bodies of meteors." A year later, *Laplace's disciple Jean-Bapiste Biot, who knew Howard's chemical analysis, rigorously confirmed reports of some 3,000 stones falling from the sky on the village of l'Aigle. Still, when members of Yale University recovered fragments of a meteorite that fell in Connecticut in 1807, President Thomas Jefferson perhaps said that he would rather believe that Yankee professors would lie than that stones would fall from heaven.

Early speculation about the origin of meteorites placed their origin in the Earth's atmosphere. A cosmic origin, as pieces of a previously existing planet destroyed by internal explosion or by collision with another celestial body, gained support from the discovery of asteroids. In 1801 the Sicilian astronomer Giuseppe Piazzi discovered an object he first thought was a star, and next a comet; soon Ceres, as he called the object, was shown to be in orbit around the Sun. A second asteroid, or minor planet, Pallas, was discovered in 1802; a third, Juno, in 1804; and a fourth, Vesta, in 1807.

Remarkably, the asteroids occupy the gap between Mars and Jupiter singled out in the so-called law published by the German astronomer Johann Bode in 1768 and by other astronomers even earlier. Distances of planets from the Sun in astronomical units were, with no theoretical or physical reasoning: $0.4 + 0(0.3) = 0.4$ for Mercury; $0.4 + 0.3 = 0.7$ for Venus; $0.4 + 2(0.3) = 1.0$ for Earth, $0.4 + 4(0.3) = 1.6$ for Mars; $0.4 + 8(0.3) = 2.8$ for the missing planet; and so on. Piazzi did not know Bode's law when he discovered Ceres, 2.8 astronomical units from the Sun.

*Comets were soon identified as another source of meteorites. The spectacular Leonid meteor shower of 1833 commanded attention, especially in North America, and suggested that the Earth had encountered a stream or swarm of objects. By the 1860s astronomers had linked comets with meteor showers, just in time to predict another Leonid shower in 1867.

It seemed unlikely that objects large enough to create craters could survive passage through the Earth's atmosphere. None weighing more than a ton had been seen to fall, and meteorites of that size made only a small hole in comparison with the crater in the Arizona desert nearly a mile in diameter and some 600 feet deep stumbled on by cattlemen in the early 1870s. In 1891 a prospector thought he had found an iron outcrop at the crater, but analysis said meteorites. The chief geologist of the United States Geological Survey investigated and reported that the

crater was volcanic, or possibly a limestone sink; meteorites could have fallen
symmetrically about the crater by chance!
Miners began hauling away the meteorites
and in 1903 Daniel Barringer, an American geologist and mining engineer, filed
claim to the land. Holes drilled in the
crater did not find the giant meteorite
Barringer sought, but did reveal fractured
rock to a depth of more than a thousand
feet and small meteoritic iron particles
containing nickel. Scientists now believed
the crater was meteoritic and calculated
that a meteorite possessing enough kinetic
energy to create the crater would have vaporized upon impact.

Large meteorites have been found. The
British explorer James Ross in 1818 encountered Greenland Eskimos with tools
made of iron chipped from a huge rock.
They would not tell Ross where the rock
was, but did reveal its site to the American
explorer Robert E. Peary in 1894. The next
year Peary carried off two of the three
great meteorites, one weighing about three
tons and the other half a ton. Two years
later he won his battle with the 34-ton "de-
moniac iron from heaven," transported it
to New York, and sold it to the American
Museum of Natural History for $40,000—
money that helped finance his trip to the
North Pole in 1909.

Early in the twentieth century new scientific techniques, including X-ray crystallography, enabled scientists to study
crystal structures and trace elements in
meteorites. Internal structure, in turn, revealed information about the temperature
and pressure under which a body formed,
and sometimes recrystallized on reheating,
perhaps during an energetic impact. Relative chemical abundances hint at where in
the solar system a body formed. New information on meteor trails followed from
British radar research. Fluctuations in
radio signals and also false echoes encountered during efforts to track German V2
rockets by their radar reflections were attributed to electrons and ions left behind in
the paths of meteors. In 1946 James Hey, a
British physicist who had helped develop
radar, established a correlation between
radar echoes and the Giacobinid meteor
shower.

Infall of meteoritic material may have
played a significant role in the Earth's biological history. The British astronomer
Fred *Hoyle attributed the origin of life,
and also of influenza epidemics, to organic
matter brought from space by meteorites.
Better supported by evidence is the meteoritic extinction of dinosaurs at the end of
the Mesozoic age. In 1978 the American geologist Walter Alvarez found at Gubbio,
Italy, a layer of clay from the Cretaceous-
Tertiary boundary rich in iridium, which is
scarce on Earth but relatively abundant in
meteorites. With his father Luis Alvarez, a
winner of the Nobel Prize in physics, Walter developed a theory of mass extinction
owing to the blockage of sunlight by a great
cloud of debris raised by a huge vaporizing
meteorite. Subsequent findings of the iridium anomaly worldwide confirmed a large
influx of extraterrestrial material about 65
million years ago, though not necessarily a
consequential mass extinction. Less certain are alleged catastrophes with a period
of 26 million years or so. In 1984 Richard
Muller, a physicist mentored by Luis Alvarez, conjectured that a star he named
Nemesis, a hypothetical companion star to
our Sun, periodically passes through the
Oort Cloud of comets surrounding the solar
system, scattering millions of them. The
few that reach the Earth bring death and
destruction.

The United States Congress has held
hearings on the hazards of near-Earth objects (NEOs) and NASA now spends a few
million dollars a year on a survey of space
objects. A British task force found that the
risk of asteroidal and cometary impacts exceeds limits of tolerability for the nuclear
power industry and the transport of hazardous goods. It estimated the damage
caused by an asteroid a kilometer in diameter colliding with the Earth every 100,000
years would, if amortized, amount to over a
hundred million pounds annually.

John G. Burke, *Cosmic Debris: Meteorites in
History* (1986). John and Mary Gribbin, *Fire
on Earth: Doomsday, Dinosaurs, and Humankind* (1996).

NORRISS S. HETHERINGTON

METEOROLOGY. The history of modern
meteorology begins with the Scientific Rev-

olution. The late sixteenth and first half of the seventeenth century saw the invention of the meteorological instruments—*thermometer, *barometer, hygrometer, wind and rain gauges—and around 1650 natural philosophers began using them to record weather observations. They immediately understood the importance of coordinating observations over as wide a space as possible. Scientific academies solicited weather diaries and organized observational networks: Leopold de' Medici, Grand Duke of Tuscany, founder of the Accademia del Cimento, and Robert *Hooke of the Royal Society of London sponsored networks of observers in the 1650s and 1660s. They were motivated in part by theories deriving from the Hippocratic treatise *Airs, Waters, and Places* that related the weather to disease; for two and a half centuries meteorologists attempted to correlate weather patterns with epidemic outbreaks and climate with public health. The application of meteorology to agriculture provided further motivation. In addition, Enlightenment meteorology attempted to rationalize traditional weather lore, including astrological meteorology, searching through recorded observations for patterns confirming traditional wisdom.

Among the few attempts at a theoretical understanding of weather phenomena were explanations of the trade winds by Edmond *Halley and George Hadley. According to them, the rising mass of heated equatorial air is replaced by an inflow of cooler air from higher latitudes. This north-south circulation is deflected, according to Halley, by the movement of the subsolar point with the earth's diurnal motion, or, in Hadley's theory, by the acceleration we now call Coriolis (*see* MECHANICS). A flow of warm air at high altitude from equator to poles completes these early pictures of the general circulation. They illustrate the role of oceangoing commerce both as a source of data and incentive for meteorology.

Early modern meteorologists were frustrated by observers' lack of discipline and by the poor quality of instruments, which rendered observations nearly useless. The late Enlightenment resolved these problems. Emerging modern states organized large networks of disciplined observers, instrument makers developed precise instruments of all types, and natural philosophers devised methods of systematic measurement. By the end of the eighteenth century meteorologists had access to large quantities of reliable weather data for the first time.

Enlightenment meteorologists, seeking weather patterns and correlations with agricultural harvests or outbreaks of disease, lacked a sense of the geographical expanse of weather events and of their development over time. Romantic natural philosophers worked out geographical and temporal syntheses. Alexander von *Humboldt's famous isothermal lines synthesized temperature observations over the globe; Humboldt integrated all the factors of climate into a unified science of the earth that he called "physique générale." Heinrich Wilhelm Brandes drew (or perhaps proposed to draw) synoptic maps of the weather over Europe for every day of 1783, tracing the progress of temperature changes across the Continent, uncovering the geographic distribution of barometric pressure, and relating wind direction to barometric differences. In the 1830s meteorologists took up the kinetics of storms. Heinrich Wilhelm Dove's "Law of Gyration" described the veering of storm winds resulting, he argued, from the conflict of equatorial and polar air currents. William C. Redfield insisted on the rotary motion of storms. James Espy introduced thermodynamic considerations, pointing to the adiabatic cooling of rising moist air and the energy of latent heat released in precipitation as the "motive" force of tropical storms.

The advent of the telegraph around midcentury made possible the nearly immediate collection of meteorological data on a continental basis; at the same time the growing importance of meteorology for agriculture and oceangoing commerce led governments to establish national weather services to coordinate observation, particularly for storm warning. The resulting inflow of data fed the systematic production of synoptic weather charts, which became important research tools. A community of meteorologists evolved, its members more consistently trained in physics and mathematics, while the discipline acquired journals and professional societies. These factors, along with the emergence of *thermodynamics after midcentury, led to quantitative treatment of Espy's supposition. A consistent body of work emerged, known as the "thermal" or "convective" theory of cyclones, that derived the kinetic energy of storms from the release of latent heat and the adiabatic cooling of rising currents of air. William Ferrel applied hydrodynamics to the process, showing that air movement caused by any chance pressure gradient will be bent into a spiral by the earth's motion, generating a barometric low and the beginnings of a storm system. Hermann von *Helmholtz and Vilhelm Bjerknes were the best known among scientists applying hydrodynamics to meteorology.

Around the turn of the twentieth century balloons, kites, and airplanes made available observations of the upper atmosphere, while aviation generated demand for detailed forecasts in three dimensions. World War I sharpened these requirements. Discrepancies in the temperature distribution above storms had led meteorologists around the beginning of the century to consider the role of air masses of differing temperatures and geographic origin in the formation of storms. The polar front theory, developed immediately after the war by Bjerknes and his Bergen (Norway) school of meteorologists, demonstrated the origin of cyclones in the encounter of cool, polar air masses with warmer air. In the 1920s the Bergen school extended the air-mass approach to weather not associated with storms.

Around the same time Lewis Richardson succeeded in computing (after the fact) a six-hour advance in the weather using numerical algorithms. The effort consumed six weeks, generated disappointing results, and convinced meteorologists of the uselessness of a computational approach. The advent of the electronic computer during World War II encouraged a new attempt at computational forecasting. John von Neumann, who selected meteorology to demonstrate the computer's usefulness, had by 1956 shown that it could generate accurate forecasts. The computer has enabled meteorologists to exploit the immense quantities of data arriving from weather satellites and a greatly increased number of observational sources in the atmosphere and at the earth's surface. Computational models of the atmosphere have since blurred the distinctions among observation, experiment, and theory.

Gisela Kutzbach, *The Thermal Theory of Cyclones: A History of Meteorological Thought in the Nineteenth Century* (1979). Robert Marc Friedman, *Appropriating the Weather: Vilhelm Bjerknes and the Construction of a Modern Meteorology* (1989). Theodore S. Feldman, "Late Enlightenment Meteorology," in *The Quantifying Spirit in the 18th Century*, Tore Frängsmyr, J. L. Heilbron, and Robin E. Rider, eds. (1990): 143–178. James Rodger Fleming, *Meteorology in America, 1800–1870* (1990). Frederik Nebeker, *Calculating the Weather* (1995).

THEODORE S. FELDMAN

METEORS. See COMETS AND METEORS; METEORITES.

MICHELSON, Albert Abraham (1852–1931), physicist.

A. A. Michelson was a master of precision optical measurement. His determinations of the speed of light and the lengths of light waves were the best of his day, and his attempt of 1887 in collaboration with Edward Morley to detect the motion of the earth through the *ether helped set the stage for Albert *Einstein's theory of *relativity. In 1907 Michelson became the first American to receive a Nobel Prize in the sciences.

Born in Strelno, Prussia (now Poland), Michelson emigrated to America with his family while still a child. He grew up in gold rush towns in California and Nevada. In 1869 he entered the U.S. Naval Academy at Annapolis and in 1875 became a physics instructor there.

While teaching optics at Annapolis, Michelson set out to repeat Jean Bernard Léon Foucault's 1850 measurement of the speed of light. He improved Foucault's rotating mirror method and published his first results in 1878. Two years later, while pursuing advanced work in Europe, he first confronted an enduring puzzle: the effect of motion on light. Many physicists believed that as the earth moved through the stationary ether, it was swept by a continual "ether wind" blowing at about 30 km/sec (18 mi/sec). This wind should have affected the speed of light waves moving in it. Calculations indicated that the effect would be very small, perhaps too small to detect. In 1880 Michelson hit on a way to measure it. In a "Michelson interferometer," a beam of light is split into two parts moving at right angles to one another; when reflected back and recombined, they produce interference fringes that provide an exquisitely sensitive gauge of any changes in the speeds or path lengths of the beams. Michelson tried the device at Potsdam in 1881 but found no shift of the fringes when he turned the interferometer: the ether wind seemed to have no effect. The result was not yet conclusive, however, for the expected shift lay at the limit of the sensitivity of his apparatus.

Michelson resigned from the Navy in 1881 and the next year began teaching at the Case School of Applied Science in Cleveland. There he and Morley, of neighboring Western Reserve University, carried out two important investigations, first confirming Armand-Hippolyte-Louis Fizeau's 1859 demonstration that moving water drags along light waves passing through it, and then in 1887 performing their famous repetition of Michelson's Potsdam experiment. Using a larger and more sensitive interferometer set on a block of sandstone 1.5 m (5 ft) square and floating in a trough of mercury, they found no shift of the fringes and no sign of an ether wind.

Michelson and Morley's null result seemed impossible to reconcile with the known facts of optics. George Francis FitzGerald in 1889 and Hendrik Antoon Lorentz in 1892 independently proposed a striking solution: perhaps motion through the ether slightly alters the forces between molecules, causing Michelson and Morley's sandstone block to shrink by just enough to nullify the effect they had been seeking. The "FitzGerald-Lorentz contraction" later became an important part of relativity theory. Although scholars have often exaggerated the influence of Michelson and Morley's experiment on Einstein's thinking, Einstein knew at least indirectly of

their result and it certainly loomed large in later discussions of his ideas. Michelson himself did not welcome the rise of relativity theory; he remained a firm believer in the ether.

In 1889 Michelson left Case for Clark University in Massachusetts. Disappointed by the negative result of his ether-drift experiments, he turned his interferometer to other uses, traveling to Paris in 1892 to measure the standard meter in terms of light waves. In 1893 he moved to the University of Chicago, where he headed the physics department until 1929. There he continued his optical work, making precision diffraction gratings and inventing the echelon spectroscope. His Lowell lectures of 1899 were published in 1903 as *Light Waves and Their Uses*; his *Studies in Optics* appeared in 1927.

Michelson spent much of the 1920s at Mount Wilson Observatory in California applying interferometric methods to astronomical problems and refining his measurements of the speed of light. In 1926 he bounced light between mountain peaks 32 km (20 mi) apart, pinning the speed of light down to within 4 km/sec (2.5 mi/sec). After Dayton Miller announced in 1926 that he had found positive evidence of ether drift, Michelson repeated a refined version of his old experiment, but again found no shift of the fringes and Miller's results came to be regarded as erroneous.

See LIGHT, SPEED OF.

Loyd S. Swenson, Jr., *The Ethereal Aether: A History of the Michelson-Morley-Miller Aether-Drift Experiments, 1880–1930* (1972). Dorothy Michelson Livingston, *The Master of Light: A Biography of Albert A. Michelson* (1973). Stanley Goldberg and Roger H. Stuewer, eds., *The Michelson Era in American Science, 1870–1930* (1988).

BRUCE J. HUNT

MILKY WAY. The seventeenth century saw little advance in understanding the Milky Way, a dense band of stars the color of milk across the sky, beyond *Galileo's confirmation of the ancient opinion that it was a congeries of innumerable stars. In 1755 Immanuel Kant, inspired by an incorrect summary of Thomas Wright's ideas about the Milky Way (1750), explained it as a disk-shaped system containing the earth. Kant conjectured that the disk structure arose in the same manner that the planets came to orbit almost in the same plane around the Sun. And the same cause that gave the planets their centrifugal force and directed their orbits into a plane could also have given the power of revolving to the stars and have brought their orbits into a plane.

Kant's manuscript perished in his printer's bankruptcy. A condensed version of his hypothesis appeared in 1763, hidden in the appendix of another book. Meanwhile (in 1761), the polymath Johann Heinrich Lambert published a similar theory. Kant had emphasized Newtonian dynamics and the process by which the world achieved its current shape. Lambert, while acknowledging the existence of Newton and gravitation, emphasized God and the harmonious order He had given the world. Unaware of Wright's, Kant's, and Lambert's ideas, the English astronomer William *Herschel began in the 1780s his own investigation of the construction of the heavens. He observed stars in a stratum seemingly running to great lengths and identified the Milky Way as the appearance of the stars as seen from the earth. To determine the position of the Sun in the sidereal stratum, Herschel counted stars in different directions. This number, he argued, should be proportional to the length of the stratum in the direction of the count. Herschel was an observer, not a theoretician. His Milky Way was an observed stratum of stars extending different distances in different directions, not a theoretical disk the result of the force of attraction.

Herschel lacked the means to measure distances. Not until early in the twentieth century could the American astronomer Harlow Shapley argue convincingly that the Milky Way was a hundred times larger than previous estimates and that the Sun lies tens of thousands of light years away from the center of the *galaxy.

Stanley L. Jaki, *The Milky Way: An Elusive Road for Science* (1972).

NORRISS S. HETHERINGTON

MINERALOGY AND PETROLOGY. Mineralogy's disciplinary status has under-

gone three distinct shifts. From the sixteenth through the early nineteenth century, it bridged chemistry and natural history. It used the laboratory techniques of the former and the principles of classification of the latter to study the whole of the mineral kingdom. During the nineteenth century, mineralogy lost this commanding position and became a subdiscipline of *geology. The study of minerals (chemicals that occur naturally in the earth's crust) separated from that of rocks (distinctive assemblages of minerals), leading to a distinction between mineralogy and petrology in the latter part of the century. Following World War II, geology came under the *earth sciences, and mineralogy and petrology were transformed by the theory of *plate tectonics and by new instrumentation.

In spite of changes in disciplinary status, from the eighteenth century to the present mineralogists have concentrated on two problems. The first, classification, has been essential for geological theory and practical applications in mining. It has also been a scientific nightmare. Classifying depends on being able to make clear distinctions at various levels of organization. In the case of animals and plants, individuals can usually be distinguished easily and most of the time species too by the test of reproductive capability. Although in the eighteenth century, Carl Linnaeus made a gallant attempt to extend these methods to minerals, it was a doomed strategy. Over the centuries, mineralogists have oscillated between using chemical composition and crystal form. Unfortunately, these do not map onto each other. Minerals of the same composition can have different forms, and minerals of the same form different compositions. Adding to the difficulties, the chemical composition of many of the commonest minerals is not fixed but allows a range of variation. Formal classifications have always been supplemented by keys to field identification using a variety of visible characters.

The second key mineralogical problem is how minerals and rocks originated. Since they occur interlocked with one another, mineralogists from the seventeenth century on have assumed that they originated as fluids and subsequently hardened in their present positions. The fluidity could have been caused by heat, water, or some combination of the two. Mineralogists have fought over the relative importance of these factors, but they have always hoped they would be able to use their knowledge of chemical reactions to reconstruct a genetic account of rock history—a geogony based on an invariable sequence of chemical reactions following from some initial state.

Mineralogy differed greatly from the historical geology better known to the public and more thoroughly studied by historians. Mineralogists have always focused their attention on the hard rocks (igneous and metamorphic). The sedimentary rocks so dear to historical geologists because of their embedded fossils have taken second place, even though mineralogists have attended to clay mineralogy and sedimentary petrology. Mineralogists have always worked closely with chemists and with crystallographers. They have found experiments and microscope work as important as fieldwork. Germans and Scandinavians dominated mineralogy partly because of the abundance of hard rocks in those countries, but also because of their distinguished traditions in chemistry and crystallography. Only in the twentieth century did they begin yielding to Canadians and Americans.

From the Renaissance through the eighteenth century, mineralogists produced one classification of rocks and minerals after another. Among the more important classifiers were Georg Bauer, better known as Agricola, the Swedish chemist Johan G. Wallerius, and Abraham Gottlob *Werner. All distinguished four major groups with different chemistries: earths, metals, salts and combustibles. Earths resisted heat and water, metals became fluid on heating, salts dissolved in liquids, and the combustible substances (coal, for example) burned. Because chemistry formed the basis for classifying rocks, the students spent as much time in the cabinet or laboratory working with chemicals and blowpipes as they did in the field. Werner drew the pessimistic conclusion that no theoretically sound principles of mineral classification were to be found. Hence he instructed his students to

begin dividing up rocks by the time of their formation (see STRATIGRAPHY).

Employed by European states as mining inspectors or surveyors, mineralogists left their laboratories to climb mountains and descend mine shafts. By the second half of the eighteenth century, these men—including Lazzaro Moro and Giovanni Arduino in Italy, Johann Lehmann in Germany, and Guillaume-François Rouelle in France—opted for an alternative approach to rock classification. They divided rocks into two main kinds: primary and secondary. Primary rocks were hard, often crystalline and the matrix in which metals and precious minerals were to be found. They made up the core of mountain chains. Secondary rocks were relatively soft and granular, layered or stratified and banked up against the primary rocks that formed the mountain cores. Often, secondary rocks contained fossils, which by then most mineralogists agreed were the indurated remains of animals and plants.

In seeking the causes for this twofold division of rocks, mineralogists found common ground with cosmogony, the study of the development of the globe. Since the seventeenth century, cosmogonists had argued from the earth's globular figure that at some time in the past it had been fluid. Fluidity could have been caused by heat, as a minority of cosmogonists had argued. Mineralogists, though, preferred water, as suggested by the chemist Johann Joachim Becher in his *Physica subterranea* (1669). They believed that a thick, chemical laden ocean had once covered the earth's surface. The primary rocks crystallized out of the ocean as the high mountain chains, a conclusion supported by the chemists' belief that crystals could be deposited only from watery solutions and not from hot melts. As the water became less saturated, and as waves and rivers wore away the mountains, the ocean began depositing the silt that solidified as the secondary, stratified rocks. This theory, *Neptunism, was most fully developed by Werner at Freiberg.

From about 1830 to 1880, mineralogists looked to new developments in chemistry to aid them with mineral classifications. Jöns Jacob Berzelius, the Swedish chemist, distinguished the silicates and aluminates—the classes of chemicals most abundant in the earth's crust for the first time. Gustav Rose offered the most comprehensive classification of minerals to date in his *Mineralsystem* (1852). For practical purposes, mineralogists continued to use external features. Friedrich Mohs, best known for developing a hardness scale for minerals, developed one of these. James Dwight Dana adapted it for an American audience. His *System of Mineralogy,* first published in 1837, must be one of the most enduring of textbooks in the history of modern science. In modified form, it was still in use in the 1960s.

In 1860, Henry Clifton Sorby invented the polarizing microscope, transforming the process of identifying minerals. Thin sections of minerals or rocks were placed on slides that could be rotated beneath polarizing lens. The characteristic color changes that were observed on rotating the slide served to identify the mineral. This new technique allowed mineralogists for the first time to see and identify mineral assemblages formerly invisible to the naked eye. It gave an enormous boost to petrology. Karl Rosenbusch used it to particularly good effect, summarizing the new results in his classic textbook, *Mikroskopische Physiographie der petrographisch wichtigen Mineralien* (1873).

In the nineteenth century, theories of the origin of rocks and minerals also became more sophisticated. Charles *Lyell suggested that besides volcanic, plutonic, and sedimentary rocks, geologists needed a fourth category, which he called metamorphic. These arose through transformation of the other classes by heat and pressure. While agreeing that gneisses, schists, and perhaps granite might be problematic, continental mineralogists continued to believe that water, perhaps under heat and pressure, perhaps containing many strong chemicals, was crucial to petrographic change. Carl Gustav Christoph Bischof summed up the state of the argument in what became the standard geochemical text, *Lehrbuch der chemischen und physikalischen Geologie* (1848). In Canada, Thomas Sterry Hunt made another stab at a chemical geogony. His theory was rejected.

At the end of the nineteenth century, mineralogists and petrologists found that new research in *thermodynamics, particularly that of J. Willard *Gibbs on the phase rule, offered an alternative way to think about mineral and rock origins. They began constructing phase diagrams for certain common rocks to clarify the sequence and manner in which the different crystals had formed. The Carnegie Institute of Washington's lavishly equipped laboratory aided this program of research. The United States could now compete with Germany. In 1928, Norman Bowen summed up then recent developments in his classic, *The Evolution of the Igneous Rocks* (1928).

The question of the origin of the rocks was pursued in the field as well as in the laboratory by two opposing camps that frequently compared themselves to the Neptunists and Plutonists of a century earlier. The minority camp, the migmatists, believed that migmatites (as they called the puzzling hard rocks of varied composition) were formed in place as circulating fluids converted extant rocks into something completely different. The majority camp, the magmatists, led by the Canadians Norman Bowen at the Carnegie Institute and Reginald Daly at Harvard, argued that they were intruded from reservoirs of molten magma beneath the earth's crust. Within this camp, heated debates raged about whether there was one magma or many, and whether magmas were homogenous or differentiated. There the matter stood at the beginning of World War II. Following the war, novel techniques in the laboratory and the field, including deep sea drilling, suggested new directions for mineralogical research.

See also CRYSTALLOGRAPHY.

Karl Zittel, *History of Geology and Paleontology* (1901). Rachel Laudan, *From Mineralogy to Geology* (1987). David Oldroyd, *Thinking About the Earth* (1996).

RACHEL LAUDAN

MOHOLE PROJECT AND MOHOROVIČIĆ DISCONTINUITY. The purpose of the Mohole Project (1957–1966) was to drill through the earth's crust to the Mohorovičić discontinuity, the seismic in-

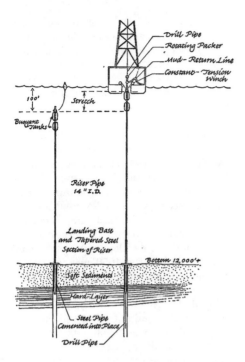

Schematic diagram for producing a hole in the bottom of the sea. The riser pipe floats on the submerged buoys when the drilling rig is not on station.

terface between the earth's crust and mantle. This boundary was discovered in 1909 by Yugoslav geophysicist Andrija Mohorovičić who noted that seismic waves returning from depth indicated there was a zone of abrupt change in the speed of seismic waves some kilometers below the earth's surface. Called Moho for short, this zone defines the base of the earth's crust and marks a change in composition. The depth of Moho varies from about 25–40 km (15–25 mi) beneath the continents, to 5–10 km (3–6 mi) beneath the ocean floor. Recently, following the acceptance of *plate tectonic theory, geoscientists have decided that changes in deformational behavior are a more significant aspect of the earth's structure than changes in composition. They divide the outer earth into the rigid lithosphere (crust and upper mantle) overlying the more deformable asthenosphere.

Project Mohole was the brainchild of AMSOC, the American Miscellaneous Society, an informal group of geoscientists formed in the 1950s. Drilling for scientific purposes had begun with efforts to determine the structure, composition, and history of coral islands. In 1877 the Royal Society of London sponsored a borehole that went down 350 m (1,140 ft) on Funafuti in the South Pacific. In 1947, pre-bomb-test drilling of Bikini reached 780 m (2,556 ft). In 1952 drilling on Eniwetak finally reached basaltic crust beneath coralline rock at a depth of over 1,200 m (4,000 ft), still well short of the Moho. In the 1950s, some countries, including Canada and the Soviet Union, proposed drilling deep holes in continental crust.

Project Mohole, funded by the U.S. National Science Foundation, was designed to drill the deep-ocean floor in water depths of thousands of meters. Although Mohole successfully drilled cores in water depths of 950 and 3,560 m (3,111 and 11,672 ft), it did not come close to reaching the Moho. Mohole, commonly seen as one of the first big-science projects in the United States, was terminated by Congress in 1966 and widely considered to have been a failure.

Nonetheless the project demonstrated that ship-based ocean drilling was feasible. It helped spawn the highly successful Deep Sea Drilling Project, begun in 1968, to drill through cover sediments on the ocean floor. Since then scientific deep-ocean drilling has become an international endeavor. Oil companies are reaching deeper and deeper objectives. Moreover, a number of countries and consortia have developed continental deep-drilling projects. None of this drilling has reached the Mohorovičić discontinuity, however.

Even so, scientists now believe that we can observe the Moho on land. By the late nineteenth century, a number of European scientists had recognized that in the Alps uplifted oceanic crust was represented by layered chert (lithified deep-sea sediments) overlying basalt (oceanic crust) overlying ultramafic rocks (high-density rocks rich in iron and magnesium), the so-called Steinmann trinity, after Gustav Steinmann. This package of rocks is called an ophiolite, and the transition from basalt to ultramafic rocks is believed to be the crust-mantle boundary—the Mohorovičić discontinuity.

Willard Bascom, *A Hole in the Bottom of the Sea: The Story of the Mohole Project* (1961). Elizabeth N. Shor, "A Chronology from Mohole to JOIDES," in *Geologists and Ideas: A History of North American Geology*, Ellen T. Drake and William M. Jordan, eds. (1985): 391–399.

JOANNE BOURGEOIS

MOHOROVIČIĆ DISCONTINUITY. See MOHOLE PROJECT AND MOHOROVIČIĆ DISCONTINUITY.

MOON. Since classical Greek times astronomers have been occupied with the orbital characteristics of the Moon, with its proposed epicycles, eccentrics, and, after Johannes *Kepler, ellipses. Concern with lunar dynamics extends to the present. Astronauts placed reflectors on the Moon's surface so that its orbital characteristics could be measured to a factor of inches using laser beams.

The study of the Moon as a physical world began in 1609, when *Galileo first observed its mountains and "maria" (seas) through a *telescope, and announced his findings in *Siderius nuncius* (1610). Since classical times, it had been assumed that the Moon was a perfectly smooth, albeit tarnished, sphere: its tarnish, like its phases, perhaps deriving from its pivotal location at the boundary between the mutable "sublunary," or earthly, realm and the "superlunary" realm of eternal perfection, as in Aristotle's *De caelo*. Galileo interpreted his lunar discoveries (as he also did his planetary and sidereal discoveries) against Aristotle. In his measurement (by shadows) of the height of the lunar Apennine Mountains (named after their Tuscan counterparts), and of certain crater formations, Galileo demonstrated the Moon to be a world with topographical features. Galileo's lunar observations may have been preceded slightly by those of Thomas Harriot and Sir Richard Lower, who also recorded mountains, "seas," and craterlike formations, but neither of them published their findings, nor did they use them to advance a particular scientific argument.

For almost forty years after Galileo's announcements few further discoveries were made in selenography (from Selene, the Greek lunar goddess), largely owing to the inability of opticians to make lenses capable of showing additional detail. Meanwhile popular science writers such as John Wilkins, William Godwin, and Cyrano de Bergerac speculated about lunar voyages and moon men.

Johannes Hevelius's *Selenographia* (1647) reopened lunar study. Using a telescope with a 3.5 centimeter (1.5 in) aperture convex object glass ground to a 3.6-meter (12-ft) radius curvature, Hevelius produced a series of drawings of the lunar surface under changing conditions of illumination, which vastly improved on Galileo's sketches. Developments in glassmaking and the development of lens-polishing machines after 1650 opened up new technological potential. Giambattista Riccioli (1651), Gian Domenico *Cassini (1679), and others produced maps displaying remarkable and faithful detail. Riccioli introduced the custom of naming lunar craters after eminent scientists.

Johann Schröter revived detailed selenographical mapping in Germany around 1790 by using large-aperture reflecting telescopes giving high magnifications. The new fascination with the Moon as a geological body from the 1830s further stimulated interest. Were the craters and "seas" the products of a once geologically active Moon? Did the supposed disappearance of the Crater Linné in 1866 indicate that a central heat might still be present? Those who sought the answers, including Schröter, Wilhelm Beer (who employed the professional Johann von Mädler), Wilhelm Lohrmann, Johann Krieger, James Nasmyth, and the members of the short-lived British Selenographical Society, were "Grand Amateurs." Nasmyth, an iron master by profession, attempted to replicate lunar features in the blast furnace, much as Robert *Hooke had tried to replicate crater formations by blowing air through molten alabaster and dropping bullets into pipe-clay. Nonetheless the nineteenth century thought that craters were *not* meteoritic in origin.

Lunar study in the twentieth century was overwhelmingly dominated by space exploration. In 1959, the Russian spacecraft *Luna 3* succeeded in photographing the "dark" side of the Moon; the pictures revealed a difference in the distribution of crater and "maria" from that seen on the side visible from the earth. Then, both before and after the American Apollo 11 mission's first landing of astronauts on the Moon in 1969, a succession of manned and robotic expeditions yielded a wealth of new information about lunar geology, magnetism, orbital characteristics, and much besides. This recent work has now itself entered the historiographical canon of astronomy. From Galileo's *Siderius nuncius* down to the present day the Moon, in its scientific, poetic, and science-fiction aspects, has probably generated a larger body of literature than any other single astronomical body.

See also CELESTIAL MECHANICS.

Robert Grant, *History of Physical Astronomy* (1852). James Nasmyth and James Carpenter, *The Moon, Considered As A Planet, A World, And A Satellite* (1874). Ewen A. Whitaker, *Mapping and Naming the Moon: A History of Lunar Cartography and Nomenclature* (1999). Scott L. Montgomery, *The Moon and the Western Imagination* (2000).

ALLAN CHAPMAN

NATURPHILOSOPHIE. Naturphilosophie was a school of thought characterized by a speculative, idealistic, and holistic approach to the study of nature that had its origins in Germany at the end of the eighteenth century. Since many of its proponents were also associated with Romanticism, Naturphilosophie is often referred to as Romantic science. In much the same way that Romantics reacted against rationalism, adherents of Naturphilosophie reacted against the traditions they saw as deriving from Francis Bacon and Isaac *Newton: the idea that the world was atomistic, that an inductive and empiricist methodology was the best way of exploring it, and that mathematics was the language of nature.

Naturphilosophie combined a version of Neoplatonism with a reading of Immanuel Kant. From the former, which they found in the works of Paracelsus and Jean Baptiste van Helmont, came the Naturphilosophen's belief that all the forces we perceive in the world are manifestations of one basic force. From the latter, they took the idea that in the construction of knowledge, the mind imposes its categories such as space, time, and cause and effect on nature. But whereas Kant emphasized that the mind cannot know the nature of things, the Naturphilosophen reinterpreted the imposition of our mental categories not as veils but as insights. They celebrated human reason for its capacity to participate in the divine reason and thereby to comprehend nature in its entirety.

Friedrich Schelling, who wrote his influential *Ideas for a Philosophy of Nature* in 1797, is regarded as the founder of the movement. For Schelling, nature consisted of opposites or polarities: positive and negative for electrical phenomena, north and south for magnetic, acids and bases for chemical. In each case, the opposing forces resolved and unified in new phenomena and new forces on a higher plane. All manifested a single underlying force and could be converted one into the other in the proper circumstances.

Although he stood apart from Naturphilosophie, Johann Wolfgang von Goethe shared many of its ideas and was held in high regard by many of its leaders. In his seminal work, *On the Metamorphosis of Plants* (1790), Goethe interpreted the organs of the individual flowering plant as modifications, or metamorphoses, of a single basic form, an idealized, primal leaf. As a group, flowering plants were variations of an ideal plant archetype, or Urpflanze. Goethe made his opposition to Newton most explicit in his study of the perception of colors, *Zur Farbenlehre* (1810).

Lorenz Oken, who worked in comparative anatomy, synthesized many of Naturphilosophie's themes in his *Elements of Physio-Philosophy* (1809–1811, English translation 1847). He influenced a generation of students at the University of Jena and abroad. Schelling's student Heinrich Steffens and Johann Wilhelm Ritter also edged Naturphilosophie in the empirical direction with their research on *geology and *electricity, respectively.

During the early nineteenth century, Naturphilosophie spread beyond Germany. It inspired some natural philosophers, particularly those working on problems in electricity, chemistry, magnetism, and anatomy. Naturphilosophie helped Hans Christian Ørsted formulate the questions that led him to the discovery of *electromagnetism (1820). Similarly, it stimulated Thomas Seebeck's research that detected thermal electricity (1822), Humphry Davy's and Michael *Faraday's exploration of electrochemical and *electromagnetic phenomena, and the concept of the conservation of energy (*see* CONSERVATION PRINCIPLES). Schelling himself believed that Faraday's work on electromagnetism confirmed his own theories.

In the life sciences, Naturphilosophie postulated that the succession of higher life forms on earth was the outcome of opposing forces present in lower forms. Although suggestive of modern evolutionary thought, this succession was not evolution in the sense of genetic descent, but rather a process of ascent toward a pre-ordained ideal, comparable to embryological gestation. Such ideas crop up in the transcendental anatomy of Étienne Geoffroy-Saint Hilaire, Henri de Blainville, Robert Knox, Robert Grant, Richard Owen, Edward Forbes, Jr., and Louis *Agassiz.

By the mid–nineteenth century, reactions to Naturphilosophie, like its fundamental theory, were sharply polarized. Within German philosophy, its influence continued, thanks largely to Georg Wilhelm Friedrich Hegel's influential *Enzyklopädie der philosophischen Wissenschaften* (1817); it re-emerged in the critique of modern science mounted by Continental philosophy in the twentieth century. In the scientific mainstream, the rise of positivism, Darwinism, and professional specialization, among other factors, led to Naturphilosophie's demise. Its influence on early nineteenth-century science was largely forgotten until the 1960s. Then, with the turn away from positivism as the dominant historiography of science, historians of science (as well as historians of ideas and literary historians) began uncovering the paths by which Naturphilosophie affected the growth of modern science.

Alexander Godevon Aesch, *Natural Science in German Romanticism* (1941). L. Pearce Williams, *The Origins of Field Theory* (1966). Trevor Levere, *Poetry Realized in Nature: Samuel Taylor Coleridge and Early Nineteenth-Century Science* (1981). Philip F. Rehbock, *The Philosophical Naturalists: Themes in Early Nineteenth-Century British Biology, Part I* (1983). Andrew Cunningham and Nicholas Jardine, eds., *Romanticism and the Sciences* (1990).

PHILIP F. REHBOCK

NEBULA. With little more to recommend himself than neat handwriting, Charles Messier began his working life as a clerk to a French astronomer, and eventually became the world's foremost hunter of nebulae. M31, the thirty-first object in his 1781 catalog of 103 nebulous-appearing objects that might be mistaken for *comets, is now known as the great spiral *galaxy in the constellation Andromeda. By 1800, Willam Herschel, with his reflecting telescope, had raised the number of known nebulae to around 2,000.

Opinion swung back and forth over whether nebulae were gaseous or stellar, clouds of luminous fluid or remote star systems. Herschel's big instruments resolved several nebulae into stars. In 1790, however, he encountered a planetary nebula (now known to be an expanding shell of gas ejected from and surrounding a very hot star) that defied resolution. Astronomers came to doubt that any nebulae were composed of stars, or constituted island universes. They came to consider even the Andromeda Nebula as a mass of nebulous matter, definitely not a stellar system. Larger reflecting telescopes constructed around the middle of the nineteenth century resolved more nebulae into groups of stars, and opinion again swung toward the concept of stellar composition of all nebulae. Even the Orion Nebula (now known to be a cloud of glowing gas lit by young stars in the process of formation) was resolved into stars. The director of the Harvard College Observatory exclaimed to the college president, "You will rejoice with me that the great nebula in Orion has yielded to the powers of our incomparable telescopes."

The end of the nineteenth century saw opinion swing yet again. Spectroscopic observations by William Huggins revealed that about a third of some seventy nebulae displayed a gaseous character. Furthermore, Huggins insisted that the nebulae with stellar spectra were composed of gas under special conditions. Astronomers persisted in an "either-or" choice, unable to imagine that both gaseous and stellar nebulae might exist, even as the number of specimens grew. The *New General Catalogue of Nebulae and Clusters of Stars* (1888, 1895, 1908) listed over 13,000 nebulae, and astronomers estimated that 150,000 nebulae were within reach of existing instruments.

The distribution of nebulae was revealing, their avoidance of the plane of the *Milky Way seemingly linking them physically to our galaxy. Later, after the extragalactic nature of many nebulae was established, the apparent zone of avoidance was attributed to obscuring gas and dust in the plane of our galaxy. As late as

1917, however, the American astronomer Edwin *Hubble could note that "extremely little is known of the nature of nebulae, and no significant classification has yet been suggested; not even a precise definition has been formulated."

The spectroscope revealed the true nature of gaseous nebulae; other nebulae, including the great spirals, are stellar systems. During the 1920s, Hubble proved that the spirals in fact lie outside our galaxy.

Edwin Hubble, *The Realm of the Nebulae* (1936). Norriss S. Hetherington, *Encyclopedia of Cosmology: Historical, Philosophical, and Scientific Foundations of Modern Cosmology* (1993).

NORRISS S. HETHERINGTON

NEPTUNISM AND PLUTONISM. The debate between Neptunists and Plutonists about the origin of the rocks of the earth's crust took place in the late eighteenth and early nineteenth centuries. The Neptunists believed that essentially all rocks had been formed in water. According to the most prominent Neptunist, Abraham Gottlob *Werner, who taught at the Mining Academy of Freiberg in Saxony, originally the earth was covered by a hot, thick, basic aqueous brew. As this cooled, the rocks that form the core of mountain chains crystallized out. Later, noncrystalline rocks were deposited as layers of strata banked up against the primary, crystalline rocks. Volcanoes were late and largely inconsequent phenomena caused by the burning of plant remains. Underlying his theory was a tradition, stretching back to Johann Joachim Becher and Georg Ernst Stahl in the seventeenth century, of chemical cosmogonies based on the assumption that processes observed in the laboratory could inform theories about mineral formations observed in the field.

The Vulcanists Nicolas Desmarest and Rudolph Eric Raspe argued that basalts, often found interbedded with strata, had actually flowed from volcanoes. Studies of extinct volcanoes in the Massif Central in France confirmed this assertion. The point was quickly accepted by the School of Freiberg, some of whose members, particularly Leopold von Buch and Alexander von

*Humboldt, carried out major studies on volcanoes in the first part of the nineteenth century.

James *Hutton is regarded as the leading representative of the Plutonist theory. According to Hutton, heat consolidated rocks at the bottom of the ocean and subsequently elevated them to form land. John Playfair, in the *Illustrations of the Huttonian Theory of the Earth* (1802), concentrated on the evidence for the theory and downplayed its natural philosophical and chemical foundations. James Hall carried out dangerous experiments with limestone heated under pressure and lived to report that it did indeed consolidate under sufficient pressure. However, neither Playfair nor Hall succeeded in convincing the geological community that the strata had been consolidated by heat.

By the 1820s, most geologists agreed that strata formed under water and that basalt and certain other igneous rocks were spewed out from volcanoes. The origin of the hard and often crystalline rocks such as granites and gneisses continued to pose a problem. In the late nineteenth century, petrologists fought about the origin of granite in what appeared to many of them as a replay of the Neptunist-Plutonist (or Vulcanist) debates of a hundred years earlier.

Following and amplifying Charles *Lyell's influential historical introduction to the *Principles of Geology* (1830), historians and geologists characterized this debate as a sterile consequence of the intrusion of nonscientific issues into geology. They criticized the Neptunists for taking the idea of a universal ocean from the Bible, and esteemed the Plutonists as excellent field geologists. A more adequate understanding of the debate recognizes that both sides supported their positions with a mixture of empirical data and theoretical arguments, and that in doing so both contributed to the clarification of the basic principles of *geology.

See also MINERALOGY AND PETROLOGY.

Roy Porter, *The Making of Geology* (1977). Rachel Laudan, *From Mineralogy to Geology: The Foundations of a Science, 1650–1830* (1987).

RACHEL LAUDAN

NEWTON, Isaac (1642–1727), one of the greatest mathematicians and natural philosophers the world has known.

Newton was born in Woolsthorpe, near Lincoln, on Christmas Day 1642. He was educated at Trinity College, Cambridge, and became a fellow in 1667. In 1669 he succeeded his teacher, Isaac Barrow, as Lucasian Professor of mathematics.

While the university closed from summer 1665 to April 1667 because of plague, Newton spent most of his time on the family farm in Woolsthorpe. He returned to Cambridge briefly from March to June 1666. Newton made himself master of the latest mathematics and created a wholly new branch, the differential and integral calculus. At about the same time, Gottfried Leibniz independently invented a similar calculus. Newton was the first to produce this new mathematics, but Leibniz was the first to publish it.

During those brilliantly creative months in plague-fearing Cambridge, Newton also set the foundations of modern optics. He analyzed the solar spectrum, revealing the phenomena of dispersion and composition of light and their causes. He invented a new type of telescope, featuring a magnifying mirror instead of a magnifying lens. Newton also devoted time to force and motion, but he did not then—as he later claimed—find that gravity extends to the Moon or discover the law of gravity.

In 1672 he published an account of his new discoveries concerning light and color. This paper gave rise to extensive criticism, to which Newton wrote careful replies. He declared that he so regretted having to reply to critics that he would never again publish his discoveries. For almost twenty years he remained faithful to this vow. He gave the lectures required by the terms of his professorship and busied himself with investigations that may seem far removed from science. They included alchemy, biblical prophecy, the interpretation of Scripture, the chronology of ancient kingdoms, but not astrology, one of the few sorts of ancient knowledge Newton thought invalid. He believed that these investigations were intimately related to his work in optics, mathematics, rational *mechanics, and celestial dynamics.

In 1679, Newton learned of Robert *Hooke's idea that orbital or curved motion could be explained by a combination of a linear inertial component along the orbit's tangent and a continual falling inward toward the center. Newton wrote that he had never before heard of this "hypothesis." But he perceived a connection between Hooke's suggestion and Johannes *Kepler's law of areas, and showed that they implied that the tendency toward the center in planetary elliptical orbits must vary as the inverse square of the distance from the Sun. He informed no one about this great breakthrough.

In 1684 Newton received a visit from Edmond *Halley, who asked for help in solving a problem that had stumped everyone in London: the force that produces planetary elliptical orbits. Newton replied that he had already solved it. He wrote up his solution in a little tract called *De motu*. While revising and expanding it, he discovered that the same force that keeps the planets in *orbit must cause perturbations in the orbital motions of other planets, the key to the great principle and law of universal gravitation.

Encouraged by Halley, Newton now began to develop his work in detail. In 1687 he published the resulting masterpiece, *Philosophiae naturalis principia mathematica* (Mathematical Principles of Natural Philosophy). Here Newton gave his new concept of mass and the principle of inertia, and his famous three laws of motion, the foundation of the new science of rational mechanics.

The first of the *Principia*'s three "books" sets forth the science of motion; the second, the conditions of fluid resistance and their consequences; and the third, the system of the world, built up from his mechanical principles in the law of universal gravity. In the second and third editions, the *Principia* concludes with a General Scholium containing Newton's famous slogan, *Hypotheses non fingo*—"I do not feign hypotheses"—referring to his disinclination (and inability) to declare a mechanical cause of gravitation.

Book three gives explanations of the tides, the motions of the Moon and of the comets, the shape of the earth, the variation of weight with change in terrestrial position, the acceleration of falling bodies, and much more besides.

A decade or so after the publication of the *Principia*, Newton moved to London, where he became warden and then master of the mint and president of the Royal Society. In 1704, he published his *Opticks*, an account of his many optical discoveries, which also contained, in the form of "Queries," hints and experiments on all sorts of physical and chemical phenomena. These sometimes contradictory queries, expanded in successive editions of the *Opticks*, helped to guide the experimental philosophy of the eighteenth century. Newton was buried in Westminster Abbey.

I. Bernard Cohen, *Franklin and Newton* (1956). Richard S. Westfall, *Never at Rest* (1983). Derek Gjertson, *The Newton Handbook* (1986). Newton's major works exist in modern formats: *Opticks*, ed. E. T. Whittaker (1952); *Principia*, trans. I. B. Cohen et al. (1999).

I. BERNARD COHEN

NEWTONIANISM. Isaac Newton joined terrestrial and celestial mechanics together in 1687 with the publication of his *Philosophiae naturalis principia mathematica*, transforming the two kinds of physics into a single system oriented around the inverse-square law of gravitational attraction. Twenty years after this feat of mathematical synthesis, Newton contributed to framing experimental physics in his *Opticks; or, A Treatise of the Reflections, Refractions, Inflections and Colours of Light*, first published in 1704 and going through four editions by 1730. The *Opticks* laid out an experimentally derived geometry of light-rays, including the use of prisms to analyze white light into the colors of the spectrum and then resynthesize them into white light. The *Opticks* also included a final section of "queries" containing all the unfinished business of Newton's career. By means of this laundry list, Newton set the agenda for his eighteenth-century followers. Most importantly, he proposed that weightless *ethers were the medium and material cause of forces and phenomena including light,

heat, *electricity, *magnetism, gravitational attraction, and animal sensation.

Over the past three centuries, "Newtonianism" has meant several things. On the model of the *Principia*, it has meant a mathematical, synthetic approach to physics, and more specifically, the confirmation and promulgation of inverse-square laws of force. Important examples are Charles Augustin Coulomb's demonstration on the "laws" of electrical and magnetic electrical attraction and repulsion (1785–1789) (*see* CAVENDISH AND COULOMB). Eighteenth- and nineteenth-century Newtonians also developed the field of *astronomy by testing Newton's law of gravitational attraction against new observations and resolving apparent conflicts. Alexis-Claude Clairaut's explanation of the motion of the lunar apogee (1749) (*see* MOON) and his accurate prediction of the return of Halley's *comet in 1759 confirmed and vindicated Newtonian astronomy. These efforts gave rise to an increasingly complex picture of the mutual gravitational influences of celestial bodies. The culmination of eighteenth and early nineteenth century Newtonian astronomy was Pierre-Simon *Laplace's *Traité de mécanique céleste* (1798–1827), in which Laplace used Newton's law of gravitation to develop a complete theory of the solar system, taking into account complexities such as the perturbations in the orbits of the planets and the satellites caused by their mutual attraction.

On the model of the *Opticks*, meanwhile, Newtonianism has meant an inductive, experimental approach to physics. Users of this meaning of the word cite Newton's promise to "feign no hypotheses." An example of a Newtonian in this sense is Benjamin *Franklin, who presented his electrical science as one founded in experimental tinkering rather than theory. His followers and historians have likened him, on that basis, to the Newton of the *Opticks*, the empirical essayer and querist. The empiricist meaning of Newtonianism has also referred, more specifically, to the use of analysis and synthesis experiments, for which Newton's investigations of white light served as the paradigm. The leading eighteenth-century example of experimental Newtonianism in this sense is Antoine-Laurent *Lavoisier's analysis of water into hydrogen and oxygen and his resynthesis of these elements into water (1785).

The *Opticks* gave rise, finally, to a third meaning of Newtonianism, the eighteenth- and nineteenth-century research program of so-called *imponderables fluids that grew from Newton's hypothesis regarding force-bearing ethers. The hypothesis that an imponderable fluid medium carried each force informed theories of electricity, magnetism, heat, and light well into the nineteenth century. An eighteenth-century example of a fluid theory was Franklin's account of electricity, according to which a weightless electrical fluid, whose particles were mutually repulsive, permeated common matter, balancing the mutual attraction of its particles. Another important eighteenth-century example was the common understanding of latent and specific heats, the one being heat fluid bound to a body so as to affect a thermometer, the other measuring the capacity of a body to hold the fluid.

In addition to the work set forth in the *Principia* and the *Opticks*, another factor shaped the meaning of Newtonianism: the contrast—partly genuine, but also over-drawn by Newton and his followers—between Newton's approach to physics and that of the French mathematician and natural philosopher René *Descartes, to whose example Newton owed the beginnings of many of his ideas. Descartes allowed his rationalism and his commitment to rigorously mechanical explanations of natural phenomena to get the better of his physics. Based upon the principle that there could be no intelligible difference between matter and space, and on the conviction that physical events must have mechanical causes in the form of pushes between bits of matter, Descartes derived a picture of the universe as a great plenum in which all things were constrained to move in vortices. Newton's followers called Cartesian physics dogmatic, misguided, and arrogant in its claims to completeness. They pointed to Newton's abstention, in the *Principia*, from assigning a mechanical cause for gravitational attraction as the epitome of empiricist open-mindedness and humility.

Leaving a gap at the heart of his system of mechanical causation, Newton allowed his disciples to fill in the metaphysics of their choosing. He himself wrote, in the queries to the *Opticks*, that natural phenomena arose not from mechanical causes, but from the will of a divine intelligence. This appeal to a final cause lying beyond the efficient ones pleased Enlightenment eulogists of Newton's mechanical system, who showed a remarkable tendency to cite its breaches. An example is David Hume's satisfaction that although Newton "seemed to draw off the veil from some of the mysteries of nature," he also demonstrated "the imperfections of the *mechanical philosophy," restoring Nature's secrets "to that obscurity in which they ever did and ever will remain" (*The History of England* [1754–1762]). Voltaire, in his *Lettres philosophiques* (1734), popularized for a French audience the contrast between Newton's heroic acceptance, and Descartes's dogmatic refusal, of obscurity.

We now have four meanings of Newtonianism: a mathematical, synthetic approach to natural philosophy (particularly one founded in inverse-square laws of force); an inductive, experimental approach to natural philosophy (particularly one founded in analysis and synthesis experiments); the attribution of forces to weightless, force-bearing ethers or "imponderable fluids;" and the appeal to final causes, manifestations of the will of a divine intelligence, as the ultimate cause of natural phenomena.

The promulgation of Newtonianism coincided with an increasing interest in natural knowledge among the literate public. Some of the first people to teach courses of experimental physics were Newton's propagandists: Francis Hauksbee and John Theophilus Desaguliers, demonstrators at the Royal Society of London, and Willem Jacob 'sGravesande, professor of mathematics at the University of Leyden, who was inspired by a meeting with Newton during a visit to London. These lecturers professed to translate Newton's physics from the language of mathematics into the language of experience, using demonstration experiments to make complicated ideas accessible to polite audiences. Popular written expositions of Newton's physics, including Desaguliers's and 'sGrave-sande's published lectures, emerged during the first third of the eighteenth century. Turning Newton's natural philosophy into a source of philosophical amusement, lecturers and authors established in the minds of their public a particular model of natural knowledge: quantitative and synthetic but also rigorously experimental; materialist and mechanist but also resting upon an underlying assumption that the ultimate causes in nature were final rather than efficient, reasons rather than mechanisms. The same model of knowledge took root in universities, academies, and technical and professional schools during the eighteenth century, beginning with the Royal Society, Cambridge University, and the University of Leiden, and spreading after about 1730 to France, Italy, Russia, and Sweden, where it mixed with continental traditions informed by the work of Descartes, Gottfried Leibniz, and others. Not only mathematicians and philosophers but doctors and engineers studied and taught Newtonian curricula by the end of the eighteenth century.

I. Bernard Cohen, *Franklin and Newton* (1956). Gerd Buchdahl, *The Image of Newton and Locke in the Age of Enlightenment* (1961). Henry Guerlac, *Essays and Papers in the History of Modern Science* (1977). Margaret C. Jacob, "Newtonianism and the Origins of the Enlightenment: A Reassessment," *Eighteenth-Century Studies* 11 (1977): 1–25. I. Bernard Cohen, *The Newtonian Revolution* (1980). Betty Jo Teeter Dobbs and Margaret C. Jacob, *Newton and the Culture of Newtonianism* (1995).

JESSICA RISKIN

NOBLE GASES. The least reactive of the chemical elements proved the most fruitful guide to their interrelations. Some twenty-five years after Dmitrii Mendeleev first worked out his *periodic table, Lord Rayleigh (John William Strutt), briefly James Clerk *Maxwell's successor as Cavendish professor of physics at Cambridge, discovered that the nitrogen he drew from the air had a specific weight greater than that of the nitrogen derived from mineral sources. He asked publicly

for ways to resolve the discrepancy and then hit on the solution himself. He read a paper, then a century old, in which Henry *Cavendish mentioned an unoxydizable residue of gas he obtained after sparking atmospheric nitrogen with oxygen.

While Rayleigh tried to collect enough of this residue to weigh it, William Ramsay, a leading British chemist alerted to the problem by Rayleigh's request, isolated the residue by more effective chemical means. It weighed enough to account for the discrepancy of one part in two hundred that had started Rayleigh's quest. By 1895 they could announce the discovery of a constituent of the atmosphere they named "argon" (from the Greek for "lazy") because it declined chemical intercourse. Ramsay then looked for other trace gases by examining air and argon liquefied by then-new cryogenic techniques (*see* COLD AND CRYONICS). Neon ("novel"), krypton ("hidden"), and xenon ("strange") quickly put in an appearance in the spectroscope and then the balance. So did helium, already named and known as the supposititious source of certain otherwise unattributable lines in the solar spectrum.

The sluggishness and aloofness of the noble gases put them in a class apart. Astonishingly, Mendeleev's chart could accommodate the five newcomers, confirming its importance and renewing its mystery. The only pinch came with argon, whose atomic weight placed it after potassium, but whose nobility placed it before. It took almost twenty years and the invention of the concept of isotope to resolve this problem of precedence. In 1919 Francis Aston at the Cavendish made the first crisp separation of isotopes using neon gas in a *mass spectroscope he had invented. Meanwhile the reversal at argon-potassium (and at cobalt-nickel and iodine-tellurium) helped to alert chemists and physicists that something other than atomic weight regulated the properties of the elements (*see* ATOMIC STRUCTURE).

The discovery of the noble gases was almost a prerequisite to unraveling the complexities of *radioactivity. Ernest *Rutherford and Frederick Soddy identified the "emanation" from thorium as a new and flighty member of the noble family, now called radon; the occurrence of a decaying nonreactive gas in their experiments provided the clue for working out their theory of the transmutation of atoms. Radium also gives off a radioactive emanation and the two similar (indeed chemically identical) noble gases offered an early example of isotopy. However, the lightest of the noble gases proved the weightiest. Helium is often found with uranium and other active ores. With the spectroscopist Thomas Royds and an apparatus made by the virtuoso glass blower Otto Baumbach, Rutherford demonstrated in 1908 that the alpha particles emitted from radioactive substances turned into helium atoms when they lost their electric charge. In 1910–1911 he showed that alpha particles acted as point charges when fired at metal atoms, and devised the nuclear model of the atom to explain the results of the scattering and to deduce that helium atoms have exactly two electrons. The replacement of atomic weight by atomic number (the charge on the nucleus) as the ordering principle of the periodic table followed. Rayleigh, Ramsay, Aston, and Rutherford all received Noble Prizes in large measure owing to their work on noble gases.

Morris T. Travers, *The Discovery of the Rare Gases* (1928; expanded ed., *A Life of Sir William Ramsay* [1956]). Isaac Asimov, *The Noble Gases* (1966). John Robert Strutt, *The Life of John William Strutt, Third Baron Rayleigh*, rev. ed. (1968).

J. L. HEILBRON

NUCLEAR MAGNETIC RESONANCE. The technique of nuclear magnetic resonance (NMR) emerged as a consequence of quantum mechanics. The orbital motions of electrons give atoms a magnetic dipole moment, which according to quantum theory can take only particular orientations in a magnetic field. One method to explore the effect involved sending a beam of atoms or molecules from a gas through a magnetic field and measuring the deflection. The so-called molecular-beam method could also measure the magnetic moment of an atomic nucleus, which resulted from quantum-

mechanical spin. In 1937 I. I. Rabi, an expert in molecular beams, proposed applying an alternating magnetic field, oscillating at radio frequency, on top of the static magnetic field. The particles in the beam would suddenly switch orientation when the frequency of precession of the spin axis matched, or resonated with, the frequency of the applied magnetic field. Rabi and his group at Columbia University used the magnetic resonance technique to produce surprising results for the magnetic moment of the proton and the quadrupole moment of the deuteron by 1939.

World War II interrupted research on magnetic resonance, but also fostered the development of new radio frequency electronics that would aid future research. After the war work with resonance led to atomic clocks, which inverted the approach and used the minute difference between quantum-mechanical energy levels as the basis for a constant frequency, and to the detection, in 1946, by groups under Felix Bloch at Stanford University and Edward Purcell at Harvard, of magnetic resonance in condensed matter. Bloch and Purcell shared the Nobel Prize in physics for 1952 for the work; Rabi had won in 1944 for the molecular-beam resonance technique.

Subsequent research brought NMR from the nuclear physics lab into diverse fields of science. In the early 1950s Bloch's group found that the chemical environment of the nucleus—the close presence of nuclei of other chemical elements or molecules—affected the resonance frequency, as did couplings between the spins of nearby nuclei. The so-called chemical shift and spin-spin coupling suggested a powerful tool for chemistry. Increasing use of NMR in the 1950s helped transform organic chemistry from painstaking test-tube analysis of chemical reactions to a quick mechanical process of structure elucidation. Solid-state physicists similarly took up NMR to reveal the internal structure of their samples.

Yet another use for NMR, perhaps the most widely known, appeared in medicine. In 1972 the chemist Paul Lauterbur proposed to use magnetic field gradients to distinguish NMR signatures from different portions of an object. The resulting scan could identify biochemical properties across a tissue sample, such as water content, blood flow, and cancerous growths. Groups in Aberdeen, Scotland, and Nottingham, England, soon scaled up NMR scanners for human use and commercial firms brought them to the medical marketplace. In the 1980s the technique acquired its current name of magnetic resonance imaging, or MRI, which skirted public fear of things nuclear.

John S. Rigden, *Rabi: Scientist and Citizen* (1987). Timothy Lenoir and Christophe Lécuyer, "Instrument Makers and Discipline Builders: The Case of Nuclear Magnetic Resonance," in *Instituting Science: The Cultural Production of Scientific Disciplines*, ed. Timothy Lenoir (1997): 239–292.

PETER J. WESTWICK

NUCLEAR PHYSICS AND NUCLEAR CHEMISTRY. Research into the atomic nucleus derived from the study of *radioactivity, which flourished starting around 1900 as a program in both chemistry and physics; chemists sorted out the different radioactive elements and their decay products, and physicists elucidated the nature of the emitted rays. The physicist Ernest *Rutherford, who would later deride most science outside physics as stamp-collecting, won a *Nobel Prize in chemistry for his radioactivity research. However, his nuclear model of 1911 (*see* ATOMIC STRUCTURE) definitively divided the physical from the chemical in the study of radioactivity. It established the nucleus as the seat of radioactivity. In 1919 Rutherford achieved the artificial disintegration of the nucleus by bombarding nitrogen atoms with alpha particles from radioactive substances. Hydrogen nuclei (protons) were knocked out of the nucleus by the impact of the alphas.

Most nuclear models through the 1920s built nuclei out of protons and electrons, since their opposite electric charges would bind them together; also, the apparent expulsion of electrons from the nucleus in beta decay argued for their inclusion. But the presence of both protons and electrons defied attempts at detailed descriptions, and the nucleus seemed to reveal deficiencies in quantum mechanics. *Bohr believed, characteristically, that the problem required bold steps and suggested that conservation of energy and momentum failed at the nuclear level. Wolfgang Pauli (*see* HEISENBERG AND PAULI) instead proposed a new nuclear particle, later dubbed the neutrino, to save the energy balance in beta decay.

Another new particle solved the problem and reconciled nuclear models with quantum theory. In 1920 Rutherford suggested that protons and electrons could combine to form a neutral particle, a "neutron," which might reside in the nucleus. The hypothetical neutron attracted little attention, but James Chadwick, a protégé of Rutherford's at the Cavendish Laboratory in Cambridge, looked for it through the 1920s. In 1932 Chadwick found it in radiation, studied by Frédéric Joliot and Irène Joliot-Curie, emitted by beryllium when it was exposed to alpha rays. Physicists soon accepted the neutron as a single elementary particle and new models of the nucleus, notably one proposed by Werner *Heisenberg, gradually accommodated the neutron alongside protons and removed electrons from nuclei. The neutrino and neutron, along with the positron (detected, like the neutrino, in 1932), were followed in the 1930s by other new particles, such as the *cosmic-ray mesons. The floodgates opened to the proliferation of so-called *elementary particles that emerged from the heads and machines of physicists in ensuing decades.

The neutron provided a useful projectile with which to probe the nucleus, since its neutral charge experienced no electrical repulsion. Physicists at the time were developing devices to accelerate charged particles to energies high enough to penetrate the electrical barrier of the nucleus. In a corner of the Cavendish, John D. Cockroft and Ernest T. S. Walton had built a high-voltage apparatus, which they used in 1932 to fire a proton into a lithium nucleus and disintegrate it into two alpha particles. Simultaneously, Ernest Lawrence (see BLACKETT AND LAWRENCE) in Berkeley was developing a circular particle *accelerator as an easier route to high energies. Then in 1934 Joliot and Joliot-

Curie bombarded aluminum with alpha particles from a polonium source and found that they created an isotope of phosphorous, which decayed radioactively. The process provided a way to produce radioactive isotopes in the lab. A group under Enrico *Fermi in Rome quickly extended the results in a systematic bombardment of elements with neutrons; when they got up to uranium, their results suggested that neutron capture produced new transuranic elements.

The artificial production of radioisotopes, whether using rays from natural radioactivity or accelerated particles, required chemistry to disentangle the decay processes and identify short-lived parent species and their daughter isotopes by comparison to stable elements with similar chemical behavior. Nuclear research labs of the 1930s restored the sort of collaboration between physicists and chemists that had marked radioactivity research thirty years earlier. Lawrence, who recruited a group of chemists to his cyclotron program, noted the indistinct boundary between the existing disciplines and wondered whether to call the field nuclear physics or nuclear chemistry. An international community emerged among the major centers in nuclear science: the Cavendish under Rutherford; Lawrence's group in Berkeley; Joliot and Joliot-Curie in Paris; Fermi's group in Rome; Bohr's institute in Copenhagen; a group under Igor Vasilyevich *Kurchatov in Leningrad; and the Riken laboratory in Tokyo under Yoshio Nishina.

A momentous collaboration formed in Otto Hahn's chemistry institute in Berlin, where the physicist Lise *Meitner worked with chemists Hahn and Fritz Strassmann. The Berlin group began studying the decay modes of the postulated transuranics produced by neutron bombardment of uranium; after several years of analysis, Hahn and Strassmann convinced themselves that the suppositious transuranics behaved chemically like elements much further down the periodic table. The chemical evidence implied that uranium could split into two lighter elements in a nuclear version of cell fission, but the conclusion challenged the results of Fermi's group and seemed to contradict the nuclear theory developed by physicists. As chemists, Hahn and Strassman hesitated to take such a bold step. They appealed to the physicist Meitner, in exile from the Nazis, who with her nephew and fellow physicist Otto Robert Frisch showed how to reconcile fission with nuclear physics.

Physicists and chemists alike recognized the importance of nuclear fission after its announcement in January 1939, and both disciplines would play central roles in the military and industrial application of nuclear energy in World War II and afterwards. The discovery required the sort of interdisciplinary collaboration possible in Hahn's institute, which helps explain why physicists alone failed to notice the effect earlier. Fission exemplifies the interplay of chemistry and physics within a common field of nuclear science.

The contributions of nuclear scientists to the war ensured public prestige and government funding for their postwar programs, which attracted new practitioners and allowed the construction of more and larger particle accelerators. The devices encouraged the separation of a new field of *high-energy physics from nuclear physics, but also supported thriving research. Nuclear physicists recognized the existence of several different mesons and incorporated the various types into the theory of nuclear forces. Based on evidence of the stability of certain isotopes, Maria Goeppert Mayer postulated an alternative nuclear structure based on shells of nucleons instead of the liquid drop; theorists later reconciled the competing models in a rotational scheme combining single-particle states with surface oscillations. Nuclear chemists, especially a group under Glenn Seaborg, meanwhile were using cyclotrons, nuclear reactors, and even nuclear bombs to produce new transuranic elements up to and beyond element 100 in the period table. Both nuclear physics and nuclear chemistry earned state support by continuing as prime contributors to the military and industrial development of nuclear energy in the cold war.

Roger H. Stuewer, ed., *Nuclear Physics in Retrospect* (1979). William R. Shea, ed., *Otto Hahn and the Rise of Nuclear Physics* (1983). Abraham Pais, *Inward Bound: Of Matter and Forces in the Physical World* (1986). Finn Aaserud, *Redirecting Science: Niels Bohr, Philanthropy, and the Rise of Nuclear Physics* (1990). Laurie Brown and Helmut Rechenberg, *The Origin of the Concept of Nuclear Forces* (1996).

PETER J. WESTWICK

O

OCEANOGRAPHY, as a distinct scientific discipline, and "oceanography," as the standard term for the study of all of the marine sciences, both date from the late nineteenth century, when the first major expeditions were undertaken specifically to explore the physics, chemistry, biology, and geology of the world's oceans. The pre-history of oceanography begins centuries earlier.

The first text devoted exclusively to marine science was the *Histoire physique de la mer* (1725) of Count Luigi Ferdinando Marsigli, a military man and founder of the Academy of Sciences in Bologna. From studies of the Gulf of Lyons, Marsigli assembled information about water temperature, salinity, specific gravity, tides, waves, currents, depth contours, and marine plants and animals. The diligence required in these efforts convinced Marsigli of the limitations of individual research in marine science; larger scale results required teams of investigators and government support—the hallmarks of oceanography since the nineteenth century.

The second half of the eighteenth century witnessed an acceleration of marine research. Enthusiasm for, and advances in, chemistry underpinned the chemical analysis of sea water by leading chemists, including Antoine-Laurent *Lavoisier and Torbern Bergman. Salinity was also studied during voyages of exploration, such as the Danish expedition to Arabia Felix, and, especially, James Cook's expeditions in the Pacific. Meanwhile, ocean currents and circulation patterns engaged the curiosity not just of sailors but of scientists such as Benjamin *Franklin, chronicler of the Gulf Stream, and Benjamin Thompson, Count Rumford, whose heat experiments led him to attribute ocean circulation to differences in water density, a theory finally accepted after a long delay.

Studies of currents and tides intensified in the early nineteenth century, exemplified by British naval surveyor James Rennell's *An Investigation of the Currents of the Atlantic Ocean* (1832). Also in the 1830s, John Lubbock and William Whewell, Lubbock's mentor at Trinity College, Cambridge, reduced tidal phenomena to mathematical analysis. And the British Admiralty helped to install tidal gauges around the southern coast of England. The limitations of the data for establishing general tidal patterns, and for drawing what Whewell dubbed "cotidal lines," led him to develop an international scheme for the collection of information about tides. The nascent British Association for the Advancement of Science soon took over the program. At the U.S. Naval Observatory, Matthew Fontaine Maury assembled data on winds, currents, and other oceanographic phenomena from ships' captains, and published the results in his textbook, *The Physical Geography of the Sea and Its Meteorology* (8 editions, 1855–1861).

The British Association also played a key role in marine biological enterprises during the middle decades of the nineteenth century. The chief investigator was Edward Forbes, a marine naturalist and paleontologist for the Geological Survey and a native of the Isle of Man. Excursions of small boats to ascertain the depth and distribution of bottom-dwelling species led to summer-long cruises in British waters, continued aboard Admiralty ships in the late 1860s by naturalists Charles Wyville Thomson and William B. Carpenter. The success of these efforts, and the increasing curiosity about life and conditions in the deep oceans, gave rise to the most ambitious oceanographic endeavor of the era, the expedition of HMS *Challenger* (1872–1876). Wyville Thomson headed a team of five scientists and an artist during this three-and-a-half-year circumnavigation of the globe. They oversaw dredging and trawling at more than 360 stations. International authorities in the various subspecialties analyzed the data and the specimens collected. The resulting *Challenger Reports* (1885–1895), published in fifty volumes, remain the founding benchmark of oceanographic science.

Although the supremacy of its navy allowed England to take a leading role initially, other nations used the precedent to launch important oceanographic enterprises during the late nineteenth and early twentieth centuries. Germany focused initially on the North Sea, but its S. S. *Gazelle* also operated in the Atlantic at the same time as the *Challenger*, and its S. S. *National* (1889) carried out a global Plankton Expedition. Alexander Agassiz headed American cruises in the Atlantic aboard the *Blake*, and in the Pacific aboard the *Albatross*. France, Denmark, Italy, and Russia had also launched projects by the turn of the twentieth century. This period saw the creation of the first marine biological laboratories, the prototype being the Stazione Zoologica created at Naples in 1873 by Anton Dohrn.

The two world wars provided an unprecedented stimulus to physical oceanography at the expense of marine biology. The deployment of submarines and their detection by sonar brought new urgency to studies of bathymetry and the relation of temperature, salinity, and bottom sediments to acoustic transmission. And an intimate knowledge of waves, currents, and surf conditions would become crucial later on to the success of amphibious landings. By World War II, a productive, if sometimes stormy, collaboration had emerged between civilian oceanographers and naval officers. In the United States, the principal centers of this collaboration were the Woods Hole Oceanographic Institution in Massachusetts, headed by Columbus O'-Donnell Iselin, and the Scripps Oceanographic Institution in California, under the direction of Harald U. Sverdrup. In the midst of the war, Sverdrup, with co-authors Martin W. Johnson and Richard H. Fleming, published the first modern textbook of oceanography, *The Oceans: Their Physics, Chemistry, and General Biology* (1942).

After 1945, the Cold War relentlessly drove the growth of oceanography. As early as 1950, the U.S. Navy defined Soviet submarines patterned after advanced German designs seized at the war's end as the greatest maritime threat to the security of the United States. Understanding the ocean environment became critical to antisubmarine warfare. The Office of Naval Research and the Bureau of Ships quickly and regularly made resources available to address the threat, funding work by hundreds of scientists and institutions around the country. The results quickly outstripped administrative efforts to bring disciplinary coherence and recognition to oceanography, slowing the creation of degree programs and formal technical education at major universities. However, this massive investment made at a dizzying pace resulted in amazingly comprehensive ocean surveys, the understanding and exploitation of the deep sound channel, fundamental advances in sonar, the very rapid development of ocean acoustics as a field of study, the creation of the Navy's ocean surveillance system (SOSUS), the quieting of nuclear and conventional submarines, and the possibility of submerged missile launching. In addition, these circumstances effectively launched the careers of Roger Revelle, Walter Munk, J. Lamar Worzel, Dale Leipper, Waldo Lyon, Henry Stommel, Alan Robinson, and many others. The new sophistication of oceanography made it possible for American nuclear attack submarines to detect and shadow their Soviet missile-carrying counterparts. The deep ocean quickly became the front line in the Cold War.

The Russians followed suit, responding to American determination for the same reasons. While definitely competitive in terms of theoretical understanding and scientific capability, the material resources of the former Soviet Union did not permit it to keep pace. However, its scientists made very significant contributions to understanding antisubmarine acoustics, the study of the Arctic region, ocean surveying, and the construction of advanced oceanographic vessels.

Twentieth-century oceanography achieved a new, mathematically rigorous understanding of the coupling of atmospheric and oceanic phenomena, and of the climatic implications of oceanographic events such as El Niño. But after World War II, the discipline turned increasingly toward questions of marine geology and geophysics. The leading catalyst here was the continental-drift hypothesis proposed by Alfred Wegener in his book *The Origin of Continents and Oceans* (1915, first English translation

1924). Rejecting notions of continental stability and *isostasy, Wegener proposed that the present configuration of the continents, and other phenomena from *stratigraphy, paleontology, and biogeography, could be accounted for by assuming the gradual movement of the continents horizontally over the face of the globe. The theory gained few adherents until the 1960s, by which time new lines of evidence helped bring about the *plate-tectonics revolution. Evidence came from studies of paleomagnetism and polar wandering carried out by P. M. S. *Blackett, S. Keith Runcorn, and their colleagues in Britain; heat flow from mid-ocean ridges, by British geophysicist Edward Bullard; seismological activity along mid-ocean ridges, by Americans Maurice Ewing and Bruce Heezen; and gravity anomalies, by the Dutch geophysicist Felix Andries Vening-Meinesz and the American Harry H. Hess. In 1960, Hess proposed the hypothesis, subsequently known as sea-floor spreading, that would explain all of these phenomena. In the mid-1960s, the British geophysicists Frederick Vine and Drummond Matthews confirmed the hypothesis by analyzing patterns of magnetic anomalies around mid-ocean ridges, and the *Glomar Challenger* drilled directly into the Mid-Atlantic Ridge. J. Tuzo Wilson's 1965 concept that the earth's surface consists of several rigid but mobile plates put the finishing touch on plate tectonics.

William A. Herdman, *Founders of Oceanography and Their Work* (1923). Margaret Deacon, *Scientists and the Sea 1650–1900: A Study of Marine Science* (1971; 2d ed. 1997). A. Hallam, *A Revolution in the Earth Sciences: From Continental Drift to Plate Tectonics* (1973). Susan Schlee, *The Edge of an Unfamiliar World: A History of Oceanography* (1973). Eric L. Mills, *Biological Oceanography: An Early History, 1870–1960* (1989). Walter Lenz and Margaret Deacon, eds., *Ocean Sciences: Their History and Relation to Man*, Proceedings of the Fourth International Congress on the History of Oceanography, Hamburg, September 1987 (1990). Philip F. Rehbock, *At Sea with the Scientifics: the Challenger Letters of Joseph Matkin* (1993). Gary E. Weir, *An Ocean in Common: American Naval Officers, Scientists, and the Ocean Environment* (2001).

PHILIP F. REHBOCK AND GARY WEIR

OPPENHEIMER, J. Robert. See KURCHATOV, IGOR VASILYEVICH, AND J. ROBERT OPPENHEIMER.

OPTICS AND VISION. During the early seventeenth century Johannes *Kepler, echoing the earlier views of Leonardo da Vinci, likened the eye to a camera obscura, a black box containing a pinhole opening through which an image of external objects is projected on the back wall. Kepler worried that the analogy implied that the eye must invert the images it observes, but his work on light, optics, and the camera obscura forced him to accept the inference. He concluded that light forms images on the back wall of the eye, now called the retina, and not on the crystalline humor, now called the lens, as most scholars of the period thought.

Kepler argued that light as a passive entity followed the laws of geometry. His *Astronomiae pars optica* (1604) and *Dioptrice* (1611) dealt with the refraction of light, whose "law"—a relation between the angle of incidence i and the angle of refraction r—was given in 1621 by Willebrord Snel, professor of mathematics at the University of Leyden, who busied himself with astronomical and triangulation studies. Snel did not print his result; René *Descartes, who may have seen it in manuscript, published it in 1637 as $\sin r = (\sin i)/n$, where n, the index of refraction, is a constant. Christiaan *Huygens elaborated upon the law of refraction in his *Traité de la lumière* of 1690 by assuming that light was undulatory, rather than corpuscular, in nature. Like Kepler, Huygens believed that refraction arises because light moves more slowly in a denser than in a rarer medium.

Descartes had accounted for vision by analogy. According to him, the colors we perceive, like the impulses transmitted by a blind man's stick, result from pressure. He took color to be a secondary quality, not a property of external objects but the mind's interpretation of the pressure registered on the optic nerve. Nicolas Malebranche deepened Descartes's theory of vision in his *De la recherche de la verité* (1688). Sight cannot ascertain the truth of things in themselves; its purpose is to facilitate our navigation through the world.

Further to the impugning of vision, George Berkeley (*Essay towards a New Theory of Vision*, 1709) undermined the idea that the concepts the mind creates by working on sense impressions reliably indicate the nature of things.

Isaac *Newton's grand discovery, that sunlight is made up of rays of different refrangibilities, transformed the study of light and vision. Newton showed in 1672 and at length in his *Opticks* of 1704 that sunlight passed through a glass prism yielded a spectrum of rays of different colors refracted at characteristic angles; that the rays of a given color all had the same refrangibility (and so could not be further divided by a prism); and that the differently colored rays, if reunited by a second prism, again produced white light. Newton supposed that the different rays were made up of particles; but to explain why some rays striking glass are refracted and others reflected, as well as to model interference phenomena like the colors of the plates and "Newton's rings," he also supposed that the particles interacted with a pervasive subtle matter or *ether. The emission and motion of the particles set the ether in vibration, and the particles penetrated or rebounded from a surface in accordance with the phase or "fit" of vibration of the ether there.

The different refrangibilities of rays of different colors make single lenses cast colored images. Newton thought, mistakenly, that this evil could not be cured, and he turned his attention to the construction of reflecting telescopes. In the 1750s the instrument maker John Dollond proved Newton wrong by manufacturing achromatic doublets made of crown and flint glass, which compensated one another's dispersion. Drawing upon Dollond's idea of an achromatic doublet, Joseph Fraunhofer, a Bavarian skilled artisan and optician working in the secularized Benedictine monastery of Benediktbeuern, improved upon a complex glass-stirring technique first developed by Pierre Louis Guinand, a Swiss bell pourer. With this method, Fraunhofer was able to manufacture flint glass of unprecedented homogeneity in the 1810s and 1820s. In addition to this crucial technological development, Fraunhofer reckoned

that he could use the dark lines of the solar spectrum (later called the Fraunhofer lines), which his superior glass prisms produced, to demarcate precise portions of the spectrum. By altering the ingredients of his glass samples, he adjusted the refractive and dispersive properties in order to produce a second lens that would correct for the chromatic aberration produced by the first. After Fraunhofer's death in 1826, the British firm Chance Brothers of Birmingham and the French company Feil of Paris took over the world's optical glass market from Bavaria. But during the 1880s, the physicist Ernst Abbe and chemist and glassmaker Otto Friedrich Schott manufactured apochromatic lenses for the Carl Zeiss Company in Jena, Germany, which would monopolize optical glass and equipment production until World War II.

During the eighteenth century prominent natural philosophers like Leonhard *Euler and Benjamin *Franklin challenged Newton's particulate model of light, but their opposition, based on qualitative considerations such as the improbable loss of matter from the sun that the model implied, did not make an effective alternative. The corpuscular theory reached its pinnacle after 1800 in the school of Pierre-Simon de *Laplace, which developed a quantitative theory of optical phenomena based on distance forces between light particles and the particles of ponderable matter. They managed to incorporate the polarization of light, which Huygens had known in the case of birefringent crystals and Étienne Louis Malus had discovered in 1810 in light reflected from glass, into their scheme by supposing that light particles had different properties on different "sides."

The remarkable success and high patronage of the corpuscular theory did not preserve it from an attack launched around 1800 by the English physician Thomas *Young, who drew upon an analogy between sound, which was understood to be a wave motion in air, and light. Young perceived a further, and more useful, analogy between diffraction patterns and the interference of water waves. In 1807 he presented the persuasive demonstration (now famous as the Young double-slit experiment) in which light from a common source passes through two parallel narrow slits in an opaque screen. The superposition of the two transmitted beams on a surface beyond and parallel to the screen produces dark and light stripes rather than clear images of the slits.

Young's initiative was continued by a better mathematician than he, Augustin Fresnel, whose memoirs, composed between 1815 and 1827, challenged his compatriots who championed the corpuscular theory (see CARNOT AND FRESNEL). One of Fresnel's techniques was to treat each point of the wave front as a source of secondary waves, a technique introduced by Huygens. By adding the contributions of those secondary waves in accordance with Young's principle of interference, Fresnel could determine the intensity of light in a diffraction pattern. The Laplacian Simeon-Denis Poisson deduced from Fresnel's equations the paradoxical result that a bright spot should appear at the center of the geometrical shadow of a disk illuminated by a beam of light. Experiment decided for Fresnel.

To incorporate polarization into the theory, both Young and Fresnel independently suggested that the vibrations constituting light are perpendicular to the direction of travel, not, as in the model proposed by Huygens and, originally, Young, in the direction of travel, as is the case of sound in air. With this addition, the wave theory bested the corpuscular theory at its strongest point. All efforts to find a mechanical model for the ether failed, and the assumption that one might exist came into increasing conflict with *electromagnetic theory as developed by James Clark *Maxwell and demonstrated by Heinrich Hertz (see HELMHOLTZ AND HERTZ). The nature of light and the fate of its medium were resolved—for the twentieth century at least—by the theory of *relativity and *quantum physics.

Young contributed to the study of vision as well as the theory of light by criticizing the custom of separating consideration of the physical and mental states involved in the perception of color. He argued that the eye, being far less complex than the mind, simplifies the information provided by a particular scene and channels it to the brain, which in turn paints the scene. He demonstrated that the most sensitive points of the retina, which are connected directly to the brain, can detect only the three primary colors—blue, red, and yellow. The optic nerve is composed of filaments, portions of which correspond to a primary color. The brain mixes the sensations to create all the possible colors. Young's theory was ignored until Maxwell and Hermann von *Helmholtz elaborated upon his work in the mid-nineteenth century. Individuals suffering from color blindness (a disorder first recognized during Young's lifetime, see DALTON, JOHN) have an abnormally low number of retinal cones, which detect color.

With the help of the opthalmoscope he invented in 1851, Helmholtz demonstrated that the optic nerve itself is insensitive to light. Sensory nerves apparently merely transmitted stimuli between the end organs and the sensorium. Between 1852 and 1855, Helmholtz studied color mixing and vision. He based his further work on Young's theory of three distinct modes of sensation of the retina, which he had previously rejected. He adopted the interpretation in Maxwell's paper, *Experiments on Colour, as Perceived by the Eye* (1855), which stated as Young's theory the proposition that, although monochromatic light stimulates all three modes of sensation, only one or two color responses will prevail in the resulting mixed color. This reading made one of the foundations for Helmholtz's *Handbuch der Physiologischen Optik* of 1860. In this classic work, Helmholtz developed Young's notion of three distinct sets of nerve fibers, although he observed that three distinct and independent processes might take place in each retinal fiber.

Binocular vision presented another set of physical-psychological problems. Charles Wheatstone's stereoscope, invented in 1838, played a critical role in the debate over how perception transforms two flat, monocular fields into a single, binocular field illustrating objects in relief. One group, led by David Brewster, argued for a theory of projection according to which the mind imagines lines starting at two stimulated points on the retinas and projecting outward to their point of intersection at the observed object. Another theory, proffered by the German physiologist Johannes Peter Müller, argued that retinas possess pairs of corresponding or identical points, each pair providing only one point in the unified field of vision. Wheatstone's stereoscope seemed to disprove Müller's views.

Physiologist Ewald Hering challenged Helmholtz's work on vision, declaring that four primary colors—red, blue, green, and yellow—formed the psychological basis for all color sensation. Hues arrange themselves in opposing pairs: red/green and blue/yellow. Helmholtz characterized the controversy as a clash between opposing epistemologies, nativism (Hering's position) and empiricism (his own view). At a deeper level, Hering opposed reckoning the processes of perception in analogy to functions of the human mind. He viewed vision as immediate impressions on the eyes and so rejected theories of projection in favor of a theory of identity similar to Müller's. Against Wheatstone and Brewster, Hering demanded that corresponding pairs of retinal points determine visual directions and depth perception. Helmholtz argued that the processes governing our spatial perception are psychological in nature and thus conditioned by learning and experience.

The controversy lasted well into the 1920s. Modern research affirms both the Young-Helmholtz three-color theory and Hering's four-color theory. The two theories complement one another. While the trichromatic approach explains how the eye detects and perceives color, Hering's explains how color information is encoded and sent to the brain via the nerve pathways. Research during the 1960s and 1970s confirmed that the eye contains three types of color sensors, the photoreceptors, composed of red, green, and blue cones, so-called because of their absorption of light at those wavelengths.

Lael Wertenbaker, *The Eye: Window to the World* (1981). G. N. Cantor, *Optics after Newton: Theories of Light in Britain and Ireland, 1704–1840* (1983). Jed Z. Buchwald, *The Rise of the Wave Theory of Light: Optical Theory and Experiment in the Early Nineteenth Century* (1989). Margaret Atherton, *Berkeley's Revolution in Vision* (1990). Gary Hatfield, *The Natural and the Norma-*

tive Theories of Spatial Perception from Kant to Helmholtz (1990). R. Steven Turner, *In the Eye's Mind: Vision and the Helmholtz-Hering Controversy* (1994). Myles W. Jackson, *Spectrum of Belief: Joseph von Fraunhofer and the Craft of Precision Optics* (2000).

MYLES W. JACKSON

ORBIT. An orbit is the path of a celestial object around another body, as in planets, comets, and asteroids moving around the sun, and satellites circling their planets. The key historical advances in the study of orbits came when Johannes *Kepler deduced that the orbit of a planet was an ellipse, and Isaac *Newton realized that a planet was bound to its elliptical path by a gravitational *force that was proportional to the product of the mass of the planet and the mass of the Sun, and inversely proportional to the square of the distance between them.

Prior to Kepler, planetary paths were thought to be a combination of circles and the importance of an "orbit," or single uncompounded curve, had not entered astronomical thought. Kepler used the observations of Mars made by Tycho *Brahe. Fortunately Mars has an orbital eccentricity of 0.093, nearly six times that of the earth. In *Astronomia nova* (1609), Kepler stated his first two laws of planetary motion: (1) planets move in elliptical orbits, the sun being at one of the foci; (2) the line joining the planet to the sun sweeps out equal areas in equal times. Kepler's third law, buried in his *Harmonice mundi* (1619), made the square of the orbital period proportional to the cube of the semi-major axis. The accuracy of the planetary ephemerides published in Kepler's *Rudolphine Tables* (1627) underlined the importance of the three laws, even though their physical foundations were unknown.

Newton's *Principia mathematica* (1687) established the physics of orbits. All orbiting bodies submit to gravity, even comets. Newton described how to calculate their orbits from three accurate observations separated in time by a few weeks. He used this method to calculate the orbit of the great comet of 1680 (*see* COMETS AND METEORS). Edmond *Halley (1656–1742) fol-

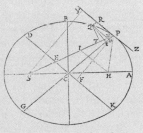

[50]

S E C T. III.

De motu Corporum in Conicis Sectionibus excentricis.

Prop. XI. Prob. VI.

*Revolvatur corpus in Ellipsi: Requiritur lex vis centripetæ tenden-
tis ad umbilicum Ellipseos.*

Esto Ellipseos superioris umbilicus S. Agatur S P secans Ellipseos tum diametrum D K in E, tum ordinatim applicatam Q v
in x, & compleatur parallelogrammum Q x P R. Patet E P æ-
qualem esse semi-
axi majori A C, eo
quod acta ab altero
Ellipseos umbilico
H linea H I ipsi E C
parallela, (ob æ-
quales C S, C H)
æquentur ES,EI,a-
deo ut EP semisum-
ma sit ipsarum P S,
P I, id est (ob pa-
rallelas H I, P R &
angulos æquales I P
R, HP Z) ipso-
rum P S, P H, quæ
conjunctim axem totum 2 A C adæquant. Ad S P demittatur
perpendicularis Q T, & Ellipseos latere recto principali (seu
$\frac{2 B C \ quad.}{A C}$) dicto L, erit L x Q R ad L x P v ut Q R ad P v;
id est ut P E (seu A C) ad P C: & L x P v ad G v P ut L ad G v;
& &

Part of Isaac *Newton's demonstration that a
body can describe an ellipse under an inverse-
square centripetal force.

lowed by calculating the orbits of a further
twenty-three comets. Both Newton and
Halley assumed that the cometary orbits
were parabolic. But when Halley found
that the comets of 1531, 1607, and 1682
had very similar orbits, he inferred that he
was dealing here with a single comet on an
elliptical, periodic orbit. Halley understood
that the gravitational influence of Jupiter
and Saturn perturbed comets and slightly
changed their periods.

In 1798, Henry *Cavendish (1731–1810)
used a torsion balance to provide the first
measurement of the density of the earth.
Knowing this quantity, an astronomer

could calculate the mass of a large body,
provided a much smaller one orbited
around it. Previously, only ratios could be
deduced—for example, that of the Sun's
mass to the earth's from the orbital periods
and semi-major axes of the moon and the
earth. The production of accurate orbital
parameters from diverse and imprecise observations, greatly improved by the work of
Carl Friedrich *Gauss, made possible
ephemerides of the first asteroids. The
greatest success of orbital perturbation
analysis came in the 1840s when John
Couch Adams and Urbain Jean Joseph Le
Verrier independently used observations of
the orbit of Uranus to predict the position
of a planet beyond the confines of the
known solar system. This work led to the
discovery of Neptune in 1846. The detailed
analysis of the orbits of binary stars about
their common centers of mass was carried
out by Sir William *Herschel between 1782
and 1804. Many other analyses followed,
but not until the measurement of stellar
parallax—after 1840—would these orbital
results be converted into individual star
masses.

William Herschel discovered the movement of the solar system as a whole in the
direction of the constellation Hercules in
1783. In fact, as uncovered through Harlow
Shapley's work of 1917 on the distances to
globular clusters, the solar system orbits
around the center of the *galaxy. Once
again, Kepler's third law could be used.
This time, it provided the mass of our
galaxy interior to the sun's orbit.

Forest Ray Moulton, *An Introduction to Ce-
lestial mechanics*, 2nd rev. ed. (1914). Fred
Hoyle, *Astronomy* (1962). Eric M. Rogers,
*Astronomy for the Inquiring Mind: The
Growth and Use of Theory in Science* (1982).
Bruce Stephenson, *Kepler's Physical As-
tronomy* (1987).

DAVID W. HUGHES

P

PALEONTOLOGY, the study of the remains of living beings and their traces in the rocks, emerged as a distinct area of investigation in the late eighteenth and early nineteenth centuries. By then, most mineralogists and naturalists concurred that fossils were the remains of living beings. The German naturalist Johann Friedrich Blumenbach went further, and argued that in the past many animals and plants had become extinct while new and different ones had been created in their place. Two of his followers, Ernst von Schlotheim and Georges Cuvier, realized this meant that extinct species and genera could be used to identify and correlate stratified rocks. In an important study of the geology of the region around Paris published in 1812, Cuvier, used fossil quadrupeds for the purpose. Schlotheim embarked on the massive task of classifying the less glamorous but much more widely distributed invertebrate fossils as a preliminary to using them in *stratigraphy. Not long after, in 1816, the English surveyor William Smith published his *Strata Identified by Organized Fossils*. By the 1830s, British geologists, ignoring the Continental tradition, claimed that Smith should be credited with the discovery that fossils could be used to identify strata.

Whatever the merits of this dispute, geologists agreed that fossils offered the best way to determine when a stratum was formed and to correlate strata over large distances. They busied themselves collecting, identifying, cataloguing, and producing monographs on the fossils most useful in stratigraphy. Overwhelmingly these were marine invertebrate fossils, usually shells of one kind or another, but also corals, trilobites, ammonites, belemnites, and so on. Geologists also studied fossil plants, especially abundant in the economically important coal-bearing formations. With the possible exception of fishes, vertebrate fossils were too rarely preserved to be of use.

In the 1870s, paleontology matured as a scientific discipline with its own specialist journals, meetings, and societies. Although university departments employed paleontologists, the new specialty derived the bulk of its institutional support by being an ancillary to stratigraphy, which was of great economic importance first in the mining industry and later in the petroleum industry as well. Jobs were offered in national geological surveys and later in laboratories run by oil companies. Because of the small diameters of the rock cores brought up in drilling for oil, paleontologists turned in earnest to the study of fossils visible only under the microscope, such as ostracods, foraminifera, and pollen grains.

For stratigraphic and economic purposes, questions about the meaning of fossils—how life had been created and evolved, how dinosaurs had lived, how man had evolved—were irrelevant. Most important debates within professional paleontology dealt with matters that bore directly on stratigraphy. Was a single fossil species or genera adequate to identify a formation, or did the paleontologist have to take whole assemblages of fossils into account? Did whole fossil faunas and floras change abruptly and simultaneously, as Cuvier claimed? Or did they change slowly and gradually, as Charles *Lyell contested? Were fossils reliable as indicators of the relative age of rocks? From the early nineteenth century on, geologists had recognized that different extinct species, just like different living species, lived in different environments. If so, was it possible that fossils varied with the environment in which they had been deposited, and not with time? By the late nineteenth century, geologists used the concept of facies to describe these environments, and began to factor environmental factors into their stratigraphic investigations.

In two areas, paleontologists concerned themselves with issues outside stratigraphy. One was paleo-ecology. In the late nineteenth century, German and Russian scientists began using fossils to reconstruct past climates, past environments, and past shorelines. They studied the environments of living marine invertebrates to shed light on the environment in which trilobites might have lived. They used the distribution of fossil coral reefs to infer what earlier climates were like.

The other area was the Cambrian boundary, below which paleontologists had failed to find any traces of life. Above it, fossils appeared abruptly as rather specialized invertebrate animals. In the mid-twentieth century, paleontologists detected microscopic unicellular animals and sedimentary structures built up of colonies of algae and bacteria (called stromatolites). The reason for the puzzlingly abrupt change, they concluded, was that until the beginning of the Cambrian almost all life on earth consisted of simple unicellular bacteria, algae, and protozoans.

Following World War II, as funding within universities and new and powerful instruments became more readily available, paleontologists began asking a wider range of questions. Paleo-ecologists used the ratios of stable isotopes of oxygen in fossils to trace temperature fluctuations in past oceans. Other paleontologists used the trace amounts of nucleic compounds preserved in fossils to shed light on evolutionary affinities and relationships. More traditional paleontology used the scanning electron microscope to subdivide strata from the Mesozoic to the present in quite remarkable detail.

Wilfred Norman Edwards, *The Early History of Palaeontology* (1967). Martin Rudwick, *The Meaning of Fossils* (1972). Eric Buffetaut, *Short History of Vertebrate Palaeontology* (1987). Richard Fortey, *Fossils: The Key to the Past* (1991).

RACHEL LAUDAN

PARALLAX. Parallax is the difference in apparent direction of an object seen from two different places. Stellar parallax, a measure of the distance from the earth to a star, is the angle subtended by the radius of the earth's orbit at the star. The absence of a measurable parallax embarrassed Copernicans, who insisted that the earth circled the sun. They had to argue that the great distance of the stars made the parallax angle too small to observe. *Galileo proposed a variation on the direct parallax measurement, substituting the easier-to-measure relative positions of stars. Making the assumption that all stars are equal in luminosity, and thus that a faint star was more distant than a bright star, he hoped to detect an alteration in their relative positions over a period of time.

Robert *Hooke, curator of experiments for the Royal Society of London, noted in 1674 that "whether the earth move or stand still hath been a problem, that since Copernicus revived it, hath much exercised the wits of our best modern astronomers and philosophers," and he well appreciated the fame that would accrue to the first person to prove that the earth moved. His reports to the Society intimated success, but attracted little attention, perhaps because no one believed them. In the 1720s, the English astronomer James Bradley, attempting to verify Hooke's claim, discovered stellar *aberration, the apparent displacement of a star in the direction in which the earth is moving.

Improved instrumentation (*see* TELESCOPE) ultimately brought success. Friedrich *Bessel at the Königsberg Observatory and Friedrich Struve at the Dorpat Observatory (now Tartu in Estonia) and then at the new Pulkovo Observatory near St. Petersburg, working on different stars, observed parallax in 1838 and 1840, respectively. Meanwhile, the Scottish astronomer Thomas Henderson, luckily having chosen a much nearer star to study, detected a parallax with less accurate measurements than Bessel's and Struve's. Henderson was the first to begin his measurements but the last to report them; Struve published first, but Bessel most convincingly.

In addition to the stellar, or trigonometric, parallax, there is a spectroscopic parallax or distance. An empirical correlation between spectral characteristics and absolute magnitudes of stars (the Hertzsprung-Russell diagram), once established, can subsequently be used to estimate the distances to other stars. Statistical parallax is an estimated distance to a group of stars. Assuming that the group members have random movements, the average of the measured radial velocities (line-of-sight components, corrected for the observer's movement) equals any other component, including that perpendicular to the line of sight. This average perpendicular velocity, combined mathematically with its corresponding average observed angular change (proper motion), yields an average distance.

Norriss S. Hetherington, *Science and Objectivity: Episodes in the History of Astronomy* (1988). Norriss S. Hetherington, *Encyclopedia of Cosmology: Historical, Philosophical, and Scientific Foundations of Modern Cosmology* (1993).

NORRISS S. HETHERINGTON

PERIODIC TABLE. Chemists of the early nineteenth century had to rely on indirect methods, analogies, and simplifying assumptions to determine the relative atomic weights of elements, and thus did not reach a consensus on them. Gradually, however, methods improved, and at the international Karlsruhe Congress of 1860 Stanislao Cannizzaro advocated what is, in essence, the system still in use today (*see* ATOM AND MOLECULE).

Even before agreement on a single set of atomic weights, several theorists noticed that chemical and physical properties reappeared in a regular, periodic fashion in the various elements. Among these predecessors of the periodic law were Johann Wolfgang Döbereiner, John Newlands, William Odling, and Jean-Baptiste-André Dumas. Many others proposed partial periodic classifications of the elements.

Dmitrii *Mendeleev reaped the success of these endeavors. (The German chemist Lothar Meyer pursued a closely parallel path independently, and published his similar contribution shortly after Mendeleev's.) Mendeleev arranged the elements horizontally according to increasing atomic weight, and started a new row below the first whenever similar properties in the elements reappeared. The resulting semirectangular table of atomic weights showed many intriguing regularities. The horizontal rows ("periods" or "series") and the vertical columns ("groups" or "families") revealed a "periodic law." For example, Mendeleev's arrangement placed the alkali metals (sodium, potassium, and the then

recently discovered rubidium and cesium) in a single vertical group with a marked family resemblance. The set of next-heavier elements to each of these alkali metals formed a second family group, the alkaline earth metals (magnesium, calcium, strontium, and barium).

This example understates the magnitude of the problems Mendeleev faced, for the elements simply failed to order themselves neatly. (Sometimes the greatest genius requires *ignoring* a certain number of anomalies, while discerning the larger pattern hidden within the data.) Lithium, for instance, fell into the alkali metal group by weight order, but it seemed to have more family resemblance to magnesium than to sodium, the element directly below it in the table. By the same token, beryllium seemed to resemble aluminum more than its putative family member magnesium. In certain other cases two adjacent elements seemed to come in a different order by weight from that dictated by the family groupings. For the first anomaly, Mendeleev provided an extenuating rationale; for the second, he did not hesitate to violate the weight order and reverse the two elements in the chart, hence preserving the periodicity of properties.

Despite the problems, Mendeleev had sufficient confidence in the periodic system he published in 1869 to hazard predictions based on it. Leaving vacant places in his chart in certain critical instances, he predicted the properties of undiscovered elements. For example, he left a space between calcium and titanium, and two between zinc and arsenic. Here Mendeleev used a strategy that converted yet another anomaly—holes in his periodic table—into potentially powerful evidence for the validity of his discovery. But of course there he risked the danger that his system would fail the test of prediction.

The first of Mendeleev's predicted new elements, discovered by Paul Lecoq de Boisbaudran in 1875, was named "gallium" in honor of France. Gallium's atomic weight of about 70 came close to Mendeleev's prediction of 68, its density of 5.9 grams per cubic centimeter virtually coincided with the prediction, its valence and oxide pattern were as expected, and a long list of observed chemical properties also matched what Mendeleev had forecast. Four years later scandium filled another space, and in another seven years, germanium. Three times Mendeleev had triumphed, not only with the fact of the discoveries, but with the details of physical and especially chemical properties as well.

These confirmed predictions caught the attention of the scientific world. For the first few years after Mendeleev announced his periodic system, it received almost no notice in journals and textbooks, and much less agreement as to its utility. After the discovery of gallium, however, textbook accounts of the periodic law began to appear, and by the 1880s they had become ubiquitous.

Further developments provided both challenges to, and support for, Mendeleev's system. Chemists and then physicists confirmed the weight inversions that Mendeleev had insisted upon to preserve periodicity (see ATOMIC STRUCTURE). A dozen or more chemically similar "rare earth elements" discovered in the last quarter of the nineteenth century presented a more worrisome problem in that they did not fit into any periodic system. Eventually chemists grouped them together in an aperiodic category as "lanthanides." In the 1890s William Ramsay and Lord Rayleigh discovered the inert gases argon, helium, neon, krypton, and xenon. They fit almost perfectly by weight before the alkalis, and soon after the turn of the century chemists decided to create an extra group for them at one end or the other of the periodic chart (see NOBLE GASES).

The development of atomic physics in the early twentieth century provided an independent method for assigning positions for elements in the periodic table: the measurement of their "atomic numbers." At the same time the study of *radioactivity revealed a number of apparently elementary bodies that did not fit into the table: the solution was to enlarge the concept of element. A group analogous to the lanthanides, the actinides, has been added to accommodate transuranic elements.

A. J. Ihde, *The Development of Modern Chemistry* (1964). J. W. Van Spronsen, *The Periodic System of the Chemical Elements* (1969). E. G. Mazurs, *Graphic Representations of the Periodic System during One Hundred Years*, 2d ed. (1974). W. H. Brock, *The Norton [Fontana] History of Chemistry* (1992). Stephen G. Brush, "The Reception of Mendeleev's Periodic Law," *Isis* 87 (1996): 595–628.

A. J. ROCKE

PHOTOGRAPHY. The technical history of photography comprises three parallel developments: camera negatives, monochrome positive prints, and recording color. The practice of photography was empirically led throughout its first hundred years, frequently outstripping the ability of contemporary science to explain the various phenomena.

Since the late Renaissance, it had been known that sunlight darkens salts of silver. In 1725, Johann Heinrich Schulze used sunlit stencils to cast images on suspensions of silver salts. Carl Scheele showed in 1777 that the violet rays of the prismatic spectrum were most effective in decomposing silver chloride, the dark product being finely divided silver. During the 1790s, Thomas Wedgwood sun-printed "profiles" of objects onto paper and leather moistened with silver nitrate, but could not prevent their obliteration by daylight. Images in the *camera obscura* were too faint to make any impression, but Wedgwood successfully obtained "copies" of specimens projected by the solar microscope.

The earliest extant camera photograph was not produced in silver, but by *heliography*, a copying process invented by Joseph Nicéphore Niépce in the 1820s. The process places a thin coating of bitumen on a pewter plate, selectively hardens it by sunlight, then dissolves it by oil of lavender to bring out the image. In 1827, Niépce captured the first photograph—now in the Gernsheim Collection of the University of Texas—by a heliographic camera exposure estimated to have taken several days. Heliography was better suited to providing etching-resists for photomechanical printing plates. Attention returned to silver; in 1837, Louis Jacques Mandé Daguerre discovered, fortuitously, that mercury vapor could develop camera images on iodized surfaces of silver-plated copper, the chemical prerequisite, iodine, having been discovered by Bernard Courtois in 1811. The *daguerreotype* process, first publicized in 1839, enjoyed widespread commercial success until photography on paper and glass replaced it in the mid-1850s.

Meanwhile, William Henry Fox Talbot had independently devised *photogenic drawing paper* by 1835. He had noted that the light-sensitivity of silver chloride, precipitated within the paper's fibers, diminished with excess salt, and had thereby discovered the first method for fixing silver images. Talbot's earliest camera negative (1835) is in the National Museum of Photography, Film and Television in Bradford, England. On hearing in 1839 of Talbot's innovation, Sir John *Herschel, who originated the terms "photography," "the negative," and "the positive," demonstrated that unchanged silver chloride in a photograph dissolved in a solution of hyposulfite of soda (sodium thiosulfate). Herschel's "hypo-fixing" superseded Talbot's salt-fixing, and remains in use today. In 1839, Talbot also noted the greater sensitivity of silver bromide—now the chief constituent of all modern photographic materials—made possible by Antoine Jerome Balard's isolation of bromine in 1826.

In 1840, Talbot made his third, crucial discovery: that an invisibly weak dormant picture in silver iodide could be brought out by gallic acid, effectively increasing the speed of his camera photography a hundredfold—from hours to minutes. He named this process *calotype*, and patented it in 1841. Although science could not account for these phenomena of latency and development, Talbot had set photography on the path of continuous refinement for the next 150 years. A quest was mounted for shorter camera exposures and higher resolution. The opacity and texture of paper were avoided by suspending the silver halide in organic binders, making "emulsions" (an inaccurate term, but universally employed): hens' egg-white (Claude Félix Abel Niépce de Saint-Victor, 1847), collodion (Frederick Scott Archer, 1851), and finally gelatin (Richard Leach Maddox, 1871). Emulsions were coated on

transparent supports ranging from glass plates (Niépce de Saint-Victor, 1847) and waxed paper (Gustave Le Gray, 1851) to the flexible, but dangerously flammable early plastic, celluloid (cellulose nitrate), which permitted the design of roll-film cameras, introduced by George Eastman in 1888. Modern safety films employ polymer bases of cellulose triacetate (1923) or polyethylene terephthalate (1955). Parallel improvements were made in the optical design of camera lenses, notably achromats (Charles and Vincent Chevalier, 1828), large aperture lenses (Josef Petzval, 1841), rapid rectilinear lenses (J. H. Dallmeyer and H. A. Steinheil, 1866), and Zeiss anastigmats (Paul Rudolph and Ernst Abbe, 1890).

Pure silver halides respond only to blue light and the ultraviolet, whose discovery by Johann Wilhelm Ritter in 1801 represents the first contribution of photography to science. To render tonally balanced negatives, emulsions must react to the entire visible spectrum. Sensitizing with dyes, introduced in 1873 by Hermann Wilhelm Vogel, extended the response to green (orthochromatic plates, 1884), and then red wavelengths (panchromatic plates, 1904), reaching the near-infrared by the 1930s. Sensitometry, the photometric study of emulsion response, was originated by Ferdinand Hurter and Vero Charles Driffield in 1890, accompanied by extensive chemical exploration for better developers, such as hydroquinone (William de Wiveleslie Abney, 1880).

The emulsion binder in universal use by 1900, animal gelatin, displayed great variability in speed. In 1926, Samuel Edward Sheppard detected the cause: traces of sulfur-containing substances, arising from the animals' diet, could sensitize silver halides. Modern emulsion technology now uses pure gelatin, with controlled addition of sulfur and gold compounds as sensitizers. Understanding the latent image became possible with the foundation of solid-state physics and chemistry during the 1920s, especially Yakov Ilyich Frenkel's theory of ionic conductivity and A. H. Wilson's electronic band theory. In 1938, Sir Nevill Francis Mott and R. W. Gurney put forward a mechanism for the formation of the latent image that has found wide acceptance. Because of the granular structure of the developed image, photographic speed is linked inversely to resolution. This trade-off was improved in the 1980s by controlled growth of silver halide crystals with a tabular habit, increasing their surface-to-volume ratio. The speed of modern negative emulsions is approaching the theoretical limit.

As Talbot realized in 1835, printing positives is an essential procedure to rectify the reversed tonality and handedness of camera negatives. Talbot's photogenic drawing paper served at first, but papers coated with albumen emulsion (Louis Désiré Blanquart-Evrard, 1850) displaced Talbot's *salted paper prints*. Silver images suffer from sulfiding, causing them to fade, but gold-toning (1855) mitigated the deterioration, and *albumen prints* remained the chief photographic medium until 1895.

Since speed is not paramount for printing positives, substances less sensitive to light than silver salts were employed in the quest for image permanence. Following the discovery of dichromates (Louis Nicolas Vauquelin, 1798), their light-sensitivity on paper (Mungo Ponton, 1838) led to light-hardening of dichromated gelatin (Talbot, *photoglyphic engraving*, 1852). The addition of artists' pigments to the gelatin matrix, as inert image substances, permitted the development of the *carbon process* (Alphonse Louis Poitevin, 1855; Adolphe Fargier, 1861; Sir Joseph Wilson Swan, 1864).

In 1842, Sir John Herschel discovered that iron (III) citrate was light-sensitive and could yield images in gold, silver, mercury, or Prussian blue—the *cyanotype* or blueprint, the first reprographic process. William Willis patented the analogous platinum process in 1873, and by 1900 his company's *platinotype* paper dominated the market. But demand for platinum as a catalyst in the growing chemical industry brought steep price rises; moreover, the introduction of roll-film cameras, whose small formats required enlargement onto the much faster silver-gelatin development papers, made platinotype commercially unviable by the 1930s. It remains today, together with carbon printing, a minority fine-art practice, yielding images of archival permanence. For the remainder of the twentieth century, silver-gelatin enlarging papers became the commercial norm.

Photography in natural colors was first achieved by James Clerk *Maxwell (1831–1879) in 1861, using optical filters to separate three primary colors and to synthesize them additively. Ducos du Hauron published details of a three-color subtractive process, *heliochrome*, in 1869. Gabriel Lippmann, who received the only *Nobel Prize (physics, 1908) awarded for photographic innovation, devised a unique interference system for recording color photographs in 1891. More successful commercially was the additive *autochrome* process of the brothers Auguste and Louis Lumière (1907), which used a mixture of starch grains dyed with three primary colors to filter the light falling on a panchromatic emulsion. Color photography progressed with research in synthetic organic chemistry of dyestuffs, notably at the Eastman Kodak Company, which produced Kodachrome in 1935: a triple-layered silver emulsion, sandwiched with subtractive primary dyes.

The scientific value of photography for faithful analogue recording is self-evident; further, by disclosing information imperceptible to the eye, photography permitted new discoveries. Early examples include the "chronophotographic" studies of animal and human locomotion (Eadweard Muybridge and Etienne-Jules Marey, 1870s); recording of shock waves (Ernst Mach, 1884); high-speed photography by stroboscopic flash (Harold Eugene Edgerton, 1932); and time-lapse photography (John Ott, 1940s).

The ability of photography to accumulate a weak optical signal over time makes its enhanced recording sensitivity particularly valuable to astronomy and optical *spectroscopy. By photographically recording galactic spectra, Edwin *Hubble (1889–1953) discovered in 1929 the "red shift," which implied that the universe was expanding. The response of photographic emulsions to *X rays has found applications since 1895 in medicine and forensic science. The x-ray diffraction patterns for elucidating crystal structure, and electron diffraction patterns from gaseous molecules, were first recorded photographically. Antoine Henri Becquerel owed his discovery of *radioactivity in 1896 to the sensitivity of emulsions to charged particles; the ensuing technique of *autoradiography* finds application in biological studies and metallographic analysis. Emulsions can register the trajectories of *cosmic rays, and particle physics employs photography to record events in its cloud and bubble chambers.

With the advance of modern electronics, photoelectric devices are replacing silver-gelatin emulsions. Digitally-processed electronic imaging has reduced the role of photography in science and the lens-based media. Yet the photograph will endure in the art and archives of humanity.

Josef M. Eder, *History of Photography*, trans. Edward Epstean (1945). Helmut and Alison Gernsheim, *The History of Photography* (1969). T. H. James, ed., *The Theory of the Photographic Process* (1977). William Crawford, *The Keepers of Light: A History & Working Guide to Early Photographic Processes* (1979). Eugene Ostroff, ed., *Pioneers of Photography: Their Achievements in Science and Technology* (1987). Larry J. Schaaf, *Out of the Shadows: Herschel, Talbot, & the Invention of Photography* (1992). Anne Thomas, ed., *Beauty of Another Order: Photography in Science* (1997). Sidney F. Ray, *Scientific Photography and Applied Imaging* (1999).

MIKE WARE

PHYSICS. The development of physics provides a particularly strong example of the evolution of natural knowledge into modern physical science. Its traditional meaning of systematized, bookish knowledge about the entire physical world persisted until the eighteenth century; as late as 1798 the Académie Royale des Sciences of Paris could announce, as the subject of a prize competition in physics, "the nature, form, and uses of the liver in the various classes of animals." To be sure, the assumptions behind the question differed in an essential respect from the implications that would have been drawn a century or so earlier, when "physiologia" was often used as a substitute for "physica." This essential difference was the expectation that

competitors would base their answers on experiments, at least some of which they would perform themselves.

Around 1800 "physics" in its old, inclusive sense, modified to imply experiment and some measure of research, lost out to a new division of science bearing the same name. This decisive historical fact is often veiled by the use of "physics" to refer to natural knowledge through the ages. The new physics of 1800 restricted itself to the inorganic world and, within it, to subjects open to investigation by such instruments as the *air pump, electrical machine, balance, *thermometer, and calorimeter. In place of the organic world, physics took on subjects previously the property of mathematics, especially *mechanics and *optics, and strove to quantify the fields that had distinguished physics during the eighteenth century: *electricity, *magnetism, heat, and *pneumatics, which had garnered attention as particularly suitable for catchy demonstration experiments to large and varied audiences.

The leaders in this reformulation are often grouped together as the school of Pierre-Simon *Laplace. They applied the mathematical approach developed for the theory of gravity to *imponderable fluids supposed responsible for the phenomena of heat, light, electricity, magnetism, *fire, and flame. The scheme of imponderables was unified by this common approach; but each fluid functioned independently and irreducibly in its theory. Neither the fluids nor the forces they were supposed to carry linked to one another in any fundamental or ontological sense. Rather, the Laplacian school and its fellow travelers deployed the fluids and forces in an instrumentalist way. It thus promoted the dismemberment of science and the dropping of scruples against instrumentalist mathematical descriptions.

The physics of 1800 differed in scope from later physics by including *meteorology and parts of subjects shared with chemistry, like atomism, pneumatics, and *thermodynamics. It fell far short of twentieth-century physics in eschewing models of the microworld apart from very general assumptions about the molecular constitution of bodies. A major exception to this generalization was the attachment of atomic weights to chemical atoms. The French-dominated physics of 1800 also differed from physics in 1900 or even in 1850 by having a vigorous opponent, *Naturphilosophie, which depreciated mathematics and dismemberment and tried to reintegrate physics on general philosophical principles.

The basis of integration proved to be mechanics, which the Naturphilosophen disliked for its mathematics, strictness, and sobriety. This integration, which characterizes the so-called classical physics of the later nineteenth century, began around 1800 with the renewal of a quantitative wave theory of light. By 1840 most physicists (to use a word then just invented) associated light with the vibrations of a world-filling *ether conceived as a medium obeying the laws of mechanics. That substituted a mechanical system for the imponderable fluid of light. As this conception crystallized, the fluid of heat vaporized; by 1860, the mechanical theory of heat and the kinetic theory of gases had replaced the old imponderable caloric. This process brought two fundamentally new concepts and techniques into physics: statistics, in the form of probability calculations about the herd behavior of molecules, and irreversibility, in the form of the thermodynamic quantity *entropy.

That left the electrical and magnetic fluids, four in all, a positive, a negative, an austral, and a boreal. The discoveries of connections between electricity and magnetism beginning in 1820 led to the representation of magnetism as electricity in motion, a development completed by the demonstration of their identity in the theory of *relativity. Thus, by 1860, the main imponderables of 1800 had either vanished altogether (caloric and the magnetic fluids), metamorphosed into a mechanical system (light), or remained with enhanced properties (electrical fluids). A further diminution occurred with the unexpected discovery that *electromagnetism could be represented as a disturbance in the same ether whose vibrations made up light; or, stated phenomenologically, that light and other radiations were manifestations of electromagnetism, a representation confirmed by the production of radio waves in 1887.

Since the ether had been understood as a mechanical medium, physicists, particularly those trained at the University of Cambridge, tried to model electromagnetic phenomena in mechanical terms. It remained only to bring matter into the system. A way opened with the discovery that whirlpools or vortex rings in a fluid with the sorts of properties usually ascribed to the ether would last forever. These rings could link and separate in accordance with other goings-on in the ether; in short, they could behave in many ways like chemical atoms. Thus arose the grand and remote program of complete mechanical reduction, which few physicists, perhaps, expected to see realized.

The possibility of concocting this program arose from the very substantial progress that mechanics itself had made during the middle third of the eighteenth century. The powerful mathematical theories of the behavior of fluids and solids then developed stood ready for exploitation by theorists of the ether. Also the more general formulations of mechanical principles were extended still further and adapted to electromagnetism and the kinetic theory of heat. The apparent success of the generalized mechanics in describing the phenomena of electromagnetism, light, and heat confirmed the possibility of mechanical reduction without the necessity of exhibiting an explicit model or picture. That was comforting. Of course, the scheme could be turned around and electromagnetism taken as primary. Many distinguished physicists around 1900 played with the idea of electrodynamic reduction (see ELECTRON).

The extraordinary progress of classical physics was supported by new institutional arrangements that made research as well as teaching the business of professors. The physics research of the eighteenth century, insofar as it existed at all, was primarily an easy-going affair associated with academies. Toward the end of the ancient régime and, increasingly after 1830, individual university professors were conceded the right, and then given the duty, to undertake research, usually at their own expense, as part of their jobs.

Their institutions provided space, a collection of teaching apparatus, and perhaps a mechanic to keep it in order. Beginning around 1870, with the foundations of the Cavendish Laboratory in Cambridge and the Physics Institute of the University of Berlin, the idea that a professor's university should furnish him with the instruments he needed and a place to work with collaborators and advanced students steadily gained ground. (Here physics followed chemistry, whose first important university facilities for teaching and research go back to the 1830s.) The instruments employed in the institutes at first were largely home made. By the end of the century, however, most of the important equipment—electrical parts, air (or vacuum) pumps, refrigerators (see COLD AND CRYONICS), batteries—came from commercial suppliers.

The recognition of the need for institutionally supported research both to advance science and to train students owed much to the demands and opportunities of the new electrotechnology and the perceived need to give doctors, engineers, and science teachers a grounding in physics. By 1900 every major university and higher school in Europe and the United States had a physics institute. By far the largest output of papers on physics came from academics. Most of them worked in Britain, France, Germany, or the United States, which then were investing the same proportion of their gross domestic product, and enlisting the same proportion of their populations, in the physics enterprise. An example of their investment, and an indication of the widening of occupational opportunities beyond the universities, was the establishment of national (or, in the case of France, international) bureaus of standards.

This enterprise reached a climax in the symbolic year 1900, when the world's physicists gathered in Paris for their first and last general conclave. They then awakened, as one participant said, to the news that the keystone of their grand synthesis might have been found. This was the electron, which, it appeared, might be the unique building block of both matter and electricity, the clue to chemical binding, the explanation of the *ion, and the root of *X

rays and *radioactivity. These considerations inspired the design of several model atoms based on electromagnetic theory. None worked well beyond the limited range of the phenomena for which it was devised, for example, *magneto-optic effects and the scattering of rays from radioactive substances. Until around 1910 all these models lacked a principal ingredient that would make the theory of *atomic structure into the leading sector of physics in the 1920s. The necessary ingredient was the quantum, introduced into the theory of black-body *radiation in 1900. Its fundamental importance for physics had only just been recognized when the world's physicists flung themselves into world war.

The performance of physicists during World War I demonstrated their indispensability to modern society. New forms of government, foundation, and philanthropic support became available for scholarships, fellowships, and research. The worldwide activities of the Rockefeller Foundation were especially fruitful, since it favored the transfer of *quantum physics from Europe to the United States and of American instrumentation, particularly cyclotrons, abroad. Basic physics, having finished provisionally with radiation and atoms (see QUANTUM ELECTRODYNAMICS), delved into the nucleus; here the United States, with sizable disposable income despite the depression, led the way into big science. American physics strengthened further as a result of the emigration of European refugees driven from their positions by fascist regimes. Nazi science policy stultified fundamental physics, in which Jews had been disproportionately prominent, in Germany and its occupied territories. As Germany declined, Japan and the Soviet Union developed their own capacities for training advanced students and set up research institutes often with liberal funding from military or industrial sources. The Soviet capacity centered on the institutes of the Akademiia nauk (Soviet Academy of Sciences) and laboratories supported by the Commissariat for Heavy Industry; the Japanese, on the imperial research institute RIKAN.

As physics spread geographically it also spread intellectually to territory newly conquered by the quantum. The study of *cosmic rays, the behavior of bodies at low temperatures, the *solid state, and the nucleus (see NUCLEAR PHYSICS) advanced prodigiously during the 1930s. The ascendancy of the electron was challenged or shared by the deuteron, meson, neutrino, neutron, positron, and proton (see ELEMENTARY PARTICLES). Molecular physics began to yield to new methods of calculation. Machines and ideas spun off into neighboring and even distant sciences, for example, *astronomy (e.g., stellar energy) and biology (e.g., radioactive tracers). The principles of *complementarity and uncertainty, invented to domesticate a mismatch between ordinary intuitions and the formalism of quantum mechanics, found applications to vitalism, psychology, philosophy, and theology.

Despite cries for a moratorium on research and movements for social responsibility among scientists, which gained some strength during the depths of the depression, the public applauded the science-based luxuries it soon found to be necessities: commercial radio, the all-electric kitchen, the long-distance telephone, air mail, air travel, and, above all, the automobile. As this easily extendable list suggests, the public did not (and does not) distinguish between science and technology. The confusion was compounded by the multiplication of industrial research laboratories, some of which, especially in the United States, permitted their technical staffs considerable leeway in choice and conduct of research. These laboratories helped to keep up demand for physicists so that, except for the loss of a year or two's crop of new graduates, the depression had no effect on recruitment, at least in the United States.

The demand became so great during World War II that physics students who had not completed their degrees were employed on advanced research projects. Once again physicists proved their value: operations research, radar, the proximity fuse, cryptography, and the atomic bomb. The great winner in the conflict, the United States, rewarded its physicists after the war by supporting them in the manner to which they had become accustomed. The infusion of public dollars, intended to keep

scientists well disposed toward government, as well as to encourage work related to weaponry, exaggerated a style already characteristic of American science: pragmatic, instrumentalist, democratic, and gigantic. The combination proved potent: countries that wanted to compete had no option but to submit to Americanization, especially in the world's most prestigious science, *high-energy physics.

The United States assisted participation and competition by helping in the recovery of science in Germany (after picking some plums in Operation Paperclip), in Japan (after throwing its cyclotrons into the sea), and in Europe (by validating CERN, the Centre Européen pour la Recherche Nucléaire, which brought together nations no longer able to compete alone). The recovery called into existence new mechanisms of national support and international collaboration. As a result, Europe and Japan caught up with, and even outpaced, the United States, which, by canceling the Superconducting Super Collider in the 1990s, showed that it no longer had the will, if it had the means, to dominate the world in high-energy physics.

While the particle physicists enjoyed their limelight, their colleagues working on a smaller scale discovered or invented things of greater importance to the public than strangeness, parity nonconservation (see SYMMETRY), the Eight-Fold Way (see THEORY OF EVERYTHING), or the unification of the fundamental forces except for gravity. Radar technologies helped academic laboratories to produce the laser and airports to handle jet-setters. Research into the solid state returned with silicon chips, the computer, and high-temperature superconducting magnets. Color television, direct-dial international telephony, satellite communications, modern banking, the credit card, and so on, demonstrate that physics warmed by cold war can do wonders.

The wonders have included nuclear power and therewith the pollution that arises from the spent fuel and radioactive parts of reactors, which began to spread worldwide in the late 1950s under the American program, Atoms for Peace. The pollution, as well as the involvement of physicists and other scientists in advanced weapons projects, fuelled an antinuclear movement and wider attacks on science. The end of the Cold War redirected some of this sentiment against the diffusion of high-tech industry to developing countries. Nonetheless, the spread of Western science, which began before World War I in the colonies of the European powers and produced important indigenous physics communities in the Soviet Union and Japan before World War II, continued into Latin America, the Near East, and continental Asia after 1945. An example of this diffusion of postwar physics, American style, is the synchrotron completed in Brazil in 1996.

The spread of physics to new cultures had a parallel within the old physics-producing countries in the increasing inclusion of groups whose participation had previously been marginal or restricted. Anti-Semitism lost its force in most of the world. Women gained easier access to training and employment. Opportunities were extended to minority groups. Acquaintance with physical principles or gadgetry diffused widely in the general culture along with the computer, high-tech weapons, and space exploration.

The possibility of an encompassing and final theory arose at the intersection of *cosmology and particle physics. Theories of the largest and smallest structures in the universe combined to make a Big Bang. Crucial evidence favoring the doctrine came from the characteristic postwar specialty, radioastronomy (see ASTRONOMY, NON-OPTICAL). The doctrine of the Big Bang and the expectation by some theoretical physicists that a complete and unified theory of the physical world might be achieved during their lifetimes provoked the interest of diverse theologians and the Vatican. In a limited region of vast extent, therefore, physics has returned to the sort of questions with which its predecessor, natural knowledge, had been engaged.

Abraham Wolf, A History of Science, Technology, and Philosophy in the Sixteenth and Seventeenth Centuries (1950). Abraham Wolf, A History of Science, Technology, and Philosophy in the Eighteenth Century (1952). Paul Forman, J. L. Heilbron, and Spencer Weart, "Physics circa 1900," Historical Studies in the Physical Sciences, vol. 5 (1971). Richard S. Westfall, The Construction of Modern Science (1971). J. L. Heilbron, Electricity in the Seventeenth and Eighteenth Centuries. A Study in Early Modern Physics (1979; 1999). Thomas L. Hankins, Science and the Enlightenment (1985). Tore Frängsmyr, J. L. Heilbron, and Robin E. Rider, eds., The Quantitative Spirit in the Eighteenth Century (1990). Peter Dear, Discipline and Experience. The Mathematical Way in the Scientific Revolution (1995). John Krige and Dominique Pestre, eds., Science in the Twentieth Century (1997). Robert D. Purrington, Physics in the Nineteenth Century (1997). Lisa Jardine, Ingenious Pursuits (1999). Helge Kragh, Quantum Generations. A History of Physics in the Twentieth Century (1999).

J. L. HEILBRON

PLANCK, Max (1858–1947), German physicist and spokesman for science.

Planck came from a family of lawyers and pastors. His marriage to the daughter of a banker placed him comfortably within the professional classes of the rising German empire. His special interest in *physics, *thermodynamics, appealed to his love of order and generalization. In 1885 he joined the university in his home town of Kiel as an assistant professor (Extraordinarius) in theoretical physics, then a small new subdiscipline. Four years later he obtained a similar position at the University of Berlin, where he remained until his retirement in 1928.

Planck's reputation among physicists rests on his solution to a problem about radiation inside a closed cavity. Thermodynamical arguments indicated that the energy falling to a given frequency (color) in cavity radiation at equilibrium depended only on the color and the temperature of the enclosure's walls, and that the "radiation formula" describing each color's energy must contain two constants of universal significance.

James Clerk *Maxwell's equations described electromagnetic radiation in general, and *entropy fixed the equilibrium state. To obtain a theory that agreed with the rapidly improving measurements (the energy distribution of cavity radiation had the practical interest of serving as a universal standard for the efficiency of electric light bulbs), Planck turned to the definition of entropy that Ludwig *Boltzmann had employed to describe the behavior of a gas. To bring the definition to bear, Planck calculated as if the sum of the energies of the resonators associated with a given frequency f consisted of a very large number of very small elements hf. (The resonators were fictional oscillators that coupled matter to the radiation field.) The key step in the derivation for Planck was thus an *extension* of the concepts of the mechanical world picture to radiation, not a *limitation* on resonator energy.

Planck's scientific reputation and sense of duty soon brought him to public attention. In the mid-1890s he defended a Jewish physicist whom the government wanted to expel for socialist activities. He supported the right of women—though only unusual women, like Lise *Meitner—to study science at the university. He gave unstintingly to his profession, especially as an editor of the *Annalen der Physik*, Germany's leading physics journal. In this capacity he welcomed and promoted Einstein's theory of *relativity. Planck's famous and severe attack in 1908 on the sensationalist epistemology of Ernst Mach marked a new direction in his efforts to steer the course of physics.

Planck's first important administrative job was to help govern the Berlin Akademie der Wissenschaften. During World War I he prevented the academy from expelling foreign members belonging to enemy countries. His repudiation of his signature of the Manifesto of the Ninety-Three Intellectuals, a declaration in support of the German invasion of Belgium, further exemplifies his wartime balance and courage. He was the only one of the ninety-three to recant publicly. After the war he worked energetically to rebuild German science and took on the presidency of the *Kaiser-Wilhelm-Gesellschaft (KWG), whose many institutes occupied the vanguard of German scientific and technological research.

The Nazi takeover presented Planck with an acute and continuing crisis of conscience; in the end, he decided to retain his positions of influence. He was able to help a

few people, but his staying eventually tarnished his reputation. In 1938 he was forced from his remaining posts. Toward the end of World War II, allied bombing destroyed his house and fear of the Russians drove Planck and his wife into the woods, where American soldiers rescued them. Planck still had a role to play. The British decided to revive the KWG under a new name. Planck acted briefly as president of the restructured organization, which took as its name the Max-Planck-Gesellschaft für die Förderung der Wissenschaften. It is now one of the world's premier research networks.

Planck's private life, begun as an idyll, became a tragedy. His first wife, the mother of his two sons and twin daughters, died in 1908. His eldest son was killed in World War I. The twin daughters died two years apart, in 1917 and 1919, both in childbirth. These catastrophes wiped out whatever pleasure might otherwise have come to him from the belated award of the Nobel Prize for physics in 1918. Planck's misery reached its apex when his second son was executed in 1944 for complicity in the plot to assassinate Hitler. Against that the loss of his material possessions in the destruction of Berlin, the last step in the annihilation of the strong, orderly, cultured Germany whose rise to European dominance had been the pride and guide of Planck's early manhood, signified nothing.

T. S. Kuhn, *Black Body Radiation and the Quantum Discontinuity* (1978). J. L. Heilbron, *The Dilemmas of an Upright Man: Max Planck as Spokesman for German Science* (2d edition, 1999).

J. L. HEILBRON

PLANET. The word "planet," derived from the Greek for "wanderer," at first referred to Mercury, Venus, Mars, Jupiter, and Saturn, which wandered amid the fixed stars of the ecliptic plane. The ancients almost universally believed that the planets rotated around the earth, although from the relationship of the orbital retrogrades of Mars, Jupiter, and Saturn to the terrestrial year, Nicholas *Copernicus argued in 1543 that they rotated around the sun. Between 1543 and 1687 the planets became agents of radical intellectual change in astronomy and physics. What was their center of rotation? After the abandonment of Aristotle's crystalline spheres around 1600, what force made the planets move? Were their orbits circular or elliptical, and how did their speed relate to their distance from the center of motion? Planetary motion lay at the heart of the work of Johannes *Kepler and Isaac *Newton; the problems it posed led to the invention of gravitational physics.

Until *Galileo first looked at planets through the telescope and discovered that they subtended disks (indicating that they were spherical worlds), astronomers had envisaged the planets as points of light. The telescope inaugurated physical astronomy. Galileo discovered that Venus showed phases; that Jupiter was slightly compressed at its poles; and that Saturn, as well as being spherical, also displayed peculiar appendages that appeared and disappeared. In 1655, Christiaan *Huygens, using a much more powerful telescope than Galileo's, resolved these appendages into a ring, and in 1675 Gian Domenico *Cassini discovered a division in them.

Following technological breakthroughs that made possible long refracting telescopes with object glasses of diameters up to 15 cm (6 ins), astronomers discovered the Syrtis Major and Polar Caps of Mars (Huygens) and Jupiter's belts and spots (Cassini and Robert *Hooke). Astronomers also obtained accurate timings of the orbital periods of the outer planets using long refracting telescopes, especially in conjunction with Huygens's pendulum clocks after 1658. New planetary satellites came into focus: Saturn's Titan (Huygens, 1655), followed by Lapetus (1671), Rhea (1672), and Tethys and Dione (1684), all discovered by Cassini. In 1668 Cassini produced tables for the motions of Jupiter's moons, which enabled cartographers to fix the respective longitudes of places on terra firma though not yet of ships at sea.

The first great wave of physical planetary discovery came to an end around 1690 as the long refracting telescope exhausted its research potential. New planetary discoveries awaited the superior resolving power of the relatively large speculum-mirror reflecting telescope, initially in the hands of Sir William *Herschel, and, after 1800, the achromatic refractor.

Herschel's discovery of Uranus on the night of 13 March 1781 heralded a new era of planetary astronomy. It suggested that more planets and satellites might be in space awaiting discovery. Herschel himself discovered Uranus's moons Oberon and Titania (1787) and two new Saturnian satellites, Enceladus and Mimas (1789). Then, from the southern latitude of Palermo, Sicily, on 1 January 1801, Giuseppe Piazzi discovered the first asteroid, or minor planet, Ceres, using a Dollond achromatic refractor (see TELESCOPE).

During the nineteenth century planetary astronomy made enormous progress. As in the past, advancing instrument technology made it possible. Three new asteroids quickly appeared: Pallas (Heinrich Olbers, 1802), Juno (Karl Ludwig Harding, 1804), and Vesta (Olbers, 1807). Then, after a gap of thirty-eight years, Karl Hencke, a German amateur, discovered Astraea, and, almost in cascade, came hundreds more, 450 in all by 1900. The asteroids fascinated astronomers, eliciting speculation about a former planet between Mars and Jupiter gravitationally destroyed in the early stages of the solar system, and giving support to the empirical Titius-Bode law of planetary distribution (1772). This so-called law, developed by Johann D. Titius and Johann E. Bode in Germany, is an empirical sequence of numbers—four, seven, ten, sixteen, twenty-eight, fifty-two, and one hundred—that correspond to the proportionate distance of the first five planets and the asteroid belt around the Sun. The Titius-Bode law lay at the theoretical heart of the greatest discovery in planetary dynamics of the nineteenth century: the detection of Neptune in 1846. Both John Couch Adams in Cambridge and Urbain J. J. Le Verrier in Paris used the law to model some of the parameters that lay at the heart of their independent predictions of the position of Neptune though, ironically, Bode's number sequence does not hold good for Neptune. This discovery, however, would not have been possible without a combination of fundamental advances in gravitational mathematics and superior tables of the motion of the known planets made at the Greenwich, Königsberg, and other observatories.

Planetary astronomy was advanced not only by professional scientists in major universities and observatories, but also by self-funded individuals. These "grand amateurs," like the Liverpool brewer William Lassell, who built the most advanced reflecting telescopes of the 1840s to 1860s to discover planetary satellites, and the spectroscopist Sir William Huggins, funded research from their private means. Less wealthy people, like the more modest amateurs in the British Astronomical Association and other bodies worldwide, devoted themselves to monitoring visible changes on the planets.

A succession of interplanetary space probes after the Russian *Venera* flight to Venus (1970) fundamentally transformed our knowledge of the planets (*see* SATELLITE). Mercury, Venus, and Mars turned out to be rocky worlds with widely different sorts of atmospheres. Jupiter, Saturn, Uranus, Neptune, and Pluto (the latter discovered by Clyde Tombaugh in 1930) are made of frozen gases. The American spacecraft *Voyager II* in the late 1980s produced breathtaking fly-past views of the outer planets and discovered several new satellites, rings, and an abundance of surface detail. The future of planetary exploration probably lies in the development of increasingly sophisticated robotics instrumentation operated through space vehicles.

John F. W. Herschel, *Outlines of Astronomy* (1849). Robert Grant, *History of Physical Astronomy* (1852). Michael Hoskin, ed., *Cambridge Illustrated History of Astronomy* (1997).

ALLAN CHAPMAN

PLANETARY SCIENCE. The term "planetary science" dates from the 1950s. It applies physics, astronomy, chemistry, geology, biology, atmospheric sciences, and oceanography to discrete bodies in the solar system. Previously the study of the planets had been known as "solar system astronomy" or "solar system science." That might have been a better name, because the field takes the whole solar system, including *comets, *meteorites, asteroids, and planetary satellites as its object of study (*see* PLANET).

Like the parallel discipline *earth science, planetary science emerged in tandem with the new technologies whose development had been spurred by World War II, rocketry and computers in particular. In 1958, the *International Geophysical Year began, in large measure to take advantage of these technologies. In the same year the Soviet Union launched *Sputnik* and the space race began (*see* SATELLITE). In October 1958, the United States established the National Aeronautics and Space Administration (NASA), to carry on and extend the work formerly done by the National Advisory Committee for Aeronautics (NACA) and other government bodies.

The term "planetary science" first appeared in the journal *Science* in 1959 in a job advertisement put out by the Goddard Space Center. In the same year, the first specialist journal, *Planetary and Space Science*, began publication. In 1962, the American Geophysical Union set up a section on planetary sciences. The journal *Earth and Planetary Science Letters* appeared in 1966. Existing institutions like the Houston Lunar Science Institute, university departments, and the journal *Meteoritics* all added "and Planetary Science" to their names.

*Copernicus, *Galileo, Christiaan *Huygens, Gian Domenico *Cassini, and William *Herschel had asked questions about the configuration of the solar system (cosmology), its mechanics, and its origin (cosmogony), and little by little discovered smaller and more distant bodies in the solar system. Galileo, William Gilbert, and Thomas Harriot mapped the Moon in the seventeenth century. Michael Florent van Langren published the first large full-Moon map in 1645, though his projected series of maps showing the Moon in its different phases never appeared in print. The first full-Mars map was published in 1840. The advent of large telescopes and photography vastly improved maps of the Moon and Mars in the remaining years of the century.

Meanwhile physicists, geophysicists, and geologists as well as astronomers pursued many aspects of what would now be planetary science. They asked how the Earth differed from neighboring planets and why. In 1801, the Italian astronomer Giuseppe Piazzi detected the first asteroid. By the end of the century, Maximilian Wolf at the University of Heidelberg had invented a technique for discovering new asteroids by the streaks they left on photographic plates. Astronomers thus discovered the asteroid belt. They also discussed the origin of craters on the *Moon and other planetary bodies. In 1803, Jean-Baptiste Biot confirmed that certain stones in Normandy really had fallen from the sky, thus establishing the extraterrestrial origin of meteorites.

After a period in which interest in the solar system waned, two American astronomers, Gerard Peter Kuiper and Harold C. Urey, renewed interest in the subject in the 1940s. Kuiper discovered the carbon dioxide atmosphere on Mars and a disk-shaped region of minor planets (now called the Kuiper belt) outside Neptune's orbit, which he proposed as the source of certain types of comets. He pioneered the development of infrared astronomy, helped identify landing sites for the first manned landing on the moon, and edited two influential works, *The Solar System* (1953–1958) and *Stars and Stellar Systems* (1960–1968). Urey synthesized his investigations of the distribution of elements in the solar system in his *The Planets, Their Origin and Development* (1952).

Since 1960, planetary science has developed rapidly with the help of new optical and radio *telescopes. In 1990, the Hubble Space Telescope reached a position high above the distorting effects of the Earth's atmosphere. Project Apollo, which culminated in 1969 with the first human moon landing, the *Pioneer* and *Voyager* spacecraft that explored the Moon and other parts of the solar system, and the *Viking* and *Mars Pathfinder* spacecraft that investigated Mars enabled new kinds of data collection, whether by humans on the moon, by robots on the moon, or by *photography and sampling of these and more distant planets. Planetary mapping, aided by radar techniques, proceeded apace. Mercury and Venus, whose surfaces had been difficult to study, Mercury because of its small size and proximity to the Sun, and Venus because of its dense atmosphere, have now been mapped.

New specialties have emerged, such as astrogeology, astrobiology, planetary atmospheres, planetary tectonics, and planetary physics. Topics of active investigation include planetary origins, the structure and composition of planets, vulcanism and tectonic activity, the atmospheres and magnetic fields of planets, and the planets of Jupiter. Public interest in planetary science, though not as high as in the 1960s, is still fueled by dramatic photographs, press coverage, and fascination with perennial puzzles like the canals of Mars and the possibility of life elsewhere in the universe. New discoveries and up-to-date information are posted on the NASA site on the World Wide Web. Planetary scientists, who until the end of the second millennium had been concerned almost exclusively with objects within our own solar system, are beginning to pursue the increasing evidence of planets in other parts of the universe.

See also ASTRONOMY, NON-OPTICAL.

Ormsby M. Mitchel, *The Planetary and Stellar Worlds: A Popular Exposition of the Great Discoveries and Theories of Modern Astronomy* (1892; new ed., ed. I. Bernard Cohen, 1980). Dominick A. Pisano and Cathleen S. Lewis, *Air and Space History: An Annotated Bibliography* (1988). Stephen G. Brush, *A History of Modern Planetary Physics*, 3 vols., (1996). Ronald E. Doel, *Solar System Astronomy in America: Communities, Patronage, and Interdisciplinary Science, 1920–1960* (1996). James H. Shirley and Rhodes W. Fairbridge, eds., *Encyclopedia of Planetary Sciences* (1997).

JOANNE BOURGEOIS

PLASMA PHYSICS AND FUSION. A plasma in physics refers to an ionized gas. Scientists first recognized the importance of plasmas in studies of the propagation of radio waves for the nascent radio industry of the early twentieth century. In seeking ways to send radio signals over long distances they realized that waves seemed to bounce off a conducting layer in the earth's atmosphere, allowing signals to travel far beyond the horizon. One of these researchers, Irving Langmuir of the United States, in the 1920s designated the atmospheric matter "plasma" and investigated its properties in gas discharges in the laboratory.

In the late 1920s physicists began to apply new theories of *atomic structure and quantum mechanics to the energy source of stars. In 1929 Robert Atkinson and Fritz Houtermans predicted that the nuclei of light atoms such as hydrogen, the primary constituent of the sun, could fuse through quantum tunneling, and that the resultant atoms would weigh less than the original constituents. Albert *Einstein's mass-energy relation suggested that fusion would release vast amounts of energy, enough to power the stars. Hans *Bethe and others developed the theory of stellar fusion in the 1930s, elucidating the chains of nuclear reactions by which fusion built up heavier chemical elements and calculating the reaction rates and energy release. Astrophysicists were then incorporating plasmas into theories of stellar structure and thus merged the study of plasma with fusion.

Physicists at the time recognized the potential of fusion for a new energy source, but the high temperatures required to produce it seemed out of reach of available technology. Although World War II and the coincident discovery of nuclear fission diverted attention from fusion, they would eventually provide the motivation and means to attain it. Nuclear fission and the subsequent development of nuclear bombs brought stellar conditions down to earth and offered a way to ignite fusion. During the war scientists working on the atomic bomb project in the United States discussed thermonuclear weapons, or the hydrogen bomb (named after its fuel), with explosive force orders of magnitude beyond fission bombs. Both the United States and the Soviet Union would pursue the hydrogen bomb in the Cold War; in the meantime, work on fission bombs advanced knowledge about plasma behavior, and the complicated hydrodynamic calculations for bomb physics spurred the development of electronic computers, which would then aid the development of fusion weapons.

Research into controlled fusion revived in 1951 with the help of Juan Perón, the dictator of Argentina. A few years earlier

Perón had set up a laboratory for Ronald Richter, an expatriate German with a scheme for controlled fusion power. In 1951 Perón announced Richter's successful production of power from a fusion reactor. The news made headlines in major newspapers, and though American and European scientists quickly discounted Richter's results, they did start thinking more seriously about the problem of fusion reactions.

One physicist so inspired was Lyman Spitzer, Jr., who was familiar with plasmas from his background in *astrophysics and who had just joined a group at Princeton University working on the crash program to build the hydrogen bomb in the United States. Spitzer devised a device to contain a plasma at high temperatures and obtained the support of the Atomic Energy Commission for the work. Commission laboratories at Los Alamos, New Mexico; Livermore, California; and Oak Ridge, Tennessee, soon followed suit. Controlled fusion seemed to offer unlimited power from an abundant fuel without the lingering *radioactivity of nuclear fission reactors, and also provided a peaceful application of nuclear research to balance the fearful implications of nuclear weapons. It did not lack for support: by the late 1950s the United States was spending tens of millions of dollars a year on fusion research.

Other countries joined what became an international race for controlled fusion. British scientists led by George P. Thomson began investigating fusion soon after the war, and the British government sponsored a major fusion effort at its nuclear research laboratory at Harwell. In the Soviet Union, Igor Kurchatov, Igor Tamm, Andrei Sakharov, and other scientists in the nuclear weapons project took up fusion research in the early 1950s (*see* KURCHATOV, IGOR VASILYEVICH, AND J. ROBERT OPPENHEIMER, and SAKHAROV, ANDREI, AND EDWARD TELLER). The connection to nuclear weapons kept work in each country secret until Kurchatov revealed the Soviet program on a visit to Harwell in 1956; an international conference on atomic energy in Geneva in 1958 opened up the field for good. Japan, France, Italy, and other nations also entered the race, but the high cost of fusion experiments spurred efforts at international collaboration.

Most fusion reactors used various configurations of magnetic fields to bottle up the charged particles of the plasma, which at the high temperatures involved proved difficult to control. Fusion research engaged scientists from diverse fields: astrophysics, cosmic ray physics, accelerator engineering, gas discharges, and weapons physics. But no unified framework emerged from this eclectic background, and the initial optimism of the early 1950s soon gave way to realization of the technical difficulty of the endeavor—skeptics compared it to trying to push all the water to one side of your bathtub with your hands—and lack of knowledge about the basic behavior of plasmas. In the late 1950s fusion researchers instead turned to the underlying theory of magnetohydrodynamics, although work on fusion reactors continued under more empirical techniques.

In the mid-1960s Soviet scientists provided a new impetus with their development of the tokamak, which combined linear and toroidal configurations of previous devices in a single toroidal device. In 1968 a Soviet team under Lev Artsimovich revealed the attainment of temperatures of around ten million degrees and confinement times of about a millisecond in a tokamak. The tokamak thereafter became the preferred device for fusion, but the conditions it produced remained far below those required for fusion. Only after decades of technical refinements did a tokamak at Princeton University provide the first definitive success in late 1993 and 1994, when it confined a plasma of hydrogen isotopes at 300 million degrees Celsius for about a second to produce 10 million watts of power. The Princeton tokamak, however, still consumed more power in heating and confining the plasma than it produced.

The development of lasers in the 1960s suggested another route to fusion. Focusing high-energy lasers on a stationary solid pellet of hydrogen isotopes could compress and heat the pellet enough for fusion. Laser fusion—a form of inertial confinement—offered a way around the difficult problems posed by confining hot moving plasma with magnetic fields, and several nations started laser fusion programs. But laser fusion also presented daunting technical problems, especially the manufacture of laser optics capable of the high energies necessary. Connections with nuclear weapons persisted in laser fusion, since it also offered a way to model miniature nuclear explosions, and secrecy began to return to fusion research. In the 1990s the United States began building the billion-dollar National Ignition Facility at Livermore to substitute laser fusion for full-scale nuclear tests.

Still another route to fusion energy was announced in Utah in March 1989 by B. Stanley Pons and Martin Fleischmann, who claimed to have obtained fusion at room temperature in a cheap and simple electrochemical experiment. The announcement set off a frenzy of popular discussion of limitless energy, but attempts to replicate the experiment and to adjust theory to accommodate the results failed. In addition, the disciplinary background of Pons and Fleischmann in chemistry did not inspire confidence in the physicists who dominated the fusion community, nor did their mode of announcement, in a press conference instead of through peer-reviewed publication. Cold fusion quickly joined N rays, polywater, and other famous nondiscoveries in the history of science.

Joan Lisa Bromberg, *Fusion: Science, Politics, and the Invention of a New Energy Source* (1982). John R. Huizenga, *Cold Fusion: The Scientific Fiasco of the Century* (1992). Gary Taubes, *Bad Science: The Short Life and Weird Times of Cold Fusion* (1993). Richard F. Post, "Plasma Physics in the Twentieth Century," in *Twentieth Century Physics*, eds. Laurie M. Brown, Abraham Pais, and Sir Brian Pippard, vol. 3 (1995): 1617–1690. T. Kenneth Fowler, *The Fusion Quest* (1997).

PETER J. WESTWICK

PLATE TECTONICS. The theory of plate tectonics, proposed in the 1960s, asserts that the creation, motion, and destruction of a small number of rigid plates, thin in relation to the earth's diameter, shape the earth's surface. Quickly termed a revolution, the switch to plate tectonics was one of the most exciting scientific developments of the mid-twentieth century.

The discoveries that stimulated scientists to propose plate tectonics came from paleomagnetism and *oceanography. At the end of the 1950s, a small but influential group of physicists based at the universities of London and Newcastle and at the Australian National University were studying paleomagnetism. They became convinced that to explain the apparent global wandering of the magnetic pole over geological time, they had to assume that the continents had moved relative to one another. They saw this as new evidence for the theory of continental drift, still widely discussed in Britain and Australia because it had been advocated in 1945 in the *Principles of Physical Geology* by the distinguished geologist Arthur Holmes.

Meanwhile, oceanographers had been surveying the ocean floor and measuring heat flow and gravitational and magnetic anomalies. They discovered a global system of mid-ocean ridges. These enormous mountain chains had some peculiar physical characteristics, such as patterns of magnetic anomalies and a median rift valley with high heat flow. In the early 1960s, Harry Hess of Princeton University suggested that these were tensional cracks through which lava welled up, created new sea floor, and spread. His conjecture of sea floor spreading was quickly corroborated by two confirmed predictions. In 1963, Fred Vine and Drummond Matthews of Cambridge University predicted that magnetic anomalies observed on either side of the mid-ocean ridges recorded global magnetic reversals preserved in the solidified lava. Physicists had dated global magnetic reversals on the continents using *radioactivity and so had a magnetic time scale. It was only necessary to find parallel zebra stripes of anomalies on either side of the ocean ridges. In 1965, J. Tuzo Wilson predicted that if sea floor spreading occurred, scientists should be able to detect seismically a new kind of fault that he named "transform." In 1966, scientists at the Lamont Doherty Geological Observatory found evidence supporting both predictions.

If the sea floor was spreading, where was the new material being accommo-

dated? Could it be that the earth was expanding? Scientists gave this possibility serious consideration. It was quickly displaced, however, by the theory of plate tectonics independently conceived by Jason Morgan at Princeton and Dan McKenzie at Cambridge in 1967 and 1968, respectively. They proposed that rigid plates, each perhaps a hundred km thick, covered the earth's surface. They, and not continents and oceans, were the important structural surface features. Created at the mid-ocean ridges, they moved apart until they sank into and were consumed in "subduction zones" signaled by intense earthquake activity and negative gravity anomalies. Abstract mathematical models of plate movements agreed well with field observations. By the early 1970s, almost all earth scientists, except in Russia, had accepted plate tectonics.

Such a rapid shift to an account of the earth so radically different from previous orthodoxy stimulated wide interest. Earth scientists published in the wide scientific press, appeared on television programs, and revised school textbooks. Once their immediate euphoria waned, many earth scientists suffered a crisis of confidence. Were they wrong to have resisted the theory of continental drift for half a century? And if science proceeded by the patient accumulation of facts, as most of them believed, was it scientific to switch in just a few years from believing that the continents stayed in place to believing that they moved?

Many earth scientists, particularly younger ones, wondered how their predecessors could have rejected continental drift and derided it as pseudoscientific when it had been supported by some evidence from similarities of paleontology and lithology on the two sides of the Atlantic and from the fit of the continents. Their reaction misreads history. Continental drift, like other theories put forward when the geological synthesis proposed by Eduard Suess in *Face of the Earth* (1883–1904) collapsed in the early years of the twentieth century, had been given a serious hearing. It was widely accepted in South Africa and viewed with an open mind by geologists in the British Isles and Australia. In the

1950s, some American geologists mocked it in their undergraduate classes largely because they believed its proponents lacked evidence. Plate tectonics, with its confirmed predictions, had much stronger evidential support; moreover, it was a different theory. The introduction of plates made continental movement an incidental theoretical consequence and not the key theoretical claim.

Earth scientists still had to face the fact that the speed with which they accepted plate tectonics did not sit well with their image of science as the gradual accumulation of facts. Casting around for an alternative picture of science, they came across Thomas Kuhn's *Structure of Scientific Revolutions* (1962). By the late 1960s, J. Tuzo Wilson and Allen Cox were describing plate tectonics as a Kuhnian revolution—an attribution still debated by historians and philosophers of science.

Allan Cox, ed., *Plate Tectonics and Geomagnetic Reversals* (1973). Ursula B. Marvin, *Continental Drift: The Evolution of a Concept* (1973). Anthony Hallam, *Great Geological Controversies* (1983). H. W. Menard, *The Ocean of Truth* (1986). Homer Legrand, *Drifting Continents and Shifting Theories* (1988). John Stewart, *Drifting Continents and Colliding Paradigms* (1990). Naomi Oreskes, *The Rejection of Continental Drift* (1999).

RACHEL LAUDAN

PNEUMATICS. The discovery of the different gas types during the third quarter of the eighteenth century caused a revolution in physical science. The new field of pneumatics made large demands on experimental technique and apparatus, and required unusual accuracy in calculating the weights of small quantities of matter. It played an important part in quantifying physical science and in forging fruitful connections among the branches of natural knowledge from anatomy (as in the work of Luigi *Galvani) to chemistry (Joseph Priestley and Antoine-Laurent *Lavoisier), meteorology (Jean André *Deluc and John *Dalton), and physics (Alessandro Volta, *see* GALVANI AND VOLTA).

The English clergyman Stephen Hales, who had learned Newtonian experimenta-

tion at Cambridge around 1700, pointed the way to pneumatics in his *Vegetable Staticks* of 1727. Hales described many ways of fixing "air" in, and liberating it from, vegetable and other matter. He collected liberated air over water in a "pneumatic trough" of his invention, measured its quantity, and studied its quality; but, although he handled several chemically distinct gases, he regarded them all as the same basic substance. The variety and quantity of substances from which he drew his "air," however, supported his conclusion, which he expressed in the Newtonian style as a query: "may we not with good reason adopt this now fixt, now volatile *Proteus* among the Chymical principles... notwithstanding it has hitherto been overlooked and rejected by Chymists, as no way entitled to the Denomination?"

Hales studied fixed air while following up his interest in the mechanics (physiology) of plants; Joseph Black came to the problem as a medical student concerned with kidney stones. For his doctoral thesis of 1754 he examined the air (carbon dioxide) released from magnesium alba (magnesium carbonate) when heated or treated with acid. He determined that it differed from common air in its inability to support combustion and respiration, and occurred fixed in the limestone implicated in urinary calculi. Novel airs then began to rise promiscuously. In 1766 Henry *Cavendish identified a special "inflammable air" (hydrogen) as a product of metals dissolved in acids. In 1772 Priestley, teacher, divine, and experimental philosopher, inspired by reading Hales, announced the new species "nitrous air" (NO) and hydrochloric acid gas; and in 1774–1775 he introduced "eminently respirable air" (oxygen), the peculiar portion of ordinary air that maintains life. In 1776 his correspondent Volta discovered a second inflammable air (methane) while gas hunting in a swamp.

In the early 1780s Cavendish, Lavoisier, and the inventor James Watt discovered that inflammable and eminently respirable air made water when sparked together. The rationale for the spark originated in Priestley's test for the respirability of gases: mix nitrous air with a sample under test and determine the contraction of the volume; the greater the diminution, the better the sample. (For oxygen the maximum contraction would be a third: $2NO + O_2 = 2NO_2$.) Volta had substituted inflammable gas for nitrous air and added the spark to speed up the process. He and other devotees of the new pneumatics devised "eudiometers" to test air by sparking. Thus they set up for themselves one of the grandest of all discoveries in physical science, the counterintuitive realization that gases that support combustion or burn freely combine to make the enemy of fire, water, and deprive it of its ancient right to be considered an element.

The discovery of the gas types led to a sweeping reformation of chemistry. It impelled natural philosophers to study the effects of heat on gases, which strengthened the caloric theory (see IMPONDERABLES) and supported measurements later important in *thermodynamics. It had practical consequences before the end of the eighteenth century in the application of eudiometry to ventilation, in the craze of ballooning initiated by the Montgolfier brothers, and in the use of laughing gas (nitrous oxide) as an anesthetic.

See also CAVENDISH, HENRY, AND CHARLES-AUGUSTIN COULOMB.

J. G. Crowther, *Scientists of the Industrial Revolution: Joseph Black, Henry Cavendish, Jospeh Priestley, James Watt* (1962). Henry Guerlac, *Essays and Papers in the History of Modern Science* (1977). J. L. Heilbron, *Weighing Imponderables* (1993).

J. L. HEILBRON

POLAR SCIENCE. Exploration of the Arctic and Antarctic, regions above 66E° 32′ North or South latitude, has been one of the most difficult and perilous chapters in the history of scientific investigation. Mortality has often been high, scientific returns often meager. The best known expeditions—those of Cook or Peary to the North Pole, and those of Scott, Amundsen, or Shackleton to the South Pole—are as remarkable for their lack of scientific interest as they are for their human drama. Most scientific information has come from scientists whose names barely register even among historians of science.

Much exploration of the North American and Eurasian Arctic from the sixteenth to the nineteenth centuries was driven by the search for Northwest and Northeast Passages—conjectural open-ocean northern routes to the Pacific. Equally important were pressures to exploit fisheries, whaling, and sealing grounds, and attempts to extend national sovereignty. Even where no obvious economic benefit lay, the enhancement of national prestige stimulated competition. The quest for personal glory and adventure also drove many of the most celebrated explorers.

Polar science includes *geography and *cartography, the study of flora and fauna, Arctic ethnology, and *geology and *paleontology; it is in this way like science in other latitudes. Its unique contribution has come in the study of ice and snow, weather and *climate (including paleoclimate), and ocean circulation. Sea ice covers 10 million square miles (7 percent of the surface) of the sea; glacier ice covers about 6.5 million square miles (11 percent of the earth's surface), most of it in Greenland and Antarctica; and this mass of ice has a profound effect on the earth's thermal regime and on ocean and atmospheric circulation.

Modern polar science began with the first International Polar Year (1882–1883). It was the brainchild of Lt. Carl Weyprecht of the Austrian navy, who had discovered Franz Josef Land (80E°N, 60E°E) in 1874 as part of the search for the Northeast Passage, a typically nineteenth-century geographic enterprise under aristocratic patronage. Weyprecht argued in 1875 that geographical finds like his would be unimportant unless they enlarged scientific inquiry. He urged a year of coordinated scientific observations (using the same instruments) by manned scientific stations ringing the North and South poles. This idea, supported by the newly created International Meteorological Congress, led to an International Polar Congress in 1879. A plan for eleven Arctic and four Antarctic stations was largely carried out. It involved 700 scientific workers and produced thirty-one quarto volumes of data. Analysis of this and other data by Vilhelm Bjerknes and his group at Bergen during World War I led to the discovery of the Polar Front and to the theory of "frontal weather."

Danish, Norwegian, and Swedish scientists played the most prominent role in polar science until quite recently. Denmark's sovereignty over Greenland (and Iceland) in the first part of the century occasioned more than thirty major expeditions. Notable among them were studies of the Greenland ice cap by the Swede Nils Nordenskiöld in the 1870s, the Norwegian Fridtjof Nansen in the 1880s, and the German Erich von Drygalski in the 1890s. Nansen's traverse of the inland ice (the first) led him to formulate a program of glaciological and geophysical investigation that guided studies of Greenland's ice (and later of Antarctica's) for almost fifty years. Shortly after, Nansen's specially built ship, the *Fram*, drifted for three years (1893–1896) locked in the Arctic sea ice between Eurasia and the North Pole, and provided data concerning ocean currents and the layering of Arctic waters. Using this, Nansen's coworker, the Swedish scientist Vagn Ekman, worked out a number of important problems in dynamic oceanography. His results, and the theoretical work of Vilhelm Bjerknes, spurred still more investigations by Johan Sandstrom and Bjørn Helland-Hansen on movement of water masses and on geostrophic currents, laying the foundations of physical oceanography.

Polar science was curtailed during World War I and, except for the relentless Danes, remained in limbo into the 1920s. In the early 1930s, interest in transatlantic flight led the United States, France, Britain, and Germany to set up overwintering stations in Greenland. The most important of these scientifically was Alfred Wegener's expedition of 1930–1931. In spite of Wegener's death in 1930, the year-long scientific program at a mid-ice station was completed. Using explosion *seismology, it demonstrated that Greenland was a basin weighed down by the ice cap itself and pioneered a technique later used to map Antarctica's geology. The world economic collapse of the later 1920s cut back plans for, and delayed the publication of results of, a second International Polar Year in 1932–1933.

Antarctic science, as opposed to Antarctic exploration, effectively began with the *International Geophysical Year of 1956–1957, which established fifty-five scientific stations in Antarctica. There followed an international treaty to keep the continent open as a scientific laboratory. This also disposed of the disputes over ownership of the Southern continent that emerged in the first half of the century. Interest in Antarctica intensified with the formulation of the geological theory of plate tectonics in the 1970s, since the continent was a key portion of the giant proto-continent Pangaea. Techniques for oxygen isotope dating have made Greenland and Antarctic ice cores an indispensable part of the climate record of the last 200,000 years. In recent years, Antarctica has proven an important resource in astronomy and astrobiology partly because iron meteorites, accumulated over millions of years, were not depleted there by aboriginal hunters seeking raw materials for weapons, and partly because the dry valleys of Antarctica provide extreme environments useful in planning for research on Mars. Investigations of ozone depletion in the atmosphere also have focused on Antarctica.

Jeanette Mirsky, *To the Arctic! The Story of Northern Exploration from Earliest Times to the Present* (1934; 1997). G. E. Fogg, *A History of Antarctic Science* (1992).

MOTT T. GREENE

PULSARS AND QUASARS. Exotic astronomical objects were first detected in the 1960s by their radio emissions, though most quasars (quasi-stellar radio sources) are radio quiet. Pulsars are pulsating radio sources.

In 1960, astronomers identified the radio source 3C48 (number 48 in the Third Cambridge Catalogue) with a star-like object. Three years later, they did the same for 3C 273, which had emission lines at unusual wavelengths. Maarten Schmidt at the Mount Palomar Observatory in Southern California recognized the mysterious spectral lines as lines from common elements shifted far toward the red. One mystery solved led to another. Assuming the red shifts arose from the expansion of the universe, astronomers had to endow quasars with tremendous speeds and distances. To be visible at vast distances, a quasar would have to be enormously bright, a thousand times brighter than all the stars in our galaxy. But a quasar's rapid variation in time would require that its energy be produced in a small volume. No known nuclear process can yield the observed energy output from a small volume.

The need for new physical explanations for new astronomical phenomena expanded further. In 1967, Anthony Hewish and Jocelyn Bell at Cambridge University, looking for rapid variations in the radio brightness of quasars, discovered a rapidly pulsating radio source. The radiation had to be from a source not larger than a planet if the signal could spread so quickly across the object to trigger bursts of radiation. Hewish won a *Nobel Prize for the discovery, though his student Bell made the actual observation.

The period of this pulsar, 1.3 seconds, was so regular that Hewish and Bell briefly thought it might be an interstellar beacon or radio lighthouse built by an alien civilization. Hence the name they gave the source LGM 1—LGM for Little Green Men.

A few months before the discovery, Francis Pacini published a theoretical paper showing that a rapidly rotating neutron star with a strong magnetic field could act as an electric generator and emit radio waves. (Once thermonuclear sources of energy are exhausted, stars of less than 1.4 solar mass shrink until they become white dwarfs; more massive stars continue contracting into even more dense stars composed of neutrons.) Most millisecond pulsars have white dwarf companions, and their amplified spins may be somehow attributable to the accretion of mass from the companion star.

The cosmological hypotheses for quasars—that their red shifts are associated with the expansion of the universe and that they are at great distances from the earth—are generally accepted. Observations by Halton Arp at Palomar, however, suggested possible physical connections between a few quasars and nearby galaxies. A committee at Palomar judged Arp's controversial research to be without value, and terminated his observing time in 1983. Quasars probably are powered by the energy released when matter falls into a gigantic rotating *black hole. Why so many quasars have red shifts around 2 remains to be explained.

See also ASTRONOMY, NON-OPTICAL; CHANDRASEKHAR; COSMOLOGY; RELATIVITY.

F. G. Smith, *Pulsars* (1977). Ajit K. Kembhavi and Jayant V. Narliker, *Quasars and Active Galactic Nuclei: An Introduction* (1999).

NORRISS S. HETHERINGTON

Q

QUANTUM ELECTRODYNAMICS

(QED) is the *quantum field theory describing the interaction of charged particles with photons. It represents positive and negative electrons by a quantized field satisfying the *Dirac equation in the presence of an electromagnetic (e.m.) field; charged spin 0 particles, such as pi mesons, by quantized field operators satisfying the Klein-Gordon equation; and the electromagnetic field by quantized field operators satisfying *Maxwell's equations. The source terms in these Maxwell equations are the charge-currents arising from the matter field in the presence of the quantized e.m. field. The small dimensionless constant $\pi = 2\pi e^2/hc = 1/137$, where e is the electronic charge, h Planck's constant, and c the velocity of light, measures the coupling between the charged matter field and the electromagnetic field. Since α is so small, the coupled equations are usually solved "peturbatively," that is, as a power series expansion in α. This perturbative approach has had amazing success in calculating extremely fine details in atomic spectra, accounting for the electromagnetic properties of electrons and muons, and predicting with precision the outcome of collisions between high-energy positive and negative electrons. As Toichiro Kinoshita and Donald Yennie, two theorists who have carried out some of the most extensive and difficult calculations testing the limits of QED, wrote in 1990, "it is inconceivable that any theory which is conceptually less sophisticated could produce the same results."

Richard *Feynman's great contribution to QED was a technique by which perturbations could be visualized and calculated by straightforward diagrams. The diagrams indicate both why and how certain processes take place in particular systems. In Feynman's approach, as generalized by Freeman Dyson, each quantized field (and associated particle) is characterized by a "propagator" represented in a Feynman diagram by a line, which if internal connects to two vertices, and if external, that is, if corresponding to an incoming or outgoing particle, connects to a single vertex. Each interaction is represented by a vertex characterized by a coupling constant and a factor describing the interaction between the fields. For a given process relatively simple expressions occur in the lowest order of perturbation theory. The diagrams that correspond to higher order contributions contain closed loops and entail integrations over the momenta of the propagators involved in the loops. In almost all cases these integrals diverge because of contributions from large momenta.

The anomalous magnetic moment of a (quasi-free) electron means the deviation from the value predicted by the Dirac equation. According to Dirac's theory, the electron has an intrinsic magnetic moment accompanying its spin, the value of which when expressed in the form $g_e = eh/4\pi mc$ is given by $g_e = 2$. The electron's anomalous magnetic moment is defined as $a_e = (g_e - 2)/2$.

Julian Schwinger's computation of a_e in 1947 constituted a landmark in the postwar developments of QED. It confirmed the experimental value that had been obtained by Isador Isaac Rabi and his associates and also the ideas of mass and charge renormalization in the low orders of QED. Since that time, both the experiments and the theory have been improved by several orders of magnitude and have provided the most precise and rigorous tests for the validity of QED. To date, the best theoretical and experimental values of the anomalous magnetic moment of the electron agree to ten significant figures—(in units of $eh/4\pi mc$) 1.001 159 652 17 (theoretical) against 1.001 159 652 19 (measured).

Richard Feynman, *QED: The Strange Theory of Light and Matter* (1985). Toichiro Kinoshita, ed., *Quantum Electrodynamics* (1990). Silvan S. Schweber, *QED and the Men Who Made It: Dyson, Feynmen, Schwinger, and Tomonaga* (1994).

SILVAN S. SCHWEBER

QUANTUM FIELD THEORY.

When initially formulated quantum mechanics described non-relativistic systems with a finite number of degrees of freedom. The extension of the formalism to include the interaction of charged particles with the electromagnetic field treated quantum mechanically brought out the difficulties connected with the quantization of systems with an infinite number of degrees of freedom. The effort to make the quantum theory conform with special *relativity disclosed further difficulties. To address both sets of problems, Ernst Pascual Jordan, Oskar Klein, Eugene Wigner, Werner *Heisenberg, Wolfgang Pauli (see HEISENBERG AND PAULI), Enrico *Fermi, and others developed quantum field theory (QFT) during the late 1920s. P. A. M. *Dirac had taken the initial step in 1927 with a quantum mechanical description of the interaction of charged particles with the electromagnetic field, which he described as an (infinite) assembly of photons, that is, of massless spin 1 particles. Dirac considered "particles" (whether they had a rest mass or, like photons, had none) to be the "fundamental" substance. In contrast, Jordan insisted that fields constituted the "fundamental" substance.

The history of theoretical elementary particle physics until the mid-1970s can be narrated in terms of oscillations between the particle and field viewpoints epitomized by Dirac and by Jordan. QFT proved richer in potentialities and possibilities than the quantized-particle approach. By the mid-1930s the imposition of special relativity on QFT had produced genuinely novel features: the possibility of particle creation and annihilation, as first encountered in the quantum mechanical description of the emission and absorption of photons by charged particles; the existence of anti-particles; and the complexity of the "*vacuum. " The latter was now seen to be not a simple substance but the seat of fluctuations in the measured observables, which fluctuations are the larger the smaller the volume probed.

Fermi's theory of beta-decay (1933–1934) was the important landmark of field theoretic developments of the 1930s. The process of β-decay—in which a radioactive nucleus emits an electron (β-ray) and increases its electric charge from Z to $Z+1$—

had been studied extensively during the first decade of the century (see RADIOACTIVITY). In 1914 James Chadwick found that the energy of the emitted electrons varied continuously up to some maximum at which conservation held to the accuracy of the measurements in the experiment. By the end of the 1920s no satisfactory explanation of the continuous β-spectrum had been found and some physicists, in particular Niels *Bohr, proposed giving up energy conservation in β-decay processes. In December 1930 Pauli, in a letter addressed to the participants of a conference on radioactivity, countered with "a desperate remedy." He suggested that "there could exist in the nuclei electrically neutral particles that I wish to call neutrons [later renamed neutrinos by Fermi], which have spin 1/2.... The continuous β-spectrum would then become understandable by the assumption that in β-decay a [neutrino] is emitted together with the electron, in such a way that the sum of the energies of the [neutrino] and electron is constant."

Fermi heard about Pauli's hypothesis for the first time at the Solvay Congress of 1933. He soon formulated a theory of β-decay that marked a change in the concept of "elementary" processes. Fermi supposed that electrons do not exist in nuclei before their emission, but that (to quote his version of 1934) "they, so to say, acquire their existence at the very moment when they are emitted; in the same manner as a quantum of light, emitted by an atom in a quantum jump, can in no way be considered as pre-existing in the atom prior to the emission process. In this theory, then, the total number of the electrons and of the neutrinos (like the total number of light quanta in the theory of radiation) will not necessarily be constant, since there might be processes of creation or destruction of these light particles."

Both Fermi's theory of β-decay and *quantum electrodynamics (QED) made clear the power of a quantum field theoretical description. In particular, they indicated that the electromagnetic forces between charged particles could be understood as arising from the exchange of virtual photons between the particles—virtual particles because they do not obey the energy-momentum relation that holds for free photons. When one of the charged particle emits a (virtual) photon of momentum k, it changes its momentum by this amount. When the second charged particle absorbs this virtual photon, it changes its momentum by k. This exchange is the mechanism of the force between the interacting particles. The range of the force generated is inversely proportional to the mass of the virtual quantum exchanged. Zero-mass photons generate electromagnetic forces of infinite range. Spin-zero quanta of mass m generate forces with a range of the order h/mc. This insight led Hideki *Yukawa to postulate that the short-range nuclear forces between nucleons could arise from the exchange of massive spin 0 bosons. Another important lesson learned from QED and Fermi's theory of β-decay was the protean nature of particles. When interacting with one another "particles" can metamorphose their character and number: in a collision between an electron and its anti-particle, the positron, the electron and positron can annihilate and give rise to a number of photons.

By the late 1930s, physicists understood the formalism of quantum field theory and its difficulties. All relativistic QFTs have the mathematical problem that the calculations of the interactions between particles give infinite, that is, nonsensical results. The root cause of these divergences was the assumption of locality, the assumption that the local fields—fields definable at a point in space-time point—whose quanta are the experimentally observed particles interact locally, i.e. at a point in space time.

Local interaction terms implied that in QED photons will couple with (virtual) electron-positron pairs of arbitrarily high momenta, and that electrons and positrons will do the same with (virtual) photons, in both cases giving rise to divergences. The problem impeded progress throughout the 1930s and caused most of the workers in the field to doubt the correctness of QFT. The many proposals to overcome these divergences advanced during the 1930s all ended in failure. The pessimism of the leaders of the discipline—Bohr, Pauli, Heisenberg, Dirac, and J. Robert Oppenheimer (see KURCHAKOV AND OPPENHEIMER)—was partly responsible for the lack of progress. They had witnessed

the overthrow of the classical concepts of space-time and had themselves rejected the classical concept of determinism in the description of atomic phenomena. They had brought about the quantum mechanical revolution. They were convinced that only further conceptual revolutions would solve the divergence problem in quantum field theory.

The way to circumvent the difficulties was indicated by Hendrik Kramers in the late 1930s and his suggestions were implemented after World War II. These important developments stemmed from the attempt to explain quantitatively the discrepancies between the empirical data and the predictions of the relativistic Dirac equation for the level structure of the hydrogen atom and the value it ascribed to the magnetic moment of the electron. These deviations had been observed in reliable and precise molecular beam experiments carried out by Willis Eugene Lamb, Jr., and by Isidor Isaac Rabi and coworkers at Columbia, and were reported at the Shelter Island Conference in the fall of 1947. Shortly after the conference, Hans Albrecht *Bethe showed that the Lamb shift (the deviation of the $2s$ and $2p$ levels of hydrogen from the values given by the Dirac equation) was of quantum electrodynamical origin, and that the effect could be computed by making use of what became known as "mass renormalization, " the idea that had been put forward by Hendrik Kramers.

The parameters for the mass m_o and for the charge e_o that appear in the equations defining QED are not the observed charge and mass of an electron. The observed mass m enters the theory through the requirement that the energy of the physical state corresponding to an electron moving with momentum p be equal to $(p^2+m^2)^{1/2}$. The observed charge e enters through the requirement that the force between two electrons at rest separated by a large distance r satisfy Coulomb's law e^2/r^2. Julian Schwinger and Richard *Feynman showed that the divergences encountered in the low orders of perturbation theoretic calculations could be eliminated by re-expressing the parameters m_o and e_o in terms of the observed values m and e, a procedure that became known as mass and charge renormalization. In 1948 Freeman Dyson working at the Institute for Advanced Study in Princeton proved that these renormalizations could absorb all the divergences arising in scattering processes (the S-matrix) in QED to all orders of perturbation theory. More generally, Dyson demonstrated that only for certain kinds of quantum field theories can all the infinities be removed by a redefinition of a finite number of parameters. He called such theories renormalizable. Renormalizability thereafter became a criterion for theory selection.

The idea of mass and charge renormalization, implemented through a judicious exploitation of the symmetry properties of QED, made it possible to formulate and to give physical justifications for algorithmic rules to eliminate all the ultraviolet divergences that had plagued the theory and to secure unique finite answers. The success of renormalized QED in accounting for the Lamb shift, the anomalous magnetic moment of the electron and of the muon, the scattering of light by light, the radiative corrections to the scattering of photons by electrons, and the radiative corrections to pair production was spectacular.

Perhaps the most important theoretical accomplishment between 1947 and 1952 was providing a firm foundation for believing that local quantum field theory was the framework best suited for the unification of quantum theory and special relativity. Perspicacious theorists like Murray Gell-Mann also noted the ease with which symmetries—both space-time and internal symmetries—could be incorporated into the framework of local quantum field theory. Gauge invariance became a central feature of the quantum field theoretical description of the electromagnetic field. Subsequently, the weak and the strong forces were similarly described in terms of gauge theories.

See also ELEMENTARY PARTICLES; QUANTUM ELECTRODYNAMICS; QUARK; SYMMETRY.

Abraham Pais, *Inward Bound* (1986). Olivier Darrigol, *From C-Numbers to Q-Numbers: The Classical Analogy in the History of Quantum Theory* (1992). Silvan S. Schweber, *QED and the Men Who Made It* (1994). Michael E. Peskin and Daniel V. Schroeder, *An Introduction to Quantum Field Theory* (1995). Helge Kragh, *Quantum Generations: A History of Physics in the Twentieth Century* (1999).

SILVAN S. SCHWEBER

QUANTUM PHYSICS. The proximate origin of the quantum theory was a perplexing paper published by Max *Planck in 1900. In it he showed that the formula he had proposed for the empirically determined spectral density of blackbody radiation could be derived by setting the energy of the collection of charged harmonic "resonators" (which he used to represent atoms capable of emitting and absorbing electromagnetic radiation) of frequency v equal to an integral multiple of hv. Here h stood for a new physical constant necessary to fit the empirical spectrum and v for the frequency of the resonator. The derivation required recourse to Ludwig *Boltzmann's probability calculation for the *entropy of a gas. It appears that, in adapting it to the blackbody problem, Planck did not recognize that he had made a break with the physics he had used to describe radiation.

In any case, Planck had full confidence in the representation of the electromag-

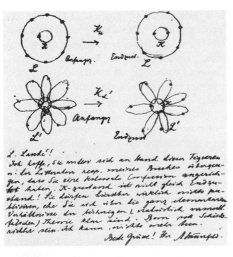

Depiction of the quantum process that, according to Arnold Sommerfeld, gave rise to certain X-ray lines. Sommerfeld drew the diagram to persuade Alfred Landé to abandon a different model, which was also wrong.

netic field given by James Clerk *Maxwell and Hendrik Antoon Lorentz. The unification of light with electromagnetism, the demonstration of the "reality" of electromagnetic waves by Heinrich Hertz (see HELMBOLTZ AND HERTZ), and the description given by the Maxwell-Lorentz equations of a multitude of wave phenomena was for Planck, and for almost all of his contemporaries, convincing evidence of the continuous nature of radiation. Albert *Einstein entertained doubts. Having scrutinized the statistical mechanical foundations upon which Planck based his derivation of his formula for the spectral density of blackbody, Einstein concluded in 1905 that a few phenomena, like the photo-electric effect, could be explained easily if "the energy of monochromatic light consists of a finite number of energy quanta of magnitude $h\nu$, localized at various points of space [that] can be produced or absorbed only as units." At about the same time, Einstein realized that Planck's radiation theory required a radical discontinuity in the energy content of the individual resonators; with his "heuristic hypothesis" concerning the photo-effect, Einstein extended the discontinuity to the free electromagnetic field, and to the interaction between light and matter.

Einstein's explanation in 1907 of the observed deviation at low temperature of the specific heat of simple solids from their classical value of $3Nk$ (N = the number of molecules in a gram, k = "Boltzmann's constant," a second universal constant from the blackbody formula) corroborated the quantum hypothesis. In Einstein's model of a solid the potential that an atom experiences near its equilibrium position is the same for all the atoms of the solid. Hence for small vibrations near their equilibrium point all the atoms oscillate with the same frequency ν . Quantization implies that each oscillator can only have an energy equal to $\varepsilon_n = nh\nu$ and Planck's formula gives, in the limit where $h\nu$ is small in comparison with kT (T = temperature), the specific heat $3Nk$. At low enough temperatures, where the limit does not hold, characteristic deviations from the classical value occur, which Walther Nernst and others detected around 1910.

In his doctoral thesis on the electron theory of metals (1911), Niels *Bohr concluded that atoms constructed according to the principles of classical physics could not represent the magnetic properties of metals. Working in Ernest *Rutherford's laboratory in Manchester just after Rutherford proposed the nuclear model of the atom, Bohr seized upon it because its radical mechanical instability made it a promising candidate for repair by a quantum hypothesis (see ATOMIC STRUCTURE). Bohr stabilized the Rutherford atom by supposing that it could exist in various "stationary states" constrained by certain quantum rules but otherwise governed by the laws of classical mechanics. However, the laws of mechanics do not hold for the transition of the system between two stationary states during which the atom radiates a quantum of energy $h\nu$ equal to the difference in energy between the two states. On Bohr's theory, radiation is not emitted (or absorbed) in the continuous way assumed by Maxwell-Lorentz electrodynamics.

Bohr's first postulate, which (in one of its forms) limited the validity of classical mechanics in the atomic domain, restricted the angular momentum of each atomic electron to an integral multiple of $h/2\pi$. The second postulate, which denied the validity of classical electrodynamics for radiative processes in atoms and made the frequencies of atomic spectral lines different from the orbital frequencies of the electronic motions, required surrendering the classical connection between the frequency ν of the emitted radiation and the mechanical frequency of the electron in its orbit.

With the help of these quantum rules Bohr accounted for the phenomenological regularities that had been discerned in the hydrogen spectrum, in particular, the Balmer formula for transitions to the $n = 2$ level, and also, and more dramatically, for the spectrum of ionized helium (1913–14). During World War I, Arnold Sommerfeld generalized Bohr's postulates to elliptical electron orbits and then to motions in three dimensions. He recorded his success in calculating regularities in doublet and triplet spectra, in the Zeeman effect, and in x-ray spectra in a long book, *Atombau und Spektrallinien* (first edition 1919), with which all physicists interested in quantum and atomic physics during the early 1920s began their work.

In the early 1920s Bohr gave a phenomenological explanation of the periodic table based on the occupancy by electrons of Coulomb-like orbits in multi-electron atoms. Thereafter, many theorists tried to justify Bohr's explanation, but, except for Wolfgang Pauli's formulation of the exclusion principle early in 1925 (see HEISENBERG AND PAULI), none of their efforts provided a stable foundation for the dynamics of atoms. They were seminal, however, in that they made manifest the problems a more complete quantum mechanics would have to solve.

In 1917 Einstein took what in retrospect was an important step toward this mechanics. Still flirting with the corpuscular nature of radiation, Einstein introduced the concept of the probability for the spontaneous emission of a light quantum by a "molecule" in a transition from one state to another. The concept allowed an easy derivation of Planck's blackbody formula. In 1923, Arthur Holly Compton's experiment on the scattering of *X rays by electrons indicated that the shift in the wave length of the scattered X ray and the recoil energy of the electron could be derived on the assumption that the X rays acted as particles with energy $h\nu$ and momentum $h\nu/c$ (c the velocity of light). The positive result of the Compton experiment led Einstein to declare that there are "two theories of light, both indispensable, and...without any logical connection." The corpuscular viewpoint accounted for the optical properties of atoms, whereas macroscopic phenomena like diffraction and interference required the wave theory of light. The two theories coexisted without any resolution during the early 1920s.

Another important guide to a more powerful quantum physics was the correspondence principle Bohr refined between 1913 and 1918. It stated that the frequencies calculated by Bohr's second postulate (during "quantum jumps") in the limit where the stationary states have large quantum numbers that differ very little from one another will coincide with the frequencies calculated with the classical theory of radiation

from the motion of the system in the stationary states. Bohr's assistant Hendrik Kramers cleverly applied the correspondence idea to compute the intensity and polarization of the light emitted from simple atoms. Kramers and Werner *Heisenberg extended the same idea to the dispersion of light and worked out ways to translate classical quantities involving a single stationary state into quantum mechanical quantities involving two or more states. Max Born, Heisenberg's teacher at the University of Göttingen, called for a "quantum mechanics" for calculating with the quantum mechanical quantities directly. That was in 1924. In less than a year Heisenberg provided him with one. Its guiding principles were satisfaction of Bohr's correspondence principle (in the appropriate limit the theory should yield the classical results); recognition that the troubles of the "old quantum theory" arose primarily from breakdown of the kinematics underlying classical dynamics; and restriction of the theory to relations between observable quantities.

Born, Heisenberg, and a fellow student of Heisenberg's, Pascual Jordan, soon developed the new mechanics into an elaborate mathematical formalism. They built a closed theory that displayed strikingly close analogies with classical mechanics, but at the same time preserved the characteristic features of quantum phenomena. Their work laid the foundations of a consistent quantum theory but at the price of relinquishing the possibility of giving a physical, visualizable picture of the processes it could calculate. Hence the relief felt by Planck, Einstein, and Lorentz when Erwin *Schrödinger, who followed a route entirely different from Heisenberg's, began to publish his wave mechanics in 1926. It seemed to avoid the unconventional features of Heisenberg's formulation and rested on more traditional foundations and easier calculations: variational principles, differential equations, and the properties of waves.

Schrödinger had followed up insights and suggestions by Louis de Broglie and Einstein. In 1923 de Broglie published an idea that was the obverse of Einstein's attribution of particle properties to wave radiation—to endow discrete matter with wave properties. By following sometimes fanciful analogies and the principle of *relativity, de Broglie associated a wave of frequency v and wavelength λ with a particle of momentum p and energy E according to $v = E/h$, $1/\lambda = p/h$. He thus extended the particle-wave duality of radiation to matter. Knowing the wavelength, Schrödinger soon found an appropriate differential equation for a wave of amplitude Ψ. He interpreted the Ψ function as describing a real material wave and considered the electron not a particle but a charge distribution whose density is given by the square of the wave function. In a short paper dated June 1926, Born rejected Schrödinger's viewpoint and proposed a probabilistic interpretation for the Ψ function. He stipulated that the wave function $\Psi(x, t)$ determines the probability of finding the electron at the position x at time t. In 1927 two different sets of experimentalists—George P. Thomson (the son of Joseph John *Thomson) in Britain and Clinton Davisson and Lester Germer in the United States—detected diffraction patterns from an electron beam.

Several physicists proved in 1926 that wave mechanics gave the same numerical answers as the "matrix mechanics" of Born, Heisenberg, and Jordan. Together they are known as quantum mechanics. In contrast to classical physics, which contained no scale and was assumed to apply both in the micro and macro domain, quantum mechanics asserted that the physical world presented itself hierarchically. As P. A. M. *Dirac emphasized in the first edition of his *Principles of Quantum Mechanics*, Planck's constant allows the parsing of the world into microscopic and macroscopic realms.

The conquest of the microrealm during the first years after the invention of quantum mechanics stemmed from the confluence of two factors: the belief in an approximately stable ontology of electrons and nuclei, and the formulation of the dynamical laws governing the motion of electrons and other microscopic particles moving with velocities small compared to the velocity of light. Approximately stable meant that electrons and (non-radioactive) nuclei, the building blocks of atoms, mole-

cules, and simple solids, could be treated as ahistoric objects with physical characteristics independent of their mode of production and lifetimes effectively infinite. These electrons and nuclei behaved as if they were "elementary," almost point-like objects specified only by their mass, their intrinsic spin, electric charge, and magnetic moment. In addition, the members of each species were indistinguishable. Their indistinguishability implied that an assembly of them obeyed characteristic statistics depending on whether they had integral or half odd integral spin (measured in multiples of $h/2\pi$). Bosons (particles with zero or integral spins) can assemble in any number in a given quantum state. Fermions (particles with half odd integral spins) do not share a quantum state. A one-particle quantum state can be characterized either by the position and the spin state of the particle or by its momentum and its spin state. Thus no two identical Fermions can be at the same position if they have the same spin. More generally, the wave function describing a system of identical bosons remains unchanged under the interchange of any two particles, whereas that describing fermions changes sign.

The quantum mechanical explanation of chemical bonding resulted in a unification of physics and chemistry. In 1929, following the success of nonrelativistic quantum mechanics in explaining atomic and molecular structure and interactions, Dirac, a main contributor to these developments, declared that "the general theory of quantum mechanics is now almost complete." Whatever imperfections still remained were connected with the synthesis of the theory with the special theory of relativity. But these were of no importance in the consideration of atomic and molecular structure and ordinary chemical reactions. "The underlying physical laws necessary for the mathematical theory of a large part of physics and the whole of chemistry are thus completely known, and the difficulty is only that the exact application of these laws leads to equations much too complicated to be soluble."

Most chemists and many physicists do not share Dirac's reductionist philosophy and its implication that quantum physics is the most fundamental of the sciences. As emphasized by Phillip Anderson, who specialized in condensed matter theory, "the reductionist hypothesis does not by any means imply a 'constructionist' one: The ability to reduce everything to simple fundamental laws does not imply the ability to start from those laws and reconstruct the universe. In fact, the more the elementary particle physicists tell us about the nature of the fundamental laws, the less relevance they seem to have to the very real problems of the rest of science, much less to those of society. The constructionist hypothesis breaks down when confronted with the twin difficulties of scale and complexity."

See also QUANTUM ELECTRODYNAMICS; QUANTUM FIELD THEORY.

Thomas S. Kuhn, *Blackbody Theory and the Quantum Discontinuity, 1894–1912* (1978). Abraham Pais, *Inward Bound* (1982). Olivier Darrigol, *From C-Numbers to Q-Numbers* (1992). Mara Beller, *Quantum Dialogue: The Making of a Scientific Revolution* (1999). Helge Kragh, *Quantum Generations* (1999).

J. L. HEILBRON

QUARK. During the 1950s and 1960s progress in classifying and understanding the phenomenology of the ever increasing number of hadrons (strongly interacting microscopic particles) came not from fundamental theory but by shunning dynamical assumptions in favor of *symmetry and kinematical principles that embodied the essential features of a relativistic quantum mechanics.

In 1961 Murray Gell-Mann and Yuval Ne'eman independently proposed classifying the hadrons into families based on a symmetry later known as the "eightfold way." They realized that the mesons (hadrons with integral spins) grouped naturally into octets; the baryons (heavy hadrons with half integral spins) into octets and decuplets. The "eightfold way" can be represented mathematically in three dimensions, a property that led Gell-Mann, and independently George Zweig, to build hadrons out of three elementary constituents. Gell-Mann called these constituents "quarks" (from a line in James Joyce's *Finnegans Wake*, "Three quarks for Muster Mark!"); Zweig called them aces. The elaboration of the quark scheme is briefly indicated here as an indication of the methods and madness of elementary particle physics.

To account for the observed spectrum of hadrons Gell-Mann and Zweig defined three "flavors" of quarks (generically indicated by q), called up (u), down (d), and strange (s), each with spin 1/2 but differing in two other quantum numbers (isotopic spin and strangeness) that defined them. Ordinary matter contains only u and/or d quarks. ("Strange" hadrons would contain strange quarks.) The three quarks had two other features: a baryonic mass of 1/3, and an electrical charge of 2/3 (for the u) and of –1/3 (for the d and s), those of the proton. This last feature startled physicists who had no experimental evidence for any macroscopic object carrying a positive charge smaller than a proton's or a negative charge smaller than the electron's.

Since a relativistic quantum mechanical description implies that for every charged particle there exists an "anti-particle" with the opposite charge, Gell-Mann and Zweig provided for antiquarks (generically denoted by $\bar{q}$) having an electric charge and strangeness opposite to those of the corresponding quarks.

Quarks bind together into hadrons as follows. An up and an antidown quark make a positive meson; two ups (with electrical charge 4/3) and a down (with electrical charge –1/3), a proton. All baryons can be made up of three quarks, all mesons of one quark and one antiquark. That, however, did not provide quite enough possibilities so quarks had to have another attribute, which, in the playful quark nomenclature, was called "color." Quark color comes in three varieties (sometimes taken to be red, yellow, and blue), each of which can be "positive" or "negative." Quarks carry positive color charges and antiquarks carry negative ones. The observed hadrons have no net color charge.

In the late 1960s the Stanford Linear Accelerator (SLAC) could produce electrons of sufficiently high energy to probe the internal structure of protons. If the proton's charge were uniformly distributed, penetrating electrons would tend to go through protons without being appreciably deflected. If, on the other hand, the charge was localized on internal constituents, then—in analogy to Ernest *Rutherford's demonstration of the atomic nucleus—an electron that passed close to one of them would be strongly deflected. The SLAC experiments showed this effect, which prompted Richard *Feynman to suggest that the proton contained pointlike particles with spin 1/2, which he called "partons." The partons soon were assimilated to the quarks, although they (the partons/quarks) appeared to be too light and too mobile to make up protons. These difficulties were eventually resolved.

The discovery in November 1974 of the J/ψ meson gave further support for the quark picture and reason to accept a fourth quark with a new flavor, "charm" (denoted by c), whose existence had been proposed by Sheldon Glashow and others in 1964 and, with greater insistence, in 1970. The J/ψ (its two names resulted from its simultaneous discovery by two different groups) appeared to be a bound state of c and $\bar{c}$. The discovery of November 1974 revolutionized high-energy physics by establishing the representation of hadrons as quark composites. As the number of hadrons grew, however, the scheme had to be extended by the addition of the "bottom" (or "beauty," b) quark in 1977 and the "top" (t) quark in

1994. Each successively discovered quark has a larger mass than its predecessors: on a scale in which the u weighs 1, the d weighs 2, the s 36, the c 320, the b 960, and the t 34,800. They all have spin 1/2, partake in the strong, electromagnetic, and weak interactions, and come in pairs: up and down (u, d), charm and strange (c, s), and top and bottom (t,b). The first member of each pair has electric charge 2/3 and the second –1/3. Each flavor comes in three colors.

If hadrons are made up of fractionally charged quarks, why have fractionally charged particles not been observed? Even if a plausible mechanism could be devised for confining quarks, what reality can be attached to them as constituents of hadrons if they can never be observed empirically? Quantum chromodynamics (QCD), which emerged a decade after the introduction of quarks, explained how quarks could be so strongly bound that they could never escape, while nevertheless behaving as quasifree particles in deep inelastic scattering.

K. Gottfried and V. Weisskopf, *Concepts of Particle Physics* (1986). M. Riordan, *The Hunting of the Quark: A True Story of Modern Physics* (1987).

SILVAN S. SCHWEBER

QUASARS. See PULSARS AND QUASARS.

R

RADIOACTIVITY. Studies in radioactivity produced the scientific research fields of *nuclear physics, *cosmic-ray physics, and *high-energy physics, and also *nuclear physics and chemistry, nuclear medicine, and nuclear engineering. Beginning in 1940 with the isolation of neptunium and plutonium, the creation of short-lived transuranium elements extended the *periodic table of the chemical elements into new territory.

That radioactivity has rested in the shared disciplinary terrains of physics and chemistry is indicated by the shared Nobel Prize in physics awarded to its discoverers Henri Becquerel, Marie Sklodowska *Curie, and Pierre Curie, and Ernest *Rutherford's receipt of the Nobel Prize in chemistry in 1908 for his (and Frederick Soddy's) elucidation of the mechanism of the radioactive disintegration of atoms.

Becquerel discovered that uranium salts emit radiation by accident while investigating whether naturally phosphorescent minerals produce the *X rays discovered by Wilhelm Conrad *Röntgen in late 1895. By 1897 Becquerel and others had demonstrated that uranium radiations carry electrical charge, a physical property that led Marie Curie to apply the quartz electrometer (then recently invented by her husband and his brother Jacques-Paul Curie) to a variety of minerals in search of the property that she termed "radioactivity." She

Marie *Curie at work in her laboratory in Paris toward the end of World War I.

A German advertisement for a radioactive toothpaste for cleaning the teeth, whitening the smile, and invigorating the gums.

identified polonium and radium as radioactive elements in 1898 and André-Louis Debierne discovered actinium in 1899. They and others found that thorium, like uranium, is radioactive.

Rutherford distinguished two kinds of radiation, which he called "alpha" (distinguished by its ready absorption) and "beta" (about one hundred times more penetrating). At the Curies' laboratory in 1900, Paul Villard discerned a gamma radiation from radioactive substances that appeared to behave exactly like X rays. By 1900, Becquerel, Rutherford, and others had established that the beta rays consist of negatively charged particles similar if not identical to *electrons. In 1903 Rutherford showed that the alpha rays also carried a charge and so had to be understood as a stream of particles. By 1908 Rutherford and his colleagues had decisive evidence that the "alpha" radiations are helium ions. Rutherford and Soddy recognized in 1903 that thorium continuously produces a new radioactive gas (radon), and a second radioactive substance (chemically identical

to radium) from which a longer line of radioactive substances descends. They settled on the hypothesis that radioactive decay is accompanied by the expulsion of an alpha or beta particle from the decaying atom and that the decay can be expressed in terms of a half life, defined as the time during which half of the mass of a radioactive element is transformed into a new substance. Rutherford's continuing study of alpha particles and their interactions with matter prompted the experiments that led to his invention in 1910–1911 of the nuclear atom.

By 1912 some thirty radioactive elements had been identified and distributed into series beginning with uranium (238), actinium (227) or uranium (235), and thorium (232). In 1913 Soddy, Alexander Russell, and Kasimir Fajans independently developed a generalized radioactive displacement law. In 1911 Soddy had noted that loss of an alpha particle produced a chemical element two places to the left of the original in the periodic table. Similarly, Russell remarked that beta decays lead to the next element in the periodic table. Soddy coined the word "isotope" for chemically identical elements differing in atomic weight. Several studies confirmed in 1914 that atomic weights of lead derived from radioactive ores varied from each other and from the established value of 207.2.

Early methods for detecting radioactivity were both electrical and visual, relying initially on the electrometer and on microscopically observed scintillations of light caused as alpha particles strike zinc sulfide. Electrical methods of detection improved in 1928 when Hans Geiger and Walther Müller succeeded in making a reliable counter using the technique with which Geiger and Rutherford had counted alpha particles before the war. C. T. R. Wilson's *cloud chamber, invented in principle in 1899 but not perfected until after World War I, also provided a means of "seeing" nuclear events. Using the older scintillation technique, Rutherford proposed in 1919 that collision of an alpha particle with nitrogen gas resulted in disintegration of the nitrogen atom and expulsion of a long-range hydrogen atom (proton). After automating a cloud chamber, P. M. S. *Blackett in 1924 fired alpha particles into nitrogen atoms and obtained dramatic photographs showing the path of a proton ejected from a recoiling nitrogen nucleus and the capture of the alpha particle by the nitrogen nucleus, creating an isotope of oxygen.

Francis Aston, who began his career as an assistant to Joseph John *Thomson, developed a *mass spectrograph that provided a photographic record of particles separated by their masses. Aston and others found evidence for isotopes not only among the radioactive and heavy elements, but also among the light elements. The concept of the isotope thus became generalized for all chemical elements.

Although John Douglas Cockroft and Ernest T. S. Walton, and Ernest O. Lawrence and Milton Stanley Livingston, had built particle *accelerators to provoke nuclear reactions, alpha particles from radium caused the first radioactivity artificially induced by humans. That was in 1934, when Frédéric Joliot and Irène Joliot-Curie produced a radioactive isotope of phosphorus from aluminum.

Rays from naturally occurring sources continued to serve nuclear physicists and chemists even in the era of the cyclotron. Fermi and his collaborators irradiated every element they could find with neutrons derived from a radon-beryllium source, discovered the efficacy of slow neutrons in inducing nuclear reactions, and, mistakenly, believed that they had made transuranic elements by shining neutrons on uranium. Lise *Meitner, Otto Hahn, and Fritz Strassmann also used neutrons from natural sources in the experiments from which, by the end of 1938, Hahn and Strassmann obtained the results whereby Meitner (by then a refugee in Sweden) and her nephew Otto Robert Frisch deduced the existence of nuclear fission.

During the course of the Manhattan Project for the development of uranium and plutonium bombs, the health hazards of radioactivity increasingly came to the fore. Radiations that had been touted since the early 1900s as a general curative and a specific agent against cancer were demonstrated to cause leukemia and other cancer-related diseases. Facial creams and mineral waters, as well as watch dials and curios containing uranium salts, disappeared from store shelves and health resorts in the 1950s as international movements against the atmospheric testing of nuclear weapons gained force from evidence of the hazards of nuclear debris or "fallout." After enthusiasm in the early 1950s for the use of nuclear energy not only as a commercial power source but also for explosives in dam-building and road-building, public suspicion of radioactivity curtailed nuclear energy projects in the United States and Great Britain, although not in France or the Soviet Union.

In chemotherapy and in medical tests radioactive isotopes continue to serve as tagging or tracer devices for studying the metabolism or pathways of iodine, barium, and other elements in the body. In his first efforts in the 1930s to get large-scale funding for his accelerator program at the University of California at Berkeley, Lawrence emphasized medical applications. Large philanthropies such as the Rockefeller Foundation increasingly turned their funding priorities to medical research in the 1930s. Lawrence and his brother John Lawrence, who served as director of the university's medical physics laboratory, argued for the medical benefits of the production of radioactive isotopes in the accelerator. One result, phosphorus-32, was used in early attempts to treat leukemia.

Radioactive isotopes had uses beyond medicine. Samuel Ruben, W. Z. Hassid, and Martin Kamen at Lawrence's radiation laboratory used carbon-11 to study the chemistry of carbon dioxide in the photosynthesis of barley and the green algae chlorella. In 1941 Ruben and Kamen identified the radioactive isotope carbon-14 produced from nitrogen in the Berkeley accelerator. Melvin Calvin followed up at Berkeley by using a combination of paper chromatography and radiochemical techniques to unravel in detail the mechanism of photosynthesis.

Willard F. Libby attracted more public attention when he showed in 1946 that living matter contains carbon-14 produced by the collision of cosmic ray neutrons with atmospheric nitrogen, which enters the carbon dioxide and carbon monoxide metabolism. Since the quantity of carbon-14 decays after death, Libby's discovery made a brilliant new means of dating very old organic remains.

Since World War II, particle accelerators and nuclear reactors have entirely superseded natural elements as the sources of radioactive materials for laboratory experiments and medical procedures.

See also ATOMIC STRUCTURE; PERIODIC TABLE.

Aaron J. Ihde, *The Development of Modern Chemistry* (1964). T. J. Trenn, *The Self-Splitting Atom: A History of the Rutherford-Soddy Collaboration* (1977). Lawrence Badash, *Radioactivity in America: Growth and Decay of a Science* (1979). Roger H. Stuewer, "Artificial Disintegration and the Cambridge-Vienna Controversy," in *Observation, Experiment and Hypothesis in Modern Physical Science*, eds. P. Achinstein and Owen Hannaway (1985): 239–307. J. L. Heilbron and R. W. Seidel, *Lawrence and His Laboratory: A History of the Lawrence Berkeley Laboratory* (1989). Susan Quinn, *Marie Curie: A Life* (1995). Ruth Lewin Sime, *Lise Meitner: A Life in Physics* (1996). S. Boudia and X. Roque, eds., "Science, Medicine and Industry: The Curie and Joliot-Curie Laboratories," special issue of *History and Technology* 13 (1997): 241–354. Jeffrey Hughes, "Radioactivity and Nuclear Physics," in *Modern Physical and Mathematical Sciences*, ed. Mary Jo Nye (2003) [*The Cambridge History of Science*, vol. 5].

MARY JO NYE

RADIOASTRONOMY. See ASTRONOMY, NON-OPTICAL.

RADIUM, named for its spontaneous emission of ionizing radiation, was introduced to the world in 1898 by Marie and Pierre *Curie in Paris. This new element had revealed its presence solely by its radioactivity, the property that enabled the Curies, assisted by the chemist Gustave Bémont, to follow it through a series of chemical separations performed on pitchblende, a uranium ore.

The route to radium's discovery began with Marie Sklodowska Curie's doctoral research. While testing various minerals for radioactivity, Curie noticed that some uranium ore samples emitted more radiation

The popular reaction to radium as depicted on the cover of the British magazine Punch.

than expected from their uranium content. She deduced that these samples contained a new radioactive element. Pierre Curie joined her in a search for the unknown substance, which yielded two new elements, polonium in July 1898 and radium in December.

These findings met with both astonishment and skepticism: astonishment, because radium appeared so highly radioactive, and because radioactivity had never been used to detect an element; skepticism, for similar reasons. *Radioactivity, thought to be of no great importance, had been discovered barely two years earlier. Dubious reports and fanciful speculations about all sorts of invisible radiations circulated widely at the turn of the century.

Most chemists did not regard evidence from the new, insignificant, and poorly understood phenomenon of radioactivity as convincing. Although the spectroscopist Eugène-Anatole Demarçay had identified a new spectral line in the Curies' purified pitchblende sample, acceptance into the *periodic table required establishment of a distinct atomic weight. Marie Curie achieved this feat in 1902, after laborious chemical purifications. Radium took its place in the alkaline earth group above barium. Later it was assigned an atomic number, 88. In 1910 Marie Curie and André-Louis Debierne isolated metallic radium. Several isotopes of radium were identified after isotopy was recognized around 1913.

Radium's strong radioactivity made new experiments possible, which led to further discoveries and insights. Demand increased for this scarce element, which at first could only be obtained from the Curies and from the German chemist Friedrich Giesel, who barely missed discovering it. Radium's prodigious, seemingly endless energy output defied explanation. After scientists adopted the transmutation theory of radioactivity (1903-1906), researchers viewed radium as one of the decay products of uranium.

Eventually the greatest demand for radium came not from scientists but from physicians, who proposed to use the destructive effects radium wrought on living tissues to destroy tumors. The hope of a cure for cancer launched a prospecting fever and fueled a burgeoning radium industry, first in Europe, then in North America, and, from the 1920s, in central Africa. The worldwide search for uranium and its daughter radium led to investigations of environmental radioactivity and *cosmic rays.

The promise of a cure for a dreaded disease coupled with the unusual circumstances of radium's discovery and its remarkable powers stoked the public imagination. Radium became a metaphor for the magic elixir. During the 1920s and 1930s the well-meaning and charlatans alike hawked patent medicines and household products purportedly containing radium. Several towns built their economies on "radium water" spas. Radium was also used in phosphorescent paints. These became especially popular for watch dials, but fatal for many dial painters who accidentally ingested the paint. After the health hazards of radioactivity were recognized, most markedly after the first atomic bomb explosions in 1945, radium's image took on a sinister cast. Radium nevertheless continued to find medical and industrial uses.

Alfred Romer, ed., *Radioactivity and the Discovery of Isotopes* (1970). Susan Quinn, *Marie Curie: A Life* (1995).

MARJORIE C. MALLEY

RAINBOW. The diaphanous multicolored arcs in the sky with pots of gold at their ends puzzled humankind from the cloudy era of Noah to the daylight of René *Descartes. From Genesis 9:13 we know that God set the rainbow in a cloud as a covenant that he would not again drown the creatures of the earth. Natural philosophers kept their eyes on the cloud and not on the covenant. As early as the time of Aristotle they knew that the Sun, the sinner, and the bow's center all lie on a straight line, and supposed that the colors arose from rays from the Sun (or eye) reflected in the clouds. Aristotle considered the bow in his *Meteorologica*, which contains most of his terrestrial physics; and subsequent learned discussions about it, up through Descartes's *Météores* (1637), typically occurred in or around commentaries on Aristotle's book. When Descartes took up the problem as part of his demonstration of the superiority of his natural philosophy to the physics taught in the schools, the Aristotelian approach had been refined to place the reflection in individual raindrops rather than in the cloud as a whole, and the origin of the rays in the Sun rather than in the eye.

Descartes demonstrated superiority by calculating. He derived the angle at which an observer (at O in the first diagram) sees the uppermost part P of the bow. Aristotle could not do it and would have thought it unimportant to try; like Noah, he attended to the qualitative features of phenomena. One quantitative feature, however, was known to the Aristotelian school: the angle ϕ through which the Sun's rays that make the bow are turned at reflection by the raindrops is the sum of the altitudes of the sun α and of the bow at P, β: $\phi = \alpha + \beta$. The angle ϕ is always around 42°. If God showed Noah a rainbow when the Sun stood higher than 42°, it was a true miracle. For, as Descartes demonstrated, the angle of 42° follows directly from the values of the refractive indexes of air and water.

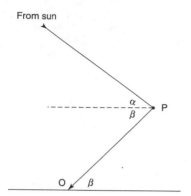

The relation between the altitude of the sun a and of the top of the bow β at P as seen by an observer at O.

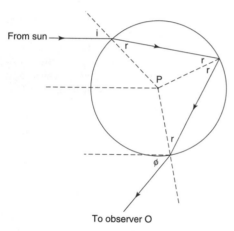

The path of a solar ray through a droplet at P incident at angle i, refracted at angle r, and turned through angle ϕ.

Willebrord Snel, professor of mathematics at the University of Leyden, established by measurement around 1620 the exact amount by which a ray of light traveling in air bends, or refracts, on entering water. With this law, Descartes could trace the path of a light ray reflected by a raindrop, as in the second diagram. He had no room to wiggle: geometry and the number 1.5, the refractive index of water relative to air, fixed ϕ. Since every angle of incidence i gives a different value for ϕ, this geometrical apparatus might not seem an advance. But, by tracing many rays, Descartes showed that for a certain range of values of

i (around 60°), ϕ stays very close to 42°. In a word, the droplets focus the rays. Explaining the colors and the pale secondary bow, with colors reversed, sometimes seen above the primary, required more words. Descartes ascribed the secondary to rays suffering two reflections in the drop. As for the colors, he suggested that in bouncing around the drop, light rays pick up spin, as if they were tennis balls sliced by a racquet, and that the different spins create the various colors when received by the eye (see MECHANICAL PHILOSOPHY). Isaac *Newton replaced this account with his revolutionary doctrine that each of the colors that he thought made up white light had a different index of refraction. On his theory, the drops focus the rays of different colors at slightly different angles ϕ.

Newton gave his theory of light and colors in a volume entitled *Opticks* (1704). Descartes explained the rainbow in his *Principles of Philosophy* (1644) as well as in his *Météores*. The difference in venue suggests the distance between the natural knowledge of the Continent around 1640, when the Aristotelian approach still provided the main competition to innovative natural philosophers, and that of England around 1700, when the claims of the Cartesians were the mark. As the most striking early example of the geometrization of a physical (as opposed to an astronomical) phenomenon, the rainbow as elucidated by Descartes and Newton represented a new covenant: that God would enable humankind to discover the numbers, weights, and measures that were employed in creation.

Carl B. Boyer, *The Rainbow from Myth to Mathematics* (2d edition, 1987).

J. L. HEILBRON

RAMAN, Chandrasekhara Venkata (1888–1970), physicist.

Raman's ancestral village lay on the banks of the River Kaveri in the Tanjore (Thanjavur) district of Southern India. The art, literature, and especially music that flourished in the region in the late seventeenth century are still palpable. Raman's mother, Parvati Ammal, was the daughter of a distinguished Sanskrit scholar. Raman's father, Chandrasekharan, a lecturer in physics, mathematics, and physical geography, was an avid reader and a connoisseur of Carnatic music. Raman developed an early and enduring interest in both music and science.

He graduated with a B.A. degree from the Presidency College of the University of Madras at the age of sixteen, ranked first in the university, and won gold medals in English and physics. In 1907 he obtained an M.A. degree in physics with the highest honors. Colonial contingencies prevented Raman from pursuing an academic career since only Indians who held advanced degrees from British universities were considered for such positions. So he joined the Financial Civil Service in 1907 as assistant accountant general and was posted to Calcutta. Before moving to Calcutta, he married Lokasundari, an accomplished musician.

In Calcutta, Raman came into contact with the Indian Association for the Cultivation of Science (IACS), which had been founded in 1876 by the physician Mahendralal Sircar to foster discussions and public exposition of new scientific developments. The laboratories of IACS provided an unexpected opportunity for Raman to pursue experimental research. He worked in the IACS laboratories before and after his daytime job as an accountant. His researches in optics, *acoustics, and musical instruments, and in particular his studies on the violin, won wide acclaim.

In 1917 Raman accepted the Palit Chair in Physics at the University of Calcutta even though the salary levels in the university were considerably less than in the Civil Service. The move to the university brought Raman a group of students and coworkers and greatly expanded the scope of his research. In 1921, on a voyage through the Mediterranean, he was fascinated by the deep blue color of the sea and conducted experiments on board the ship with the help of a small *telescope fitted with polarizers, analyzers, and a diffraction grating. He rejected Rayleigh's explanation that the color of the sea was "simply the blue of the sky seen by reflection," and showed instead that it arose from scattering of light by the water molecules.

On his return to India, Raman and his associates began a systematic study of molecular scattering of light in fluids and solids irradiated with sunlight. By placing complementary light filters in the path of the incident and scattered radiation, they showed the persistence of a weak secondary radiation in the scattered spectrum containing frequencies not present in the incident beam. Raman and K. S. Krishnan further established that the secondary radiation evident in the scattered spectrum of aromatic and aliphatic liquids was strongly polarized. The discovery of the polarized secondary radiation, now known as the "Raman effect," was immediately recognized for its exceptional importance. It provided yet another proof of *quantum physics and offered a powerful new tool for investigating the internal structure of molecules and the chemical composition of substances.

The Raman effect can be envisaged as the transfer of energy associated with inelastic collisions between molecules and incident photons. The incident photon can either impart some of its energy to the molecule, raising it to a higher energy state, or collect energy from the molecule, leaving it in a lower energy state. The frequency shifts observed in Raman scattering thus correspond to the vibrational and rotational energies of the irradiated molecules. Raman's researches into the scattering of light won him numerous honors. He was knighted by the British government in 1928 and received the Nobel Prize for physics in 1930.

In 1933 Raman moved to Bangalore to accept the directorship of the Indian Institute of Science. At the Institute, where he also headed the physics department, Raman established a thriving research program. His work with N. S. Nagendra Nath on the scattering of light by ultrasonic waves in a liquid is particularly noteworthy as it explains both the appearance of diffraction bands and the variation in their intensity as a function of the amplitude of the sound wave.

In the early 1940s, Raman took issue with the Born-von Kármán theory, which predicted quasi-continuous second-order Raman spectra of crystals. Experiments in

Raman's laboratory, however, showed discrete, line-like structures in the spectra of rock salt and diamond. The resulting controversy between Raman and Max Born spurred new interest in the field. Raman's own laboratory took the lead in exploring different aspects of lattice dynamics, including enumeration and calculation of normal mode frequencies, as well as the absorption, emission, and Raman spectra of a number of crystals. Leon Van Hove resolved the controversy by showing that the quasi-continuum frequency spectrum of a crystal would have singularities (or line-like features) for a subset of the normal mode frequencies.

Raman founded the *Indian Journal of Physics* (1926), The Indian Academy of Sciences (1934) with its monthly *Proceedings*, and the Raman Research Institute (1948). He trained more than 100 physicists, many of whom went on to important positions in universities and research institutes in India. For Raman, science was the exploration of the beautiful and wondrous in nature. Sounds of music, radiance of light, and the vibrancy of colors enthralled him, and he sought to understand their physics. His physics, in turn, was marked profoundly by his aesthetics.

G. Venkataraman, *Journey into Light: Life and Science of C. V. Raman* (1988). Abha Sur, "Aesthetics, Authority, and Control in an Indian Laboratory: The Raman-Born Controversy on Lattice Dynamics," *Isis* 90 (1999): 25–49.

ABHA SUR

RELATIVITY, a theory and program that set the frame of the physical world picture of the twentieth century. The article on Albert *Einstein describes the genesis of the theory in the mind of its inventor; the present article concerns its content and reception. Four stages may be distinguished: the invention of the misnamed "special theory of relativity" (SRT), which covered only phenomena observed in bodies moving among themselves with constant relative velocities, in 1905; the recognition of the equivalence of a body's inertial mass (the measure of its resistance to change of velocity) and gravitational mass (the measure of the pull of gravity on it) in 1907; the re-

moval of the restriction to constant velocities in the general theory of relativity (GRT) in 1915; and the application of GRT to *cosmology especially in recent decades.

The infelicity of the term SRT involves both the "S" and the "R." It is not "special" but limited, whence the French term "restricted relativity." Again, what distinguishes SRT from classical *mechanics is not "relativity"—the idea that the laws of motion look the same in all inertial frames (reference frames moving with respect to one another with constant velocity)—but rather the startling concept that all observers measure the same velocity for light in free space. This distinguishing principle is absolute, not relative. It disagrees with the principle of relativity if, as many physicists believed around 1900, radiation is reducible to mechanics. Einstein insisted on both the absolute and the relative—the constancy of the light velocity and the equivalence of inertial frames—and, in consequence, had to surrender common intuitions of space and time. One forgone concept was absolute simultaneity.

In the thought experiment Einstein often described, an observer on an embankment sees a light flash from the middle of a passing railroad carriage equipped with a mirror at either end. According to relativity, an observer seated at the center of the car would see the light returned from the mirrors simultaneously. But the observer on the embankment would see the flash from the forward mirror after that from the rear: the light has farther to go to meet and return from the forward than from the rear mirror, and the speed of light by hypothesis is the same in both directions. On a ballistic theory of light, the return flashes occur simultaneously for both parties: to the traveling observer, the light has the same speed in both directions; to the stationary one, the speed in the forward direction exceeds that in the opposite direction by twice the speed of the carriage. Further arithmetic shows that holding to the absolute and the relative simultaneously required that meter sticks and synchronized clocks moving at constant velocity with respect to a "stationary" observer appear to him to be shorter and tick slower than they do to an observer traveling with them.

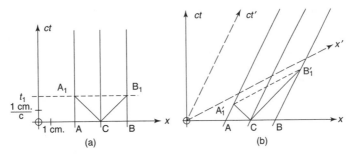

(a) The points A and B, at rest in the system x, t, describe "world lines" parallel to the ct- axis; a light signal sent from C reaches them simultaneously (at time t_1) along the "light lines" CA_1, CB_1. (b) If A and B move along x at the speed v, they describe world lines inclined to the x axis at an angle $\tan^{-1}(vt/x)$ and meet the signals from C at A_1', B_1', that is, at different times measured in the x, t system; since relativity requires the signal to be simultaneous in the system x', t' in which A and B are at rest, the x' axis must be parallel to $A_1'B_1'$.

Einstein showed that the form and magnitude of these odd effects follow from the stipulation that the coordinates of inertial systems are related by a set of equations that he called the Lorentz transformation. Hendrik Antoon Lorentz had introduced them as a mathematical artifice to make the principle of relativity apply to Maxwell's equations of the electromagnetic field (see ELECTROMAGNETISM). Einstein now insisted that the Lorentz transformation apply also to mechanics in place of what came to be called the "Galileo transformation" that guaranteed relativity to Newton's laws of motion. The Galilean, which transforms only space, reads, for relative motion at a constant velocity v, $x' = x - vt$, $t' = t$, where the primed letters refer to a coordinate system moving along the x axis of the unprimed system; the Lorentz transformation replaces these relations with $x' = \gamma(x - vt)$, $t' = \gamma(t - vx/c^2)$, $\gamma = (1 - v^2/c^2)^{-1/2}$. When v is negligibly small in comparison with c, the equations have the same form, but not the same meaning, as the Galilean transformation between the same variables. Since the Galilean transformation supports the Newtonian expressions for force, mass, kinetic energy, and so on, replacing it required reworking the formalism of the old mechanics. This labor, in which Max *Planck and his student Max von Laue played leading parts, produced expressions differing from the Newtonian ones by factors of γ. A new form of Newton's second law emerged that satisfied the

demand of relativity ("remained invariant") under the Lorentz transformation.

As an afterthought, also in 1905, Einstein argued that energy amounts to ponderable mass and vice versa. The relativistic expressions for the momentum and kinetic energy require that, for conservation of momentum to hold, the mass m of an isolated system of bodies must increase when the system's kinetic energy E decreases. The increase occurs at a rate (change of mass) = (change of energy)/c^2, $\Delta m = \Delta E/c^2$. Einstein thus united the previously distinct principles of the conservation of energy and of mass. From a practical point of view, the equivalence of mass and energy as applied to nuclear power is by far the most important consequence of relativistic mechanics.

Relativity had an enormous appeal to people like Planck, who regarded the surrender of common-sense expectations about space, time, and energy as a major step toward the complete "deanthropomorphizing of the world picture" begun by *Copernicus. The mathematician Hermann Minkowski declared in a famous speech to the Society of German Scientists and Physicians in 1908 that "space by itself, and time by itself, are doomed to fade away into mere shadows, and only a kind of union of the two will preserve an independent reality." He interpreted the Lorentz transformation as a geometrical rotation in his four-dimensional space, whose points represented "world events" and whose lines represented the histories of all the particles in the universe. "The word relativity postulate ... seems to me very feeble," he said, "to express the postulate of the absolute world," that is, Einstein's theory as geometrized by Minkowski.

Einstein's compulsion to remove the "all too human" from physics and his desire to overcome the limitation of SRT brought him to a more profound generalization than Minkowski's. Taking the equivalence of inertial and gravitational masses as his guide, he worked out by 1911 that gravitational forces should affect electromagnetic fields (for example, by giving radiation potential energy) and deduced that the sun would bend the path of a ray of starlight that passed close to it. But the major conquest wrested from the equivalence of the masses was the elimination of gravity: all freely falling bodies in the same region experience the same acceleration because the presence of large objects distorts the space around them and the bodies have no alternative but to follow the "geodesics"—straight lines in the curved space in which they find themselves. In "flat space-time," without distorting masses, bodies move in inertial straight Euclidean lines. Falling bodies apparently coerced to rejoin the earth under a gravitational force in Euclidean space in fact move freely along geodesics in the local warp in the absolute four-dimensional space-time occasioned by the earth's presence. By 1915 Einstein had found the mathematical form (tensors) and the field equations describing the local shape of space-time that constituted the backbone of GRT, and had added two more tests: an explanation of a peculiarity in the orbit of Mercury and a calculation of the effect of gravity on the color of light.

By 1910 or 1911 German theorists had accepted SRT and a few physicists elsewhere recognized its importance. Arnold Sommerfeld grafted it onto Niels *Bohr's quantum theory of the atom in 1915–1916, with spectacular results (the explanation of the fine structure of helium). Physicists demobilizing from World War I interested in *atomic structure or *quantum physics perforce had to learn relativity and, ultimately, to find ways of employing SRT systematically in quantum mechanics (see QUANTUM ELECTRODYNAMICS). And then the positive results of the eclipse expedition organized by Arthur Eddington and other English astronomers to test Einstein's prediction of the deflection of starlight engaged the public imagination. It appeared that a lone pacifist had by pure thought bettered Newton while most of the world's scientists had devoted themselves to war. Einstein traveled, quipped, became a favorite of newspaper reporters around the world and the bête-noire of anti-Semites back home. He spent the rest of his life, in Berlin and in Princeton, trying to generalize the general by bringing electromagnetism within GRT. But neither his efforts nor those of the few other theorists who thought the game worth the candle managed to reduce electricity and magnetism to bumps in space.

Application to cosmology at first seemed more promising. A solution to the field equations by Aleksandr Friedmann in 1922 indicated the possibility of a finite expanding universe. The concept agreed with measurements of red shifts in *galaxies. The notion of the origin of the universe in a compact space, or "cosmic egg," was bruited by Georges Lemaître in 1927. The resultant Big-Bang universe, with origin in time (initially set at two billion years ago from the accepted value of the Hubble constant), gained some acceptance but little development until the 1960s. Then discoveries in astronomy (quasars, *pulsars, supposititious *black holes, the cosmic background radiation), laboratory demonstrations of the gravitational red shift and tests with rockets and atomic clocks, and advances in particle physics and the mathematics of gravity made GRT fashionable. Relativistic astrophysics now has the panoply of journals, textbooks, meetings, and money that mark a flourishing science. Although gravity remains outside the unified forces of the *Standard Model, the pursuit of the vanishingly small and the ineffably large depend upon GRT for clues to the origin and evolution of the universe.

See also ASTRONOMY, NON-OPTICAL; COSMOLOGY.

Max Born, *Einstein's Theory of Relativity* (1962). Abraham Pais, *"Subtle Is the Lord...": The Science and the Life of Albert Einstein* (1982). Helge Kragh, *Cosmology and Controversy: The Historical Development of Two Theories of the Universe* (1996). Helge Kragh, *Quantum Generations. A History of Physics in the Twentieth Century* (1999).

J. L. HEILBRON

RÖNTGEN, Wilhelm Conrad (1845–1923), discoverer of X rays.

The son of a German draper and his Dutch wife, Röntgen was born in the Rhineland fifty years before he made his capital discovery. Those who knew him as a young man did not expect to see him become a German professor. His parents left the Rhineland and German citizenship for the Netherlands when Röntgen was three. He did not quite complete his secondary education at a technical school in Utrecht because he was expelled for refusing to identify a schoolmate who had caricatured a teacher. By not finishing high school he did not qualify for enrollment in a German university.

Röntgen went to the then-new Swiss Polytechnic in Zurich, which did not require a high school diploma. He enrolled in 1865 and graduated three years later as a mechanical engineer and protégé of the professor of physics, August Kundt. Since the Polytechnic did not grant doctoral degrees, Röntgen took one under a professor who also taught at the University of Zurich, an itinerary later repeated by Albert *Einstein. When Kundt took a chair at the University of Würzburg, Röntgen went along as his assistant in the hope of habilitating, that is, putting his foot on the first rung of the academic ladder. The university of which he was to be the glory denied his request to habilitate because his Swiss Ph.D. did not compensate for the want of a high school diploma.

Röntgen habilitated at Strasbourg, where Kundt was called to teach in the German university then (in 1872) newly seized from the French, and where the professors determined the qualifications for habilitation within their faculties. Röntgen could thus begin a teaching career that brought him Kundt's old chair at Würzburg in 1888. There he practiced the scrupulous, exact, careful experimental physics he had learned from Kundt. German academic physicists admired him for his work and his colleagues at Würzburg esteemed him for his mandarin qualities. In 1894 they made him their rector.

Röntgen's great discovery may be understood as a reward for his administrative service. When he left the rectorate in 1895, he decided to refresh himself with a new research topic. He began with an experiment designed by Philipp Lenard, who had replaced a bit of the glass wall in a standard discharge tube with a thin aluminum window in order to permit *cathode rays to pass from the tube into the laboratory (*see* ELECTRON). In November 1895, almost certainly using a Lenard tube, Röntgen made the observation that led to his grand discovery: a detecting screen lying upon a table fluoresced far beyond the range of the cathode rays studied by Lenard.

After making the fateful observation, Röntgen holed up in his laboratory working out the properties of what he called "a new sort of rays." One was the capacity to photograph the bones in a living human hand. Here the staid Röntgen displayed a bit of showmanship. Among the pictures circulated with the announcement of his discovery was one of his wife's hand. The mystery of the rays and of their discoverer, and their obvious application to medicine, fascinated physicists, physicians, and the general public. No less puzzling to the wider world was Röntgen's refusal to patent his discovery, to interest himself in the application of his rays, or to improve the method of their production. He received nothing for his gift to humanity except academy memberships, medals, and prizes, including the first Nobel Prize in physics (1901).

Röntgen did not tout himself or relish the many honors he received. But he was proud as well as shy, and felt bitterly the unfounded allegations that not he, but an assistant, had first noticed the effects of *X rays, and that his grand discovery was but an extension of Lenard's work.

Röntgen's status suited him to a more distinguished university than Würzburg.

In 1900 he became professor of experimental physics and the director of the physical institute at the University of Munich. Most of his time went to administration. He took no part in the demonstration by Max von Laue and others at Munich that X rays are light of very high frequency.

Ever the loyal civil servant, Röntgen was as shocked by the discovery that his Kaiser and his government had lied systematically about the military situation during World War I as he was by the final collapse. He suffered greatly after the war from political unrest, scarcity of food, rampant inflation, and loneliness. He died in 1923, some say of intestinal cancer, others say of malnutrition. In his will he responded to the innuendoes about his discovery by directing that all his papers relative to the early history of X rays be burned.

Otto Glasser, *Wilhelm Conrad Röntgen and the History of Röntgen Rays* (1933). W. Robert Niske, *The Life of Wilhelm Conrad Röntgen* (1971). Elisabeth Crawford, *The Beginnings of the Nobel Institution: The Science Prizes, 1901–1915* (1984). Albrecht Fölsing, *Wilhelm Conrad Röntgen: Aufbruch ins Innere der Materie* (1995).

J. L. HEILBRON

Rutherford's heraldic crest as Baron Rutherford of Nelson. The figures represent Hermes Trismegistus (an allusion to alchemy) and a maori warrior (a reference to Rutherford's native New Zealand); the Latin phrase signifies "to inquire into the elements of things."

RUTHERFORD, Ernest (1871–1937), the dominant personality in the early investigation of radioactivity and nuclear physics, born the son of a farmer in New Zealand, died an English baron.

Supported by fellowships, Rutherford graduated from Canterbury College, New Zealand, well educated in mathematics and physics, adept at the then-new art of wireless telegraphy, and armed with an economical style of writing and reasoning. He received an "exposition," a scholarship founded on the proceeds from the Great Exposition of London of 1851, to study abroad. Rutherford chose Cambridge, England, where he planned to perfect a wireless detector under the guidance of Joseph John *Thomson at the Cavendish Laboratory. In December 1895, a few months after Rutherford's arrival in England, Wilhelm Conrad *Röntgen discovered *X rays. Thomson invited Rutherford to join him in studying the ionization the rays produced

when passing through gases. As they finished this work, news of *radioactivity arrived. Rutherford applied the techniques that he and Thomson had developed to the ionization produced by uranium rays. He found that the rays had two distinct fractions, a short-range, or alpha type, and a long-range, or beta type.

With Thomson's support, Rutherford obtained a professorship in physics at McGill University in Montreal, Canada. With McGill's excellent equipment and the help of several colleagues—R. B. Owens (McGill's professor of electrical engineering), Frederick Soddy (briefly lecturer in chemistry, a chance visitor from England), and Otto Hahn (a determined visitor from Germany)—Rutherford discovered the emanation, radon, and its active deposit (with Owens); invented the disintegration theory of radioactive decay (with Soddy); and monitored the sequence of decay products (with Hahn). Rutherford set forth the principles of radioactivity in a textbook, first published in 1904, which defined the field for decades.

In 1907 Rutherford left Montreal for the professorship of physics at the University of Manchester, England. There he made discoveries at least equal to those of his McGill years and gathered a research group superior to anything imaginable in Canada. The findings included the demonstration of the identity of alpha particles and ionized helium atoms, a theory of the scattering of alpha particles, and the nuclear model of the atom. In 1908 Rutherford received the Nobel Prize in chemistry, for which he expressed surprise. He had expected to win one in physics. His transformation into a chemist, he said, was the fastest and oddest disintegration he had ever witnessed.

The research group in Manchester included Niels *Bohr, who combined Rutherford's nuclear model into the theory of *atomic structure that guided microphysics for a decade; György Hevesy, who developed the technique of radioactive tracers and helped define the idea of isotopes; and Henry Moseley, whose work on characteristic X rays established the concept and utility of atomic number (see ATOMIC STRUCTURE). Both Bohr and Hevesy won Nobel Prizes for their work; Moseley would have had one too if he had not died in World War I.

Rutherford helped to mobilize British scientists during the war and investigated acoustical methods of detecting submarines. He led a delegation of British and French scientists to Washington to show what science adapted to war had wrought in Europe and to encourage development in the United States. After the war Rutherford strove to moderate the treatment of German scientists, whom many allied scientists wanted to ostracize from international meetings and research projects.

In 1919 Rutherford succeeded Thomson as Cavendish professor. Again he surrounded himself with a powerful research group: James Chadwick, with whom he pursued his wartime discovery that alpha particles from radioactive substances can transmute nitrogen into oxygen, and who on his own detected (in 1932) the neutral particle (neutron) whose existence Rutherford had predicted a decade earlier; John Douglas Cockroft and Ernest T. S. Walton, who made the first *accelerator that disintegrated an atom with a particle beam (also in 1932); C. T. R. Wilson, a former fellow student from Cambridge, who invented the *cloud chamber; P. M. S. *Blackett, who used the cloud chamber to discover the positive electron (again, in 1932); Pyotr Leonidovich Kapitsa, who made the world's most powerful magnet; and Francis Aston, Thomson's last collaborator, who demonstrated experimentally the agreement between apparent atomic and true isotopic weights. All of these men won Nobel Prizes.

Rutherford's elevation to life peer in 1931, together with his other honors, scientific accomplishments, and international connections, made him an ideal figurehead for the Academic Assistance Council, a nongovernmental organization set up in 1933 to rescue scholars driven from their positions by the Nazis. During Rutherford's presidency the council found jobs in Britain for over two hundred émigrés and helped broker positions for many more elsewhere.

Rutherford died suddenly of complications arising from a strangulated hernia. He is buried in Westminster Abbey near Isaac *Newton, and in the *periodic table of elements at number 104.

A. S. Eve, *Rutherford* (1939). Ernest Rutherford, *Collected Papers* (3 vols., 1962–1965). Lawrence Badash, ed., *Rutherford and Boltwood: Letters on Radioactivity* (1969). David Wilson, *Rutherford, Simple Genius* (1983). John Campbell, *Rutherford, Scientist Supreme* (1999). J. L. Heilbron, *Rutherford, A Force of Nature* (2003).

J. L. HEILBRON

S

SAGAN, Carl. See HAWKING, STEPHEN, AND CARL SAGAN.

SAKHAROV, Andrei (1921–1989), theoretical physicist, key figure in the Soviet thermonuclear program, prominent advocate of human rights; and **Edward TELLER** (b. 1908), theoretical physicist, leading figure in the U.S. hydrogen bomb program, staunch proponent of position-of-strength policy toward the Soviet regime.

Although Teller and Sakharov have in common the fathering of the hydrogen bomb and both were prominent political and public figures, their lives on opposite sides of the Iron Curtain exhibited striking differences.

Teller was born in Budapest, Austria-Hungary, to a family of assimilated Jews of the upper middle class. After a course in chemical engineering he went to study physics in Germany (under Werner *Heisenberg) and Denmark (under Niels *Bohr). In 1935 he emigrated to the United States to work at George Washington University. In the 1930s he made a few significant contributions to theoretical physics, including the Jahn-Teller effect, which concerns crystal symmetry arising from interactions between electrons and nuclei, and turned out to be very important for material science.

Teller was involved in the American atomic project from its very beginning before the United States entered World War II. In the late 1940s, preoccupied with an idea of the hydrogen bomb, he helped convince the U.S. government that the weapon was feasible and indispensable for national security in the Cold War. He also made a key contribution to the first successful design of a thermonuclear weapon (tested in 1952). Teller, together with Ernest O. Lawrence (see BLACKETT AND LAWRENCE), urged the creation of a second nuclear-weapons laboratory, now the Lawrence Livermore Laboratory. As a prominent government adviser on nuclear policy, he opposed the ban of nuclear testing, championed peaceful uses for nuclear explosives, and promoted strategic antiballistic missile defense.

Teller pushed the hydrogen bomb against the opposition of J. Robert Oppenheimer (see KURCHATOV AND OPPENHEIMER) and the General Advisory Committee of the Atomic Energy Commission that Oppenheimer chaired. During the commission's hearing over its decision to revoke Oppenheimer's security clearance (1954), Teller testified that he regarded Oppenheimer loyal to the United States, but that he "would like to see the vital interests of this country in hands which I understand better, and therefore trust more." The great majority of the American scientific community regarded his testimony as an unacceptable violation of ethics and ostracized Teller for life. Nonetheless, in 1962 the U.S. government awarded Teller the Enrico *Fermi prize "for leadership in research on thermonuclear reactions, and for his efforts to strengthen national security and to insure the peace."

Sakharov was born in Moscow to a family of the Russian intelligentsia. World War II interrupted his study of physics in Moscow University. After two years of work at a munitions factory, in 1945 he went on to graduate study in theoretical physics under Igor Tamm. Among his first results was an idea of muon-catalyzed fusion. In 1948 the government assigned Tamm's group, including Sakharov, to check the feasibility of an H-bomb design that, unknown to Sakharov, had been developed in part through espionage. In a few months he invented a brand-new design realized in the first Soviet thermonuclear bomb (tested in 1953). In 1951 he pioneered a research of controlled thermonuclear fusion that led to the tokamak reactor. His was the main contribution to the full-fledged H-bomb tested in 1955.

In 1958 Sakharov calculated the number of casualties that would result from an atmospheric test of the "cleanest" H-bomb: 6,600 victims for 8,000 years per megaton. "What moral and political conclusion must be drawn from these numbers?" he asked. Sakharov was proud of his contribution to the 1963 Test Ban Treaty, which saved the lives of many people who would have perished had testing continued in the atmosphere. In the 1960s he started his return to pure physics. The most successful consequence was his explanation in 1966 of the disparity of matter and antimatter in the universe, or baryon asymmetry.

The major turn in Sakharov's political evolution took place in 1967, when the antiballistic missile defense (ABM) became a key issue in U.S.-Soviet relations. Sakharov wrote the Soviet leadership to argue that the moratorium proposed by the United States on ABM would benefit the Soviet Union and that otherwise the arms race in this new technology would increase the likelihood of nuclear war. The government ignored his memorandum and refused to let him initiate a public discussion of ABM in the Soviet press. Sakharov felt compelled to make his views public in an essay "Reflections on Progress, Peaceful Coexistence and Intellectual Freedom," published in samizdat (underground self-publishing in the Soviet Union) and in the West in the summer of 1968. The secret father of the Soviet H-bomb emerged as an open advocate of peace and human rights.

Sakharov was immediately dismissed from the military-scientific complex. He then concentrated on theoretical physics and human rights activity. The latter brought him the Nobel Prize for Peace in 1975 and internal exile from 1980 until 1986, when the new Soviet leader Mikhail Gorbachev released him. Upon his return from exile Sakharov enjoyed three years of freedom and seven months of professional politics as a member of the Soviet parliament. Those were the last months of his life.

Although Sakharov opposed Teller over nuclear testing and the ABM problem, he considered the attitude of his American colleagues toward Teller to be "unfair and even ignoble." Sakharov had reason to know that Americans who considered Teller's position "anti-Soviet paranoia" did not understand the Soviet regime. Teller had had inside information from two of his physicist friends, pro-socialist Lev Landau and Laszlo Tisza. They had witnessed the great Soviet terror of 1937 that destroyed, among much else, the Kharkov Physics In-

stitute, one of the best scientific centers in the USSR, and killed people dedicated to science and devoted to their country. Teller concluded that "Stalin's Communism was not much better than the Nazi dictatorship of Hitler," and never changed his mind.

Sakharov underwent a conversion. For many years he lived intoxicated by socialist idealism. He later said that he "had subconsciously ... created an illusory world to justify" himself. Totalitarian control over information enabled Soviet propaganda to brainwash even the best and brightest. Sakharov wanted to make his country strong enough to ensure peace after a horrible war. Experience brought him to a "theory of symmetry": all governments are bad and all nations face common dangers. In his dissident years he realized that the symmetry "between a normal cell and a cancerous one" could not be perfect, although he kept thinking that the theory of symmetry did "contain a measure of truth." That did not close his mind: "We should nevertheless continue to think about these matters and give advice to others that is guided by reason and conscience."

For both theoreticians the statement that "the future is unpredictable" was meaningful far beyond quantum physics. It made them feel personally responsible for the future of humanity.

Herbert York, *The Advisers: Oppenheimer, Teller and the Superbomb* (1976). Andrei Sakharov, *Memoirs*, trans. R. Lourie (1990). Andrei Sakharov, *Moscow and Beyond, 1986 to 1989*, trans. A. W. Bouis (1991). *Andrei Sakharov: Facets of a Life* (1991). David Holloway, *Stalin and the Bomb: the Soviet Union and Atomic Energy, 1939–1956* (1994). Edward Teller with Judith Shoolery, *Memoirs: A Twentieth-Century Journey in Science and Politics* (2001). Gennady Gorelik with Antonina W. Bouis, *The World of Andrei Sakharov* (2005).

GENNADY GORELIK

SALAM, Abdus (1926–1996), Pakistani physicist who together with Sheldon Glashow and Steven Weinberg won the Nobel Prize in physics in 1979 for work on the unification of the weak and the electromagnetic forces.

Salam was born into a religious family with a long tradition of learning in Jhang

Maghiana, a small market town in the then-undivided Punjab province, now in Pakistan. His father, Choudhari Mohammed Hussain, a teacher, rose to head clerk in the Department of Education of the district. A very precocious child, Salam could read and write at four and perform lengthy multiplication and division. An outstanding student, Salam won a scholarship to the University of the Punjab. In 1946, after completing his studies in Lahore, he went to St. John's College, Cambridge, again on a scholarship, and emerged with a double first in mathematics and physics (1949), the Smith's Prize for the most outstanding predoctoral contribution to physics (1950), and a Ph.D. in theoretical physics (1951).

Salam's Smith Prize paper offered a resolution of the difficult problem of the overlapping divergences in the proof of the renormalizability of the S-matrix in *quantum electrodynamics and gained him an international reputation. He returned to Pakistan in 1952 and became head of the Mathematics Department of the Punjab University with the intention of founding a school of research. But there were no postgraduate studies at Punjab University, no journals, and no possibility to attend any scientific conferences. He had to make a choice between physics or Pakistan. In 1954 he returned to Cambridge as a lecturer and committed himself to making it possible for young Ph.D.s to remain active and productive scientists while working in their own communities. In 1960 he conceived of the idea of an International Centre for Theoretical Physics (ICTP). With funds from the international community the ICTP in Trieste started operations in 1964. Salam instituted "associateships" that allowed deserving young physicists to spend their vacations at ICTP in close touch with the leaders in their field of research. They could thus overcome their sense of isolation and return to their own country reinvigorated. Earlier, in 1957, P. M. S. *Blackett had invited Salam to Imperial College, London, to found and head a Theoretical Physics Group. He remained at Imperial as professor of physics for the rest of his life.

To explain the experiments that demonstrated the violation of parity in weak interactions, Salam proposed in 1957 that the spin of neutrinos always points in the direction opposite to that of their momentum. The work that won him the 1979 Nobel Prize, which he shared with Sheldon Glashow and Steven Weinberg, advanced a model for the unification of the weak and electromagnetic forces. He had earlier worked with Jeffrey Goldstone and Weinberg on spontaneous *symmetry breaking and with John Ward on theories of the weak interactions. In 1971 Jogesh Pati and Salam proposed that the strong (nuclear) forces might also be included in this unification. In 1974 Salam and his lifelong collaborator John Strathdee introduced the idea of superspace, a space with both commuting and anticommuting coordinates, which underlay all research on supersymmetry at the end of the twentieth century.

Salam's scientific achievements reflect only one facet of his personality. He also devoted his life to nurturing international cooperation and bridging the gap between developed and developing nations. He believed that the eradication of this disparity demanded that every country become the master of its own scientific and technological destiny. The first step, the ICTP, has been a major forum for the international scientific community and a model for similar establishments in various countries. Since 1965 over 60,000 scientists from 150 countries have taken part in its activities. Salam expounded his vision of science and technology in the Third World in his book, *Ideals and Realities* (1984).

Salam won the Atoms for Peace Prize (1968), the Einstein Medal (1979), and the Peace Medal (1981) as well as the Nobel Prize; received over forty honorary degrees; and earned a knighthood for his services to British science (1989). He died in Oxford after a long, debilitating illness. He was a devout Muslim, whose religion did not occupy a separate compartment of his life. He wrote, "The Holy Quran enjoins us to reflect on the verities of Allah's created laws of nature; however, that our generation has been privileged to glimpse a part of His design is a bounty and a grace for which I render thanks with a humble heart."

Abdus Salam, *Ideals and Realities,* 2d ed. (1987). Abdus Salam, *Science in the Third Word* (1989).

SILVAN S. SCHWEBER

SATELLITE. The first artificial satellite appeared in the serial novel *The Brick Moon,* published by Edward E. Hale in 1869 in *Atlantic Monthly* magazine. The satellite, intended to be used as a navigational aid, was built of bricks to withstand the heat generated during its motion through the atmosphere. The launch system consisted of two gigantic fly wheels. An accident during the launch dispatched the brick moon prematurely, and the construction workers on board with their families organized a pleasant life in their new home planet.

The advent of rockets made the idea of artificial Earth satellites a concrete possibility. In February 1945, with the German V2 missiles still hitting London, science fiction writer and amateur scientist Arthur C. Clarke suggested that V2 rockets could be used to launch communication satellites. He proposed the use of geostationary satellites to achieve a global telecommunications system (*Wireless World,* October 1945). Intercontinental ballistic missiles (ICBM), a part of the V2 legacy in the Cold War framework, would eventually provide the technical basis for satellite launchers.

Strategic reconnaissance was the main objective of satellites from the military point of view. A spy satellite flying over foreign territory, however, violated international law, which guaranteed national sovereignty over air space. Hence the nations able to do so launched nonmilitary, scientific satellites to set a legal precedent of the "freedom of space" for subsequent military space activities. The *International Geophysical Year (IGY) offered a framework to establish the precedent.

The IGY, planned to run from 1 July 1957 to 31 December 1958, eventually had the support of sixty-six nations. One of its scientific objectives was to gain information about upper atmosphere phenomena by the use of balloons and sounding rockets. In October 1954, the IGY Special Committee recommended that governments try to launch Earth satellites for scientific purposes. Both the United States

and the Soviet Union undertook to meet the challenge.

The Soviet Union placed the project within its ICBM military program and developed the future R-7 (Semyorka) missile as a launch vehicle for its satellite. The Dwight D. Eisenhower administration, on the contrary, wanted to stress the scientific image of the venture. It rejected the U.S. Army's Orbiter, based on the Jupiter missile developed by former V2 designer Wernher von Braun, in favor of the Naval Research Laboratory's Vanguard, based on the Viking sounding rocket originally designed for upper atmosphere research.

The Soviet path proved more successful. On 4 October 1957, a modified Semyorka rocket launched *Sputnik 1,* the first artificial "fellow-traveler" of the earth. (Russian astronomers had used "sputnik" to designate any hypothetical small natural satellite of the earth.) It was an aluminum sphere 58 centimeters (22 inches) in diameter weighing 83.6 kilograms (184 pounds) that circled the earth once every 96.3 minutes. Its radio emitters sent its familiar "beep-beep" sound all over the world for 92 days.

One month later, on 3 November, the Soviet Union put up *Sputnik 2* to celebrate the fortieth anniversary of the October Revolution. It weighed more than 500 kilograms (1,100 pounds) and carried the first living being into space, the dog Laika, wired up for medical and biological studies. Space technology had not developed enough to return Laika to Earth alive.

The competition proved too strong for Vanguard. The satellite had been reduced to a small sphere of 1.5 kilograms (3.3 pounds) and the launcher not fully tested when, on 6 December 1957, Vanguard rose four feet off its pad at Cape Canaveral, Florida, and slumped back to Earth in a ball of thunder and flame. American pride was restored on 31 January 1958 when von Braun's Jupiter-C rocket put the *Explorer 1* satellite into orbit. It carried a cosmic-ray counter designed by James A. Van Allen that revealed the radiation belt trapped in the earth's magnetic field eventually named after him. Later in 1958 the Eisenhower administration created the National Aeronautics and Space Administration (NASA).

During the 1960s space was an important field of political confrontation between the two superpowers. A superior space program might indicate a superior ideology, a more efficient political institution, a greater industrial capability, and a stronger armed force. Both countries undertook ambitious programs of manned space flight aimed at capturing the public imagination. In April 1961, Soviet cosmonaut Yury Gagarin became the first human being to visit outer space. In July 1969, the United States succeeded in landing the first astronauts on the Moon.

In the following decade, the Soviet Union undertook to establish staffed space stations in Earth's orbit. Seven Salyut stations went up between 1971 and 1982, followed in 1986 by the large Mir station. The United States' space shuttle program culminated in the first launch of the shuttle Columbia in April 1981.

In the early 1960s, a number of Western European nations started civilian space programs. Europe's first satellite, the Italian San Marco 1, was carried aloft by an American rocket in December 1964. The French Astérix, launched by a French rocket, followed in November 1965. The United Kingdom collaborated with NASA in the Ariel scientific satellite program, whose first launch occurred in 1962. Western Europeans joined together in the European Space Research Organization (ESRO), whose first satellite, Iris, flew on an American rocket in 1968. ESRO's success prompted the creation of the European Space Agency (ESA) in 1975. Eastern Europe realized a parallel collaborative effort through the Intercosmos program (1969–1971).

Japan and China joined the list of space nations in 1970. The former launched its first satellite (Ohsumi) in February, the latter the China-1 satellite in April. The first Indian satellite, Aryabata, rose from a Soviet range in April 1975.

Artificial Earth satellites opened new frontiers of experimental research in many fields. It became possible to study the upper atmosphere and the ionosphere by in situ measurements. Astronomers could investigate the electromagnetic spectrum from celestial objects in wavelengths absorbed by the atmosphere (infrared, ultraviolet, X- and gamma rays). Satellites and space probes made it possible to study the structure and properties of the magnetosphere and its interaction with the solar wind and the interplanetary plasma. The *Moon and other bodies of the solar system became objects of important geophysical studies. The advent of space stations and space laboratories on board the shuttle (Spacelab) opened new opportunities in the life sciences and material sciences. The Century of Space Science contains a list of all scientific satellites launched between 1957 and 2001.

Satellites have had important civilian applications, particularly in telecommunications, meteorology, navigation, and Earth observation. Following a number of experimental satellites (Score, Echo, Telstar), the era of commercial satellite telecommunications began in April 1965 with Early Bird (Intelsat 1), the first communications satellite in geostationary orbit. Four years later, three Intelsat 3 satellites realized Clarke's vision of global space communications. Several communications satellite systems besides the Intelsat global network were set up, notably the Soviet Union's Molnyia system.

*Meteorology from space started with NASA's Tiros satellite program (1960–1965). This was followed by the Environmental Science Services Administration (ESSA), National Oceanic and Atmospheric Administration (NOAA), and Geostationary Operational Environmental Satellite in the United States (GOES); Meteor in the Soviet Union; Geostationary Meteorological Satellites (GMS-Himawari) in Japan; and the European Space Agency's Meteosat. The use of satellites made it possible to establish a world meteorological service under the aegis of the World Meteorological Organization (WMO).

The United States and the Soviet Union developed navigation satellites for military use. In February 1978, the United States launched the first Navstar satellite of the Global Positioning System (GPS), capable of providing the position of any moving object with a precision of a few meters (see GEODESY). The Soviet Glonass system (1982) had similar capabilities. Both are also used in the civilian sector.

The United States paved the way for remote sensing satellites aimed at surveying Earth and ocean resources. The first *Landsat* satellite flew in July 1972, the seventh and last in 1999. The Soviet Union, France, Japan, India, and the European Space Agency have all had Earth observation programs.

The most ambitious project of the space age's first half century was and is the International Space Station, now under construction 400 kilometers (250 miles) above the earth's surface through a joint effort of the United States, Russia, the European Space Agency, Italy, Japan, and Canada.

Homer E. Newell, *Beyond the Atmosphere: Early Years of Space Science* (1980). Walter A. McDougall, *...The Heavens and the Earth: A Political History of the Space Age* (1985). Roger D. Launius, *NASA: A History of the U.S. Civil Space Program* (1994). T. A. Heppenheimer, *Countdown: A History of Space Flight* (1997). Asif A. Siddiqi, *Challenge to Apollo: The Soviet Union and the Space Race, 1945–1974* (1998). John Krige, Arturo Russo, and Lorenza Sebesta, *A History of the European Space Agency, 1958–1987* (2000). Johann Bleeker, Johannes Geiss, and Martin Huber, eds., *The Century of Space Science* (2001).

ARTURO RUSSO

SCHRÖDINGER, Erwin (1887–1961), theoretical physicist, discoverer of the equation governing the wave representation of quantum mechanics.

Schrödinger's mother's father, Alexander Bauer, held the principal chair of chemistry at the Vienna Polytechnic. His mother's mother was English and Protestant, and through her relations Erwin spoke English from childhood. Erwin's father, Rudolf Schrödinger, a non-practicing Catholic, one of Bauer's students, operated half-heartedly an inherited linoleum and oilcloth factory and store. His free time he devoted first to painting and etching, and then to microscopic botany, being in his later years a mainstay of the Zoological-Botanical Society of Vienna.

Erwin was a bright and beautiful only child in the most favorable familial circumstances, attached affectionately and intellectually to his father, doted upon by his mother, nurses, maids, and mother's un-

married sisters. Educated at home until he entered the Gymnasium in 1898, where he was then always first in his class, Erwin continued to live in the spacious parental apartment in central Vienna until he married in 1920.

His marriage endured, but almost from the outset Schrödinger sought through extramarital sexual conquests to re-create his childhood conditions of abundant feminine love and to create the religio-mystical experience of mergence with non-self. In youth nonreligious, and in adulthood disdainful of "official Western creeds," Schrödinger, following Arthur Schopenhauer, had become, by his early thirties at the latest, strongly attached to the Eastern concept of a cosmic intelligence in which every individual soul participates.

Schrödinger entered Vienna University in 1906, receiving his Ph.D. in 1910 with an experimental dissertation on humidity as a source of error in electroscopes. The core of Schrödinger's education was Friedrich Hasenöhrl's extended cycle of lectures over the various fields of theoretical physics transmitting the outlook of Ludwig *Boltzmann. The views of both Boltzmann and Ernst Mach were incorporated in the lectures of the professor of experimental physics, Franz Exner, whose assistant Schrödinger became upon returning in autumn 1911 from his obligatory year of military service. (He would serve throughout World War I, 1914–1918.) His researches now were chiefly theoretical, applying Boltzmann-like statistical-mechanical concepts to magnetic and other properties of bodies. The results were not notably successful—Schrödinger's physical intuition profited little from his familiarity with instruments—but gained him the advanced doctorate (Habilitation) in 1914. By contrast, Schrödinger's spectroscopic studies of human color perception (in particular his own) complemented his development (1918–1920) of a theory, based on the Machist concept of elementary sensations of color, that almost represented the facts. This was Schrödinger's only significant original contribution while he remained in Vienna, where all his work was prompted by the topics and problems of his teachers and friends.

The war ended in a complete economic collapse of Austria, ruining Schrödinger's family and forcing him to pursue his career in the wider German-language world of Central Europe. Between spring 1920 and autumn 1921, Schrödinger took up, successively, positions at the Jena University, the Stuttgart Technical University, the Breslau University, and the Zurich University. Finally safe and secure in Swiss employ, he collapsed after the stresses of the previous two years, which had also seen the deaths of both his parents and grandfather Bauer.

A seven-month rest cure in Arosa in 1922 restored Schrödinger physically and opened the only period in which he contributed importantly to the mainstream of theoretical physics. This creative phase culminated in 1926 with the publication of several lengthy papers introducing his differential equation for the quantum-mechanical treatment of atomic systems, and demonstrating its power by applying it successfully to several standard problems. (In his fifty-year career, Schrödinger's median annual output of research papers amounted to forty pages; in 1926 he published 265.) Schrödinger achieved this "wave" equation, arguably the single most important contribution to theoretical physics in the twentieth century, by a creative union of his Viennese statistical-mechanical concerns with the elaborations of the Bohr theory of the atom then the vogue in Germany. The special advantages of his situation in Switzerland, which brought him into close contact with a mathematician of exceptional power, Hermann Weyl, and which permitted him to consider sympathetically the work of the Frenchman Louis de Broglie, enabled Schrödinger to seize the opportunity inherent in de Broglie's attribution of a wave process to material particles.

The wave mechanics of 1926 brought Schrödinger in 1927 the succession to Max *Planck in the chair of theoretical physics at the Berlin University. Schrödinger remained in it until the summer of 1933, accomplishing little. Although he was acceptable to the Nazis, they were not acceptable to him. He resigned to hold a succession of positions at the universities of Oxford, Graz, and Ghent, landing (after hasty, anxious escapes) in October 1939 in Dublin as senior professor at the new Dublin Institute for Advanced Studies. There Schrödinger remained nearly to the end of his career, returning to Vienna in 1956 in poor health and retiring two years later.

At the Dublin Institute Schrödinger devoted himself largely, but unsuccessfully, to the problem then also occupying Einstein, also unsuccessfully, at the similarly named Princeton institute: the creation of a field theory uniting gravity with electromagnetic and nuclear forces. As a contribution to the Dublin Institute's series of public lectures, Schrödinger, who was an engaging speaker, delivered several in February 1943 under the title *What Is Life?* In these popular-scientific lectures Schrödinger, who had only a very slight knowledge of the literature on the physical bases of life, dragged his audience into and then out of a series of blind alleys, leaving them at the end just about where he began. Nonetheless these lectures, printed the following year, achieved an immediate and great reputation with both physicists and biologists, and rank still today as one of the most overrated scientific writings of the twentieth century.

Paul A. Hanle, "The Coming of Age of Erwin Schroedinger: His Quantum Statistics of Ideal Gases," *Archive for History of Exact Sciences* 17 (1977): 165–192. Paul A. Hanle, "Indeterminacy before Heisenberg: The Case of Franz Exner and Erwin Schroedinger," *Historical Studies in the Physical Sciences* 10 (1979): 225–269. Walter J. Moore, *Schroedinger: Life and Thought* (1989), abridged as *A Life of Erwin Schroedinger* (1994). Lily E. Kay, *The Molecular Vision of Life* (1993). Lily E. Kay, *Who Wrote the Book of Life?* (2000).

PAUL FORMAN

SEISMOLOGY, a branch of *geophysics, examines the behavior and products of elastic (seismic) waves traveling within the earth. Earthquakes are the most significant generators of these waves. Other sources include volcanic eruptions, explosions (including nuclear explosions), and meteorite impacts. Trucks, trains, and thunder produce seismic "noise." The discipline of seismology covers documentation of events and effects (observations, maps,

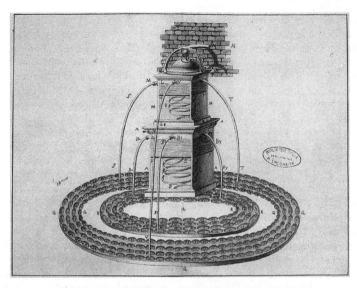

Seismic recorder invented by Atansio Cavalli (1785): shaking the wall causes the mercury-filled basins to oscillate on their springs; mercury, jetting out to a distance determined by the severity of the jolt, is caught in little cups moved around by clockwork.

catalogs), instrumentation and analysis, theory and application. Because large earthquakes can be detected worldwide, the science is international.

The great Lisbon earthquake and tsunami (tidal wave) of 1755 mark the beginning of the systematic study of earthquakes in western science. For the next century or so, studies of earthquakes consisted primarily of observations of how earthquakes behaved, their geological effects, catalogs of historical events, and continued speculation about their causes. By the early eighteenth century, inertial seismoscopes in Europe included pendulums and bowls of liquid mercury. Earthquake catalogs and other regular observations of earthquakes were widely compiled. By 1840 enough information had accumulated for Karl Ernst Adolf von Hoff to produce the first global historical catalog of earthquakes.

Around 1850 seismology began to take shape as a separate field of inquiry. Robert Mallet studied the great Italian earthquake of 1857 and wrote the landmark *First Principles of Observational Seismology*. Johann J. Noggerath in 1847 first used isoseismals to map earthquake intensity, Luigi Palmi-

eri produced an electromagnetic seismograph first used in 1856, and M. S. de Rossi and François Alphonse Forel cooperated to publish the first widely used, standardized intensity scale in 1883. By the end of the century, scientists were designing the first reliable seismographs (recording instruments). They sought ways to reduce friction between the recording needle and recording paper, to damp triggered motion in the instrument, and to eliminate local disturbance of the mechanism. In 1892 John Milne in Japan produced the first compact, simple (though still not entirely accurate) seismograph. Boris Golitsyn's galvanometric seismograph, perfected in 1911, suspended the pendulum in an electromagnetic field. Largely independently of the data, scientists developed the basic theory of elastic wave behavior. Simeon-Denis Poisson worked out a theory for the primary and secondary (P and S) waves; George Gabriel Stokes and Lord Rayleigh made further contributions; and the effort culminated with the publication of theoretical work on surface and other seismic waves by Horace Lamb in 1906 and A. E. H. Love in 1911.

By the early twentieth century, international cooperation and the standardization of observations worldwide, including accurate timing mechanisms and travel-time tables, allowed the development of new standard scales of magnitude and detailed maps of global seismicity. Using Milne's seismograph scientists in the British empire had set up the first uniform, international network. In the 1930s Charles Francis Richter and Beno Gutenberg developed a standard scale to measure the relative sizes of earthquake sources, commonly called the Richter scale. It is one of several magnitude scales in use today. Gutenberg and Richter also wrote textbooks that became standards in the field. Seismology figured among the founding six sections at the first meeting of the International Union of Geodesy and Geophysics (IUGG) in 1922. The International Association of Seismology and Physics of the Earth's Interior (the IASPEI), a branch of the IUGG, continues to coordinate international seismological research.

The twentieth century saw major advances in geophysics as a result of the accumulation of seismic data and analysis. In the first half of the century seismologists and geophysicists such as Gutenberg, Richter, Hugo Benioff, Inge Lehmann, Harold Jeffreys, and Francis Birch delineated the interior, layered structure of the earth. By the end of the century investigators were mapping heterogeneity in the earth's mantle and its boundaries using a method called seismic tomography. A second important spin-off followed from the more accurate location (both geographically and within the earth) of earthquake sources and the differentiation of fault motions, known as focal mechanisms. This information played a major role in the development of the theory of *plate tectonics. A third area of importance was the investigation of the structure of the crust itself. During World War II Maurice Ewing and others developed the technology to make seismic investigations offshore. This allowed the mapping of layered strata beneath the sea floor. The investigation of the earth's crustal structure, at ever greater depths and resolution, has continued to the present day, particularly in the field of seismic stratigraphy.

From the 1920s, seismology has been put to practical uses like the search for oil and gas in the subsurface by using artificial sources of elastic waves, primarily dynamite explosions. Beginning in the 1950s, seismology has been used for monitoring nuclear testing and for understanding other seismic events caused by humans such as those triggered by pumping fluids into or out of the ground. Volcanic seismology has improved to the point that it can help predict volcanic eruptions. While seismologists still cannot predict earthquakes with accuracy, they have done much to explore their preconditions. As human population has increased, more funding has been devoted to earthquake preparedness, including the design of structures, the education of the public, and hazard analysis. In places where the historical record of earthquakes is too short to analyze earthquake probability, workers in the field of paleoseismology use geological evidence to document prehistoric earthquakes.

*Satellite and digital technology, notably the Internet, have had major effects on seismology. In 1984 a consortium of American universities founded the Incorporated Research Institutions for Seismology (IRIS) to develop, deploy, and support modern digital seismic instrumentation. By the year 2000, IRIS had more than ninety member institutions and four major programs: the Data Management System (DMS), the Global Seismographic Network (GSN), a program for the study of the continental lithosphere (PASSCAL), and an education and outreach program.

Charles Davison, *The Founders of Seismology* (1927). Beno Gutenberg and Charles Francis Richter, *Seismicity of the Earth and Associated Phenomena* (1954). Charles Francis Richter, *Elementary Seismology* (1958). Benjamin F. Howell, Jr., *An Introduction to Seismological Research: History and Development* (1990). Bruce A. Bolt, *Earthquakes and Geological Discovery* (1993). Stephen G. Brush, *Nebulous Earth: The Origin of the Solar System and the Core of the Earth from Laplace to Jeffreys* (1996).
JOANNE BOURGEOIS

SOLAR PHYSICS. As with many topics in the history of astronomy, solar physics started with Isaac *Newton and his *Principia mathematica* (1687). Using his new theory of gravity, the contemporary values for both the Earth-Sun and Earth-Moon distances, and the length of the year and the month, Newton calculated the Sun/Earth mass ratio. Because the Earth-Sun distance was inaccurate, Newton obtained a mass ratio that was eight times too small. By the time of the second edition of *Principia* (1715), however, a better value for the Earth-Sun distance gave the astonishing result that the Sun was about 330,000 times more massive than the earth, and 110 times larger.

Newton also concerned himself with the source of the solar energy. In his *Opticks* (1704), he often treated the "rays" of light as corpuscular. On this view, the Sun, in emitting light, lost mass. If the loss were not made good, the Sun should dwindle. For Newton, the great comet of 1680 came to the rescue. He calculated that it had passed within 250,000 km of the solar surface. Newton suggested that comets that came closer would fall into the Sun and thus provide it with new fuel and light.

Worries about the solar energy source resurfaced in the nineteenth century with the realization that the earth, and thus the Sun, had to be much older than the 6,000 years allowed by Bishop Ussher, who in the seventeenth century had calculated from Biblical chronology that creation had occurred in the last week of October 4004 B.C. Measurements by Claude-Sevais-Mathias Pouillet in 1837 of the flux of radiation passing the earth indicated that the sun had a luminosity of about 3.8×10^{26} watts. The suggestion that the sun compensated for this continuous energy loss by accreting *meteorites was discarded on the realization that a mass equivalent to about 86 percent the mass of the *Moon would have to hit the Sun each year, a gain that would have resulted in an unobserved annual increase in the length of the year of 1.5 seconds. In 1854, Hermann von *Helmholtz suggested that the sun was contracting and thus converting potential energy into radiated energy. Even though the required rate amounted to only 91 m per year, the accumulated reduction of the Sun's diameter of 50 percent over 5 million years was re-

garded as untenable. The discovery of *radioactivity in 1895 opened the possibility of another source of solar heating. The solution to the problem came closer with Albert *Einstein's deduction of the energy equivalent of mass (*see* RELATIVITY). The direct conversion of solar mass into energy became more plausible in the early 1920s when *mass spectroscopy disclosed that four hydrogen atoms outweighed one helium atom. In the mid-1920s, spectroscopic analysis by Cecilia Payne-Gaposkin indicated that about 75 percent of the solar mass was in the form of hydrogen. Astrophysical modeling of the solar interior by Arthur Eddington made the Sun gaseous throughout and fixed the central temperature high enough, at 15 million degrees, for collisions between nuclei to lead to fusion and energy release. The physicist Hans *Bethe worked out the nuclear chemical equations governing the conversion of hydrogen into helium in 1939. It became clear that the sun could produce energy at its present rate for a further 5,000 million years.

The relationship between solar astronomy and physics also appears in the problem of surface temperature. The sun's radiant power was known, but not its law of cooling. Father Pietro Angeli Secchi assumed Newton's exponential formula for cooling held and calculated in 1861 a surface temperature of 10,000,000° K. At the same time, J. M. H. E. Vicaire used the empirical power law proposed by Pierre Louis Dulong and Alexis Thérèse Petit and obtained a value of 1750° K. The introduction of Stefan's law (bodies radiate as the fourth power of their temperature) in 1879 and Wien's law in 1896 pointed to 5,770° K.

Pieter Zeeman's observation of the splitting of sunspot spectral lines (*see* MAGNETO-OPTICS) enabled George Ellery Hale to measure the magnetic fields in these temporary refrigerated dark regions on the surface (1910). These fields change polarity every cycle, indicating that solar activity varied with a period of around 22 years. Harold D. Babcock and his son Horace W. Babcock used a magnetograph in the late 1940s to show how the magnetic fields of sunspots related to the general solar magnetic field. By the 1960s, the

Babcocks had linked the 11-year sunspot cycle, the 22-year magnetic cycles, the magnitude of the general solar magnetic field, and the way in which the solar spin-rate varied with latitude.

Solar flares in the vicinity of bipolar spot groups lead to the ejection of clouds of plasma from the overlaying solar corona. Richard C. Carrington and Edward Sabine in the mid-nineteenth century had related the occurrences of these solar phenomena to the observations of aurorae and geomagnetic storms on Earth. It takes a day or two for the charged particles to travel between the Sun and the earth. Eugene N. Parker put forward a hydrodynamical model that explained how the continuously expanding high-temperature corona produces a solar wind that expands past the planets into interstellar space (1958).

Robert B. Leighton's high-precision Doppler spectrometry of 1960 measured the velocity of the solar surface along the line of sight and revealed that the whole solar surface oscillates vertically. Analysis of the modes of this oscillation allow astrophysicists to determine the conditions inside the sun.

C. A. Young, *The Sun* (1881). Giorgio Abetti, *The Sun, Its Phenomena and Physical Features* (1938). A. Jack Meadows, *Early Solar Physics* (1970). Peter V. Foukal, *Solar Astrophysics* (1990). Karl Hufbauer, *Exploring the Sun: Solar Science since Galileo* (1991).

DAVID W. HUGHES

SOLID STATE (CONDENSED MATTER) PHYSICS. The publisher's note of *Galileo's last work, *Dialogues Concerning the Two New Sciences* (1638), informs the reader that the author considered "the resistance which solid bodies offer to fracture by external forces a subject of great utility, especially in the sciences and mechanical arts, and one also abounding in properties and theorems not hitherto observed." By the 1660s Robert *Hooke had discovered the proportionality of stress and strain. Notwithstanding the amazing volume of empirical data amassed by engineers and craftsmen up to the mid-seventeenth century concerning the elastic properties of materials, only in Galileo's and Hooke's work did characteristic features of what

came to be known as solid state physics surface: specific assumptions to simplify calculations, which, nevertheless, express underlying physical processes and mechanisms, together with the introduction of "constants" specific to each substance.

Jakob Bernoulli initiated the mathematical study of elasticity, which his nephew Daniel Bernoulli and Leonhard *Euler developed in the eighteenth century. By the middle of the nineteenth century the systematic experimental and theoretical investigations of these and other phenomena of the solid state were closely associated with engineering. Slowly the study of solids independently of the problems encountered in their practical use led to the establishment of the first laboratories devoted to work in solid state physics. The main areas of study have been elasticity, crystal structure, strength of materials, thermal and electrical conductivity in metals, thermoelectricity, optical, magnetic, and electric properties of solids, the Hall effect, low and high temperature *superconductivity, incandescent lamps, semiconductors, transistors, and computer chips. The amazingly successful explanations of these phenomena rested on general theories such as electrodynamics and *quantum physics and many new specific concepts. The development of calibration instruments, the needs for *standardization, the extensive use of *x-ray diffraction, and the attainment of temperatures close to absolute zero helped further to consolidate the study of solids.

In 1827 Augustin-Louis Cauchy considered solids as continua and proposed a theory of elasticity involving the use of a large number of disposable parameters that could not readily be determined by experiment. The study of crystals brought about a significant change in the treatment of solids. Louis Pasteur discovered in 1848 that tartaric acid can have two distinct crystal forms, each polarizing in a different direction. He inferred that crystals acted as an aggregate of "unit cells" rather than of atoms. The shift from a strict atomic viewpoint facilitated the introduction of symmetry techniques and the understanding of a number of properties as deviations from symmetry.

In 1900 Paul Drude, using the methods of the kinetic theory of gases, showed that the quotient of thermal conductivity and electrical conductivity was proportional to the absolute temperature. This derivation of the empirical Wiedemann-Franz law rested on the assumption that the conduction electrons in a metal could be considered as a free gas. Hendrik Antoon Lorentz proceeded to refine the calculations of Drude by taking into consideration the statistical distribution of the electron velocities and their collision with what he considered to be positively charged atomic cores. His results, which differed from Drude's by a factor of 1/2, also explained other properties such as normal and anomalous dispersion of light, rotation of polarization, and the Zeeman effect (see MAGNETO-OPTICAL EFFECTS).

The experimentally determined dependence of specific heats on temperature at low temperature could not be understood with the model of the free electron gas in solids. It appeared necessary either arbitrarily to reject the gas equation for free electrons or to make the number of free electrons much smaller than the number of atoms in the metals. By 1905 Walther Nernst formulated his heat theorem (see ENTROPY; THERMODYNAMICS) and surmised that as temperature goes to absolute zero, the specific heat of a body must approach a limiting value independent of the nature of the body. Experimental corroboration in 1910 of the behavior predicted by Nernst helped to consolidate Albert *Einstein's treatment in 1907 of specific heats using Max *Planck's radiation law.

Undoubtedly the most decisive developments concerning solid state physics resulted from the advent of quantum mechanics in 1926. The electrical conductivity of metals, paramagnetism, diamagnetism, ferromagnetism, magnetoresistance, the Hall effect, the behavior of semiconductors, and superconductivity found a satisfactory explanation within the framework of quantum mechanics. Theoretical solid state physicists faced two particularly vexing problems in their attempts to solve the wave equation in a periodic crystalline potential: the numerical solution of the Schrödinger equation for the problem and the establishment of the proper expression for the potential used in the equation, which involved approximating interelectron effects.

Long before the extensive use of digital computers, physicists developed effective numerical solutions. They handled interelectron effects by likening a solid to a periodic array of associated nuclei and their core electrons immersed in a sea of valence or conduction electrons. The treatment of conduction electrons in solids as nearly free particles occupying a series of energy bands that correspond to the electronic shells of atoms turned out to be so successful that the success itself required an explanation. Why did the electrostatic interactions among the electrons not restrict their freedom? Lev Landau showed in 1950 that a system of strongly interacting fermions (particles that, like the electron, have half-integral spin) can be regarded as a collection of "quasi-particles" resembling free fermions. This idea underlay the development of the excitation model of the solid that formally solved the puzzle of how supposedly highly correlated electrons can act as if free.

The understanding of the magnetic properties of matter remained elusive for a long time. Empirical methods and a mass of data about magnets provided sufficient information for the construction and manipulation of magnetic and magnetized materials. By 1903 Joseph John *Thomson had come to consider *magnetism as a property of atoms and to attribute both paramagnetism and diamagnetism to the motion of the atomic electrons under their reciprocal repulsions and the externally applied field. Pierre *Curie's systematic experimental treatment of magnets and his empirical law that the magnetization of a paramagnetic body is proportional to the intensity of the magnetic field divided by the absolute temperature became the background for Paul Langevin's theory of 1905. This theory, based on ideas of Thomson and Lorentz's electrodynamics, derived para- and diamagnetism from magnetic moments arising from motion of the atomic electrons. In the late 1920s Werner *Heisenberg showed that the exchange interaction between electrons might be the key to an understanding of the success of Langevin's approach.

In 1926 Wolfgang Pauli (*see* HEISENBERG AND PAULI) calculated the paramagnetic susceptibility quantum mechanically on the assumption that an electron gas consists of free fermions. In contrast to the prediction of the classical treatment, Pauli found that at low temperatures the susceptibility approached a constant. Only electrons in metals within a certain small range of energy can be aligned by the magnetic field, an effect that dramatically decreases the magnetic susceptibility in metals.

The electron theory of metals was systematically developed after 1928, the year that Felix Bloch defended his doctoral dissertation under Heisenberg at the University of Leipzig. Bloch assumed that the electrons did not act on one another and that they moved freely through a lattice (the metal). A perfect lattice of identical atoms would give an infinite conductivity; electrical resistance resulted from lattice imperfections or ionic motion. Bloch also proved that if the electron Fermi distribution was in equilibrium and at rest with respect to the lattice, and if it had the same temperature as that at which the lattice vibrated, then the electrons and the vibrations would be in equilibrium. In this case the motion of electrons did not have any consequences for the thermal motion in a solid. From these considerations Bloch derived a temperature dependence for electrical resistivity.

Hans *Bethe's doctoral dissertation with Arnold Sommerfeld was another turning point of the electron theory of metals. He showed that electrons with negative potential energy in a metal have a larger kinetic energy inside it than outside, and concluded that the crystal shortened their wavelengths. This explained the positions of the maxima in electron-diffraction experiments, which did not agree with the predictions of the previous theory. Two additional notions associated with Bethe's close friend Rudolph Peierls—holes and band gaps—became important in understanding the conduction processes in metals. In his attempt to deal with the anomalous Hall effect, Peierls found that the Hall constant in the limit of a slightly

filled energy band had the same value as that derived by the classical electron theory. In a nearly full band, it again had the value derived by the classical electron theory, but now for carriers of positive charge equal in number to the unfilled states in the band. These vacancies or holes in an otherwise full band behaved as positively charged bodies. Since, as Peierls showed, electrical conductivity vanished in the case of completely occupied bands, the holes (the negative electrons in Paul *Dirac's *quantum electrodynamics) became an indispensable notion for understanding the behavior of electrical insulators.

In 1934 Eugene Wigner and Frederick Seitz calculated the bands of sodium, a monovalent metal. They modeled the crystal as a network of identical cells surrounding single metallic ions and thus considered the conduction electron in each cell to be influenced only by its "own" ion's field. This calculation initiated other studies of band structures of real materials.

The nearly free electron model could not account for low temperature phenomena such as superconductivity and superfluidity or for phase transitions and critical (or collective) phenomena. Here interactions between the electrons became important. Not until after World War II, however, were the necessary many-body methods developed. New concepts like elementary excitations (phonons, spin waves, quasi-particles), macroscopic wave functions (long range order), order parameters and changes of symmetry in phase transitions, collective modes, low-lying excitations above the ground state, Bose-Einstein condensation, pairing, and broken symmetries played an important role in understanding these interactions. The articulation of the notion of macroscopic quantum effects, first formulated by Fritz London in 1936, brought about a deeper understanding of these phenomena.

Electronic conductivity of semiconductors and the mechanical properties of metals are controlled by minute additives of foreign atoms or by irregularities in crystalline structure. Small concentrations of defects in a largely undisturbed lattice have strong effects on macroscopic crystal phenomena such as optical properties and electrical conduction. Semiconductors (flawed crystals) became a separate class of materials and the objects of intensive research, much of it done in industrial laboratories. In one of the largest of them, Bell Laboratories in the United States, researchers invented the transistor. The contact of two crystals, one with a minute excess of electrons and the other with a predominance of holes, formed the so-called p-n junction, the basis of the transistor. The need for purer materials, the understanding of structure-dependent properties, and the introduction of specific impurities defined the possibilities of semiconductors, and led to the elucidation of their most important property, the rectifying contact that made the transistor practicable. Lasers soon followed.

Research and development in many universities and industrial research laboratories, the direct and intense interest of the military, and the profitability of the new materials and inventions based on the understanding of the many properties of the solids played a decisive role in shaping the characteristics of solid state physics. These characteristics derived from many schools of thought and experimental practices from all over the world.

The only person to have been awarded the Nobel Prize twice for the same science was John Bardeen, who shared the physics prize for his work on semiconductors and again for his work on superconductivity, in 1956 and 1972, respectively. This unique honor reflects solid state physics' idiosyncratic inheritance of diverse phenomena, intriguing problems, and miscellaneous methods.

Marcus Fierz and Victor F. Weiskopf, eds., *Theoretical Physics in the Twentieth Century: A Memorial Volume to Wolfgang Pauli* (1960). Crosbie S. Smith, *A History of Metallography* (1960). John G. Burke, *Origins of the Science of Crystals* (1966). Lillian Hoddeson, George Baym, and Michael Eckert, "The Development of the Quantum Mechanical Electron Theory of Metals, 1926–33," *Reviews of Modern Physics* 59 (1987): 287–327. Lillian Hoddeson, Ernest Braun, Jurgen Teichmann, and Spencer Weart, *Out of the Crystal Maze, Chapters from the History of Solid State Physics* (1992).

KOSTAS GAVROGLU

SOUND. See ACOUSTICS AND HEARING.

SPACE AND TIME. The evolution of the modern understanding of space and time, which is closely related to the formation and development of physical sciences, can be divided into three stages dominated by Newton's absolute space and time, Minkowski's spacetime, and Einstein's spatial and temporal structures constituted by gravitational fields interacting with material bodies or other physical fields.

Absolutism. The scientific revolutionaries of the seventeenth century rejected the scholastic view of space and time as accidents of substance along with most other fundamental tenets of Aristotelianism. Against this view, which left no room for a void and assumed that time was the same everywhere at once, the revolutionaries flirted with atomism and other systems that gave space an independent existence.

Here, René *Descartes took an intermediate position. Although he rejected the notion of space as an immaterial, infinite, immobile container with indistinguishable parts, he did allow it an independent, even a material, existence, by characterizing it by its extension and identifying it with matter. He thus rejected a vacuum in favor of a plenum, and deduced that motion can be transmitted only by impact, that motion of a body can be measured only relative to other bodies, and that the total motion in the universe is conserved. Conservation of motion suggested to him a principle of inertia, according to which uniform rectilinear motion is equivalent to rest, but rest requires for its definition an immobile frame of reference, which, alas, cannot be supposed within the ceaselessly moving Cartesian universe.

In forming his theory of motion, Isaac *Newton recognized the importance of Descartes's principle of inertia, transformed it radically, and provided it with new conceptual foundations. By conceiving the inertia of a body not as an expression of the conservation of its motion, but as its inertness measured as its mass, Newton purchased ground for taking force, an external mover of inert matter, as a primitive entity existing independently of bodies. To give meaning to the revised notion of inertia and to make the revised law of inertia self-consistent, Newton promoted the atomists' void into a primitive entity.

If motion occurs only relatively to other bodies, immaterial and immobile space as a frame of reference is dispensable; but if absolute motion exists, then the frame of reference has to be taken as a primitive concept (absolute space). Newton adduced the motion of water in a rotating vessel as evidence of a centrifugal force generated by a rotation in absolute space. He described this space eloquently as an entity that, in its own nature, without relation to anything external, remains always similar and immovable. In the same manner, without much argument, Newton defined true time as absolute time, which, of itself and from its own nature, flows equably without relation to anything external. He understood absolute space and absolute time as attributes of God, one expressing divine omnipresence, the other divine eternity.

Newton's absolutist view of space and time came under criticism from Christiaan *Huygens and Gottfried Leibniz. Huygens tried to interpret rotation as a relative motion of the parts of the rotating body, driven to different sides and in different directions, and argued that this relative motion gave the appearance of centrifugal force in Newton's bucket experiment. But this argument failed because in a rotating coordinate system, parts rest but the centrifugal force does not disappear. Leibniz employed his principles of the identity of indiscernibles and sufficient reason to dismiss Newton's absolute spatial and temporal relations and to insist, characteristically, that in Newton's homogeneous absolute space, God would have had no reason to create the world in the way he did rather than in infinitely many other ways. But since Leibniz had to accept rotation as an example of absolute motion, and offered no relational theory to accommodate it, his metaphysical arguments did not carry much force for his contemporaries.

In the two centuries after Newton, natural philosophers accepted absolute space and time as the bedrock of physical theory. The only significant challenge came from

Immanuel Kant. In his influential teaching, space and time are imposed by us on the world as the ground or possibility of our intuitions of it. Kant's a priori view of space collapsed with the discovery of non-Euclidean geometry in the mid-nineteenth century. But his transcendental arguments about space and time as necessary prerequisites for experience was revived by Niels *Bohr and many others concerned to anchor quantum physics on classical observables in space and time (see COMPLEMENTARITY).

Spacetime. Newton defined absolute space in terms of the resting center of gravity of the world. However, for the validity of *mechanics, any "inertial system," that is, any body moving uniformly with respect to absolute space, could serve as a reference system. The question of inertial systems in mechanics was entangled with the question of the *ether, the carrier of electromagnetic waves, in the late nineteenth century. Since physicists tended to identify the ether with absolute space, they expected to be able to detect the effect of motion relative to it. The negative result of the Michelson-Morley experiment (1887) posed a puzzle, explained away by the hypothesis, suggested by George Francis FitzGerald and Hendrik Antoon Lorentz, that moving bodies, owing to their interactions with the ether, contract along their line of motion. Lorentz's explanation (1895, 1902), also proposed by Joseph Larmor in 1900, made use of a quantity they called "local time," different for different observers, which they regarded as a mathematical artifice. When local time was taken to be the real time for a moving observer, first by Henri Poincaré in 1902 and then by Albert *Einstein in 1905, the absolutist notion of a single universal time collapsed, and absolute simultaneity could no longer be defined. Times and locations can be defined meaningfully only in accordance with the states of motion of inertial systems. The relation between the space and time coordinates in two inertial systems in relative motion can be obtained mathematically from the principle of *relativity (physical laws take the same form in all inertial systems) and the postulate of the constancy of the *speed of light, first suggested by Poincaré in 1902, without resorting to the contraction hypothesis.

In 1905, Poincaré noted that the mathematics (Lorentz transformations) that relate spatial and temporal intervals of inertial systems to another mixed spatial and temporal coordinates, but left the formula for the spacetime intervals between events the same in all coordinate systems. Poincaré observed further that this formula behaved as if it represented a four-dimensional analogue of a line in three-dimensional space, so that the Lorentz transformations could be pictured as four-dimensional analogues of ordinary rotations. These observations suggested a complete change in the ideas of space and time to Hermann Minkowski. He conceived the relative spaces and times of inertial systems as projections of an absolute four-dimensional spacetime manifold, the true and independent stage for physical events to occur, onto the three-dimensional space of the observer.

Absolute spacetime has richer structures than absolute space had. Most important among them is the light cone, defined at each point by the events that can be causally related to the observer and those that lie absolutely elsewhere and absolutely elsewhen. Minkowskian spacetime, together with its kinematic and causal structures, has replaced Newtonian absolute space and time and played a foundational role in all forms of relativistic dynamical theories, including quantum mechanics and quantum field theories, except for the general theory of relativity and its variations.

Dynamical Spatial and Temporal Structures. Non-Euclidean geometries make possible use of intrinsic local variations of curvature to designate positions in space without resorting to a material coordinate system and thus opened the way to a new version of absolute space. But Bernhard Riemann observed in 1854 that since the structure of physical space had to be determined by physical forces, the new notion of absolute space could not be sustained. Einstein vigorously pursued Riemann's idea in developing his general theory of relativity. Here Einstein's work shows the influence of Ernst Mach, whose program of freeing science from metaphysics included ridding the world of the concept of absolute space. In 1883 he rejected Newton's crucial bucket experiment with the argument that the centrifugal forces on the water arose because of its relative motion with respect to the mass of the earth and the other celestial bodies. Mach thus replaced absolute space with the cosmic distribution of matter, which would determine the inertia of bodies and the spatial structures of local inertial systems, and thus provide dynamics with a relationist foundation.

Einstein's general relativity (1915) has a spatial structure (curvature or metric) that varies with the distribution of matter. But an interpretation of general relativity along Mach's lines ran into trouble with the discovery of the "vacuum solution" to the theory's equations, which showed that spatial structures exist in the absence of matter. Further reflection showed that any description of the properties and state of matter necessarily involves a metric as an indispensable ingredient, and thus presupposes the existence of spatial structure. Thus, although Einstein initially liked Mach's idea, which, in 1918, he raised to "Mach's Principle," he later (1953) rejected it. In his final formulation dynamical (gravitational or metric) fields, but not masses, determined the spatiotemporal structures that grounded the dynamic behavior of everything in the world. Spacetime as a quality of the field had no independent existence.

There seems to be unanimity that spatiotemporal structures are not conventional, but specified or constituted by metric fields or their variations. Serious disagreements nevertheless persist over the ontological status of spacetime. Substantivalists ascribe spatial-temporal positions and structures directly to the individual points of a spacetime manifold, and only in a derivative sense to physical entities occupying points of the manifold; relationists claim that the spacetime characteristics of a physical entity belong to it in a primary and underived sense.

Some ardent substantivalists argue that an immovable spacetime substratum as a primitive existence has to be presumed if we wish to ground absolute motions and field theories. The relationist counters that absolute motions can be measured by deviations from geodesic motions and that chirality (right- or left-handedness), as Kant had realized, cannot be understood by reference to the points in absolute space. It is an intrinsic spacetime characteristic of physical entities and belongs to them in a primary sense.

Hans Reichenbach, *The Philosophy of Space and Time* (1958). Lawrence Sklar, *Space, Time, and Spacetime* (1974). John Earman, Clark Glymour, and John Starchel, eds., *Foundations of Space-Time Theories* (1977). Michael Friedman, *Foundations of Space-Time Theories* (1983). Robert Torretti, *Relativity and Geometry* (1983). Julian Barbour, *Absolute or Relative Motion? Vol. 1: The Discovery of Dynamics* (1989). John Earman, *World Enough and Space-Time* (1989). Max Jammer, *Concepts of Space*, 3d enlarged ed. (1993). Julian Barbour and Herbert Pfister, eds., *Mach's Principle: From Newton's Bucket to Quantum Gravity* (1995).

TIAN YU CAO

SPACE SCIENCE. The term "space science" came into use in the late 1950s just after the Soviet Union launched *Sputnik I*, but it had antecedents in the 1930s, when astronomers climbed mountains to observe the heavens and meteorologists and physicists sent instruments aboard high-flying balloons to study *cosmic rays.

After World War II, captured German V-2 missiles and their immediate successors, like the Navy's *Viking* or the long line of Soviet missiles based upon V-2 technology, carried probes into near space to examine Earth's upper atmosphere, the nature of cosmic rays, the Sun's high-energy spectrum, and the particles and fields contained within Earth's magnetic system. After *Sputnik*, space science research relied on rockets powerful enough to send nuclear warheads ballistically to another continent or satellites into orbit. Space science identified its programs in terms of the capabilities of specific transport vehicles—balloons, aircraft, sounding rockets, *satellites, and space probes.

In Britain, most of the activity in the 1950s centered on the Gassiot Committee of the Royal Society and on less formal splinter groups at major university centers. The society's members had established interests in the physics of the upper atmosphere and had followed closely the progress of the rocketry groups in the United States. Starting in 1955, they championed what became known as the *Skylark*, a sounding rocket on the scale of the American *Aerobee*, and eventually gained access to space post-Sputnik with the Ariel series of scientific satellites sent up by American vehicles. Meanwhile other British launchers were being developed—*Blue Streak* (based on preexisting military systems) and *Black Knight* (developed out of research programs)—but they did not survive as scientific launchers once Britain decided to work within an international structure.

In the United States, those who instrumented the V-2s between 1946 and 1951 came from disciplines that traditionally had not inquired into the natural phenomena they now addressed with rockets. They were practical, tool-making, problem-solving physicists and engineers experienced in building and maintaining long-range radio networks, rugged and reliable high-speed optical systems, proximity fuses for artillery shells, and radiation detectors for atomic tests. These were the skills needed to make an instrument perform delicate observations in the violently hostile realm of the rocket.

A second generation of practitioners, typically trained in the university groups that had access to rockets in the 1950s, did postdoctoral work in military laboratories. They tended to identify more with the disciplines they could address with the instruments they built than with the objects of their handiwork. Starting in the 1960s, leading academic scientists like Lyman Spitzer, Jr., at Princeton, Leo Goldberg at Harvard, John Simpson at Chicago, James van Allen at Iowa, Fred Whipple at Harvard, Charles Hard Townes at Berkeley, Joshua Lederberg at Wisconsin, William Dow at Michigan, and Joseph Kaplan at UCLA supported graduate students and assistants on contracts from NASA, the Air Force, and the Navy to develop instruments and techniques to pursue science from space.

Graduate students in astronomy were attracted to groups conducting solar physics from rockets, and became specialists in methods most suited to studying the Sun from space. Many went on to satellite-based research in the 1960s and to manned orbiting platforms in the 1980s and 1990s. As the generational cohorts established stronger and stronger interdisciplinary ties with traditional areas of research, they migrated more freely within their subject matter disciplines; generally they no longer moved from one scientific discipline to another but were attracted to problems within their discipline where they could exploit their expertise with rocket and satellite technologies.

Space science thus came to lie at the intersection of three elements: a technical capability (the use of rockets and satellites as platforms to make observations of any accessible phenomenon) with a scientific interest (framing problems that can be addressed by observations from rockets and satellites) and a military or commercial need (creating a capability to use and manage space for communication, weapon delivery systems, reconnaissance, and command and control). At the intersection, expensive, government-sponsored technologies made research in space possible.

Scientific *satellite development roughly followed the growth of the launch capabilities of the vehicles. The very first American scientific satellites, those typical of the long-lived Explorer series, were single-purpose instrument packages weighing thirty to one hundred pounds. They contained primitive telemetry systems, onboard data storage, and rudimentary temperature stabilization. By the early 1960s, spin-stabilized "observatory" class satellites (the Orbiting Solar and Geophysical Observatories) and the multifunctional but unstabilized Ariel series were flying. These satellite series coexisted throughout the 1960s and offered access to a wide range of electromagnetic phenomena from the Sun, Earth's magnetosphere, and the interplanetary medium. Dedicated sounding rockets and a few experimental high-energy satellites began to detect nonsolar x-ray sources: the first fully dedicated x-ray mapping satellite,

Uhuru, was launched only in 1970 as the forty-second in the Explorer series.

The first of the OAO series appeared in 1968. With the OAO, in a weight class of thousands rather than hundreds of pounds, lead times crept up during the 1970s from a few to many years and began to slow the pace of training and advancement in the participating disciplines. Astronomers have called OAO-2 the first true observatory in space because the scale and resolution of its instrumentation complemented those available on the ground. It operated from December 1968 to January 1973 and could achieve a pointing accuracy as good as 1 minute of arc with a stability of some seconds of arc providing the capability to secure sustained photometric and spectroscopic data from tiny celestial sources.

In the post-Apollo era, two drivers of the American space program both propelled and severely limited the continuing development of science satellites. The primary driver was NASA's preoccupation with establishing a permanent human presence in space. As the Apollo program wound down in the wake of the successful lunar landings between 1969 and 1972, NASA decided that national goals could best be met in a reusable launch system, the Space Shuttle. Accordingly it concentrated access to space in the Shuttle bay, severely reducing or eliminating altogether suborbital and orbital programs based upon conventional launch vehicles. This had the effect of requiring even the smallest packages to be rated for human space flight, vastly increasing costs and lead times for development and testing.

At the same time, NASA's propensity for mission-based, rather than problem-based, programming drove up the scale of successive satellites. While the OAO's were flying and scientific groups tried to keep alive programs in the Explorer and smaller observatory and planetary probe classes, NASA and a vocal faction within the scientific community set their sights on a "great observatory" class of satellites, instruments that would fill the Shuttle bay and offer truly high resolution and broadband access to the faintest of celestial sources. This older class had been wildly successful beginning in the 1970s. By the end of the century, all the planets save Pluto had been visited, mapped, revisited, and, in the cases of Mars and Venus, investigated by landers. In the 1980s and 1990s, however, the costs of these probes were competing directly with the great observatories like the Hubble Space Telescope, the Compton Gamma Ray Observatory, the Advanced X-Ray Astronomy Observatory (AXAF, renamed *Chandra*), and the Space Infra-Red Telescope Facility (SIRTF). Pressure mounted under competition from satellites of the European Space Agency and Japan.

For these reasons, as well as a lack of public enthusiasm for the continuation of NASA programs at the levels enjoyed in the Apollo era, American promoters of space research increasingly sought out new and more substantial modes of international cooperation at all levels, from Explorer to great observatory class. Few nations have the resources to conduct science from space. The United States from the start entertained an international program, mainly to launch instruments, and in some cases, satellites, for other countries. The Soviet Union soon followed, and all eventually recognized the need for the creation of a new international body, now the Committee on Space Research (COSPAR), that would coordinate international work. Many countries, including Great Britain, created new organizational structures to communicate with the international body. Thus, COSPAR formed and quickly assumed far broader international responsibilities.

At the same time, the United States announced that it would provide launch vehicles for COSPAR countries as its contribution to international cooperation. Initially Western Europe, Great Britain, and the Commonwealth countries dominated COSPAR, with a strong representation from the United States and hardly any from the Soviet Union and Eastern Bloc. This was not satisfactory to member countries of ICSU, and eventually the Soviet Union joined COSPAR, which thus became an important forum for international cooperation, eventually as an agency within the United Nations. Its existence, and the American offer of launch vehicles, weakened British resolve to purchase its own

systems. The United States sealed the arrangement when it announced at a meeting of COSPAR in March 1959 that it would provide launch systems without charge.

The emergence of the European Space Research Organization (ESRO), and out of that the European Space Agency (ESA) in the 1960s, even with Britain's initial reluctance to join in, marked the completion of the overall structure for space science in Europe and the Americas. Led by France, ESA produced a competitive launch vehicle, *Ariane*, which prompted creation of similar-scaled vehicles in China and Japan. Thus at the beginning of the twenty-first century there exist five competing national—or, in the case of ESA, transnational—sources for placing satellites into orbit. Before the multinational armada of space probes that met Halley's Comet in the spring of 1986, the United States and Soviet Union were the only countries actively supporting interplanetary probes, landers, and orbiters. By 2000 at least four of the five major launching programs were considering new probes, orbiters, and landers.

Homer E. Newell, *Beyond the Atmosphere: Early Years of Space Science* (1980). James A. van Allen, *Origins of Magnetospheric Physics* (1983). Richard Hirsh, *Glimpsing an Invisible Universe: The Emergence of X-Ray Astronomy* (1983). Harrie Massey and M. O. Robbins, *History of British Space Science* (1986). Robert W. Smith, *The Space Telescope: A Study of NASA, Science, Technology and Politics* (1989). Joseph N. Tatarewicz, *Space Technology and Planetary Astronomy* (1990). Karl Hufbauer, *Exploring the Sun: Solar Science Since Galileo* (1991). David H. Devorkin, *Science with a Vengeance: How the Military Created the U.S. Space Sciences After World War II* (1992). Ronald Edmund Doel, *Solar System Astronomy in America: Communities, Patronage, and Interdisciplinary Science, 1920–1960* (1996). *A History of the European Space Agency, 1958–1987. Vol. 1: The Story of ESRO and ELDO, 1958–1973*, ed. John Krige and Arturo Russo; *Vol. 2: The Story of ESA, 1973–1987*, ed. John Krige, Arturo Russo, and Lorenza Sebesta (2000).

DAVID DEVORKIN

SPECTROGRAPH, MASS. SEE MASS SPECTROGRAPH

SPECTROSCOPY is the science on the borders of chemistry and physics that studies the properties of matter by analyzing, usually prismatically, the light it emits when rendered incandescent. Spectroscopy's progress has depended on the development of the necessary equipment. It uses physical methods to study chemical phenomena. Not until a chemist and a physicist collaborated in working on spectra did spectroscopy begin to yield useful chemical knowledge.

While working to improve optical glass, Joseph von Fraunhofer found that flame spectra were characterized by discrete bright lines (see OPTICS AND VISION). He also found a number of dark lines crossing the continuous spectrum of the sun and noted that their positions did not change with intensity. These dark lines, subsequently named after him, still bear the letters he used to designate them.

While spectral lines facilitated the calibration of optical instruments, their meaning eluded satisfactory explanation for many years. The physical interpretation of the lines played a major role in the wave-particle debate over the nature of light that raged, especially in Great Britain, during the 1820s and 1830s. William Henry Fox Talbot suggested in 1826 that spectral lines might be used for chemical analysis. This idea, however, was not pursued, largely because the generally poor quality of glass prisms, and the impurities present in chemical substances, made it difficult to achieve replicable results. Attempts in the 1840s and 1850s to analyze the spectra of electric sparks also came to little, although the theory behind them favored the development of spectroscopy.

From his work on photochemistry in the 1850s, Robert Bunsen, professor of chemistry at the University of Heidelberg, became convinced that the light emitted from flames was uniquely characteristic of the chemical elements present. He pursued this idea with the school's professor of physics, Gustav Kirchhoff. Together they showed conclusively, in 1859, that a chemical element emitted a unique characteristic spectrum that could be used for chemical analysis.

In 1860–1861, using what was then known as spectrochemical analysis, Bunsen detected and then chemically isolated two hitherto unknown chemical elements, cesium and rubidium, which occurred in trace quantities in mineral waters. In 1861, William Crookes in London discovered the chemical element thallium using spectrochemical methods. These discoveries placed the new method on a secure evidential basis. Furthermore, they helped popularize knowledge of the new method and to arouse widespread interest. In 1865 August Wilhelm von Hofmann demonstrated spectroscopy to Queen Victoria at Windsor. Thus publicized, the method quickly became established as an invaluable laboratory technique.

During the same period, Bunsen also collaborated with Kirchhoff to show experimentally that the bright yellow lines characteristic of sodium corresponded with Fraunhofer's dark D lines in the solar spectrum. Kirchhoff provided a thermodynamic explanation of the coincidence. This extension of chemical analysis to the sun and stars (entities that the French philosopher Auguste Comte had pointed to in 1835 as examples of things forever unknowable) led to the new science of *astrophysics. In the ensuing decades, spectroscopic observations allowed astronomers to develop theories of the evolutionary sequence of stars. Measurements made in the late 1890s on the spectrum of cavity radiation prompted Max *Planck's quantum theory. Somewhat later, using the measurements and analyses of the distribution of lines emitted by particular elements, physicists began to investigate the internal composition of matter, leading to Niels *Bohr's theories of *atomic structure. The discovery of the diffraction of *X rays in 1911 led to high-frequency spectroscopy with crystals rather than glass prisms as analyzers, and to important information about the deeper reaches of atoms.

The obvious benefit of the spectroscope to a wide range of scientific and technical activities prompted its commercial manufacture by a large number of instrument makers throughout Europe. The instrument was refined and developed during the late nineteenth century by substituting diffraction gratings or hollow prisms filled with carbon bisulphide for the glass prism.

Some spectroscopes had prisms so arranged that they appeared to resemble telescopes. The basic principles and uses of the spectroscope did not undergo any fundamental change until the invention of the *mass spectrograph in 1919.

See also THERMODYNAMICS AND STATISTICAL MECHANICS.

Frank A. J. L. James, "The Establishment of Spectro-Chemical Analysis as a Practical Method of Qualitative Analysis, 1854–1861," *Ambix* 30 (1983): 30–53. Frank A. J. L. James, "Of 'Medals and Muddles'. The Context of the Discovery of Thallium: William Crookes's Early Spectro-Chemical Work," *Notes and Records of the Royal Society of London* 39 (1984): 65–90. Frank A. J. L. James, "The Discovery of Line Spectra," *Ambix* 32 (1985): 53–70. Myles W. Jackson, *Spectrum of Belief: Joseph von Fraunhofer and the Craft of Precision Optics* (2000). Klaus Hentschel, *Mapping the Spectrum: Techniques of Visual Representation in Research and Teaching* (2002).

FRANK A. J. L. JAMES

SPEED OF LIGHT. See LIGHT, SPEED OF.

SPUTNIK. Shortly before midnight on 4 October 1957, the Soviet Union launched the world's first artificial Earth *satellite, a 184-pound sphere about 22 inches in diameter dubbed *Sputnik* (loosely translated as "fellow traveler"). It was made of aluminum alloy and contained two zinc-oxide batteries, a thermal regulation system, and a radio that transmitted temperature and pressure data and signaled its presence to the world with a "beep-beep" sound. A month later, *Sputnik II* went up, weighing 1,121 pounds, packed with a maze of scientific instruments and signaling back the condition of the live dog named Laika it was carrying.

The prime mover behind the satellite program was Sergei Korolev, an imaginative engineer and efficient organizer and the chief designer in the Soviet effort to develop intercontinental ballistic missiles (ICBMs). A longtime enthusiast of space exploration, Korolev urged in 1954 that one of the missile rockets be used to launch an Earth-orbiting satellite. In mid-1955, the United States announced that it intended to launch

such a satellite during the *International Geophysical Year, scheduled to begin in mid-1957. Early in 1956, having proposed to Soviet policymakers that the nation beat the Americans to the punch, Korolev and his allies obtained operational authorization for the satellite project. *Sputnik* was launched less than two months after the Soviets first successfully tested an ICBM.

The Soviet achievement stunned the West. The *Sputnik*s demonstrated that the Soviets possessed the rocket and guidance capability for ICBMs, and that by putting a live dog on board, they were well on the way toward putting a man into space. President Dwight Eisenhower refused to panic. He knew from secret intelligence, particularly the information gathered by overflight surveillance of the Soviet Union, that the United States held the lead over the Soviets in intercontinental rocketry. (The United States launched its own satellite soon thereafter.) Eisenhower nevertheless accelerated the nation's program to build and deploy ICBMs and authorized the deployment of short-range missiles in Italy and Turkey. *Sputnik* also prompted him to strengthen American science and its role in policymaking by creating the President's Science Advisory Committee. And it led the government in 1958 to establish the National Aeronautics and Space Administration (NASA) to oversee and coordinate all the nation's nonmilitary activities in space research and development.

Eisenhower thought of the new agency more as a consolidation of existing agencies than as a bold departure, remarking to his cabinet that he did not want to pay to learn "what's on the other side of the Moon." However, Congress, resolute in its sense of Cold War competition, pushed the president hard. "I want to be firstest with the mostest in space," a member of the House of Representatives declared.

Sputnik's tiny "beep-beep" precipitated a space race between the United States and the Soviet Union, and among its achievements were the first exposure of the far side of the *Moon to human observation via the Soviet *Luna 3* in 1959, and the first landing of men on the Moon through the American Apollo program in 1969.

See also SPACE SCIENCE.

Robert A. Divine, *The Sputnik Challenge: Eisenhower's Response to the Soviet Satellite* (1993). Asif A. Siddiqi, *Challenge to Apollo: The Soviet Union and the Space Race, 1945–1974* (2000).

DANIEL J. KEVLES

STANDARDIZATION is ancient and ubiquitous, because a community requires it, or at least functions better when standards of weights, measures, currency, morals, and the like exist. The nineteenth and twentieth centuries, however, saw a great extension and acceleration of movements toward standardization—in science, technology, industry, and society in general.

In a narrow sense, standardization refers to the establishment of specifications for the measurement, design, and performance of scientific and technological processes and products, enforced either voluntarily or by some authority. In a broader sense, the term "standardization" may be used to characterize wide-ranging historical movements in the nineteenth and twentieth centuries, on par with terms like "nationalism" and "liberalism."

The origins of the modern movements of standardization may be traced to the eighteenth century, when a quantifying and systematizing spirit began to grow among the Enlightenment *philosophes*. They took many of their cues from Isaac *Newton's triumphant system of the world. As he quantified and systematized the solar system in the seventeenth century, so they would extend his techniques to all fields and endeavors, from chemistry to forestry to public health to bureaucracy. Prominent results of this spirit included Linnaeus's system of natural classification, the great *Encyclopédie* of Diderot and d'Alembert, and the beginnings of the metric system in France.

Though the metric system did not always receive unqualified support from French revolutionaries in the 1790s because of the royalist connections of some of its proponents and the usual academic rivalries, the system did have the advantage of suppressing the standards and measures of the old order. Napoleon, in turn, did his part to spread the system by mandating its use in the states he conquered. After Napoleon, metric-based weights and measures reform became a key lever that the new centralized bureaucratic states of Europe used to establish their control over trade and taxes.

The same period saw the beginnings of the revolutions in transportation and communications centered around the railroad and the telegraph, which spurred movements for standardization on a number of levels. As railroads expanded from local to regional to national lines, mismatches in track gauge often occurred. Locals often worked against standard gauges to protect their commercial interests. Eventually, however, the forces encouraging railroad system integration overcame local barriers and inertia. In the United States the increase in east-west trade, the military urgencies of the Civil War, and the first transcontinental railroad forced standardization of gauges.

The expansion of the railroads also precipitated standardization throughout whole economies by encouraging the mass manufacturing and mass marketing of standardized products. The long-distance, high-volume trade in grain, for example, meant that buyers no longer could personally inspect and oversee each purchase. The new impersonality required the establishment of standardized units and grades of grain.

Similar developments during the second half of the nineteenth century may be observed in manufactured products, especially in the United States. Increased speed and other innovations in continuous-process manufacturing enabled the mass production of items such as matches, cigarettes, soap, dressed meat, beer, and canned goods in standard sizes and grades. The celebrated American system of manufacture—the use of standardized, interchangeable parts to enable the mass production of durable goods—had its origins in mid-nineteenth-century small-arms production, from which it spread to sewing machines, bicycles, and typewriters. Overall the implementation of the system was spotty and its full impact came only with the rise of the automobile industry in the twentieth century.

Railroads also acted to standardize time-keeping. For its schedules each railroad usually chose a standard based on the local time of the principal city it served, and thus, by means of its schedules, extended the use of that time to the surrounding communities. The expansion of the railroads initially caused a multiplication of time standards and confusion where railroads met. But lofty proposals from scientists to establish an international system of time, much debated in the 1870s, had little appeal to businesses and railroads. In the United States a mixed system of quasi-official standard time zones and a patchwork of local times persisted until 1918, when Congress mandated a national system.

These developments—the coming of the railroads, the telegraph, mass production, and mass distribution—unintentionally worked to erase local cultures and create national ones in the Western world. More intentionally, nationalistic leaders used the new public education, conscription, and mass media of the late nineteenth century to turn "peasants into Frenchmen" or whatever nationality was at stake.

Nor did scientific, technological, and cultural standardization stop at national borders. Organizations like the International Telegraphic Bureau (founded 1865), the International Metric Union (founded 1875), and the Universal Postal Union (founded 1878) represented an international cooperative trend that prompted or forced reluctant states to yield sovereignty over what historically had been internal matters. The proliferation of international congresses in the second half of the nineteenth century, bringing together devotees of subjects ranging from the usual academic fields to interests such as world peace, the abolition of tobacco, and liturgical chanting, marked the trend on the nongovernmental level.

Scientists were in the vanguard scheduling and promoting such congresses, and standardization was a topic that they often raised and debated. They continued to agitate for the universal acceptance of the metric system and also sought to systematize standards of measurement in other fields, such as *electricity. The needs of the rapidly expanding telegraph industry provided the initial push around 1860 for the development of a system of electrical units and standards. At a series of international electrical congresses between 1881 and 1904 electrical practitioners and scientists, including William *Thomson (Lord Kelvin), Hermann von *Helmholtz, John William Strutt (Lord Rayleigh), and Henry Rowland, worked to construct an international system of such units, based on the ohm, volt, and ampere. In addition to nationalistic rivalries, they struggled with the tension between scientific coherence and practical application. Defining magnetic units that had natural relationships to electrical units and also had convenient magnitudes for engineering work turned out to be impossible. Scientists also were enamored over the increasing capabilities of their instruments for precision measurement and the determination of the next decimal point. Most practitioners, meanwhile, felt it to be wasted effort.

A more subtle impediment was raised by the different legal and scientific cultures of the participants. The Germans and the English not only had varying legal approaches to setting standard weights and measures, they often were at odds on issues as fundamental as what constituted an acceptable measurement and the proper way to calculate, report, and interpret the errors in their results. By their very nature, standards are social creations of networks of people and institutions, and they therefore require shared values and a foundation of trust. Constructing that foundation of trust—in instruments, materials, methods, and measurers—is a delicate and often frustrating iterative process.

In the end the forum of the international electrical congress proved inadequate to the task. The organizations that ultimately and ironically succeeded in establishing an international system of units were the bureaus of standards of Germany, England, and the United States. Though often rivals, the Physikalisch-Technische Reichsanstalt (founded 1887), National Physical Laboratory (founded 1899), and National Bureau of Standards (founded 1901) had common interests in standard-setting and keeping. Their cooperative work in the first decade of the twentieth century, which also involved France's Laboratoire Central d'Électricité (founded 1888), led to a well-organized system by 1912.

World War I greatly expanded and accelerated the movements for standardization. The war motivated attempts to rationalize the industrial effort on the home front and eventually forced a transformation of the economies of the combatants to a more efficient, or at least more orderly, command structure. As one result, the war and its aftereffects spawned national standardizing agencies in nineteen countries in Europe, North America, and Asia between 1916 and 1924. Their work and the work of similar organizations ranged widely and deeply. In any given country, thousands of committees and subcommittees worked to develop standards of testing, dimensions, and quality for all products imaginable: from bolts to mattresses to biological and medical substances. The effort extended all the way to the shop floor, where Frederick W. Taylor and his followers in "scientific management" sought to determine and prescribe the most efficient motions of workers.

Nor did the standardizers exempt their own skills and conduct, though they had an ulterior motive. The second half of the nineteenth century saw many attempts to construct new scientific disciplines and engineering professions. A key means used to demarcate professionals from lesser practitioners was the establishment of standards of knowledge, skill, and behavior via university degrees and professional exams and codes of conduct. The creation of such standards requires that knowledge itself be organized and standardized. The standardizers helped to accelerate this ancient and ubiquitous process. Its end points today may be found in textbooks, teaching apparatus, competency exams, and encyclopedic companions.

Norman F. Harriman, *Standards and Standardization* (1928). David A. Hounshell, *From the American System to Mass Production, 1800–1932* (1984). Tore Frängsmyr, J. L. Heilbron, and Robin E. Rider, eds., *The Quantifying Spirit in the 18th Century* (1990). Meinolf Dierkes and Ute Hoffmann, eds., *New Technology at the Outset: Social Forces in the Shaping of Technological Innovations* (1992). Theodore M. Porter, *Trust in Numbers: The Pursuit of Objectivity in Science and Public Life* (1995). M. Norton Wise, ed., *The Values of Precision* (1995). Jed Z. Buchwald, ed., *Scientific Credibility and Technical Standards in Nineteenth and Early Twentieth Century Germany and Britain* (1996).

LARRY R. LAGERSTROM

STANDARD MODEL. During the 1980s, physicists who worked on *elementary particles came to agree that matter consists of three pairs of leptons—very light and even weightless particles—and their antiparticles (of which *electrons and their corresponding neutrinos are the exemplars) and three pairs of *quarks and their antiparticles (of which protons, neutrons, and other heavy particles, or "baryons," are made). So much for the bricks. The mortar that holds the quarks together (the "strong force") comes in eight kinds of "gluons" (*see* TERMINOLOGY); the cement that binds leptons to one another and to quarks (the "electroweak force") consists of the photon (the electrical part of the force) and three other adhesives, W^+, W^-, and Z^0 (the weak part). The detection of the W and Z particles and the top quark in 1982–1983 completed the experimental identification of the elements of the standard model. The successes of this model gave impetus to Grand Unified Theories (GUTs), intended to unify the strong and electroweak forces, and to dreams of *Theories of Everything (TOEs).

Some particle physicists, notably Steven Weinberg (who won the Nobel Prize in physics in 1979), have asserted that now that the guts of GUTs are in place, the final TOE will soon follow. A glance at earlier claimants to the status of "standard model" does not give cause for confidence in his prediction. To go back no further than 1800, the system of *imponderables developed by Pierre-Simon de *Laplace and his school seemed capable of describing all the phenomena then known in the same terms (though not the same language) as the current standard model: several leptons (the weightless "fluids" of electricity, magnetism, heat, light, and so on), a baryon (the particles of "common matter"), and forces of attraction and repulsion. Many natural philosophers looked forward to a unified theory that would connect the various "fluids" (leptons), a project encouraged by the discoveries of radiant heat and *electromagnetism. But with difficulties in the theory of heat (*see* ENTROPY) and new fashions in science (*see* CONSERVATION LAWS; FIELD), the imponderable fluids evaporated. A new standard model was drawn up, based on the unification of light with electromagnetism, heat with kinetic energy, and *magnetism with vortical motion, which strove to manage with one sort of material substrate, or *ether, subject to the laws of *mechanics. In the most austere of these GUTs, the "vortex atom," developed especially by William *Thomson (Lord Kelvin), James Clerk *Maxwell, and Joseph John *Thomson, all physical phenomena were to be referred to as motions of a single, perfect, incompressible space-filling medium.

The program of mechanical reduction, the standard model of the late nineteenth century, collapsed under experimental discoveries (*see* RADIOACTIVITY; X RAYS) and difficulties in the theories of radiant heat and electrodynamics (*see* QUANTUM PHYSICS; RELATIVITY). The discovery of the *electron and subsequent speculation about *atomic structure suggested that matter might be built from three ingredients: in today's language, a negative lepton (electron), a positive baryon (proton), and, after the Compton effect, a neutral photon. But the study of the nucleus (*see* NUCLEAR PHYSICS AND NUCLEAR CHEMISTRY) and *cosmic rays between the world wars, and the building of ever more powerful *accelerators after World War II, revealed many more "particles" than three. The gigantic instrumental and theoretical effort to classify and comprehend this cornucopia resulted in the standard model of the 1980s.

Like the decuplet of five pairs of twin brothers who ruled Atlantis, the standard model with its three pairs of leptons, eight gluons, and so on, might well slip into the sea. Its place may be taken by uncountable numbers of unimaginably small, wriggling, vibrating bits, as in string theory, the latest candidate for the Theory of Everything.

Yuval Ne'eman and Yoram Kirsh, *The Particle Hunters* (1983).

J. L. HEILBRON

STAR. To pretelescopic astronomers the night sky hemisphere contained about three thousand visible stars, grouped into constellations whose shapes did not vary over thousands of years. This pattern formed a backdrop against which the orbital wanderings of the planets could be traced. The most prominent star was only about a hundred times brighter than the faintest. Hipparchus of Nicaea classified star brightness into six degrees of importance, a division that has formed the basis of today's stellar magnitude system.

From Antiquity to the Renaissance there had been debate over whether the Sun, a heat-giving, golden disc subtending half a degree at Earth, had anything in common with the cold stellar pinpoints seen at night. To Giordano Bruno, the Sun and stars differed only in their distance from Earth. He also suggested that stars moved. Edmond *Halley proved as much in 1718 by finding that Sirius, Aldebaran, Betelgeuse, and Arcturus had changed position since Ptolemy drew up his catalogue fifteen hundred years earlier.

The Ptolemaic cosmos had the stellar sphere just beyond the orbit of Saturn. The Copernican heliocentric hypothesis, post-1542, had Earth moving 300 million kilometers every six months. The implied stellar parallax was eagerly sought. In 1632 *Galileo Galilei suggested that monitoring the separations and angular positions of optical double stars might provide some parallaxes. This approach eventually succeeded around 1838 when Friedrich Wilhelm *Bessel, Friedrich Struve, and Thomas Henderson observed the parallax of 61 Cygni (0.293 arc seconds), giving this star a distance of 3.4 parsecs (9.8 light years). Now astronomers could compare the solar luminosity with that of other stars. The number of stars with known distances became an effective aid to *astrophysics around 1903 when photographic measurements at the Yerkes and Allegheny Observatories in the United States reached a precision of about 0.01 seconds.

Galileo's telescopic observations of 1609 revealed that the Milky Way consisted of a myriad of faint stars. Isaac *Newton and Edmond Halley convinced themselves (between 1692 and 1720) that the starry realm was infinite, and balanced by gravitation. The discovery of the Doppler principle in 1842 revolutionized the study of stellar motion. Radial velocities could then be combined with proper motions perpendicular to the line of sight. In the second half of the nineteenth century, *photography further transformed stellar research. Photographic spectra came in 1872 and photographic cataloguing and sky mapping in 1882.

Earlier, around the turn of the century, William *Herschel had observed optically close stars and discovered that many were gravitationally associated binaries orbiting a common center of mass. Astronomers soon realized that about 50 percent of the stars were single, like the Sun, the majority of the remainder being binaries. In 1912 Henrietta Leavitt found that pulsating giant Cepheid stars have periods that are a function of their luminosity.

Early-nineteenth-century estimates of stellar surface temperatures made using Newton's law of cooling gave temperatures far too high. Better values were obtained by calculating with the radiation laws of Josef Stefan (1879) and Wilhelm Wien (1896). Stellar surface temperatures came out in the range of 3000° to 40,000°K. Stellar spectral classification blossomed. The astronomical spectroscopist William Huggins showed that stars and Earth were made of similar chemical elements.

Around 1914 Ejnar Hertzsprung and Henry Norris Russell independently plotted logarithmic graphs of stellar luminosity against surface temperature. This Hertzsprung-Russell diagram indicated two main types of stars, dwarfs (like the Sun) and giants (some one hundred times larger). The discovery of planet-sized white dwarfs and huge supergiants soon followed. The H-R diagram acted as a basis for the study of stellar evolution.

Before 1840, when most people still regarded the universe as relatively young, few worried about the source of solar (and stellar) energy. Some suggested mass accretion, others solar contraction. By the mid-nineteenth century the accepted age of the universe had increased to 400 million years and the problem of the generation of stellar energy became acute. Radioactive decay (discovered in 1896) appeared to be a possible energy provider (see RADIOACTIVITY). Albert *Einstein's equivalence of *mass and energy (see RELATIVITY) became the cornerstone of stellar energy generation when in 1919 Francis Aston, using a mass spectrometer, found that a helium atom weighed less than four hydrogen atoms. Hans Albrecht *Bethe worked out the scenario in the late 1930s. Around 1926, Arthur Eddington estimated that the central stellar temperatures of 10^7°K were high enough to enable the implied transmutation of hydrogen into helium to occur. Energy generation by the transformation of elements indicated that only a little loss in mass took place during stellar evolution. In 1944 Albrecht Otto Johannes Unsöld found that sun-like stars could "burn" nuclearly for about 10^{10} years, whereas hot, luminous O and B stars would have lifetimes as short as 10^6 years and would die very close to where they were born.

In 1921 Meghnad Saha proposed a theory of thermal ionization and excitation that made possible the calculation of stellar chemical composition from spectra. By 1925, Cecilia Payne-Gaposkin realized that hydrogen and helium comprised 99 percent of stellar material. Around 1943 Walter Baade discovered groups of younger stars rich, and groups of older stars poor, in metals. In 1957, Fred Hoyle and William Fowler showed how nuclear fusion synthesized all the other elements.

In 1924 Eddington had discovered that all star masses lay in the range of 0.05 to 100 solar masses and that the luminosity of a Main Sequence star (that is, the diagonal band of stars on the H-R diagram, all of which generate energy by converting hydrogen to helium) was approximately proportional to the fourth power of its mass. Thus stars did not evolve along the Main Sequence; they moved to the right as they ran out of hydrogen. From the early 1950s, observations of this break-off point were related to the age of stellar clusters.

Recent observations of pulsating radio stars have confirmed that stars more massive than the Chandrasekhar mass (suggested by Subrahmanyan *Chandrasekhar in 1931 to be about 1.4 solar masses) do not stabilize at the white dwarf stage (by which point the star is about the size of the earth and consists mainly of heavy elements), but continue condensing to become neutron stars, about 20 km across. Stars more massive than about 3.2 solar masses (the Oppenheimer-Volkov mass, calculated in 1939) become *black holes.

See also ASTRONOMY; COSMOLOGY; SOLAR PHYSICS.

Agnes M. Clerke, *A Popular History of Astronomy During the Nineteenth Century* (1885). Martin Johnson, *Astronomy of Stellar Astronomy and Decay* (1950). A. Pannekoek, *A History of Astronomy* (1961). Otto Struve and Velta Zebergs, *Astronomy of the 20th Century* (1962). Martin Harwit, *Cosmic Discovery: The Search, Scope, and Heritage of Astronomy* (1981). Dieter B. Herrmann, *The History of Astronomy from Herschel to Hertzsprung* (1984).

DAVID W. HUGHES

STATISTICAL MECHANICS. See THERMODYNAMICS AND STATISTICAL MECHANICS.

STEADY-STATE UNIVERSE. Steady-state theory was conceived at a movie theater in 1947. The British Astronomer Fred *Hoyle, accompanied by Hermann Bondi and Thomas Gold, Austrian-born physicists who had worked with Hoyle on radar during World War II, saw a ghost story that ended the same way it began. That inspired thoughts about a universe unchanging yet dynamic. According to Hoyle, "It did not take us long to see that there would need to be a continuous creation of matter."

A universe unchanging in density holds a philosophical advantage over a big-bang, expanding universe. If density changes, various physical laws might also change, invalidating extrapolations from the present back to a super-dense origin of the universe. Steady-state theory also enjoyed an observational advantage over big-bang theory in 1947. The estimated rate of expansion extrapolated back to an initial big bang yielded an age for the universe that was less than the estimated age of the solar system.

Scientific arguments about the steady-state theory in Great Britain turned on philosophical questions, with little appeal to observation. Religion and politics also played a part. Pope Pius XII suggested in 1952 that big-bang cosmology agreed with the notion of a transcendental creator, and was in harmony with Christian dogma—an extrapolation for which he was later criticized at the Second Vatican Council.

Soviet astronomers rejected both steady-state and big-bang cosmologies as idealistic and unsound. Hoyle associated steady-state theory with personal freedom and anti-Communism. Observational challenges to steady-state theory came from the new science of radio astronomy. Martin *Ryle at Cambridge University reported in 1955 that his survey of almost 2,000 radio sources contradicted steady-state theory. His conclusion, however, was premature. Hoyle felt bitterly that Ryle was motivated not by a quest for the truth, but by a desire to destroy steady-state theory. For his part, Ryle did not respect theoretical cosmologists. The final blow against steady-state theory came in 1965 with the discovery of cosmic microwave background radiation.

Hoyle responded that cosmic background radiation could arise from interaction between stellar radiation and interstellar needle-shaped grains of iron. But few found the response persuasive. For most purposes, the big bang had defeated the steady state by 1965.

See also ASTRONOMY, NON-OPTICAL; COSMOLOGY.

Norriss S. Hetherington, ed., *Encyclopedia of Cosmology: Historical, Philosophical, and Scientific Foundations of Modern Cosmology* (1993). Helge Kragh, *Cosmology and Controversy: The Historical Development of Two Theories of the Universe* (1996).

NORRISS S. HETHERINGTON

STELLAR ABERRATION. See ABERRATION, STELLAR.

STRANGENESS. Shortly after World War II, physicists found evidence for a flurry of new particles in cosmic rays and the products of *accelerators. In addition to the muon and π-meson, the new entities included V particles, named after the forked tracks they left in cloud chambers and nuclear emulsions, and the K meson or kaon. By the early 1950s experimenters had distinguished several different V particles (now known as Λ, Σ and Ξ). The high production rate of kaons and V particles suggested that they were governed by the strong force, but they decayed slowly with lifetimes typical of weakly interacting particles.

Physicists thought this behavior strange and sought a theory for the so-called strange particles. In 1951 Abraham Pais in the United States and a group of Japanese physicists independently proposed that kaons and V particles could be produced only in pairs, although they could still decay individually via weak interactions. The notion of associated production seemed to reconcile the strong production rates with the weak decay modes. In 1953 Murray Gell-Mann and, independently, Kazuhiko Nishijima and Tadao Nakano refined the idea, proposing that the strange particles carried a new quantum number, which Gell-Mann called the strangeness quantum number, S. Pions, muons, and nucleons would have $S = 0$, but the strange particles would have nonzero, integer strangeness: for example, $S = +1$ for kaons and $S = -1$ for the existing V particles. According to the theory, strangeness was conserved in strong interactions but not in weak ones, thus indicating what interactions were possible. As many more new particles were produced in the 1950s they were assigned strangeness values, and strangeness would prove a useful guide to the particle zoo.

"Strangeness" was a strange term. Physicists had previously appealed to Greek to name new particles, as indicated by the deuteron and mesotron or the atom itself, or to physical concepts such as spin. Strangeness represented a whimsical turn in the argot of physics that would flower in later neologisms such as quarks, flavors, and colors. This represents in part the linguistic interests of Gell-Mann. But terms such as strangeness also represent an approach characteristic of American physics, which spurned classical culture and philosophical ruminations for a more pragmatic engagement with the subject (see TERMINOLOGY).

Andrew Pickering, *Constructing Quarks: A Sociological History of Particle Physics* (1984). Laurie M. Brown, Max Dresden, and Lillian Hoddeson eds., *Pions to Quarks: Particle Physics in the 1950s* (1985).

PETER J. WESTWICK

STRATIGRAPHY AND GEOCHRONOLOGY. The principles of stratigraphy—the study of the earth's strata or layers of sedimentary rock—and of geochronology—the naming and describing, though not necessarily dating, of the periods of earth history—were established rapidly between 1810 and 1840. For the next century, stratigraphers filled in the details of the stratigraphic column with ever-greater precision. Much of this research could be put to good use by the mining industry, and from the 1920s and 1930s by the petroleum industry.

Although stratigraphy flowered in the first half of the nineteenth century, it had its roots in the seventeenth century. The Danish Cartesian Niels Stensen (or Steno), in his *Prodromus to a Dissertation on Solids Naturally Contained within Solids* (1669), considered bodies that made up the earth, particularly fossils, crystals, and strata. In any sequence of undisturbed strata, he concluded, the oldest strata would be on the bottom and the youngest on the top. This was an early version of the first of the three major principles of stratigraphy—the principle of superposition.

The second principle—that rock types or lithology usually occur in a predictable sequence—followed from the work of eighteenth-century mineralogists in the German states, Italy, France, the British Isles, and Russia. Independently of one another, they became convinced that the strata of the earth occurred in the same order everywhere. On the small scale, they knew that in an individual mining area they would find the same rocks in the same sequence in adjoining shafts. On a grander scale, they believed that around the globe the rocks could be sorted into three main groups that appeared to represent a time sequence: the primary rocks, hard and often crystalline; the secondary rocks, softer, layered, and often fossiliferous; and the tertiary rocks, the topmost and softest rocks. Unfortunately, the principle of superposition failed when strata had been disturbed after deposition, and the principle of lithological regularity broke down when rocks of the same lithology occurred more than once in the sequence.

By the second decade of the nineteenth century, the third principle of stratigraphy—that fossils can be used to identify and correlate strata—had been established. For the next century and a half, *paleontology was to be chiefly a tool for stratigraphy. Armed with these three principles, geologists between 1820 and 1840 established and named the greater part of the stratigraphic column, an accomplishment that has held up in outline to the present day. In practice, it involved one controversy after another about particular puzzles in the sequence. The British played a large part, perhaps because the strata of England are relatively straightforward.

In 1815, a mineral surveyor, William Smith, had published a map of the strata of England that although not fully correct, made a good start. Charles *Lyell gave names to the epochs of the Tertiary—Pleistocene, Pliocene, Miocene, and Eocene— and distinguished them by their proportion of still extant fossils. Adam Sedgwick renamed the older part of the Secondary formations, the Paleozoic. With Roderick Murchison, he introduced the names Cambrian, Silurian, and Devonian. The Carboniferous and the renaming of the upper part of the older Secondary Mesozoic were English suggestions. The renaming gave birth to Permian (another Murchison coinage), Triassic (a German suggestion), Jurassic (largely French), and Cretaceous (Belgian). The establishment of geochronology was a magnificent achievement. Museum panoramas and book illustrations showing the development of life forms still encapsulate what the general public understands by geology.

More important within professional geology were stratigraphic maps, topographic maps colored to show the strata that outcrop at the surface of the earth. Geological mapping developed with great speed between the 1780s and the 1830s when most of the techniques employed until World War II were at hand. As *cartography progressed and accurate topographic base maps that showed change of elevation by contours became more widely available, the task of making geological maps became easier. Stratigraphers used maps both as a record of their fieldwork and as a way to extract new information. Using them in conjunction with the stratigraphic column—a theoretical reconstruction of the strata arranged according to age—and the section—the vertical arrangement of strata along some line or traverse across the surface of the map—they could construct a mental picture of the three-dimensional structure of the strata and thus predict what would be found beneath any spot on the earth's surface.

Until the 1950s, most geological education gave priority to teaching students to construct and interpret maps, and professional stratigraphers were largely occupied with mapping the earth's surface. During the nineteenth century, they extended their mapping beyond northwestern Europe. They resolved problems about the Cambrian-Silurian boundary by introducing the Ordovician Period. American stratigraphers found that the Carboniferous did not work well for their territory, and replaced it with Mississippian and Pennsylvanian. The Canadians began trying to make sense of the afossiliferous pre-Cambrian rocks that made up much of their country.

Stratigraphers saw themselves as men who traveled widely, scaling mountains and descending mines, hammer, notebook, and map in hand, returning to their bases with packages of fossils and rocks to add to growing collections. Some were independently wealthy, but most found employment in universities and national geological surveys. In 1878, at the first International Geological Congress in Paris, they began the huge task of codifying stratigraphic nomenclature, a task that still continues.

Stratigraphers puzzled about how to reconcile the distinct breaks in the fossil record with the gradual changes predicted by evolutionary theory, writing at length about how elevation and erosion had destroyed part of the record. They worried that fossils might indicate changes in the environment of deposition rather than in the time of deposition. Walking through a gorge with sloping strata might be a walk through time or it might be a walk through space, from the deep ocean to a continental shelf to a brackish delta. With the growth of the petroleum industry, further subdivision of the stratigraphic sequence became a necessity. New tools were developed, such as well logs and microscopical examination of microfossils, particularly foraminifera.

By World War II, the intellectual excitement in stratigraphy had evaporated. It revived after the war when stratigraphy was subsumed under the *earth sciences, with their host of new concepts and sophisticated instruments.

Claude C. Albritton, Jr., *The Abyss of Time: Changing Conceptions of the Earth's Antiquity after the Sixteenth Century* (1980). Barbara M. Conkin and James E. Conkin, eds., *Stratigraphy: Foundations and Concepts* (1984). James A. Secord, *A Controversy in Victorian Geology: The Cambrian-Silurian Dispute* (1986). Martin J. S. Rudwick, *The Great Devonian Controversy* (1985). Stephen Jay Gould, *Time's Arrow, Time's Cycle* (1987). Brian W. Harland et al., *A Geologic Time Scale* (1990). David R. Oldroyd, *The Highlands Controversy* (1990). Robert H. Dott, Jr., ed., *Eustasy: The Historical Ups and Downs of a Major Geological Concept* (1992).

RACHEL LAUDAN

SUPERNOVA. Supernovae have been observed on several occasions, recently and spectacularly in 1885, when Ernst Hartwig saw a new star brighten the Andromeda galaxy by 25 percent. Six months later this supernova was ten thousand times fainter.

In 1911 the American astronomer Edward Charles Pickering differentiated between low-energy novae, often seen in the *Milky Way *galaxy, and novae seen in distant *nebulae like Andromeda. By 1919 Knut Lundmark had realized that low-energy novae occurred commonly, whereas the brighter novae, up to tens of thousands times more luminous, occurred rarely. The 1920s saw two theories of these brighter novae (named "supernovae" by Fritz Zwicky and Walter Baade in 1934), one relying on runaway instabilities in stellar interiors, the other (by Alexander William Bickerton) suggesting that collisions had occurred between stars.

Zwicky started the first supernova detection patrol in 1933; J. J. Johnson joined him in 1936. Using the new 45-cm Palomar Schmidt telescope they found twelve new supernovae in three years based on 1625 photographs of 175 extragalactic regions. The 1.2-m Palomar Schmidt came into use after 1958, and by 1977 the supernova tally had reached 450.

In 1981 Gustav Tammann estimated that around three supernovae occurred every century in the Milky Way galaxy. Most go undetected owing to obscuring interstellar material. During the last millennium local supernovae were detected only in 1006, 1054, 1572, 1604, and 1667.

The Taurus supernova of 1054 was extensively recorded in the East, being visible in daylight and reaching –5 in the magnitude scale. (The magnitude scale of stellar brightness is logarithmic: the brightest naked eye star is of magnitude 1, the faintest of magnitude 6. Hence, a negative magnitude denotes a body that is brighter than the brightest naked-eye star.) John Bevis discovered the expanding cloud of material that resulted in 1731. In 1758 Charles Messier labeled the cloud M1, the first entry in his catalogue of *nebulae. By 1937, O II, O III, N II, and S II emission lines (spectral lines emitted by excited atoms as they decay) had been found in the cloud. Owing to the large expansion velocity produced by the stellar explosion, the emission lines were particularly broad. After 1948 astronomers found several supernova radio sources (*see* ASTRONOMY, NON-OPTICAL).

In 1934, two years after the discovery of the neutron, Baade and Zwicky suggested that supernovae arose when giant stars became neutron stars. Their view became generally accepted after Jocelyn Bell Burnell discovered *pulsars in 1967. The Crab Nebula pulsar came to light in 1969. Over 120 supernova remnants have been discovered in the Milky Way. One type, exemplified by Cassiopeia A and the Veil nebula in Cygnus, has a ring-like structure. Others are irregular with a central brightening, like the Crab.

In the mid-1950s, astronomers recognized two supernova varieties. Type I are binary white dwarfs. Mass accretion, pushing the star beyond the Chandrasekhar limit (*see* STAR), triggers a wave of nuclear reactions and a flood of neutrinos, either destroying the star completely or leaving behind a neutron star. Type II results from the explosion of a young, massive giant star that has exhausted its nuclear fuel. In February 1987 a Type II supernova exploded nearby, in the Large Magellanic Cloud. The pre-nova star was a supergiant. In exploding, its brightness increased by 10^8 in a few hours. The visible energy release of 10^{44} joules was dwarfed by the 10^{46} joules of high-energy neutrinos, many of which were captured by atomic nuclei thus manufacturing elements heavier than iron. The explosion scattered these elements far and wide throughout the galaxy.

David H. Clark and F. Richard Stephenson, *The Historical Supernovae* (1977). Michael Hoskin, ed., *The Cambridge Concise History of Astronomy* (1999).

DAVID W. HUGHES

SYMMETRY AND SYMMETRY BREAK-ING. The modern scientific notion of symmetry begins with the geometric symmetries of objects, both mathematical and physical. A perfect snowflake rotated through 60° about its center is indistinguishable from its original appearance. Rotation through 90°, however, yields an appearance distinguishable from both. Rotating the snowflake transforms it relative to something external. Symmetry transformations of an object leave the initial and final states indistinguishable (at least with respect to the properties we specify as relevant). This concept of symmetry—indistinguishability under transformations—has blossomed in science over the past 400 years. Here, three developments are fundamental: the extension of the concept to "physical symmetries," the development of group theory and its scientific applications, and the increasing importance of "symmetry-breaking."

In science, the distinction between geometric and physical symmetries is the distinction between symmetries of objects and of laws. An object may fail to possess a given geometric symmetry, and still evolve in accordance with laws that do possess that symmetry. For example, a chair is not

rotationally symmetric (turn it through any angle other than 360° and the initial and final positions will be distinguishable), but since the laws of nature are rotationally symmetric in the absence of external influences, the natural behavior of the chair does not change with the direction it faces.

*Galileo made an early and famous application of a physical symmetry in the debate over the system of *Copernicus. Opponents of heliocentrism claimed that if the Earth moved around the Sun, the behavior of terrestrial objects would show it. In his *Dialogue Concerning the Two Chief World Systems* (1632), Galileo claimed that no such observations are possible, and he argued for this using an analogy with a ship. He pointed out that someone shut up in a windowless cabin on a ship would be unable to distinguish by means of any experiments carried out within that cabin whether the ship was at rest or in smooth, uniform motion. This so-called "Galilean relativity" is a symmetry of space and time; it quickly found its way to the heart of seventeenth-century natural philosophy, being used by Christiaan *Huygens in his solution to the problem of colliding bodies, and appearing in *Newton's *Principia* as Corollary V to his laws of motion. The Galilean group of symmetries also includes spatial translations and rotations, and temporal translations. The principle of relativity remains at the heart of modern physics as one of the two postulates of *Einstein's 1905 special theory of *relativity. Here, however, it belongs to a different group of space-time transformations, the Poincaré group.

The group concept emerged from developments in late eighteenth- and early nineteenth-century mathematics. In the early 1830s, Evariste Galois used discrete groups (groups consisting of a finite number of elements) to characterize polynomial equations via the structural properties of their solutions. In the 1870s, Sophus Lie set about extending Galois's theory from algebraic equations to differential equations, and this led him to the concept of continuous analytic groups (Lie groups). Felix Klein's 1872 "Erlanger Program" used group theory to characterize geometries, putting non-Euclidean geometry (so important for the general theory of relativity) on an equal footing with Euclidean geometry.

One of the first applications of group theory in science was in *crystallography. René-Just Haüy used symmetry to characterize and classify crystal structure and formation in his *Traité de mineralogie* (1801). With this application, crystallography emerged as a discipline distinct from mineralogy. From Haüy's work, two strands of development led to the 32 point transformation crystal classes and the 14 Bravais lattices, all of which may be defined in terms of discrete groups. These were combined into the 230 space groups by E. S. Fedorov (1891), Artur Schönflies (1891), and William Barlow (1894). The theory of discrete groups continues to be fundamental in *solid state physics, chemistry, and materials science, and in *quantum field theory through the CPT (charge conjugation, parity, and time-reversal) theorem.

Continuous symmetries come in two kinds: global symmetries, such as Galilean translations and rotations, and local symmetries, such as the gauge symmetry of *electromagnetism and the invariance under general coordinate transformations of the field equations of general relativity (1915). The importance of continuous symmetries in physical theories, and the power of symmetries in theory construction, was increased in 1918 when Emmy Noether proved the existence of a general connection between continuous symmetries and conserved quantities, and shed new light on the structure of theories with continuous local symmetries. Group theory and symmetries can provide powerful constraints on theories. For example, in particle physics global symmetries are used to classify particles and to predict the existence of new particles, such as the omega-minus particle (predicted in 1962, detected in 1964) via the SU(3) symmetry classification scheme. In 1918, Hermann Weyl introduced local scale symmetry to construct his unified theory of gravitation and electromagnetism, intended to succeed the general theory of relativity; this theory failed, as did the 1954 proposal of Chen Ning Yang and Robert Mills, today credited as the first

"modern" local gauge theory. Following the developments of the 1970s, however, theories with local gauge symmetry have come to dominate fundamental physics.

Symmetry breaking has become as important in modern science as symmetry itself. In 1894, Pierre *Curie highlighted the importance of symmetry breaking and asserted the so-called "Curie principle"—that an effect cannot be less symmetric than its cause. This assertion was challenged in the 1950s in two ways. First, the phenomenon of spontaneous symmetry breaking came to the attention of physicists in the context of superconductivity (and was later reapplied in the context of *quantum field theory). In fact, the symmetric solution of a symmetric problem may be unstable; in such cases, the observed stable outcome will be less symmetric than the cause, in apparent violation of Curie's principle. Nevertheless, theoretically there exists a set of equally likely effects (only one of which is observed in any given instance) that together possess the symmetry of the cause, and in this "sophisticated" sense Curie's principle survives the challenge of spontaneous symmetry breaking. The second challenge is the violation of parity, in which one possible outcome of an experiment dominates its mirror-image. This violation, predicted by Tsung Dao Lee and Chen Ning Yang in 1956, was detected soon afterward experimentally by Chien Shiung Wu and her colleagues (*see* LEE, WU, AND YANG). The law governing the weak nuclear interaction breaks the symmetry, and the Curie principle can only be saved by including the law within the cause.

During the latter half of the twentieth century, spontaneous symmetry breaking also became important in biology. Brian Goodwin describes one such application in *How the Leopard Changed Its Spots* (1994). All organisms start off as highly symmetric entities, such as a single spherically symmetric cell. As the organism grows, this highly symmetric state becomes unstable, owing either to internal stresses and strains, or to influences from the environment. Enter spontaneous symmetry breaking: the organism will move to one of a set of possible stable, but less symmetric, states. In this way, the dynamics of stability and spontaneous symmetry breaking constrain the possible general forms that an organism may take during its growth. Which of the possible states the organism moves to at each stage can be controlled internally (by a nudge from the DNA, for example) or by the environment (through temperature or a chemical). All this has radical implications for the theory of evolution. On the standard Darwinian approach, evolution is free to explore a huge variety of possibilities, constrained only in general terms by the laws of physics and chemistry. This approach leaves us several major puzzles, two of the most important being the emergence of the same general forms in different lineages, and the fact that we do not see evolution exploring all possibilities, but instead a rather limited subset. A response to this is that the domain of the "biologically possible" is highly constrained by dynamical stability, in which spontaneous symmetry breaking plays a key role.

Hermann Weyl, *Symmetry* (1952). Ian Stewart and Martin Golubitsky, *Fearful Symmetry* (1992). Klaus Mainzer, *Symmetries of Nature* (1996).

KATHERINE A. BRADING

SYMPATHY AND OCCULT QUALITY. Aristotelian physics was strong on classification (four elements, four causes, types of motion, categories of being) but weak on dynamics (generation, corruption, physical interaction). Bodies acted on one another primarily through the "manifest active qualities" of the elements predominating in their constitutions: hotness, coldness, dryness, and moistness. Thus, to take a complicated example used by Aristotle, the sun melts wax and dries mud, the different consequences of the same manifest quality (hotness) depending on the elementary makeup of the recipient body. Two other widespread attributes of matter, gravity and levity, were often treated as if they were manifest qualities, since they characterized the four elements even though they could not be reduced to the tangible qualities hotness, coldness, and so forth, to which Aristotle gave priority.

The world has many physical properties less widely encountered than gravity and levity but, like them, not easily or obviously explainable in terms of the action of manifest qualities. Later Peripatetic philosophy designated these properties "occult," because, although evident in their consequences, their causes were hidden. The exemplar of an occult quality was *magnetism. The ancients knew it as the ability of a peculiar rock to draw bits of iron to it—but why only iron? The answer lay, according to the natural philosophy taught when *Galileo was in school, in an innate sympathy or harmony between lodestones and iron. This example indicates the level of explanation that, in the seventeenth century, made "occult" a byword for nonsense. Originally an expression that aided the classification of properties whose causes were provisionally unknown, the occult became a trash heap of innate and irreducible qualities. A purge or poison, the deadly glance of the basilisk, astrological influences, the powers of talismans, and the force by which that small pesky fish, the remora, stops big ships—all operated by occult sympathies and antipathies between agent and recipient. Molière neatly satirized the level of explanation afforded by the occult in his *Malade imaginaire* (1673), in which he praises a doctor for ascribing the soporific quality of opium to an occult "dormative virtue."

The *mechanical philosophy, especially in its radical form of René *Descartes's limitation of the affections of matter to extension, shape, and motion, appealed to the scientific revolutionaries of the seventeenth century because it annihilated the complex of qualities taught by the traditional philosophy they opposed. Even manifest qualities had to go: the hot, cold, moist, and dry became secondary effects arising from the interaction of the few primary qualities of extended, moving, material bits with the human sensory apparatus. Explanations in mechanical terms, like Descartes's referral of thunder and lightning, and also rains of blood, to the precipitous fall of one cloud on another, might appear no more persuasive than magnetic sympathies; nonetheless, the corpuscular philosophy, by seeking a mechanical account of properties held by its opponents to be innate and irreducible, opened the possibility of further analysis.

Robert *Boyle's concept of the "spring of the air," for which he offered several mechanical analogies, and Descartes's representation of magnetism by a vortex of specially shaped particles, suggest the range and limitations of seventeenth-century mechanical models.

Against the rhetorical and explanatory advantages of the corpuscular philosophy, Isaac *Newton's apparent invocation of an occult sympathy—the "universal attraction" of the *Principia* (1687)—seemed retrograde to many natural philosophers enlightened by Descartes. They were both right and wrong. Newton did return to an occult quality, but in its most useful and responsible form: a widespread property of matter, exactly described, whose cause had not yet been found. Newton's famous phrase "hypotheses non fingo" ("I feign no hypotheses") meant that, as far as he was concerned, gravity would remain occult. Until we have a *Theory of Everything (*see* STANDARD MODEL), and perhaps even then, scientists necessarily will continue to invoke occult qualities.

Brian Vickers, ed., *Occult and Scientific Mentalities in the Renaissance* (1984). David C. Lindberg and Robert S. Westman, eds., *Reappraisals of the Scientific Revolution* (1990). Dennis Des Chene, *Physiologia: Natural Philosophy in Late Aristotelian and Cartesian Thought* (1996).

J. L. HEILBRON

T

TELESCOPE. Lenses for reading were available in Italy in the thirteenth century, but not until the seventeenth century did spectacle makers in the Netherlands put together a device "by means of which all things at a very great distance can be seen as if they were nearby." In 1609 *Galileo Galilei heard rumors of spyglasses, made more powerful ones, and pointed them at the heavens. In 1611 Johannes *Kepler explained the path of light rays through lenses and the formation of images. The improved Kepler telescope formed images in its focal plane, where they were viewed by a magnifying lens.

Anything placed in the focal plane of a telescope appears sharply alongside the celestial object, as the Englishman William Gascoigne noticed in about 1640 when a spider spun its web inside his instrument. Astronomers inserted cross hairs, facilitating precise alignment of telescopes on objects, and micrometers, to measure small angular distances and diameters. They also developed, though more slowly, stable, precise mountings and large arcs with precisely divided and marked scales against which the telescope's alignment could be noted when pointed at a celestial object. Still their instruments suffered from chromatic and spherical "aberrations"—fuzziness of the image—arising from the fact that different wavelengths or colors of light are refracted by different amounts, and light incident on the periphery of the lens focuses closer to the lens than does light striking near the center. To reduce the aberrations astronomers ground lenses with very long focal lengths, which led to long and unwieldy instruments. In 1757 the Englishman John Dollond perfected the achromat, a combination of glass lenses that brought rays of different colors to the same focus, enabling more precise measurements of positions of faint stars by means of shorter instruments easier to use. Inability to make large pieces of optical-quality glass, however, limited the size of refracting telescopes, in which light passes through transparent lenses.

In 1668 Isaac *Newton, having decided that chromatic aberration in lenses could not be defeated, built the first successful reflecting telescope. It employed a concave mirror to collect light and form the image. William *Herschel in England built telescopes with large reflecting metal mirrors in the 1780s and William Parsons in Ireland built the 6-foot "Leviathan of Parsonstown" in 1845. Giant reflectors, though producing spectacular observations, ultimately were disappointing: the mirrors flexed under their immense weights and tarnished quickly.

The second half of the nineteenth century saw advances in refracting telescopes, especially by the Boston firm of Alvan Clark and Sons. Their metal tubes were stiffer yet lighter than wooden telescopes. Larger pieces of optical glass were now available, from France and England, and five times the Clarks figured the lens for the world's largest refracting telescope, culminating in a 40-inch lens in 1897. It is yet to be surpassed in size. Larger pieces of optical glass are difficult to cast; heavier lenses flex more; and thicker lenses absorb more light.

Lenses also absorb strongly in the blue region of the spectrum, where *photography is most effective. A new interest in astrophysics and distant stars required a new technology. George Willis Ritchey made the photographic reflecting telescope the basic instrument of astronomical research, constructing at the Mount Wilson Observatory a 60-inch telescope in 1908 and a 100-inch in 1918. Later the Rockefeller Foundation paid for a 200-inch reflecting telescope at nearby Mount Palomar. Corning Glass Works cast the mirror in 1934 as a thin piece of Pyrex glass with a system of ribbing in the back. Grinding the lens removed five tons of material, leaving sixteen tons of curved mirror, which received its reflective coating of aluminum in 1949.

To circumvent the problems of casting and supporting large mirrors, many small mirrors can be assembled into a close array. The Keck Telescope, erected in Hawaii in

1993, has thirty-six 1-meter mirrors mounted together on a tracking structure, and the European Southern Observatory in Chile links four 8-meter and three 1.8-meter mirrors into one very large telescope. Its huge cost is shared among nine countries.

Reflecting telescopes bring only rays from stars in the center of the viewing field to a sharp focus. Given a usable field of view of 15 seconds of arc, approximately a million photographic plates would be required to cover the entire sky. In 1930 Bernhard Schmidt, an Estonian-born optician at the Hamburg Observatory, designed a reflecting telescope with a usable field of view of 15 degrees. The Schmidt telescope has a simple spherical mirror plus a thin correction plate for spherical aberration. Palomar completed a 1.2-meter Schmidt telescope in 1948.

Non-optical telescopes can detect radio and gravitational waves unblocked by the earth's atmosphere. Other non-optical telescopes rise above the earth's atmosphere. There *X rays incident on mirrors at small "grazing angles" are reflected into a detector, where their interaction with an inert gas generates countable electrons. The telescope, with several mirror surfaces nested concentrically within it, looks like a funnel. Telescopes in space also detect infrared emissions and gamma rays.

The Hubble Space Telescope enables traditional, optical astronomers to escape our atmosphere. The telescope's primary mirror is eight feet in diameter. Including recording instruments and guidance system, the telescope weighs twelve tons. It has been called the eighth wonder of the world; critics say it should be, given its cost of two billion dollars. It was as much a political and managerial achievement as a technological one; approval for it came only after a political struggle lasting from 1974 to 1977. In 1990, after overcoming a host of problems, its designers launched it into space, only to discover that an error had occurred in the shaping of the primary mirror. One newspaper reported "Pix Nixed as Hubble Sees Double." Addition of a corrective mirror solved the problem.

Over four centuries the telescope has evolved from two small glass lenses afford-able and operable by an untrained individual of no great wealth into an immense political, managerial, and technological undertaking beyond the reach of all but the wealthiest countries. Our understanding of the universe has expanded apace, as ever larger, more expensive, and technologically sophisticated telescopes range over ever more of the electromagnetic spectrum to detect ever more distant objects.

See also ASTRONOMY, NON-OPTICAL; INSTRUMENTS AND INSTRUMENT MAKING.

Henry C. King, *The History of the Telescope* (1955). Isaac Asimov, *Eyes on the Universe: A History of the Telescope* (1975). James Cornell and John Carr, eds., *Infinite Vistas: New Tools for Astronomy* (1985). J. A. Bennett, *The Divided Circle: A History of Instruments for Astronomy, Navigation and Surveying* (1987). Donald E. Osterbrock, *Pauper and Prince: Ritchey, Hale, & Big American Telescopes* (1993). Robert W. Smith, *The Space Telescope: A Study of NASA, Science, Technology, and Politics* (1993).

NORRISS S. HETHERINGTON

TELLER, Edward. See SAKHAROV, ANDREI, AND EDWARD TELLER.

TERMINOLOGY. The terms invented to denote new entities and instruments may illuminate the state of science as much as the objects of its investigation. At first, neologisms had unexceptionable derivations from ancient languages: *barometer, microscope, *telescope, *electricity. The custom continued in the nineteenth century with coinages that may not have satisfied all philologists: electrode, ion, scientist, physicist, telegraph, telephone, cesium, *electron, argon, helium, *radium. These names not only respected the convention that neologisms be based on the languages that once served the so-called Republic of Letters, but also that they indicate a characteristic feature of the object named. In the second half of the nineteenth century, however, both conventions were shaken by the terminology introduced into mechanics by the English (curl, twist) and by the nationalistic naming of new elements by their discoverers (germanium, gallium, scandium, polonium).

In the twentieth century the decline in the humanistic education of scientists and the jocularity of American physicists produced *strangeness, *quark (in its top, bottom, charmed, and colored varieties), and gluon alongside the old-fashioned quantum, proton, neutron, deuteron, positron, meson, and the playful neutrino. A similar development does not seem to have affected the biological sciences, which have kept to safe items like gene and genome, owing, perhaps, to closeness to medicine, still stuck in ancient argot.

See also COSMIC RAYS; ELEMENTARY PARTICLES; NOBLE GASES.

J. L. HEILBRON

TERRESTRIAL MAGNETISM. Practical needs, particularly of navigators, have inspired interest in terrestrial magnetism since at least the fifteenth century. Equally important have been conceptual puzzles about how to reconcile terrestrial magnetism with basic physical theory and with theories based on laboratory studies of *magnetism. Consequently interplay between field studies and experimental studies has been a regular feature of the history of geomagnetism. So has a tension between explaining the ultimate causes of geomagnetism, usually in terms of some kind of fluid movement in the interior of the earth, and surveying the spatial and temporal variations of geomagnetic declination, dip, and intensity.

Serious work on geomagnetism began in 1600 with the publication of William Gilbert's *De magnete*. By that date, navigators knew that their needles sometimes pointed at an angle to true north (declination) and that sometimes they inclined from the horizontal (dip). Philosophers generally assumed that the earth's magnetism arose through the occult properties of the mineral magnetite or by some Neoplatonic correspondence between the polestar in the heavens and the magnetic north pole on Earth.

Gilbert, a member of the Royal College of Physicians in London, discussed the five motions associated with magnetism—attraction (he called it coition), orientation, declination, dip, and rotation—as a preliminary to presenting his theory of earth magnetism. Experiments with small magnetic needles on a small spherical lodestone (called a terrella) showed that irregularities in the lodestone changed the orientation of the needles. They also demonstrated that needles parallel to the surface of the sphere at the equator gradually dipped to a vertical as they moved to the position at the poles. The earth, he concluded, was a giant lodestone with an immaterial rotating magnetic soul.

Because magnetism, including geomagnetism, seemed an exemplary occult force, mechanical philosophers had to find an alternative explanation in terms of matter in motion. In his *Principles of Philosophy* (1644), René *Descartes traced the earth's magnetism to circulating streams of corkscrew-shaped particles. From this suggestion arose the tradition, predominant until the 1820s, of attributing the earth's magnetism to subtle active magnetic fluids. Edmond *Halley in 1683 and again in 1692 proposed that the earth consisted of an inner sphere and outer shell. They rotated at different speeds and each had a north and south pole. The interactions between these four poles accounted for the variations in declination and dip. Between 1698 and 1700, Halley sailed the Atlantic, measured the variations in declination, and charted them on a pioneering map that appeared in different editions between 1701 and 1703.

Descartes's effluvial theory continued to be important until the early nineteenth century. The alternative, most fully articulated by Charles Augustin Coulomb (*see* CAVENDISH AND COULOMB) and Simeon-Denis Poisson—whose theory presented to the Paris Academy in 1826 represented the culmination of the tradition—assumed distance forces resulting from fluids locked in magnetic substances. Other important figures in the debate were Gavin Knight; Leonhard *Euler, who with others won the prize of 1746 offered on the subject by the Paris Academy of Sciences; Franz *Aepinus; and Jean-Baptiste Biot. Many researchers attempted to deal with earth magnetism though the requisite mathematics was dauntingly complex. During the 1820s and 1830s theories like Poisson's based on "austral" and "boreal" fluids were losing their luster. Christopher Hansteen revived Halley's two-axis–four-pole model in his *Investigations Concerning the Magnetism of the Earth* (1819). To look for poles, defined either as regions of maximum magnetic intensity or of vertical dip, Hansteen traveled to Siberia around 1830 and James Clark Ross went to Canada. Although the two-axis theory did not win acceptance, the reintroduction of poles as an object of investigation, the attempt to mathematize the theory, and the expeditions brought fresh ideas and evidence to geomagnetic studies.

In the same decades, Hans Christian Ørsted's discovery in 1820 that electric currents produce magnetic effects, Thomas Seebeck's discovery of thermoelectricity in 1822, and Michael *Faraday's discovery in 1831 that magnetism can produce electric currents gave rise to new questions and possibilities concerning earth magnetism. Alexander von *Humboldt, who had been fascinated by the global variations of magnetism since the 1790s, speculated about the similarities between lines of equal magnetic intensity and isothermal lines and about interconnections between geological, meteorological, and magnetic phenomena. In 1805 he reported that magnetic intensity varied across the earth's surface. To plot these variations, Humboldt encouraged the establishment of a network of magnetic observatories. By 1834 the twenty-three European observatories had detected the phenomenon of magnetic storms. In the fifth volume of his *Cosmos* (1845), Humboldt summed up the state of knowledge of magnetic variation, distribution, and storms.

In the 1830s Carl Friedrich *Gauss and his younger collaborator Wilhelm Weber took over from Humboldt as leaders in geomagnetism, tackling problems from instrumentation to basic theory. Early in the decade Gauss designed the bifilar magnetometer, developed for the first time an absolute measure of magnetic intensity, and launched his own version of the Magnetische Verein (magnetic union) to establish a network of magnetic observatories worldwide. The results from these observatories came out in six volumes, *Resultate aus den Beobachtungen des magnetischen Vereins*, between 1836 and 1841. With new data

about variations of magnetic intensity in hand, Gauss could publish his mathematical analysis of the vertical and horizontal components of earth magnetism in 1839. He analyzed the magnetic potential at any point on the earth's surface by an infinite series of spherical functions. Not dependent on a theory about the ultimate causes of geomagnetism, his method of analysis shaped theoretical work on geomagnetism for the rest of the century.

In Britain, a follower of Hansteen, Edward Sabine, fretted that Britain was letting slip the chance of contributing to the growing field of geomagnetism. In 1838 he enlisted the astronomer John *Herschel to help him raise support for a British magnetic survey. The publication of James Clerk *Maxwell's *Treatise on Electricity and Magnetism* in 1873 encouraged investigators to speak of the earth's magnetic *field, not its magnetic forces, and gave them another set of mathematical tools.

Between 1890 and 1900 geomagnetism began to take on the trappings of a separate discipline. A new generation of mathematically trained physicists, notably Arthur Schuster, continued working on mathematical analyses of the earth's field although they did not propose new comprehensive theories. With the establishment of national surveys and observatories, the amount of data available multiplied. The beginning of submarine warfare accelerated military interest in geomagnetism. International organizations expanded; in 1896 the journal *Terrestrial Magnetism* was founded. Another period of rapid breakthroughs in geomagnetism occurred in the years following World War II. In 1947, following measurements of the magnetic fields of the sun and some stars, the English physicist Patrick *Blackett suggested that magnetism (including the earth's magnetism) might be a property common to all rotating bodies. A decade earlier, Göttingen-trained physicist Walter Maurice Elsasser had published a series of papers suggesting that a self-excited magneto hydro dynamo in the earth's core created its field. For a few exciting years, scientists explored the consequences of the two theories in the hope of deciding between them. Then in 1952, after obtaining negative results from an experiment intended to detect the effects of rotation in the laboratory, Blackett himself rejected his own theory. Versions of Elsasser's theory held sway for the rest of the century.

Other major developments, interesting in their own right and for what they contributed to *plate tectonics, occurred in paleomagnetism. Already in the nineteenth century, scientists had detected remanent or fossilized magnetism. They noticed that ferrous minerals in baked clays and cooled lava flows preserved the alignment of the earth's main field as it was when they had cooled. In the late 1950s, physicists in London, Newcastle, and the Australian National University who systematically surveyed remanent magnetism found that the magnetic north pole appeared to have wandered widely over the globe in the past. They proposed various hypotheses to explain this result: their instruments created the effect, the earth's field had not always been dipolar, the continents had moved relative to one another, or the earth's magnetic poles had wandered independently. By the end of the decade, a small but influential group of scientists, Keith Runcorn prominent among them, had convinced themselves that the continents had moved. This served to give the largely discredited theory of continental drift new life.

In the 1920s and 1930s, scientists had discovered another peculiarity about remanent magnetism. In some rocks the magnetism had a polarity opposite to that of the present geomagnetic field. In the 1940s, researchers in the Carnegie Institution of Washington developed a spinning magnetometer capable of detecting weak magnetic fields. From the 1950s through the 1960s, paleomagnetists at the United States Geological Survey and the Australian National University raced to reconstruct the history of these reversals, using radioactive dating to determine their sequence. By the mid-1960s, they had constructed a fairly complete scale. It proved to be a key piece of evidence for the theory of sea floor spreading.

See also IMPONDERABLES; SYMPATHY AND OCCULT QUALITY.

Sydney Chapman and Julius Bartels, *Geomagnetism* (1940). R. W. Home and P. J. Conner, *Aepinus's Essay on the Theory of Electricity and Magnetism* (1979). William Glen, *The Road to Jaramillo* (1982). Christa Jungnickel and Russell McCormick, *Intellectual Mastery of Nature* (1986). Robert P. Multhauf and Gregory Good, *Brief History of Geomagnetism* (1987). David Barraclough, "Geomagnetism: Historical Introduction," in *The Encyclopedia of Solid Earth Geophysics*, ed. David E. James (1989).

RACHEL LAUDAN

THEORY OF EVERYTHING. By the 1920s physicists had constructed a standard model of fundamental particles and forces, which consisted of two particles, the electron and proton, interacting through the forces of *electromagnetism and gravity. Following the development of general *relativity, Albert *Einstein, Theodor Kaluza, Oskar Klein, and a few other physicists and mathematicians searched in vain for a theory that would combine the forces of gravity and electromagnetism in a unified field theory. Most physicists at the time focused instead on quantum theory and *nuclear physics, but these fields and the subsequent development of *high-energy physics introduced new forces (the weak and strong nuclear forces), new particles, and new conditions to physical theory. Physicists in the last few decades of the twentieth century thus embraced the program of unification in order to simplify the scheme of particles and forces, in search of what they called a grand unified theory, or theory of everything. A theory of everything would embrace in a single mathematical structure the different sets of equations needed to describe the actions of the four basic forces.

In the early 1970s theorists succeeded in unifying the electromagnetic and weak forces in the electroweak theory, based on an idea first advanced by Steven Weinberg and Abdus Salam in 1967. Experimental evidence of neutral currents in the early 1970s and the detection of W and Z particles a decade later supported the electroweak unification. Physicists then sought to link the electroweak theory to the recently developed theory, called quantum chromodynamics (QCD), for the strong nuclear force. Promising early attempts by Howard Georgi, Sheldon Glashow, and others accounted for the elementary particles of each theory—leptons for electroweak, quarks for QCD—as well as the carriers of the three forces, but experiments designed to test the theory, in particular its prediction of proton decay, failed to provide convincing evidence.

Physicists recognized that these "grand unified theories" included only three of the four forces. Gravity proved difficult to accommodate. The development in the mid-1980s of string theory, which treated constituents of matter not as particles but as strings, offered a candidate for complete unification. But string theory, while mathematically elegant, increasingly departed from experimentally verifiable predictions and, despite frequent intimations of imminent success, by the end of the century had failed to incorporate gravity with the other three forces.

A theory of everything did not imply an end to scientific research, but rather that the quest for fundamental knowledge had ended and all that remained was to fill in the details. Claims of completeness in physics echoed similar anticipations in the past—for instance, in the late 1920s after the formulation of *quantum physics, or at the end of the nineteenth century after the construction of classical physics. Theories of everything also assumed that elementary particle physics was the foundation for the rest of science, an assumption disputed by *solid-state physicists and chaos theorists, and likewise by biological scientists, for whom *quarks or string theory offered few clues to the meaning of life or consciousness. Some theoretical physicists strayed from science altogether into the realm of theology, and claimed that a theory of everything would give humankind a glimpse of the mind of God.

Steven Weinberg, *Dreams of a Final Theory* (1992). David Lindley, *The End of Physics: The Myth of a Unified Theory* (1993).

PETER J. WESTWICK

THERMODYNAMICS AND STATISTICAL MECHANICS. The development of the theory of heat in the first half of the

nineteenth century, which eventually led to thermodynamics, was linked with the technology of steam engines. Their operation was originally analyzed in terms of the caloric theory, which represented heat as a conserved *imponderable fluid. In 1824 the French military engineer Sadi Carnot employed the caloric theory in his analysis of an idealized heat-engine, which aimed at improving the efficiency of real engines. On the basis of an analogy with the production of work by the fall of water in a waterwheel, Carnot assumed that a heat-engine produced work by the "fall" of caloric from a higher to a lower temperature. The analogy suggested that the work produced was proportional to the amount of caloric and the temperature difference of the two bodies between which caloric flowed. Carnot proved that no other engine could surpass his reversible ideal engine in efficiency by showing that the existence of a more efficient engine would imply the possibility of perpetual motion. In 1834 a mining engineer, Benoit-Pierre-Émile Clapeyron, reformulated Carnot's analysis, using calculus and the indicator (pressure-volume) diagram. Carnot's theory was virtually ignored, however, until its discovery in the mid-1840s, via Clapeyron's paper, by William *Thomson (Lord Kelvin), and Hermann von *Helmholtz.

James *Joule's experimental work of the 1840s, which indicated the interconversion of heat and work, undermined caloric theory. His precise measurements supported the old idea that heat consists in the motion of the microscopic constituents of matter. The interconversion of heat and work, along with other developments spanning several fields (from theoretical mechanics to physiology), led to the formulation of the principle of energy conservation. In the early 1850s all these parallel developments were seen, with the benefit of hindsight, as "simultaneous" discoveries of energy conservation, which became the first law of thermodynamics.

Joule's experiments, however, presented a problem for Carnot's analysis of a reversible heat-engine based on the assumption of conserved heat. In the early 1850s Thomson and the German physicist Rudolf Clausius resolved the problem by introducing a second principle. Carnot's analysis could be retained, despite the rejection of the conservation of heat, because, in fact, it dealt with a quantity—the amount of heat divided by the temperature at which the heat is exchanged—that is conserved in reversible processes. During the operation of Carnot's engine, part of the heat dropped from a higher to a lower temperature and the rest became mechanical work.

In 1847 Thomson diagnosed another problem, also implicit in Carnot's analysis. Carnot had portrayed heat transfer as the cause of the production of work. In processes like conduction, however, heat flows from a warmer to a colder body without doing any work. Since the heat does not spontaneously flow from cold to hot, conduction resulted in the loss of potential for doing work. Both Joule and Thomson agreed that energy cannot perish, or, rather, that only a divine creator could destroy or create it. Thomson resolved the difficulty in 1852 by observing that in processes like conduction, energy is not lost but "dissipated," and by raising the dissipation of energy to a law of nature. "Real"—that is, irreversible—processes continually degrade energy and, in a good long time, will cause the heat-death of the universe. The Scottish engineer William Rankine and Clausius proposed a new concept that represented the same tendency of energy toward dissipation. Initially called "thermodynamic function" (by Rankine) or "disgregation" (by Clausius), it later (in 1865) received the name *"entropy" from Clausius, who grafted onto the Greek root for transformation. Every process (except ideal reversible ones) that takes place in an isolated system increases its entropy. This principle constituted the second fundamental law of thermodynamics, and its interpretation remained the subject of discussion for many years.

The dynamical conception of heat provided a link between mechanics and thermodynamics and led eventually to the introduction of statistical methods in the study of thermal phenomena. In 1857 Clausius correlated explicitly thermodynamic and mechanical concepts by identifying the quantity of heat contained in a gas with the kinetic energy (translational,

rotational, and vibrational) of its molecules. He made the simplifying assumption that all the molecules of a gas had the same velocity and calculated its value, which turned out to be of the order of the speed of sound. Clausius's idealized model faced a difficulty, however, as pointed out by the Dutch meteorologist C. H. D. Buys Ballot. On the model, gases should diffuse much faster than actually observed. In 1858, in response to that difficulty, Clausius attributed the slow rate of diffusion to the molecules' collisions with each other and introduced the new concept of "mean free path," the average distance traveled by a molecule before it collides with another one.

In 1859 James Clerk *Maxwell became aware of Clausius's kinetic interpretation of thermodynamics and, in the following years, developed it further by introducing probabilistic methods. In 1860 he developed a theory in which the velocities of the molecules in a gas at equilibrium distribute according to the laws of probability. He inferred from "precarious" assumptions that the distribution followed a bell-shaped curve, the so-called normal distribution, which had been familiar from the theory of errors and the social sciences. Following up these ideas, he published in 1871 an ingenious thought experiment that he had invented four years earlier to suggest that heat need not always flow from a warmer to a colder body. In that case the second law of thermodynamics could have only a statistical validity. A microscopic agent ("Maxwell's demon," as Thomson called it), controlling a diaphragm on a wall separating a hot and a cold gas, could let through either molecules of the cold gas faster than the average speed of the molecules of the hot gas, or molecules of the hot gas slower than the average speed of the molecules of the cold gas. Heat thus would flow from the cold to the hot gas. This thought experiment indicated that the "dissipation" of energy did not lie in nature but in human inability to control microscopic processes.

Ludwig *Boltzmann carried further Maxwell's statistical probing of the foundations of thermodynamics. In 1868 he rederived, in a more general way, the distribution of molecular velocities, taking into account the forces exerted between molecules as well as the influence of external forces like gravity. In 1872 he extended the second law of thermodynamics to systems not in equilibrium by showing that there exists a mathematical function, the negative counterpart of entropy, that decreases as a system approaches thermal equilibrium. This behavior was subsequently called the "H-theorem."

Furthermore, Boltzmann attempted to resolve a severe problem, pointed out by Thomson in 1874 and Joseph Loschmidt in 1876, which undermined the mechanical interpretation of the second law. The law defines a time asymmetry in natural processes: the passage of time results in an irreversible change, the increase of entropy. However, if the laws of mechanics govern the constituents of thermodynamic systems, their evolution should be reversible, since the laws of mechanics run with equal validity toward the past and the future. *Prima facie*, there seems to be no mechanical counterpart to the second law of thermodynamics.

Boltzmann eluded the difficulty in 1877 by construing the second law probabilistically. To each macroscopic state of a system correspond many microstates (particular distributions of energy among the constituents of the system), which Boltzmann ranked as equally probable. He defined the probability of each macroscopic state by the number of microstates corresponding to it and identified the entropy of a system with a simple logarithmic function of the probability of its macroscopic state. On that interpretation of entropy, the second law asserted that thermodynamic systems have a tendency to evolve toward more probable states. The interpretation came at the cost of demoting the law. A decrease of entropy was unlikely, but not impossible.

Maxwell's and Boltzmann's statistical approach to thermodynamics was developed further by J. Willard *Gibbs, who avoided hypotheses concerning the molecular constitution of matter. He formulated statistical mechanics, which analyzed the statistical properties of an ensemble, a collection of mechanical systems. This more general treatment proved to be very useful

for the investigation of systems other than those studied by the kinetic theory of gases, like electrons in metals or ions in solutions.

D. S. L. Cardwell, *From Watt to Clausius: The Rise of Thermodynamics in the Early Industrial Age* (1971). S. G. Brush, *The Kind of Motion We Call Heat: A History of the Kinetic Theory of Gases in the 19th Century*, 2 vols. (1976). Lawrence Sklar, *Physics and Chance: Philosophical Issues in the Foundations of Statistical Mechanics* (1993). Crosbie Smith, *The Science of Energy* (1998).

THEODORE ARABATZIS

THERMOMETER. The notion of a scale or degrees of heat and cold dates back at least to the second-century physician Galen, as does the idea of using a standard—such as a mixture of ice and boiling water—as a fixed point for the scale. Ancient philosophers' experiments, such as Hero of Alexandria's "fountain that drips in the sun," demonstrated the expansion of air with heat, and were known among natural philosophers of the sixteenth century. In the second decade of the seventeenth century, *Galileo, Santorio Santorio, and others began to use long-necked glass flasks partially filled with air and inverted in water to measure temperature, applying them to medical and physical experiments and keeping meteorological records. The first sealed liquid-in-glass thermometers, filled with spirit of wine, were constructed for the Accademia del Cimento in Florence in 1654 by the artisan Mariani; though not calibrated from fixed points, his thermometers agreed very closely among themselves.

The succeeding century saw experimentation with thermometric liquids, among which spirit of wine was favored for its quick response and because no cold then known would freeze it. Several natural philosophers, including Robert *Hooke, Christiaan *Huygens, and Edme Mariotte, worked out methods for graduating their instruments from a single fixed point, typically the freezing or boiling point of water. Toward the end of the seventeenth century, Italian investigators began using two fixed points, as did the Dutch instrument maker Daniel Fahrenheit in the first few decades of the eighteenth century. Fahrenheit's excellent thermometers spread his method and his preference for mercury throughout England and the Low Countries, while the dominance of France and the fame of its Académie Royale des Sciences secured the position of academician René-Antoine Ferchault de Réaumur's thermometer on the rest of the Continent.

Réaumur and his contemporaries despaired of precision in their instruments. The inconstant composition of spirit of wine; air dissolved or trapped in the liquid, whether mercury or spirits; the lack of good glass—these and the lack of motivation to precision rendered the thermometer's readings at best qualitative indications of the temperature. After the Seven Years' War, the rational bureaucratic state and industrial manufacturers generated pressure for precise measurement for cartography and navigation, enclosures and canals, and the construction of steam engines and other machinery. In England, instrument making grew from a handicraft to an operation of industrial scale, exploiting advances in glassmaking and metallurgy and serving an international clientele. The thermometer played an auxiliary role in the precise measurements of the late Enlightenment, but from about 1760, the Genevan natural philosopher Jean-André *Deluc developed methods for rigorously calibrating the instrument and for using it in exhaustive series of systematic measurements. In England, a committee of the Royal Society of London under the chairmanship of Henry *Cavendish worked out methods in 1776 of setting the upper fixed point in a water bath, methods that remain in use today. Late-eighteenth-century thermometers achieved a precision of $1/10°$ F. The chief development of the nineteenth century was the discovery that the glass of new thermometers contracted in time so that their zero point fell; in the 1880s, glasses were developed that did not experience these effects.

W. E. Knowles Middleton, *A History of the Thermometer and Its Use in Meteorology* (1966).

THEODORE S. FELDMAN

THOMSON, Joseph John (1856–1940), physicist, known as the "discoverer of the electron."

Thomson was born at Cheetham Hill near Manchester, England, the son of Joseph James Thomson, a bookseller, and Emma Swindells, who came from a textile manufacturing family. His parents intended him to become an engineer, entering him at Owens College, Manchester, at the age of fourteen. When his father died two years later, Thomson could no longer afford an engineering apprenticeship and was compelled to rely on scholarships, concentrating on mathematics and physics (taught by Thomas Barker and Balfour Stewart, respectively), in which he excelled.

In 1876 Thomson obtained a minor scholarship at Trinity College, Cambridge, to study mathematics. He was coached by Edward Routh, who gave him a thorough grounding in analytical dynamics (the use of Joseph-Louis Lagrange's equations and William Hamilton's principle of least action). This emphasis on physical analogies and a mechanical worldview is evident throughout the rest of his work. In 1880 he graduated as Second Wrangler (second place in mathematics).

Thomson's early work was dominated by his reliance on analytical dynamics to explore James Clerk *Maxwell's electrodynamics, which he had first encountered at Owens College, and then learned from William Niven at Cambridge. In 1881 he was the first to show that the mass of a charged particle increases as it moves, and suggested that the particle dragged some of the *ether with it. In 1882 he won Cambridge's Adams Prize for *A Treatise on Vortex Motion,* investigating the stability of interlocked vortex rings and developing the then-popular theory that atoms were ethereal vortices to account for the *periodic table. This work laid the foundations for all of his subsequent atomic models.

In 1884 Thomson was elected Cavendish Professor of Experimental Physics at Cambridge, becoming almost overnight a leader of British science and training a high proportion of the next generation of British physicists. He held an increasing number of positions in scientific administration, was on the Board for Invention and Research during World War I, president of the Royal Society from 1915 to

1920, and, from 1919 to 1927, an active member of the Advisory Council to the Department for Scientific and Industrial Research. His position was confirmed by a knighthood in 1908 and the Order of Merit in 1912. In 1890 he had married Rose Paget, daughter of Cambridge's Regius Professor of Physics. They had a son, George, and a daughter, Joan.

As Cavendish Professor, Thomson took up the discharge of electricity through gases. By 1890 he had developed the concept of a discrete electric charge, modeled by the terminus of a vortex tube in the ether, which guided his later experimental work. The discovery of *X rays in 1895 proved a turning point: X rays ionized the gas in a controllable manner and clearly distinguished the effects of ionization and secondary radiation. Within a year, Thomson and his student Ernest *Rutherford had convincing evidence for Thomson's theory of discharge by ionization of gas molecules. X rays also rekindled interest in the *cathode rays that caused them. With new confidence in his apparatus and theories, Thomson in 1897 showed that all the properties of cathode rays could be explained by assuming that they were subatomic charged particles. He called these "corpuscles," and guessed that they were a universal constituent of matter. They soon became known as "electrons," a term previously introduced to signify the elementary unit of electrical charge. In the following years Thomson unified his ionization and corpuscle theories into a general theory of gaseous discharge of wide applicability, for which he won the Nobel Prize in 1906.

Thomson's sophisticated model of the atom, in which thousands of corpuscles orbited in a sphere of positive electrification, went some way toward explaining the *periodic table and chemical bonding. But his discovery in 1906 that the number of corpuscles in the atom was comparable with the atomic weight (i.e., that the atom contained hundreds rather than thousands of corpuscles) raised problems with the origin of the atom's mass and its stability, to which he sought solutions through experiments on the positive ions in a discharge tube. This work led to recognition of the

H_3^+ ion and the discovery of the first nonradioactive isotopes, those of neon, in 1913, prompting the invention by Thomson's collaborator Francis Aston of the *mass spectrograph in 1919.

In 1919, following his appointment as Master of Trinity College, Thomson resigned the Cavendish Professorship. Under his leadership the Laboratory had become a place of lively debate with a social life of its own, though it suffered from financial stringency. Thomson continued to experiment until a few years before his death, laying the foundations for, among other things, plasma physics. He is buried in Westminster Abbey.

Joseph John Thomson, *Recollections and Reflections* (1936; reprinted 1975). Lord Rayleigh, *The Life of Sir J. J. Thomson* (1942; reprinted 1969). E. A. Davis and I. J. Falconer, *J. J. Thomson and the Discovery of the Electron* (1997).

ISOBEL FALCONER

THOMSON, William (Lord Kelvin) (1824–1907), natural philosopher, inventor of electrical and navigational instruments.

Born in Belfast, William Thomson was the fourth child of James and Margaret Thomson. His father taught mathematics in the Belfast Academical Institution (noted for its political and religious radicalism). His mother came from a Glasgow commercial family, but died when William was just six. Encouraged throughout his formative years by his father (mathematics professor at Glasgow University from 1832 until his death in 1849), William moved from the broad philosophical education of Glasgow to the intensive mathematical training at Cambridge University, where he came second in the Mathematics Tripos of 1845. While working in Paris for some weeks to acquire experimental skills, Thomson was elected to the Glasgow Chair of Natural Philosophy in 1846, a post he held for fifty-three years.

Thomson fashioned for himself a career that took him to the very pinnacle of British imperial science. His capacity to direct his physics toward practical ends placed him among the most eminent of Victorian scientists and engineers. His central position in a network of elite mathematical physicists—

including James Clerk *Maxwell, George Gabriel Stokes, Hermann von *Helmholtz, and Peter Guthrie Tait—gave him a leading role in the emergence of physics as a scientific, laboratory-based discipline. And his active part in geological and cosmological controversies following publication of Charles Darwin's *Origin of Species* (1859) located him in the mainstream of Victorian debates about humanity's place in nature.

Thomson's early scientific papers owed much to Joseph Fourier's *Théorie analytique de la chaleur* (1822). In an original paper written at the age of seventeen, Thomson used Fourier's mathematical treatment of heat flow to reformulate the orthodox theory of electrostatics. Replacing action-at-a-distance forces by continuous flow models, Thomson's radical approach was a principal inspiration for Maxwell's later construction of electromagnetic field theory, exemplified in his famous *Treatise on Electricity and Magnetism* (1873).

Thomson extended Fourier's treatment to the analysis of electric signals transmitted by very long telegraph wires, and advised on the optimum dimensions and operating conditions for transatlantic and imperial cables. He also constructed delicate measuring instruments (notably his "marine mirror galvanometer") for use in telegraphy. In recognition of his services to the Empire, he was created Sir William Thomson by Queen Victoria in 1866, following completion of the first successful Atlantic telegraph.

Through the engineering influence of his older brother James, William became committed in the late 1840s to Sadi Carnot's analogy between the motive power of heat and the fall of water driving a waterwheel. This representation gave Thomson the means of formulating in 1848 an "absolute" temperature scale (later named the Kelvin scale) that correlated temperature difference with work done, independent of the working substance. At the same time, Thomson pondered over the significance of James *Joule's experiments. Committed to Carnot's theory, Thomson could not accept Joule's proposition that work converted into heat could be recovered as useful work. Prompted by the competing investigations of Macquorn

Rankine and Rudolf Clausius, Thomson reconciled the theories of Carnot and Joule in 1850–1851. The production of motive power required both the transfer of heat from high temperature to low temperature and the conversion of an amount of heat exactly equivalent to the work done.

Thomson and Rankine introduced the terms "actual" energy (later "kinetic") and "potential" energy. The laws of energy conservation and dissipation became the foundation of a new "science of energy." Thomson and Tait began a monumental project to extend the energy treatment throughout physics, but completed only one volume, on dynamics, of the *Treatise on Natural Philosophy*.

Working from energy principles, Thomson calculated ages for Earth and Sun in the range of 20 to 100 million years, a time scale that contradicted the geological assumptions on which Charles Darwin had built his theory of evolution. Darwin found Thomson's challenge the most difficult he had to counter.

Having built up a university physical laboratory (the first in Britain) for research and teaching, Thomson played a leading role in developing absolute standards of electrical measurement. Closely allied to this work were extensive business interests in the patenting and manufacture of scientific, navigational, and electrical instruments. In 1892, he became the first British scientist to be made a peer, and took the title Kelvin from the tributary of the River Clyde that flowed close to the University.

See also STANDARDIZATION; THERMODYNAMICS AND STATISTICAL MECHANICS.

Crosbie Smith and M. Norton Wise, *Energy and Empire. A Biographical Study of Lord Kelvin* (1989). Crosbie Smith, *The Science of Energy. A Cultural History of Energy Physics in Victorian Britain* (1998).

CROSBIE SMITH

TIDE. See MOON.

TIME. See SPACE AND TIME.

TOMONAGA, Shin'ichirō (1906–1979), Japanese theoretical physicist.

Born in Tokyo, Tomonaga moved with his family to Kyoto in 1913 when his father became professor of philosophy at the imperial university there. His father had studied in Germany and appreciated both the potency and weaknesses of Western culture. He became a close friend of the chief priest of the head temple of the Tendai sect of Buddhism, who invited him to live in a large house on the temple ground. Shin'ichirō grew up in this house, filled with his father's books, and in the temple grounds, communing with nature and Japanese culture. He developed a deep sense of responsibility to the nation that led him to accept the burdens of academic and governmental administrative posts after 1951.

Tomonaga was a sickly child, frequently absent from school, sensitive, poor at gymnastics, and bullied by classmates. He attended a prestigious senior high school, where he became an outstanding student, and formed a lifelong friendship with his classmate Hideki *Yukawa. By middle school he had decided on a career in biology, but Albert *Einstein changed that. Einstein's visit to Japan in 1922 resulted in extensive popular accounts of the theory of *relativity that Tomonaga found unsatisfactory. He turned to a book by Jun Ishiwara and became fascinated by the four-dimensional world and non-Euclidian geometry.

Yukawa and Tomonaga both entered Kyoto University in 1923 and majored in physics. The theoretical physicist there, Kajûrô Tamaki, worked on relativity and hydrodynamics but had no interest in the old quantum theory. In their senior year, with Tamaki's encouragement, Tomonaga and Yukawa studied *quantum physics. They read the papers of Werner *Heisenberg, P. A. M. *Dirac, Ernst Pascual Jordan, Erwin *Schrödinger, and Wolfgang Pauli (see HEISENBERG AND PAULI) that had laid the foundations of the field, and explained them to each other. Shortly after their graduation, they attended the lectures that Dirac and Heisenberg gave at the Institute of Physical and Chemical Research (Riken) during their visit to Japan in 1929.

Tomonaga completed the work for his Rigakushi (bachelor's degree) in physics in 1929. With Japan in the throes of an economic depression and no prospect for a job, he decided to stay at Kyoto for graduate work. In 1931 Yoshio Nishina—who had worked closely with Niels *Bohr and Oskar Klein in Copenhagen in the 1920s and had contributed to the new quantum mechanics—lectured in Kyoto. He had come back to Japan to organize and oversee *cosmic ray and *nuclear physics at Riken. Nishina was deeply impressed by the acuity and incisiveness of the questions Tomonaga posed after the lecture and invited him to work at Riken. Tomonaga thrived in Nishina's laboratory and assumed the position of house theorist. He, Minoru Kobayasi, and Shoichi Sakata translated the second edition of Dirac's *Quantum Mechanics* into Japanese. In 1937 Tomonaga went to Leipzig to work with Heisenberg. On his return in 1939, he accepted a professorship at Bunrika (Liberal Arts and Science) University in Tokyo. His socratic teaching became legendary. Students said of him that "he was like a magician."

In 1940 Tomonaga married Ryoko Sekiguchi, the daughter of the director of the Tokyo Metropolitan Observatory. They had two sons and one daughter.

In 1943, while engaged in wartime work on magnetrons and radar devices, Tomonaga recast quantum field theories in a form that explicitly satisfied the requirements of the theory of special relativity. In the ruins of Tokyo after World War II, making use of this formalism, Tomonaga—independently of Hendrik Kramers, Hans Bethe, Julian Schwinger, Richard *Feynman, and Freeman Dyson—formulated the renormalization procedures that made it possible to circumvent the difficulties that all *quantum field theories faced. He was able to isolate and discard in a consistent manner the divergences encountered in perturbation theoretic calculations of *quantum electrodynamics and thus to perform a fully relativistic calculation of the Lamb shift.

Nishina died unexpectedly in 1951 and Tomonaga took over many of his duties on governmental committees. He became chair of the Liaison Committee for Nuclear Research of the Science Council responsible for the establishment and oversight of national research institutes in high-energy and nuclear physics. Tomonaga's technical knowledge, sagacity, and even-handedness led to an almost full-time involvement in science policy. Eventually he became president of the Science Council. He practiced a style of arbitrating among competing claims that became known as the Tomonaga method: "waiting for the fruit to ripen and fall." Tomonaga often served as the official representative of Japanese culture and science at international gatherings. As a personal statement and commitment, he actively participated in the Pugwash conferences.

Most of Tomonaga's time after 1951 went to administrative duties. He welcomed his retirement in 1969. It gave him the leisure to give lectures, write essays (collected in a volume entitled *Birds That Come to My Garden*), and nurture his friendships. At the time of his death he had completed a manuscript published posthumously in two volumes as *What Is Physics*. After his death, his friends constructed a book of testimonials under a threnody from *Man'yoshu*, an anthology of eighth- and ninth-century Japanese poems. It runs, "If I could believe that there were/Two men like you in Japan,/I would never grieve."

S. S. Schweber, *QED and the Men Who Made It* (1994). H. Ezawa, ed., *Sin-intiro Tomonaga—Life of a Japanese Physicist*, trans. C. Fujimoto and T. Sano (1995).

SILVAN S. SCHWEBER

TRANSURANIC ELEMENTS. Glenn T. Seaborg popularized the term "transuranic elements" after the declassification of information on the elements with atomic numbers greater than uranium's.

The Rutherford-Bohr model of the atom, established between 1911 and 1922, explained for the first time why the lanthanides had similar properties—namely that they possessed identical outer electronic configurations and slightly different numbers of electrons in interior subshells. Both Johannes Rydberg in 1913 and Niels *Bohr in 1921 speculated that a similar group of heavier elements might exist, and Vicktor Goldschmidt proposed that they should be named the neptunium group. Following the discovery of nuclear fission in 1939, Edwin M. McMillan and Philip H.

Abelson, working at the University of California at Berkeley, showed that neutron bombardment of uranium produced a radioactive element beyond it in the *periodic table. They followed Goldschmidt in naming it neptunium. At the same time, and using the Berkeley 150-cm (60-inch) cyclotron, Seaborg and his colleagues prepared plutonium (element 94) by the beta-decay of neptunium. McMillan and Seaborg received the Nobel Prize in chemistry in 1951 for this work. During the war, Seaborg found time to pursue further examples of synthetic transuranic elements. Americium and curium were identified between 1944 and 1945, and berkelium, californium, einsteinium, fermium, and mendelevium by 1953. Altogether Seaborg participated in the identification of nine of the fifteen actinides (as chemists called Goldschmidt's neptunides) between 1940 and 1970.

Although the chemistry of neptunium and plutonium proved analogous to uranium's, their successors did not. This anomaly forced a reexamination of their place in the periodic table. As early as 1944 Seaborg noted that elements 95 and 96 should have properties analogous to those of the lanthanides europium and gadolinium. He therefore suggested that elements 90–92 (thorium, protactinium, and uranium) should be moved from the seventh period of the periodic table to form a second rare earth (actinide) family of elements that extended from 89 (actinium) to 102 (nobelium), and subsequently to lawrencium, made in an accelerator in Berkeley in 1961 under the leadership of Seaborg's colleague Albert Ghiorso. This amendment to the periodic table proved the key to discovering the remaining elements. The preparation and confirmation of a few fleeting atoms of synthetic transactinides beyond lawrencium saw rival groups from the Soviet Union and Germany competing with the Americans. Settling the names of these synthetic elements, rutherfordium, dubnium, seaborgium, bohrium, and meitnerium, proved controversial and protracted. In the fifty years between 1940 and 1990, seventeen new elements were added to the periodic table beyond uranium.

Seaborg also speculated that beyond transuranic elements 113 and 164 there may be islands of stability containing super-heavy elements with long radioactive half-lives that would allow detailed comparisons with natural elements in their group positions within the periodic table. Besides forming an outstanding example of twentieth-century big science, the investigation of the transuranic elements has shown the continuing value and power of Dmitrii Mendeleev's periodic table.

See also ATOMIC STRUCTURE.

Glenn T. Seaborg, *The Transuranium Elements* (1958). George B. Kauffman, *"Beyond Uranium," Chemical and Engineering News* (19 November1990): 18–29.

WILLIAM H. BROCK

U

UNCERTAINTY. See COMPLEMENTARITY AND UNCERTAINTY.

UNIFORMITARIANISM AND CATASTROPHISM. William Whewell, Master of Trinity College, Cambridge, coined the terms uniformitarianism and catastrophism in 1832 in his review of the second volume of Charles *Lyell's Principles of Geology* (1830–1833) in the *Quarterly Review*. Uniformitarianism referred to Lyell's methodology and theory of the earth, catastrophism to the mainstream geological doctrines that he himself favored. No terms in the history of *geology have created more confusion.

Lyell in his *Principles* had wanted to avoid two methodological extremes—the conservative enumerative induction adopted by the Geological Society of London as its quasi-official methodology, and the method of hypothesis of the seventeenth- and eighteenth-century cosmogonists. The first inhibited theorizing; the second put no empirical checks on the free rein of the imagination. Lyell opted for the vera causa method advocated by Isaac *Newton in the *Rules of Reasoning* in his *Principia mathematica* (1687), given canonical form by the Scottish philosopher Thomas Reid and used by John Playfair in his *Illustrations of the Huttonian Theory of the Earth* (1802). According to the vera causa principle, any postulated cause has to satisfy two conditions: it must be known to exist, and it must be known to be adequate to produce the effect.

Observing both causes and their effects was difficult in geology. Causes frequently acted over long periods or out of sight in the center of the earth. Lyell modified the method to accommodate the special problems of his science. Geologists could be sure of the existence of the cause they postulated only by observing one like it in action. In practice, this meant that geologists had to restrict themselves to causes operating at present. The same applied to the adequacy of the cause. Geologists could not postulate causes that acted more forcefully than those they observed in action.

Uniformitarianism, as Lyell understood it, thus consisted of three theses. The first, that the laws of nature had not changed, was uncontroversial. All geologists accepted this constancy except in the case of the creation of new species, which Lyell, like everyone else, believed required divine intervention. The second was that the kinds of causes operating on the surface of the earth had never changed (the existence condition of the vera causa principle), and the third, that the intensity of causes had not changed either (the adequacy condition of the vera causa principle). The highly regarded astronomer John *Herschel, the mathematician Charles Babbage, and Charles Darwin, who made the vera causa method central to the extended argument of his *Origin of Species* (1859), welcomed uniformitarianism.

Almost everyone else rejected Lyell's principles. Accepting them would have meant repudiating the most successful causal theory in the discipline, Élie de Beaumont's theory of the cooling, contracting earth with its spasmodic bursts of mountain-building and species-extinction. If the earth had cooled, then presumably in the past no glaciers had gouged the earth as they now did. And presumably volcanoes would have been more active in the past. Not surprisingly, Lyell's denial that causes had ever varied in intensity seemed both arbitrary and improbable, and therefore almost all geologists rejected uniformitarianism. For the geologists' position, with its commitment to a directional theory of the earth's history, Whewell adapted the Greek word for the denouement of a drama: *catastrophe*.

Uniformitarians and catastrophists also divided along religious lines. Lyell, like Hutton before him, rejected Christianity for deism. He was delighted that his methodology supported the theory of a steady-state earth because, in his opinion, a benevolent deity would not have created an earth that contained the seeds of its own destruction. Most geologists, though, were Christians. Although they did not believe the literal truth of the Biblical accounts of the Creation and the Flood, they did believe that the earth had been created at a specific time in the past, and would come to an end at a specific time in the future. This predisposed them to a belief in a directional Earth history.

After Lyell's death in 1875, uniformitarianism seemed to have lost its luster. Then Sir Archibald Geikie, head of the Geological Survey of Great Britain, praised it in his *Founders of Geology* (1905), and, perhaps because of his endorsement, uniformitarianism became the rallying cry of geologists in the first half of the twentieth century. They had no knowledge of its roots in the old vera causa tradition. They summed it up with the slogan, "The present is key to the past," and used it in a number of not always consistent ways. They argued that the earth had suffered no dramatic upheavals, as catastrophes were now conceived; that geologists should study present processes, though they rarely did; that geologists should not resort to religious explanations; and that they should not invoke extraterrestrial causes.

As geologists came to accept the destruction of crust implied by *plate tectonics, the possibility of non-gradual evolutionary change asserted by proponents of punctuated equilibria, the fact of meteor impacts, and the likelihood that one such impact had killed the dinosaurs, catastrophism (understood as dramatic large-scale events) regained plausibility.

See also EARTH SCIENCES; EARTH, AGE OF.

Reijer Hooykaas, *The Principle of Uniformity in Geology, Biology and Theology* (1963). Rachel Laudan, *From Mineralogy to Geology* (1987). Charles Lyell, *Principles of Geology* (1830–1833), ed. Martin Rudwick (1991).

RACHEL LAUDAN

UNIVERSE, AGE AND SIZE OF THE.

At the end of the seventeenth century, Isaac *Newton's concept of an infinite universe created instantaneously by God rendered the question of the size of the universe moot and the question of its age a matter for historical rather than scientific determination. For more than two centuries after the publication of Newton's *Principia*, efforts to measure astronomical distances were confined to our solar system and a few nearby stars. Ever larger *telescopes and the development of *photography and *spectroscopy greatly increased the variety and accuracy of the data available to astronomers, but at the end of the nineteenth century, the size of our own Milky Way *galaxy was still unknown and, despite earlier speculations about the possibility of other systems, few, if any, astronomers believed that anything existed outside our own galaxy in the infinite void of the universe.

Between 1910 and 1930, new techniques for estimating interstellar distances finally enabled astronomers to determine the approximate size and shape of the Milky Way galaxy. During the same period, Vesto M. Slipher and Edwin *Hubble measured the red shift in the spectra of a number of spiral nebula, and determined that almost all of them were moving away from the earth at high radial velocities. By 1929, Hubble had also calculated the distance to several *nebula, and for the first time had provided convincing evidence that spiral nebula were clusters of stars (he did not identify them as galaxies) far beyond the borders of our own galaxy. Even more significantly, Hubble had found that the farther away the nebulae were, the greater their radial velocities away from the earth.

The key to calculating the age, and hence the size, of an expanding universe is the determination of the intergalactic velocity/distance ratio (which became known as the "Hubble Constant"). In the 1930s, Hubble's original value for the velocity/distance ratio produced an estimate of about 2×10^9 years for the age of the universe. This result briefly created the curious anomaly of a calculated value of the age of the universe that was smaller than radiometric measurements of the age of the earth. The work of Walter Baade in the 1940s and Allan Sandage in the 1950s resulted in substantial revisions in the accepted value of Hubble's Constant, and by the 1960s astronomers agreed that the universe was between 10 and 20×10^9 years old, with a corresponding size of the order of 10^{10} light-years. Also by the 1960s, the success of the "Big Bang" hypothesis had provided astronomers with a causal physical model of an expanding universe with an instantaneous beginning and a finite age and size. Since the 1950s, the rapid proliferation of new techniques and technologies has enabled astronomers to probe ever closer toward the outer edge of the universe, and in the 1990s, a more precise redetermination of the Hubble Constant was made a priority of the new Hubble Space Telescope. The results of the several independent efforts to recalculate the Hubble Constant were not consistent among themselves, however, and the late 1990s saw renewed controversy over the age and size of the universe. At the century's end, some astronomers believed that the universe's age had been determined to lie within the range of 12 to 15 × 10^9 years, while their more cautious colleagues remain unwilling to allow for a greater certainty than the earlier 10 to 20 × 10^9 years. The size of an expanding universe, of course, depends on its age, but recent cosmological theories, such as Alan Guth's inflationary model, suggest the possibility that our observable universe may be only a small part of a much larger structure.

See also COSMOLOGY; SPACE AND TIME.

Kitty Ferguson, *Measuring the Universe* (1999). John Gribbin, *The Birth of Time* (1999). Stephen Webb, *Measuring the Universe, the Cosmological Distance Ladder* (1999).

JOE D. BURCHFIELD

UREY, Harold (1893–1981), American physical chemist, discoverer of deuterium, and Nobel Prize winner.

Urey was born in Walkerton, Indiana, the son of a schoolteacher. He read zoology and chemistry at the University of Montana at Missoula, graduating in 1917 just as America entered World War I. Although he retained a passionate interest in biology for the rest of his life, the war forced him to become an industrial research chemist. He returned to Montana as a chemistry instructor at Montana State University before entering graduate school under G. N. Lewis at the University of California at

Berkeley in 1921. His doctoral thesis (1923), concerned with the *entropy of gases, introduced him to *spectroscopy and whetted his appetite for the new *quantum physics then gaining the attention of physicists and physical chemists. He spent the years 1923–1924 on an American-Scandinavian Fellowship studying with Niels *Bohr and Hendrik Kramers at the Institute for Theoretical Physics in Copenhagen. On his return to America in 1924, he taught chemistry at Johns Hopkins University in Baltimore. He married a bacteriologist, Frieda Daum, in 1926. They had four children.

In 1929 Urey moved to New York as associate professor of chemistry at Columbia University. His interest and expertise in quantum physics led him to coauthor a pioneering textbook on quantum mechanics with A. E. Ruark (*Atoms, Molecules, and Quanta*, 1930) and to become founding editor of the *Journal of Chemical Physics* in 1933. During World War II he served as director of research of the so-called Substitute Alloys Materials Laboratory at Columbia, a part of the Manhattan Project. In 1945 he became professor of nuclear studies at the University of Chicago, where, in 1952, he published *The Planets: Their Origin and Development*. From 1958 until his retirement in 1970, he was professor of chemistry-at-large at the University of California's San Diego campus.

Urey's early research and publications embraced chemical kinetics, quantum mechanics, and infrared spectroscopy. During the 1920s, Francis Aston of the Cavendish Laboratory in Cambridge, England, had pioneered, and virtually monopolized, the identification of isotopes using the *mass spectrometer. Aston's instrument had failed to detect a heavier isotope of hydrogen. In 1931, however, while working with Ferdinand Brickwedde and George Murphy, Urey found evidence for what he named "deuterium" in commercially available hydrogen at room temperature. In the following year he separated deuterium from deuterium oxide ("heavy water") by the electrolysis of ordinary water, and encouraged colleagues like Rudolf Schoenheimer to use deuterium as a radioactive tracer in metabolic studies. Urey received the Nobel Prize in chemistry in 1934 for this discovery.

Urey's unrivalled expertise in techniques of isotope separation was put to use during World War II when he helped separate uranium isotopes for the atomic bomb. He also had a hand in the separation of tritium (the third isotope of hydrogen), an ingredient of the hydrogen bomb. While willing to serve in what he saw as a just war, Urey was sternly critical of postwar secrecy and the continuation of the manufacture of atomic weapons. Joseph McCarthy's House Un-American Activities Committee accused him of having communist beliefs, which he denied.

In the 1950s, reflections on thermodynamic relations between isotopes led Urey to develop a method of estimating ocean temperatures in past geological eras by using fossil evidence. In 1952 he published *The Planets, Their Origin and Development*, in which he suggested that the earliest atmosphere of the earth must have been a reducing one. His speculation that ultraviolet light and atmospheric electrical discharges could have formed organic molecules that became the basis of life was supported by experiment by his doctoral student, Stanley Miller, in 1953. The fact that lunar meteoric craters appeared much older than terrestrial ones led him to conclude that the moon had been formed quite separately from the earth by the accretion of interstellar particles. Urey was forced to abandon his lunar hypothesis when the *Apollo* 2 mission (1969) confirmed that the moon had once been a molten mass. Nevertheless, his cosmochemical speculations provided a considerable stimulus to space exploration in the second half of the twentieth century.

H. C. Urey, "Some Thermodynamic Properties of Hydrogen and Deuterium," in *Nobel Lectures, Chemistry, 1922–1941* (1966): 333–356. F. G. Brickwedde, "Harold Urey and the Discovery of Deuterium," *Physics Today* 35 (September 1982): 34–39. S. G. Brush, "Nickel for Your Thoughts: Urey and the Origin of the Moon," *Science* 217 (1982): 891–898. K. P. Cohen et al, "Harold Clayton Urey," in *Biographical Memoirs of Fellows of the Royal Society* (1983): 623–659. J. N. Tatarewicz, "Urey, Harold Clayton," in *Dictionary of Scientific Biography*, ed. F. L. Holmes, 18 (1990): 943–948.

CURTIS WILSON

V

VACUUM. Nature abhors a vacuum. So said medieval natural philosophers, following Aristotle. Against the ancient atomists, who held that material atoms move in an infinite void, Aristotle presented several arguments for the impossibility of a vacuum: the lack of resistance would produce infinite velocities; the homogeneity of the void precluded natural motion, which for Aristotle relied on a distinction between up and down; and the void likewise prevented violent motion, which needed an external medium for continued propulsion. The plenum persisted into the seventeenth century, notably in the system of René *Descartes, who identified matter with space. But the possibility of the void received some discussion from medieval scholastics, who wondered in particular about the space beyond the stars, where God perhaps resided.

Experimental refutation of the *horror vacui* came in the seventeenth century. Mining pumps of the time operated according to the abhorrence of the vacuum, up to a point—thirty feet, in fact, above which they could not draw water. In 1644 an Italian mathematician, Evangelista Torricelli, explained the limitation by a mechanical equivalence between the weight of atmospheric air and the weight of the column of water, and demonstrated it using a glass tube, closed at one end, inverted in a basin of mercury. The mercury rose to a height one-fourteenth that to which water attained. Torricelli's new device—what we now call a *barometer—figured in a famous experiment four years later by Blaise Pascal, who initially doubted Torricelli's explanation of the barometer and thought it showed only the limits to the force of vacuum. Pascal pursued barometric experiments with a variety of liquids and glass tubes up to 14 m (46 ft) long, for the latter relying on the state-of-the-art products of the glass factory in his hometown of Rouen. In the decisive experiment, Pascal in 1648 sent his brother-in-law up a mountain in France with a barometer; the lower level of the mercury at the peak convinced Pascal and others that the weight of the atmosphere, not the vacuum inside the barometer, was forcing up the mercury in barometers and the water in mining pumps.

The famous experiment of the Magdeburg spheres (1654). A pair of hollow hemispheres easily separable under ordinary conditions could not be pulled apart by two teams of horses when put together, sealed hermetically, and exhausted of air. The experiment demonstrated the power of atmospheric pressure and of the air pump newly invented by Otto von Guericke.

Otto von Guericke soon provided an equally famous demonstration using his new air pump, a piston-driven suction pump with valves that could suck the air out of sealed chambers and thus make a vacuum. In 1657 in Magdeburg, where he was the mayor, Guericke worked his air pump on two copper hemispheres stuck together and showed that a team of horses could not pull them apart; the force of the vacuum—or, rather, of the air on the outside of the hemispheres—held them together. Robert *Boyle developed the air pump into a means of easy production of a vacuum. When Boyle placed a barometer inside a glass globe, the level of mercury descended as the pump evacuated the enclosed space until the mercury no longer stood. Boyle's account of the results, *New Experiments Physico-Mechanicall Touching the Spring of the Air* (1660), showed along the way that cats and candles could not survive in a vacuum but that electric and magnetic effects could.

The technology of the vacuum would henceforth be crucial for modern science. The fruitful program of experiment with evacuated *cathode ray tubes in the nineteenth century relied on a new generation of vacuum pumps, the first major advancements over von Guericke's original design; in particular, a pump made by German instrument maker J. H. W. Geissler using a mercury column instead of pistons, which improved residual pressures from one inch to one millimeter of mercury. Rotary pumps made by Wolfgang Gaede in Germany in the early twentieth century proved crucial for the development of vacuum tube technology in the commercial electronics industry. High-energy particle *accelerators later in the century required further advances in the production of large empty spaces.

The vacuum of physicists since the early modern period has not been empty. The imponderable fluids of electricity, light, and magnetism in Laplacian physics pervaded it, as did the ether and electromagnetic fields of Maxwellian electromagnetism. Even after Albert *Einstein banished the *ether, his postulated equivalence of energy and mass implied that matter could still intrude in empty space, and fields,

electron holes, and ghost particles continue to clog up the vacuum of modern physics.

Edward Grant, *Much Ado about Nothing: Theories of Space and Vacuum from the Middle Ages to the Scientific Revolution* (1981). Steven Shapin and Simon Schaffer, *Leviathan And Air-pump: Hobbes, Boyle, And The Experimental Life* (1985). Per F. Dahl, *Flash of the Cathode Rays: A History of J. J. Thomson's Electron* (1997).

PETER J. WESTWICK

VACUUM PUMP. See AIR PUMP AND VACUUM PUMP.

VAVILOV, Nikolai Ivanovich (1887–1943), Soviet botanist, geographer, and organizer of agricultural research, and **Sergei Ivanovich VAVILOV** (1891–1951), Soviet physicist and statesman.

The Vavilov brothers were born in Moscow into a one-time peasant family that had risen to become wealthy merchants. Both chose science after graduating, Nikolai from the Agricultural Academy and Sergei from the physico-mathematical department of Moscow University. Nikolai was among the first Russian researchers to take up the new field of genetics just before World War I. After the war that interrupted their academic pursuits, and the subsequent revolution and civil war that forced their father to emigrate, both brothers stayed in Soviet Russia and continued their scholarly careers. In 1920, Nikolai made a sensational presentation at a congress of Russian breeders, suggesting a new "law of homological series" in hereditary variation of plants. Shortly thereafter, he was appointed director of the Bureau (subsequently Institute) of Applied Botany in Petrograd (Leningrad). In this position, he became the key organizer of the Soviet system of agricultural research, which grew to a countrywide network of several hundred government-sponsored breeding stations and experimental research institutions. In 1929, these institutions were united under the newly created All-Union Lenin Academy of Agricultural Sciences (VASKhNiL), with Nikolai Vavilov as president. The same year, he was elected ordinary member of the USSR Academy of Sciences, where he directed the Institute of Genetics.

While neither he nor his younger brother ever joined the communist party, Nikolai shared many of the social and economic goals of the revolutionary modernizing regime. He endorsed the Stalinist collectivization campaign, hoping that large-scale collective farms would introduce modern scientific practices into backward agriculture. To help breed and introduce better varieties of cultivated plants, Vavilov built up, at his Leningrad institute, a comprehensive worldwide collection of seeds, which grew to include some 300,000 specimens. Many of them came from Vavilov's expeditions to remote areas of Asia, Africa, and Latin America. Another product of these travels was Vavilov's innovative and acclaimed theory of the geographical centers of origin of cultivated plants, which relied on genetic and cytological analysis. The most widely traveled Soviet scientist, Vavilov also served as president of the All-Union Geographical Society. He actively promoted the revolutionary science of genetics, attracting to his institutes some of the best Soviet and foreign researchers. Soviet genetics flourished under his patronage during the 1920s and early 1930s, second only to American genetics.

Geneticists first came under criticism in the early 1930s because of their association with eugenics, denounced by Soviet Marxists as racist. Stalinist officials often blamed agricultural failures on the inadequate application of scientific research. The charismatic but poorly educated agronomist T. D. Lysenko attracted followers with his rival brand of "Michurinist" genetics (named after a plant breeder), which, he claimed, was more practical and ideologically more sound than "formalist" and "idealistic" Mendelian genetics. No less dangerous to Vavilov personally were accusations from militants at his own institutes, who challenged the agricultural usefulness of his research program and the wisdom of spending government funds on expensive overseas expeditions. Under mounting pressure, Vavilov stepped down from the VASKhNiL presidency in 1935, but remained its vice president. The more difficult his situation became, the more outspoken his courageous, but losing, defense of genetics against Lysenkoist ideological accusations. Along with many

important agricultural officials, Vavilov fell victim to Stalinist purges. He was arrested in 1940, convicted of "wrecking" in the field of agriculture, and died of malnutrition in Saratov prison in 1943. Most key administrative positions in Soviet biological and agricultural research came under the control of Lysenko, whom many in the scientific community held responsible for Vavilov's fate. Having lost its most important spokesman and patron, Soviet genetics fell on hard times.

Sergei Vavilov lacked his elder brother's charisma, rebelliousness, and revolutionary vigor, but excelled in hard work, self-discipline, and manners. His career advanced more slowly than Nikolai's until 1929, when he became professor at Moscow University. His research in experimental optics, on luminescence and the quantum structure of light, required patience and tedious work. This bore important fruit in 1933, when Vavilov asked one of his graduate students, P. A. Cherenkov, to look for luminescence induced by rays from radioactive substances. Cherenkov noticed in the dark a very weak blue glow coming from the liquid. Vavilov recognized that the glow could not be luminescence. It was a new physical phenomenon now known as Cherenkov or Vavilov-Cherenkov radiation. In 1937, Vavilov's colleagues I. E. Tamm and I. M. Frank explained the effect theoretically as caused by *electrons traveling faster than light propagates in the liquid, an electromagnetic analogue of the acoustical shock waves produced in the atmosphere by supersonic projectiles. The discovery later found important application in high-energy physics as the basis of the Cherenkov detectors of elementary particles. In 1958 (after Vavilov's death), Cherenkov, Tamm, and Frank shared the Nobel Prize in physics.

The discovery did not contribute much to Sergei's rapid career advance during the 1930s, which rested on his administrative talent and unmatched sense of responsibility. In 1932, Vavilov was elected ordinary member of the USSR Academy of Sciences and assumed the scientific directorship of the State Optical Institute in Leningrad, where he coordinated R&D in the national optical industry, both military and civilian.

Vavilov's responsibilities included the Academy's own small physical institute (FIAN, where Cherenkov made his discovery), which turned into a major assignment in 1934 with the Academy's move from Leningrad to Moscow. Under Vavilov's directorship, FIAN developed into the nation's largest center for advanced research in fundamental physics and the home for six researchers who would later become Nobel laureates.

After the war's end in 1945, Vavilov became Stalin's choice for the Academy's president. He oversaw the major expansion of the Academy's scientific research in response to the American atomic bomb. As the country's major political representative of science and its public spokesman—in the political climate characterized by Stalin's personality cult, the outbreak of the Cold War, and ideological campaigns in sciences—Vavilov faced difficult compromises. Some were very bitter, such as the Academy's compliance with the ban on Mendelian genetics imposed in 1948 by Lysenko, and others more rhetorical, such as Vavilov's elaborate glorification of Stalin.

Vavilov received a political commission to write about the history of science. He wrote excellent historical works on optics, Isaac *Newton, and eighteenth-century Russian science. In his philosophical publications, Vavilov argued that the twentieth-century revolution in physics (theories of *relativity and the quanta) fully confirmed the predictions of Marxist-Leninist philosophy. This stance, along with Vavilov's subtle political maneuvering, helped deflect the danger of an ideological pogrom in physics. While genetics struggled to survive underground, Soviet physics continued its spectacular successes throughout the Stalin years. Decades of hard work took a toll on Vavilov's health. He died in 1951: to Stalinist leadership, an ultimately reliable and ideologically loyal scholar; to scientists, especially physicists, an effective protector in dangerous times.

N. I. Vavilov, *The Origin, Variation, Immunity and Breeding of Cultivated Plants: Selected Writings*, trans. from Russian by K. Starr Chester (1951). Andrei Sakharov, foreword to M. A. Popovskii, *The Vavilov Affair* (1984). Alexei Kojevnikov, "President of Stalin's Academy: The Mask and Responsibility of Sergei Vavilov," *Isis* 87 (1996): 18–50. B. M. Bolotovskii, Yu. N. Vavilov, A. N. Kirkin, "Sergei Ivanovich Vavilov: The Man and the Scientist: A View from the Threshold of the 21st Century," *Physics Uspekhi* 41(5) (1998): 487–504.

ALEXEI KOJEVNIKOV

VAVILOV, Sergei Ivanovich. See VAVILOV, NIKOLAI IVANOVICH, AND SERGEI IVANOVICH VAVILOV.

W

WERNER, Abraham Gottlob (1749–1817), German geologist and mineralogist.

After working for some time in an iron foundry where his father was employed, Werner studied mining at the Bergakademie (Mining Academy) in Freiberg (in 1769) and then read law for some years at the University of Leipzig. In 1774 he published his first book, *Von den äusserlichen Kennzeichen der Fossilie,* which led him back to the Mining Academy in Freiberg, where he became a lecturer and also curator of the mineral collection . He remained there for the rest of his life. In his book he laid down the foundations of his classification system, which was based on the external characteristics of minerals. However, he believed that there must also be a correlation between external characteristics and chemical composition, and in Axel Fredrik Cronstedt's *Försök til mineralogie* ("Researches on mineralogy," 1758) he found the system based on chemical composition he sought. Werner translated parts of Cronstedt's treatise from Swedish into German. He published this translation, together with his own amendments and additions, in *Axel Kronstedts Versuch einer Mineralogie* (1780). He then extended his *Mineral System* to more examples and more editions (1789, 1816, 1817). Between the first and last of these editions the system grew from 183 species to 317. Werner discovered eight of them himself and named a further twenty-six. Many of his results were published by his colleagues with his permission.

Werner was as interested in broad perspectives as in mineralogical detail. He had an organic view of the history of the earth resembling Nicolas Steno's, whose work he knew well. He presented his division of the types of rocks in a slim booklet, first published as an essay, *Kurze Klassifikation und Beschreibung der verschiedenen Gebirgsarten* ("A brief classification and description of the different types of rocks," 1786). He distinguished five types. The oldest, *uranfänglich,* was the hard deposit produced by stratification or crystallization in the universal ocean that Werner supposed once to have covered the earth. This group included the granites, which were formed first, gneiss, basalt, and nine other types. When the water began to recede, it left a series of strata of a softer type, *Flötzgebirge,* which included limestone, sandstone, and coal. Ocean swells and storms caused local variations in the strata by alternately raising and lowering the water surface. The third and fourth groups consisted of igneous rocks, which could be traced directly to active volcanoes: *rocks attributable* entirely to fire, *ächtvulkanische* (lava, pumice and volcanic ash), and ones originally composed of *Flötz* but transformed by fire, *pseudovulkanische* (cinder and porcelain jasper). The fifth group was made up of alluvial deposits of eroded material from primary rocks and *Flötz.*

Volcanic forces play only a marginal role in Werner's scheme. He was a strong force for *neptunism in its fierce fight with vulcanism. In a key point, the nature of basalt and granite, Werner's point of view was outmoded. His belief in the great universal ocean at the dawn of time opened him to the charge of biblicism. However, Werner made no reference to the Flood or even to the Bible in his writings and manuscripts. His independence from scriptural tradition appears further in his allowing at least a million years to geological history as against the 6000 years since Creation calculable from the Bible. Werner had grown up in a form of German Pietism, which was quite open-minded and recommended education in science. He eventually adopted a deistic outlook on life that to some critics seemed close to atheism. During his time in Leipzig he belonged to a little group of freethinkers.

Werner's personal view was one thing, that of his pupils another. A militant group of neptunists who claimed to be Werner's supporters provoked the great quarrel with the vulcanists. Because Werner committed so few of his original thoughts to print, people interpreted and manipulated the little that existed as they wished. His most loyal disciples were in England. Unfortunately, they wanted to buttress the story of Creation and linked that Werner's neptunism with an orthodox biblical view. However, divisions occurred among the leaders of this group—Richard Kirwan, Jean-André *Deluc, and Robert Jameson.

Besides his lectures and an oral tradition, Werner left an extensive collection of manuscripts. A very popular teacher, he devoted the last twenty years of his life to teaching and administration.

Heinrich Bingel, *Abraham Gottlob Werner und seine Theorie der Gebirgsbildung* (1934); Alexander Ospovat, "Abraham Gottlob Werner," *Dictionary of Scientific Biography,* 14 (1980), 256–264.

TORE FRÄNGSMYR

X

X RAYS. When the president of the Berlin Physical Society spoke at its jubilee in 1896, he could not manage much enthusiasm about the future of its science. Later he said that, had he known about the discovery of X rays, he would instead have expressed his joy "that the second fifty years in the life of the society had begun as brilliantly as the first." His reaction was representative: from the minute Wilhelm Conrad *Röntgen made known his discovery, physicists recognized it as a tonic to their senescent science. X rays challenged theory, abetted experiments, made a public sensation, and gave doctors a diagnostic tool of unprecedented power. Until the medical profession could provide itself with the necessary apparatus, people who swallowed pins or stopped buckshot appealed to physicists to locate the mischief.

X rays refused easy classification into the available categories. They did not bend in electric or magnetic fields and thus did not belong among charged particles; and since they could not be reflected or refracted, they failed the test for light. Most physicists supposed them to be a peculiar form of electromagnetic radiation. The peculiarities of X rays included behavior unbefitting a wave, however. As the English physicist William Henry Bragg stressed, an X ray could impart to an electron almost as much energy as had gone into the ray's creation. But if a wave, the ray should have spread out from its point of origin, diffusing its energy; how then could the entire original energy reassemble when a small section of the wave front encountered an electron? This difficulty appeared the stronger when in 1912 Max von Laue and his colleagues at the University of Munich, and in 1913 Bragg and his son William Lawrence Bragg, then a student at Cambridge, showed, respectively, that a crystal can both refract and reflect X rays. Röntgen's discovery thus appeared to have properties characteristic of a wave and of a particle.

At first physicists did not worry over the properties that conflicted with the wave model confirmed by the diffraction experiments. The wave model allowed investigations of crystal structure, pioneered by the Braggs. It also made possible determination of the frequencies of the characteristic X rays emitted by the elements. These rays resemble the visible line spectrum but are simpler, depending only on the atomic number, Z, and a "screening constant" s, indicative of the place within the atom where the electron involved in the emission of a given line ends up. The study of characteristic x-ray spectra thus helped to elucidate *atomic structure. In his influential theory of the constitution of atoms of 1922, Niels *Bohr made systematic use of x-ray data to determine the quantum numbers of atomic electrons. The doublet structure of some x-ray lines helped Samuel Goudsmit and George Uhlenbeck to construct the concept of electron spin in 1925.

Meanwhile, the American physicist Arthur Holly Compton reopened the shelved problem of the nature of X rays by his discovery in 1922 of what was soon called the Compton effect. According to Compton, a high-frequency X ray collides with an electron as if both were billiard balls. From relativity and the quantum theory Compton assigned the X ray a frequency, v; an energy, hv; and a momentum, hv/c. By assuming that energy and momentum are conserved in the collision, he obtained a relation between Δv (the ray's loss in frequency), T (the electron's gain in momentum), and the angles between the velocities of the interacting particles.

Measurement confirmed Compton's equations and lent such support to the material conception of X rays that thereafter atomic theorists felt obliged to work both particle and wave properties into their descriptions. Louis de Broglie moderated the behavior of photons (the concept Albert *Einstein introduced in 1905 for hypothetical particles of high-frequency light) by coupling them to unobservable waves; Bohr, Hendrik Kramers, and John Slater abandoned the conservation of energy in considering the relation between electron

A playful Swedish depiction of a beach holiday recorded by Roentgen photography (ca. 1900).

jumps and emitted light. Erwin *Schrödinger followed up de Broglie's lead, and Werner *Heisenberg reworked the Bohr–Kramers–Slater approach and other work of Kramers to arrive at their alternative versions of *quantum physics.

The completion of the quantum theory of the electronic cloud of the atom did not end the usefulness of characteristic high-frequency radiation in studying the fundamental structure of matter. Nature produces very hard X rays in spontaneous nuclear decay. Beginning in the 1930s, physicists analyzed the energies of these "gamma rays" for data about nuclear transformations and for help in specifying nuclear energy levels. More recently, X rays from stars have given information about stellar processes.

The diagnostic uses of X rays soon gave rise to a new profession, radiology, and, when their effect on lesions and tumors was noticed, to therapeutics against skin diseases and cancers. Demand for more penetrating and more plentiful X rays prompted a rapid development of apparatus, culminating just before World War I in the high-voltage, heated-cathode Coolidge tube. The modest gains of physics from these developments rapidly multiplied after the war, in part because of wartime electrical engineering and surplus electrical equipment. Pioneers in California—at the newly established Caltech in Pasadena and at the University of California at Berkeley—pushed x-ray generators to gigantic sizes to produce radiation that could reach deep-lying tumors. Medically they achieved little, but technically they advanced substantially the art of high-power electrical engineering in the service of science. Some of the techniques and funding for these machines supported the early development of particle *accelerators for *nuclear physics.

The capacity of X rays to peer into previously secret places has had many applications beyond medicine. They have been used to inspect welds, test materials, fit shoes, detect dental cavities, search pyramids for mummies, etch circuits, and so on. The use of characteristic x-ray spectra to analyze the elementary makeup of even minute samples of materials became a staple in chemical assays. The optimism immediately inspired by Röntgen's discovery, and reaffirmed in the presentation to him of the first Nobel Prize in physics (1901), has been justified repeatedly in the sciences, medicine, and industry, although long exposures and inappropriate therapies have claimed martyrs among physicians and patients.

See also RADIOACTIVITY.

P. P. Ewald, *Fifty Years of X Ray Diffraction* (1962). J. L. Heilbron, *H. G. J. Moseley* (1974). J. L. Heilbron and Robert W. Seidel, *Lawrence and His Laboratory* (1989). Bettyann H. Kevles, *Naked to the Bone* (1997).

J. L. HEILBRON

Y

YANG, C. N. See LEE, T. D., C. S. WU, AND C. N. YANG.

YOUNG, Thomas (1773–1829), natural philosopher, Egyptologist, physician, and man of letters. The son of a cloth merchant and banker, Young was born at Milverton, Somerset, and died at his home in Park Square, London. His strict Quaker upbringing, which stressed the importance of education, was the major formative factor of his character. Although he attended school for six years, Young was largely self-taught and he is often cited as a child prodigy, having mastered several languages by his mid-teens. Encouraged by his kinsman Dr. Richard Brocklesby, Young decided to pursue a career in medicine. He attended both the medical school in London founded by William Hunter and also St. Bartholomew's Hospital.

Like many dissenters, who were barred from Oxford and Cambridge, he attended the Edinburgh medical school (1794–1795) and then continued his medical training at the University of Göttingen (1795–1796), where he submitted his dissertation, entitled *De corporis humani viribus conservatricibus*, and graduated doctor of physic in July 1796. By this time Young had repudiated his Quaker background and was finally disowned in February 1798. He embraced the Church of England and entered Emmanuel College, Cambridge, intent on gaining a Cambridge degree that would open the way to a Fellowship of the Royal College of Physicians. He gained his Cambridge M.B. in 1803, but the degree of M.D. and the coveted Fellowship eluded him until 1808. Marrying into the minor aristocracy in 1804 further increased his distance from his Quaker upbringing and helped launch his medical career in London.

Brocklesby introduced Young to the London scientific elite, including many of the leading medical men, who were among his supporters in the Royal Society. Elected in June 1794—at the early age of twenty—he had already gained a scientific reputation from having presented a paper to the society. He closely associated himself with the Royal Society, serving as its foreign secretary from 1804 until his death, and he supported Joseph Banks and other traditionalists, who wished to maintain a cozy alliance with the aristocracy, against the reformers, who wanted active scientists to control the society.

In July 1801 he accepted the position of professor of natural philosophy at the recently founded Royal Institution. His main duty during his two-year tenure was to deliver lectures on natural philosophy; these lectures were subsequently revised and published as *A Course of Lectures on Natural Philosophy and the Mechanical Arts* (1807). A number of innovations can be traced to Young's lectures, the most important being his research on the wave theory of light, which was also published in several papers between 1799 and 1804.

Rejecting the dominant corpuscular theory of light—usually attributed to Isaac *Newton—Young developed a wave theory that attributed light to the vibrations of a ubiquitous *ether. His main intellectual sources were Leonhard *Euler and those often-overlooked passages in which Newton had entertained an ether and suggested that light is a periodic vibration. Young's primary innovation was a two-ray version of the principle of interference, which he developed from his work earlier on *acoustics. He showed how this principle could explain such phenomena as the colors of thin plates and those seen when a fiber is held close to the eye. Only in the published *Lectures* of 1807 did he apply his principle to the "two-slit" experiment that has become associated with his name. Among his several later optical contributions was the proposal that polarization could be explained by assuming that light is a transverse vibration of the ether particles. Yet despite a number of insights into optical theory, Young's interpretation of the wave theory of light attracted little interest and was subsequently superseded by Augustin Fresnel's mathematically more sophisticated theory.

Young also laid the basis of a three-color theory of color sensation, often called the Young–Helmholtz–Maxwell theory. In rejecting the seven-color theory, often attributed to Newton, Young suggested that the visible spectrum is continuous but that color vision is due to just three types of receptor, according to the three principal colors: red, yellow, and blue. Having first proposed this in a lecture in 1801, he subsequently improved it in the published *Lectures* and in an article on "Chromatics" published in 1817.

Among the other fields in which Young made brief but incisive excursions was hieroglyphics. He helped decipher the inscriptions on the Rosetta stone. However, his writings on this subject were less detailed and not as clearly focused as those of Jean-François Champollion; Young's main excursion was his article *Egypt*, which appeared in a supplement to the *Encyclopaedia Britannica* in 1819. Despite his medical practice, which was never very remunerative, Young was principally an essayist who contributed a large number of articles on diverse subjects to the *Britannica* and to numerous medical and periodical publications. Always the gentleman scholar, his contributions show him as well informed, well read, and often brilliantly insightful but lacking in depth and application.

Alexander Wood, *Thomas Young: Natural Philosopher, 1773–1829* (1954). P. D. Sherman, *Colour Vision in the Nineteenth Century: The Young–Helmholtz–Maxwell Theory* (1981). G. N. Cantor, *Optics after Newton: Theories of Light in Britain and Ireland, 1704–1840* (1983). N. S. Kipnis, *History of the Principle of Interference of Light* (1990).

GEOFFREY CANTOR

YUKAWA, Hideki (1907–1981), Japanese theoretical physicist.

Hideki Yukawa, who originated the meson theory of nuclear forces, was born in Tokyo and spent most of his life in Kyoto. The fifth of seven children, Yukawa came from a line of Japanese scholars of samurai origin, including both grandfathers and his geologist father, Takuji Ogawa, professor of geography at Kyoto Imperial University. Hideki's youthful interests centered on literature and philosophy, although he also enjoyed mathematics. Physics took precedence for him in high school, where he had an excellent teacher and a classmate, Shin'ichirō *Tomonaga. In 1932 Yukawa married Sumi Yukawa, adopting her family name as his own. In the same year he became a lecturer at Kyoto Imperial University (now Kyoto University), where he served as professor from 1939 to 1969.

Yukawa's quantum field theory of nuclear forces of 1934 proposed that the forces between nuclear particles, protons and neutrons, are transmitted through the exchange of "heavy quanta." This mechanism resembles the transmission of electromagnetic forces by the exchange of light quanta (photons). The field theory employing photons is called *quantum electrodynamics (QED). While light quanta are electrically neutral and massless, Yukawa's proposed quanta carried unit electric charge (positive or negative) and had a mass intermediate between that of the electron and the proton (hence "meson"). Using a simple relation that he discovered between the range of force and the mass of the corresponding quantum, Yukawa estimated that the range of the strong nuclear force, a distance much smaller than the size of the atom, implied a meson mass about two hundred times that of the electron.

In addition to explaining the origin of the strong forces that hold nuclei together, Yukawa's theory also gave an account of beta-decay, a form of weak interaction in which a neutron decays into a proton, electron, and a very light neutral particle, the neutrino. Through its own weak interaction, a meson produced in free space would decay in less than a microsecond. Yukawa argued that, owing to this short lifetime, free mesons should not be found in nature except where, as in *cosmic rays, there is sufficient energy ($E = mc^2$). Later physicists observed mesons produced by beams of the high-energy elementary particle accelerators of the late 1940s.

Although Yukawa published his paper in English in 1934, his theory was ignored until in 1937 American cosmic-ray physicists Carl Anderson and Seth Neddermeyer discovered charged particles of intermediate mass and both signs of charge. Their result led to worldwide recognition of Yukawa's idea. Physicists in the West, and also Yukawa and his students in Japan, extended the scope of his theory. Meanwhile, Meanwhile American nuclear experimentalists showed that the nuclear forces between any two nuclear particles, whether neutron or proton, were the same (a property called charge-independence), suggesting that an additional electrically neutral meson was required to mediate between like particles.

Over the next decade, further cosmic-ray experiments proved that the particles discovered by Anderson and Neddermeyer did not behave as mesons should, but more like heavy electrons. They belong to a family of weakly interacting particles (leptons) and are now called "muons." In 1947 cosmic-ray researchers in Bristol, England, led by Cecil Powell used a new technique to observe the charged mesons ("pions") predicted by Yukawa and their decay into muons. (Particle accelerators in California produced the neutral pi meson in 1950.) In 1949 Yukawa, then a visiting professor at Columbia University in New York City, received the Nobel Prize in physics, the first Japanese citizen to be so honored.

The success of the meson theory established a new paradigm of elementary particle physics. The current *Standard Model of elementary particle interactions, involving *quarks that interact by exchange of particles (gluons), follows the Yukawa pattern, and has replaced meson theory as an underlying fundamental theory of nuclear forces. However, Yukawa's theory is still a useful way of understanding nuclear processes especially at low and intermediate energies. Physicists now understand mesons, which come in many types, as composites of a quark and an antiquark.

Yukawa's Nobel Prize was a point of pride for the Japanese people at a time when they badly needed one. He received many additional accolades, notably the Research Institute for Fundamental Physics (now Yukawa Institute) founded in his honor at Kyoto University. Yukawa wrote many essays on science and Eastern thought and participated in world movements for peace.

Hideki Yukawa, *Creativity and Intuition*, trans. John Bester (1973). Hideki Yukawa, *"Tabibito" (The Traveler)*, trans. Laurie Brown and Rick Yoshida (1982). Nicholas Kemmer, "Hideki Yukawa," *Biographical Memoirs of Members of the Royal Society* 29 (1983): 661–676. Laurie M. Brown, "Yukawa, Hideki," *Dictionary of Scientific Biography* 18 (supplement II) (1990): 999–1005. Laurie M. Brown and Helmut Rechenberg, *The Origin of the Concept of Nuclear Forces* (1996).

LAURIE M. BROWN

INDEX

ILLUSTRATION SOURCES AND CREDITS